International Conference Physics of Neutron Stars - 2017

50 Years After

Journal of Physics: Conference Series Volume 932

St. Petersburg, Russia
10 – 14 July 2017

ISBN: 978-1-5108-5364-5
ISSN: 1742-6588

Printed from e-media with permission by:

Curran Associates, Inc.
57 Morehouse Lane
Red Hook, NY 12571

Some format issues inherent in the e-media version may also appear in this print version.

Printed by Curran Associates, Inc. (2018)

For permission requests, please contact the Institute of Physics
at the address below.

Institute of Physics
Dirac House, Temple Back
Bristol BS1 6BE UK

Phone: 44 1 17 929 7481
Fax: 44 1 17 920 0979

techtracking@iop.org

Additional copies of this publication are available from:

Curran Associates, Inc.
57 Morehouse Lane
Red Hook, NY 12571 USA
Phone: 845-758-0400
Fax: 845-758-2633
Email: curran@proceedings.com
Web: www.proceedings.com

Table of contents

Volume 932

International Conference Physics of Neutron Stars - 2017. 50 years after.
10–14 July 2017, St. Petersburg, Russian Federation

Accepted papers received: 22 November 2017
Published online: 12 December 2017

Physics of pulsar magnetospheres

Extragalactic bursts

Binary systems and accretion

International conference "Physics of neutron stars – 2017. 50 years after"

G G Pavlov[1], J A Pons[2], P S Shternin[3] and D G Yakovlev[3]

[1] Pennsylvania State University, USA
[2] Universitat d'Alacant, Spain
[3] Ioffe Institute, Saint Petersburg, Russia

The conference

This was the 11th gathering on neutron star physics in Saint Petersburg since 1988. The 2017 Conference was organized by the Theoretical Astrophysics Department of the Ioffe Institute and the Relativistic Astrophysics Department of the Sternberg Astronomical Institute. It commemorated the semicentenary of the discovery of pulsars.

During these 50 years, the field of astrophysics of neutron stars has become a broad field covering all ranges of physical scales, from microscales characteristic of strong interactions to macroscales of neutron star sizes strongly affected by General Relativity. All branches of contemporary physics are involved in the research of astrophysical phenomena related to neutron stars, and observations throughout the entire range of electromagnetic spectra have been used to constrain fundamental physical theories. Towards the end of the 20th century, electromagnetic observations were joined by direct neutrino detections from a newly born neutron star (Supernova 1987a) and, between the end of the Conference and the publication of this volume, a long-awaited gravitational wave signal from a binary neutron star merger was detected for the first time, thus promoting neutron stars to the first class of astrophysical sources observed in the electromagnetic band, neutrinos, and gravitational waves. With many more detections of binary neutron star mergers to come, the next decade of the neutron star research is expected to be developing under great influence of the gravitational wave astronomy. In addition, the launch of telescopes of the next generation (proposed, planned and already operating) will shed light on many of the current mysteries, but surely it will also unveil new ones, making the near future even more stimulating than the previous five decades.

In this conference, 130 participants from 24 countries presented their contributions on various topics related to observations and theory of neutron stars. These included, among others, observations of neutron stars as radio pulsars or high energy emitters, studies of pulsar wind nebulae, supernovae and supernova remnants, theory of pulsar magnetospheres, the neutron star equation of state, structure and evolution of neutron stars, or gravitational wave emission from neutron stars and black holes. Many manifestations of neutron stars were discussed: magnetars and high magnetic field pulsars, central compact objects in supernova remnants, accreting X-ray pulsars (including millisecond pulsars), neutron stars in low-mass X-ray binaries, X-ray bursters, and more. A significant part of the Conference contributions resulted in 54 proceedings papers collected in this volume. We hope that the volume will bring pleasant memories to the participants and become a useful reference for many astrophysicists and physicists who would like to learn the current state-of-the-art in the field.

Acknowledgments

We thank all the participants for the wonderful atmosphere during the Conference. We are especially grateful to those who prepared high-quality publications. We are deeply indebted to the 48 referees who thoroughly reviewed all the contributions collected in this volume. We also thank all the SOC members for their expert scientific advice which allowed the organizers to configure the fruitful scientific programme of the Conference, and all the LOC members for fulfilling their numerous organization duties. We particularly wish to thank Dick Manchester for giving the after-dinner talk that drove the audience through the 50 years of the pulsar investigations. We thank the official conference operator INNO-MIR (www.inno-mir.com) and personally I E Zemlyanskaya and E A Lifshits for handling the financial part of the event. We acknowledge the financial support of the Russian Foundation for Basic Research under project # 17-02-20307, Saint Petersburg Polytechnic University and the Government of Saint Petersburg.

Organizing committees

SOC

- V S Beskin *Lebedev Physical Institute, Moscow, Russia*
- C O Heinke *University of Alberta, Edmonton, Canada*
- M R Gilfanov *MPI, Germany; IKI, Moscow, Russia*
- V M Kaspi *McGill University, Montreal, Canada*
- D R Lorimer *West Virginia University, Morgantown, USA*
- S Mereghetti *INAF IASF, Milan, Italy*
- G G Pavlov *Pennsylvania State University, USA*, **co-chair**
- J A Pons *University of Alicante, Spain*
- S B Popov *Sternberg Astronomical Institute, Moscow, Russia*
- J Poutanen *University of Turku, Finland*
- N Rea *University of Amsterdam, the Netherlands*
- R W Romani *Stanford University, USA*
- M van der Klis *University of Amsterdam, the Netherlands*
- D G Yakovlev *Ioffe Institute, Saint Petersburg, Russia*, **co-chair**

LOC

- R L Aptekar *Ioffe Institute*
- D A Baiko *Ioffe Institute*
- S A Balashev *Ioffe Institute*
- D P Barsukov *Ioffe Institute*
- D M Beronya *Ioffe Institute*
- A I Chugunov *Ioffe Institute*
- A A Danilenko *Ioffe Institute*
- V A Dommes *Ioffe Institute*
- O A Goglichidze *Ioffe Institute*
- M E Gusakov *Ioffe Institute*
- A D Kaminker *Ioffe Institute*
- E M Kantor *Ioffe Institute*
- A V Karpova *Ioffe Institute*
- A A Kozhberov *Ioffe Institute*, **co-chair**
- K P Levenfish *Ioffe Institute*
- D D Ofengeim *Ioffe Institute*
- K A Postnov *Sternberg Astronomical Institute*
- A Y Potekhin *Ioffe Institute*
- Yu A Shibanov *Ioffe Institute*
- P S Shternin *Ioffe Institute*, **co-chair**
- Yu A Uvarov *Ioffe Institute*
- D A Varshalovich *Ioffe Institute*
- I E Zemlyanskaya *INNO-MIR*
- D A Zyuzin *Ioffe Institute*

IOP Conf. Series: Journal of Physics: Conf. Series **932** (2017) 011001 doi:10.1088/1742-6596/932/1/011001

Invited speakers

- **Andrea De Luca** (*INAF, Italy*)
  Central compact objects in supernova remnants and the slow magnetar in RCW 103
- **Pawel Haensel** (*CAMK, Poland*)
  Equation of state of neutron stars: Recent developments
- **Jason Hessels** (*Anton Pannekoek Institute for Astronomy, the Netherlands*)
  Fast Radio Bursts
- **Oleg Kargaltsev** (*The George Washington University, USA*)
  Observations of pulsar-wind nebulae in X-rays and other wavelengths: outstanding questions
- **Martin Obergaulinger** (*Universidad de Valencia, Spain*)
  Magnetic field amplification during core collapse
- **Alessandro Patruno** (*Leiden Observatory, the Netherlands*)
  Observations of accreting millisecond X-ray pulsars
- **Alexander Philippov** (*Princeton University, USA*)
  How do pulsars shine?
- **Juri Poutanen** (*Tuorla Observatory, Finland*)
  Ultraluminous X-ray pulsars
- **Samar Safi-Harb** (*University of Manitoba, Canada*)
  Observational diversity and evolution of neutron stars
- **Patrick Slane** (*Harvard-Smithsonian Center for Astrophysics, USA*)
  The neutron star – supernova remnant connection
- **Roberto Turolla** (*Department of Physics and Astronomy, University of Padova, Italy*)
  Polarization of neutron star emission and future X-ray missions

IOP Conf. Series: Journal of Physics: Conf. Series **932** (2017) 011001 doi:10.1088/1742-6596/932/1/011001

Conference photo

Figure 1. Photo taken by M Beznogov on July 13, 2017 in front of the Academic University

Peer review statement

All papers published in this volume of *Journal of Physics: Conference Series* have been peer reviewed through processes administered by the proceedings Editors. Reviews were conducted by expert referees to the professional and scientific standards expected of a proceedings journal published by IOP Publishing.

IOP Conf. Series: Journal of Physics: Conf. Series **932** (2017) 012001 doi:10.1088/1742-6596/932/1/012001

50 Years of Pulsars!

R N Manchester

CSIRO Astronomy and Space Science, PO Box 76, Epping NSW 1710, Australia

E-mail: `dick.manchester@csiro`

Abstract. A brief, personal, and very incomplete account of 50 years of pulsar astronomy presented at the Conference Dinner for "Physics of Neutron Stars – 2017 – 50 Years After", held in Saint Petersburg, July 2017.

1. Introduction and the discovery of pulsars

First, I would like to thank George Pavlov for the invitation to attend this meeting and to present this talk at the Conference Dinner. I have to say, I was a little surprised to be asked to give the after-dinner talk as I am not well-known as a raconteur. I could only conclude that it was because I am old!

It is true that I have been involved with pulsars one way or another since their discovery, or at least the official announcement of it by Hewish et al. in the February 24, 1968, edition of Nature. I started work at the CSIRO Parkes Observatory just 12 days before the Nature paper was published and soon became involved in pulsar research.

But first back to the real beginning. In mid-1967, Jocelyn Bell was a graduate student at Cambridge University helping to build the "Four-acre array", an array of dipoles tuned to 81.5 MHz that Antony Hewish had designed to investigate interplanetary scintillation of radio sources with the aim of identifying compact sources likely to be quasars. With the completion of the array, Jocelyn was given the job of examining the chart recordings to find the rapidly fluctuating signals from compact sources that scintillated as a result of propagation through the solar wind. She did find many of these, but also began to notice some signals that were subtly different from both the scintillating sources and radio-frequency interference. The first recorded detection of one of these "scruff" sources was on August 6, 1967, as shown in Figure 1. The break-through and the point that marks the real discovery of pulsars was when she realised that this source returned at the same sidereal time most days. This meant that it had to be outside the solar system and unrelated to terrestrial activity. A few months later, on November 28, she and Antony Hewish obtained the first high-speed recording to show that the scruff was in fact a train of pulses with a spacing of about 1.3 seconds.

Follow-up observations with the Four-acre array and other instruments at Cambridge refined the periodicity of the pulsar and showed that the signals were dispersed by their passage through the interstellar medium. A paper reporting the discovery and follow-up observations of the pulsar, then named CP 1919, and an interpretation that favoured radial pulsations of either a white dwarf or a neutron star was submitted to Nature by Hewish, Bell, et al. on February 9, 1968, and published on February 24 [2].

In 1974 Antony Hewish was awarded the Nobel Prize in Physics for the discovery of pulsars.

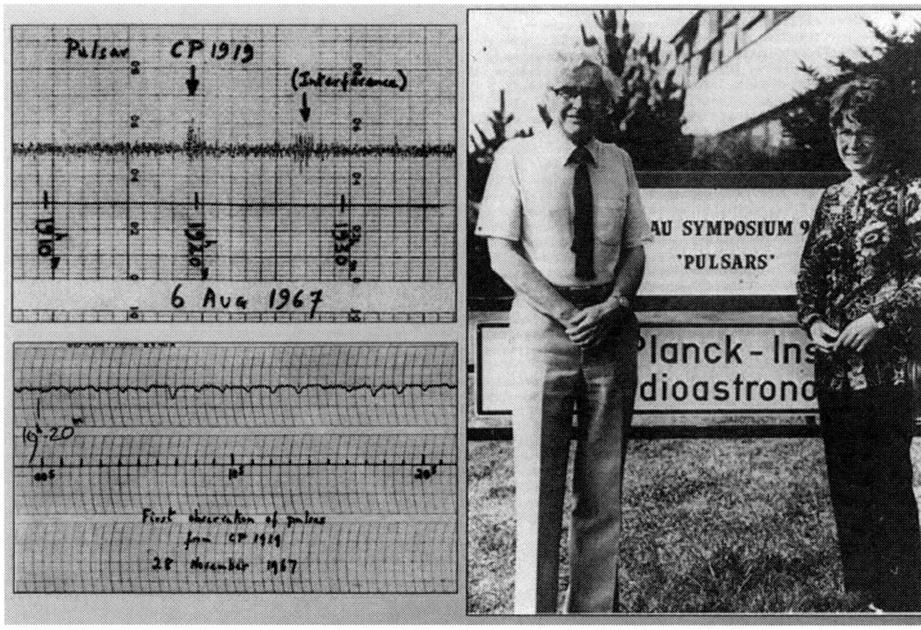

Figure 1. Left: Chart recordings from the Four-acre array at Cambridge showing the first detected signals from CP 1919 and the first recording to show that the emission was pulsed. (Reprinted with permission from Reviews of Modern Physics [1], copyright (1975) by the American Physical Society.) Right: Jocelyn Bell and Antony Hewish at IAU Symposium 95, Bonn, August 1980.

2. Pulsar observations at Parkes

Just two weeks after the publication of the Nature discovery paper a team from the CSIRO and the University of Sydney installed a remarkable set of five coaxial feeds with corresponding receivers at the focus of the Parkes 64-m radio telescope. On the morning of March 8, I was in the control room as the telescope slewed around to the position of CP 1919 and saw the wonderful burst of pulses come through on the chart recorder seconds later. By good fortune, this first observation at 150 MHz just caught a large scintillation maximum. The pulsar was not seen again with such high signal-to-noise for the rest of the session. This image was recorded for posterity on the first Australian \$50 note (Figure 2) and also published with what are still some of the best measurements of the individual-pulse spectra of any pulsar [3].

Later on that year, having learnt something about measuring polarisation with the 18-cm receiver while working with Brian Robinson and Miller Goss on observations of OH masers, I was asked to assist Radhakrishnan investigate the polarisation of the Vela pulsar. These observations led to rotating-vector model (RVM) (or magnetic-pole model as Rad preferred to call it) for pulsar polarisation (Figure 3) [4]. In follow-up observations in mid-March, 1969, we again pointed the telescope to the Vela pulsar, recording the data with a signal-averaging system that was set up to fold the data at the expected topocentric period of the pulsar. To our surprise, rather than remaining a fixed phase on the oscilloscope display, the pulse was noticeably drifting to the left, indicating an error in the predicted period. By the evening, after a while trying to understand what was wrong, Rad said that he was going to bed and left me to sort out the problem. I again checked all the instrumentation, looked at some other pulsars and eventually concluded that it must be the pulsar that had changed. I left a note for Rad and went to bed

Figure 2. The first Australian $50 note, featuring the Parkes radio telescope, a portrait of Sir Ian Clunies Ross (the first Chairman of CSIRO) and various images relating to Australian radio astronomy and biological sciences. On the left of the note is an image of the first pulsar recording made at Parkes, of CP 1919 and recorded on March 8, 1968.

myself.

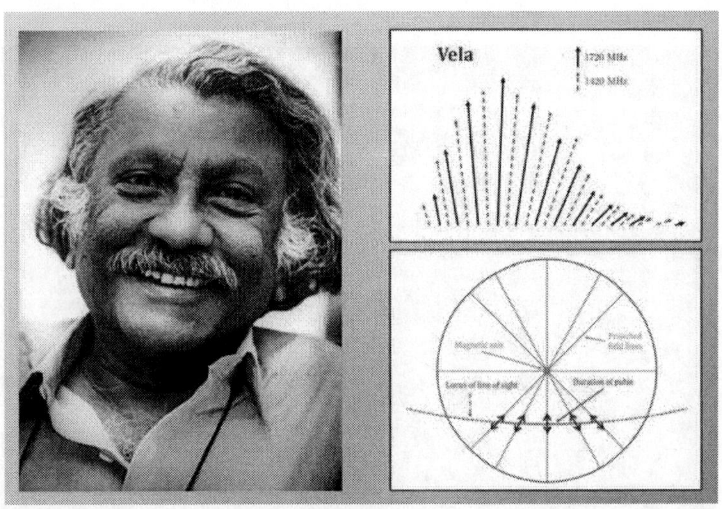

Figure 3. Left: Radhakrishnan in the year 1996 (image credit: Library, Raman Research Institute). Right: plots illustrating the frequency-independence of the position-angle swing across the Vela pulsar mean pulse profile and the interpretation of that in terms of the magnetic-pole or rotating-vector model (after Radhakrishnan & Cooke [4]). In this model, the radiation is emitted tangentially to field lines and is linearly polarised in the plane of field-line curvature as indicated by the double-headed arrows in the lower plot. (Image credit: Peter Shternin)

Observations over the next few days confirmed that it was a real decrease in the pulsar period by about 3 parts in a million - the first detection of a pulsar glitch! Rad contacted Paul Reichley and George Downs who we knew were observing the pulsar using the Goldstone antenna at the Jet Propulsion Laboratory, Caltech, and found that they too had observed the glitch. Back-to-back papers reporting the discovery were published in Nature on April 19, 1969 [5, 6].

IOP Conf. Series: Journal of Physics: Conf. Series **932** (2017) 012001 doi:10.1088/1742-6596/932/1/012001

3. Searching for pulsars

Searches for pulsars are fundamentally important. Not only does the discovery of previously unknown pulsars increase the sample for all manner of statistical studies of pulsar properties and evolution, but almost every significant pulsar search has uncovered some unexpected and interesting pulsar or class of pulsars. Examples are binary pulsars [7], pulsars in globular clusters [8], RRATs [9], pulsars with planets [10], pulsars with main-sequence binary companions [11], etc., etc..

More than 2600 pulsars are now known. Figure 4 shows the rate of pulsar discovery since 1968 by observatory (or in a couple of cases, pulsar class). Small searches (including the original Cambridge search) are grouped under "Other". This figure highlights the major contributions made by the Molonglo radio telescope in the first decade – twice in this time, more than half of the known pulsars were discovered at Molonglo – and the Parkes radio telescope. In the late 2000's, Parkes had found more than twice as many pulsars as the rest of the world's telescopes put together. Even now, the Parkes share is more than 55% of the total. The Parkes Multibeam Pulsar Survey [12] by itself has more than 830 pulsars to its credit, including recent discoveries from reanalyses of the dataset, e.g., [13]. In recent years, the *Fermi* Gamma-ray Space Telescope has been very successful in uncovering previously unknown pulsars, especially millisecond pulsars (MSPs) in so-called "redback" and "black widow" binary systems where the pulsar radio emission is often obscured or eclipsed by plasma streams from the companion star. Most of these systems have been found through radio searches of unidentified gamma-ray sources with properties known to be characteristic of pulsars, e.g., [14].

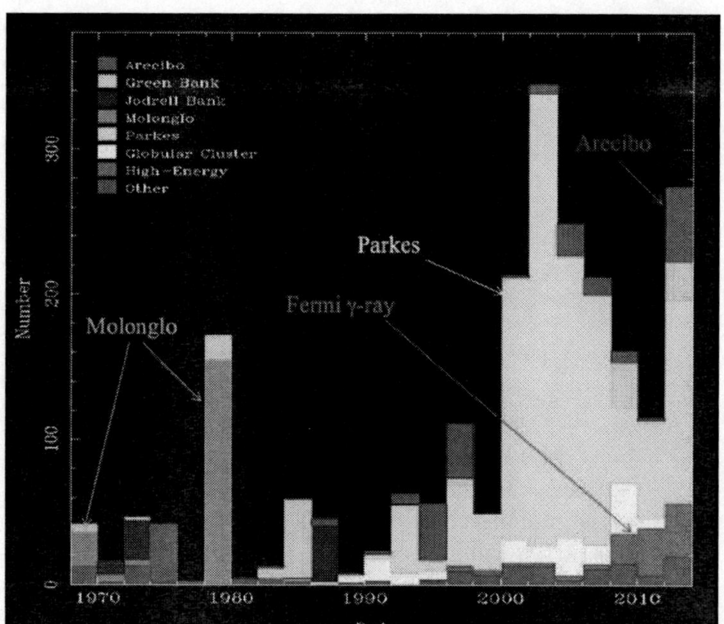

Figure 4. Rate of pulsar discovery, sorted by observatory or class. Each bar represents the number of pulsars discovered in the corresponding 2-year interval. Data from the ATNF Pulsar Catalogue [15]

People often ask me why Parkes has been so successful in pulsar discovery. There are three main reasons:

- The centre of our Galaxy passes almost overhead at Parkes and the southern part of the Galactic plane is far richer than that in the north.
- At ATNF we have had and continue to have a very skilled group of engineers and excellent co-operation with the scientific staff. This led, for example, to the development of the Parkes 13-beam receiver system, by far the most successful pulsar-finding machine ever.
- We had very experienced teams of scientists working on the pulsar survey projects at Parkes (see, e.g., Figure 5). This led to the development of efficient signal-processing systems (both hardware and software) for pulsar detection and confirmation.

Figure 5. Andrew Lyne, Joe Taylor and Dick Manchester at IAU Colloquium 177 "Pulsar Astronomy – 2000 and beyond", Bonn, August, 1999.

4. Key discoveries

The discovery of the first binary pulsar, PSR B1913+16, at Arecibo in 1974 by Russell Hulse and Joe Taylor [7] is the prime example of an unexpected and exciting find in pulsar searches. Observations of this system over the next few years provided the first accurate determinations of neutron-star masses, the first observational evidence for gravitational radiation and confirmation that Einstein's general theory of relativity gives an accurate description of motions in strong gravitational fields [16, 17]. These important results led to the award of the 1993 Nobel Prize in Physics to Taylor and Hulse. As an aside, I note that I shared an office with Joe at the University of Massachusetts from 1971 to 1974, but unfortunately was not involved in the Arecibo search!

Probably the next most important pulsar discovery was that of the first MSP at Arecibo by Don Backer and his colleagues [18]. This was a little different to the other examples of important discoveries in that it was not found in a large-scale search, but rather in a directed investigation of a highly unusual radio source, 4C21.53W. This source was known to have a compact component (it scintillated) and it is steep-spectrum and polarised, all properties consistent with it being a pulsar. Yet efforts to detect a pulsar in this direction (including one at Parkes by Nichi D'Amico

IOP Conf. Series: Journal of Physics: Conf. Series **932** (2017) 012001 doi:10.1088/1742-6596/932/1/012001

and RNM) had been unsuccessful. This changed dramatically when Backer et al. employed a sub-millisecond sampling system that revealed the 1.558 ms periodicity. Figure 6 shows a photograph of Don Backer, probably taken in the early 2000's, and images from the discovery paper. Papers suggesting that the rapid spin of the pulsar originated in an earlier phase of accretion from a companion star (since disappeared) were quickly published [19, 20]. The "recycling" idea had previously been invoked to account for the short period of PSR B1913+16 [21].

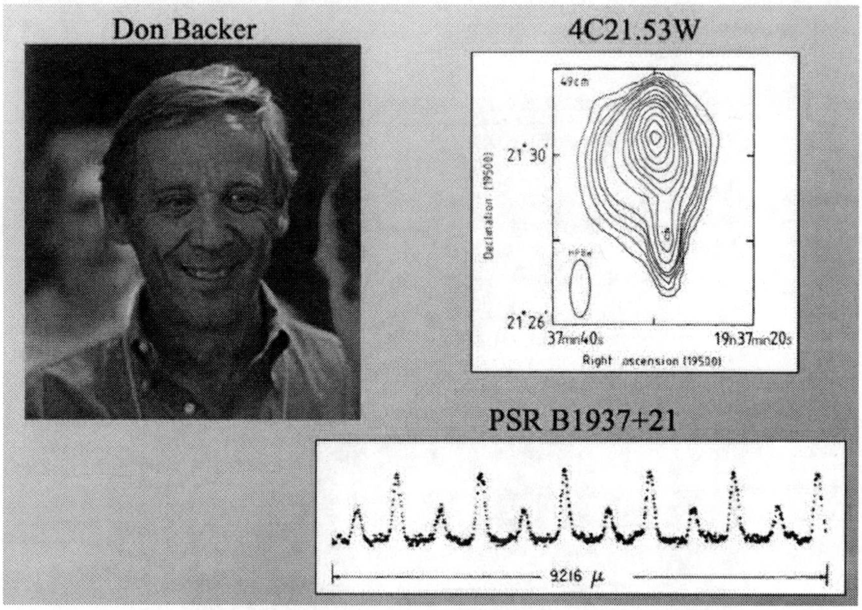

Figure 6. The discovery of the first MSP: Don Backer who led the team (image credit: Bob Rood), a 600-MHz image of 4C21.53W made with the Westerbork Synthesis Radio Telescope, and a signal-averaged recording from Arecibo showing the 1.56 ms periodicity. The cross marks the position of the compact source in the Westerbork image. (Reprinted by permission from Macmillan Publishers Ltd: Nature, [18], copyright 1982.)

Finally, I must mention the Double Pulsar, PSR J0737−3039A/B, discovered at Parkes in 2003 [22, 23]. This system is unique in a number of ways. Firstly it is a double-neutron-star (DNS) system like PSR B1913+16, but with the very short orbital period of 2.4 h. This makes it even more "relativistic" than PSR B1913+16, with a predicted (and observed) periastron advance of 16.9° yr^{-1}, four times that of PSR B1913+16. Secondly, it was and remains the only DNS system where both stars have been observed as pulsars. The B star, although younger than the A star, has a much longer pulse period, about 2.77 s. Thirdly, the orbit is viewed almost exactly edge-on, making the Shapiro delay easy to observe [24] and allowing eclipses of the A pulsar by the magnetosphere of the B pulsar [25]. Recent observations of the system have resulted in the detection of six relativistic effects, including a measurement of the orbital decay that is an order of magnitude more precise than the PSR B1913+16 determination and fully consistent with the prediction from general relativity (Kramer et al., in preparation). It never ceases to amaze me that, despite the development of many alternative theories of relativistic gravity since general relativity, Einstein seems to have got it right the first time.

5. Future prospects

The past 50 years of pulsar astronomy have been more than interesting but there is much to look forward to. Large new radio telescope such as the recently commissioned Five hundred metre Aperture Spherical Telescope (FAST) in Guizhou province, China [26] will enable the discovery and study of many more pulsars, as well as increasing the precision of observations of currently known pulsars. Further in the future, the Square Kilometre Array [27] will provide unrivalled sensitivity and resolution for essentially all radio astronomy applications. New X-ray telescopes such as eRosita, to be launched next year [28], and Athena, due for launch in 2028 [29], will provide wide-field spectroscopic imaging with unprecedented sensitivity and resolution. Existing radio telescopes will be enhanced with new wideband receivers and more sophisticated signal-processing systems. All of these and more will surely probe deeper into the secrets of pulsars, uncovering many things that are not even dreamt of yet. The future of pulsar astronomy and astrophysics is bright!

Большое спасибо!

Acknowledgments

It has been a privilege and a pleasure to have travelled through 50 years of pulsar astronomy and astrophysics and I thank all those colleagues and friends who have helped me along the way.

References

[1] Hewish A 1975 *Rev. Modern Physics* **47** 567–72
[2] Hewish A, Bell S J, Pilkington J D H, Scott P F and Collins R A 1968 *Nature* **217** 709–13
[3] Robinson B J, Cooper B F C, Gardiner F F, Wielebinski R and Landecker T L 1968 *Nature* **218** 1143–5
[4] Radhakrishnan V and Cooke D J 1969 *Astrophys. Lett.* **3** 225–9
[5] Radhakrishnan V and Manchester R N 1969 *Nature* **222** 228–9
[6] Reichley P E and Downs G S 1969 *Nature* **222** 229–30
[7] Hulse R A and Taylor J H 1975 *ApJ* **195** L51–3
[8] Lyne A G, Brinklow A, Middleditch J, Kulkarni S R, Backer D C and Clifton T R 1987 *Nature* **328** 399–401
[9] McLaughlin M A *et al* 2006 *Nature* **439** 817–20
[10] Wolszczan A and Frail D A 1992 *Nature* **355** 145–7
[11] Johnston S, Manchester R N, Lyne A G, Bailes M, Kaspi V M, Qiao G and D'Amico N 1992 *ApJ* **387** L37–41
[12] Manchester R N *et al* 2001 *MNRAS* **328** 17–35
[13] Knispel B *et al* 2013 *ApJ* **774** 93
[14] Camilo F *et al* 2015 *ApJ* **810** 85
[15] Manchester R N, Hobbs G B, Teoh A and Hobbs M 2005 *AJ* **129** 1993–2006
[16] Taylor J H, Fowler L A and McCulloch P M 1979 *Nature* **277** 437
[17] Weisberg J M, Nice D J and Taylor J H 2010 *ApJ* **722** 1030–4
[18] Backer D C, Kulkarni S R, Heiles C, Davis M M and Goss W M 1982 *Nature* **300** 615–8
[19] Radhakrishnan V and Srinivasan G 1982 *Curr. Sci.* **51** 1096
[20] Alpar M A, Cheng A F, Ruderman M A and Shaham J 1982 *Nature* **300** 728–30
[21] Smarr L L and Blandford R 1976 *ApJ* **207** 574–88
[22] Burgay M *et al* 2003 *Nature* **426** 531–3
[23] Lyne A G *et al* 2004 *Science* **303** 1153–7
[24] Kramer M *et al* 2006 *Science* **314** 97–102
[25] McLaughlin M A *et al* 2004 *ApJ* **616** L131–4
[26] Nan R, Li D, Jin C, Wang Q, Zhu L, Zhu W, Zhang H, Yue Y and Qian L 2011 *Intnl J. of Mod. Phys.* D **20** 989–1024
[27] Carilli C and Rawlings S 2004 *astro-ph/0409274*
[28] Predehl P *et al* 2016 *Space Telescopes and Instrumentation 2016: Ultraviolet to Gamma Ray* (*SPIE* vol 9905) p 99051K
[29] Ayre M, Bavdaz M, Ferreira I, Wille E, Lumb D and Linder M 2015 *UV, X-Ray, and Gamma-Ray Space Instrumentation for Astronomy XIX* (*SPIE* vol 9601) p 96010L

IOP Conf. Series: Journal of Physics: Conf. Series **932** (2017) 012002 doi:10.1088/1742-6596/932/1/012002

Pulsar timing and its applications

R N Manchester

CSIRO Astronomy and Space Science, PO Box 76, Epping NSW 1710, Australia

E-mail: dick.manchester@csiro.au

Abstract. Pulsars are remarkably precise "celestial clocks" that can be used to explore many different aspects of physics and astrophysics. In this article I give a brief summary of pulsar properties and describe some of the applications of pulsar timing, including tests of theories of gravitation, efforts to detect low-frequency gravitational waves using pulsar timing arrays and establishment of a "pulsar timescale".

1. Introduction

Pulsars are remarkably precise "celestial clocks". They are believed to be rotating neutron stars that send out beams of radio, optical, X-ray and γ-ray emission which we observe as pulses when they sweep across the Earth. About 2600 pulsars are currently known, almost all of which lie within our Galaxy.[1] The observed pulse periods range between 1 ms and 15 s, with most lying between 0.3 s and 3 s, the so-called "normal pulsars". The "millisecond pulsars" (MSPs) form a distinct group, most with periods in the range 1 ms to 10 ms. Although pulsar periods (P) are very stable and predictable, they are not constant. Most pulsars are powered by their rotational kinetic energy and, as they lose energy to relativistic particle winds and radiation, they slow down. The rate of slowdown, $\dot{P}$, and other pulsar parameters can be determined using pulse time-of-arrival (ToA) measurements made over long intervals. MSPs are also distinguished by their very small $\dot{P}$s, around five orders of magnitude smaller than $\dot{P}$s for normal pulsars, and by the fact that most are in a binary orbit with another star. These different properties result from the very different evolutionary path followed by MSPs in which a central feature is spin-up of an old neutron star by accretion from a companion star [2].

In this review, we first discuss the basics of pulsar timing in §2. Tests of theories of gravitation are discussed in §3, and pulsar timing arrays (PTAs) and their use as detectors of low-frequency gravitational radiation are described in §4. Pulsar timescales and the application of pulsar timing to navigation of distant spacecraft are discussed in §5. The main points of the review are briefly summarised in §6.

2. Pulsar timing

Measurement of a sequence of pulse ToAs over intervals ranging from hours to decades is the basis of precision pulsar timing. These ToAs are first transferred to an "inertial" reference frame, normally the solar-system barycentre, to remove the effects of rotation and orbital motion of the Earth. Radio-frequency dispersion resulting from the presence of ionised gas in the interstellar medium can be compensated for using observations in different frquency bands. Then, based on

[1] See the ATNF Pulsar Catalogue V1.56 (http://www.atnf.csiro.au/research/pulsar/psrcat) and [1]

a model for the intrinsic properties of the pulsar including its astrometric, rotational, interstellar and (if applicable) binary parameters, a predicted barycentric pulse arrival time is computed for each ToA. The difference between the observed and predicted arrival times is known as the timing "residual". Study of the time and radio-frequency variations of these residuals is the basis of all pulsar timing. Systematic variations of the residuals indicate either errors in the pulsar model or the presence of unmodelled phenomena affecting observed ToAs. Each pulsar parameter has a distinct signature in the time and frequency domains. A least-squares fit of the these signatures to the observed residuals enables corrections to the various pulsar parameters and, possibly, the measurement of previously un-measured parameters. Examination of the remaining residuals enables investigation of new and different phenomena such as relativistic binary perturbations or gravitational waves (GWs).

One of the most basic products of long-term pulsar timing is the rate of slowdown of the pulsar rotation period $\dot{P} = -\dot{\nu}/\nu^2$, where $\nu = 1/P$ is the pulse frequency. Figure 1 shows the plot of period derivative versus pulse period for the known pulsars in the Galactic disk.[2] Assuming braking of the pulsar rotation by the reaction to emission of magnetic-dipole radiation, that is, electromagnetic radiation at the pulsar rotation frequency, and assuming a dipole field structure, we can compute the magnetic field at the pulsar surface

$$B_s = 3.2 \times 10^{19}\sqrt{P\dot{P}}, \tag{1}$$

where P is in seconds and B_s is in Gauss, and the "characteristic age", that is, the time since the (back-extrapolated) pulse period was zero (see, e.g., [4])

$$\tau_c = P/(2\dot{P}). \tag{2}$$

One of the main reasons why this diagram is so useful is that different classes of pulsars generally lie in different zones on it. For example, with their short period and low spin-down rate, MSPs are located in the bottom left of the diagram. Conversely, because of their typically long periods and strong magnetic fields, magnetars are at the top right. Double-neutron-star (DNS) systems tend to lie in the gap between MSPs and normal pulsars. Young pulsars have a high probability of being detectable at high energies (optical and above) and the same applies to MSPs having a larger-than-average spindown rate. The so-called "rotating radio transients" (RRATs) (see, e.g., [5]) tend to have long periods and to lie between the magnetars and the bulk of normal pulsars. The X-ray isolated neutron stars (XINS) (see, e.g., [6]) also lie in this region.

The secular spin-down of pulsars can be described by the "braking index", defined by

$$n = \frac{\ddot{\nu}\nu}{\dot{\nu}^2}. \tag{3}$$

Different values of n are predicted for different braking mechanisms. For example, $n = 3$ for magnetic-dipole braking, $n = 5$ for braking by emission of GWs, and braking by pulsar winds which deform the magnetosphere so that fields at the light cylinder (where the co-rotation velocity equals the velocity of light) are stronger than their dipole value have $1.0 \leq n \leq 3.0$ [4]. Because of the combined effects of random irregularities in pulsar period, especially in young pulsars, and the relatively strong $\ddot{\nu}$ dependence, braking indices representing the secular braking of the pulsar have only been measured for a dozen or so pulsars [7]. Most of these are between 2.0 and 3.0 indicating some braking by a stellar wind. For example, for the Crab pulsar, the long-term average value of n is 2.342 ± 0.001 [8].

[2] Globular cluster pulsars are omitted as their observed $\dot{P}$ is often affected by acceleration of the pulsar toward the cluster centre, see, e.g., [3].

IOP Conf. Series: Journal of Physics: Conf. Series **932** (2017) 012002 doi:10.1088/1742-6596/932/1/012002

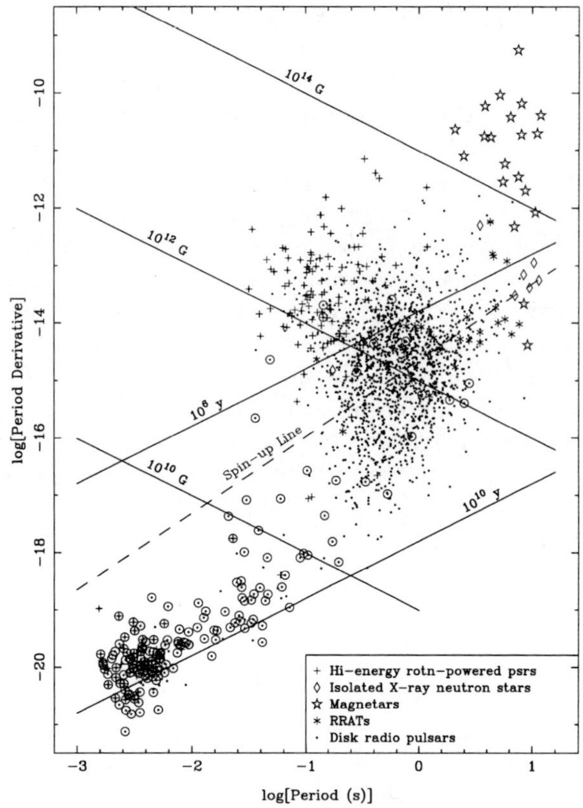

Figure 1. Plot of rate of period increase, $\dot{P}$ versus pulse period P for known pulsars lying in the Galactic disk, that is, excluding pulsars in globular clusters and extra-galactic pulsars. Different classes of pulsars are indicated by different symbols and binary pulsars are indicated by a circle around the pulsar symbol. Lines of constant characteristic age τ_c and surface dipole magnetic field strength B_s are shown as is the maximimum spin period attainable by Eddington-limited accretion of mass from a companion (see, e.g., [2]).

For young pulsars the regular slow-down is often interrupted by a sudden spin-up known as a "glitch". About 425 of these glitches have been observed in more than 140 pulsars[3] and they have relative magnitudes $\Delta\nu_g/\nu$ of between 10^{-10} and 3×10^{-5} for spin-powered pulsars. The post-glitch timing behaviour is different for different pulsars, but for large glitches generally includes a exponential relaxation toward the extrapolated pre-glitch trend followed by an enhanced linear decay of $\dot{\nu}$. This behaviour is illustrated in Figure 2.

During the linear decay phase, the apparent braking index is typically between 20 and 40 [7]. However, as Figure 2 shows, this does represent the long-term evolution of $\dot{\nu}$. Following Lyne et al. [9], Espinoza et al. [7] fitted a template to the post-glitch changes in $\dot{\nu}$ to derive braking indices of 1.7 ± 0.2, 1.9 ± 0.5 and 2.2 ± 0.6 for PSRs B0833−45, B1800−21 and B1823−13, respectively. These are all much less than the magnetic-dipole value of 3.0 and less than other well-measured secular braking indices. It should be noted, however, that Akbal et al. [10] have independently derived a braking index for the Vela pulsar (PSR B0833−45) of 2.81 ± 0.12 by fitting to the values of $\dot{\nu}$ just before a glitch rather than the post-glitch variations fitted by [7].

3. Tests of gravitational theories

The discovery of the first binary pulsar, PSR B1913+16, by Hulse and Taylor at Arecibo Observatory in 1974 [11] opened up multiple new fields of pulsar research. Important among these was tests of relativistic theories of gravitation. PSR B1913+16 was the first of the class of DNS systems (now numbering 15) that have short orbital periods, typically about a day, and relatively high eccentricities. Together with the large system mass, these properties imply large

[3] ATNF Pulsar Catalogue Glitch Table

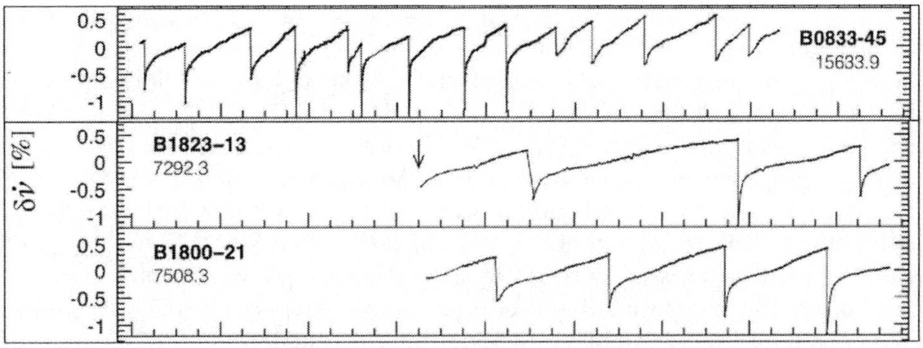

Figure 2. Time dependence of spin-down rate $\dot{\nu}$ for three young pulsars that have multiple glitches. The horizontal axis is time over the Modified Julian Date range 40000 to 57500, a total of about 48 years. The number below the pulsar name is the mean absolute spin-down rate in Hz s^{-1}. (Adapted from [7]).

orbital velocities. For example, at periastron, the orbital velocity v of PSR B1913+16 and its companion are about 300 km s^{-1} or 0.1% of the velocity of light. Since lowest-order relativistic effects go as $(v/c)^2$, they are easily detectable as orbital modulations of the observed pulsar period. Within a few years, the first two relativistic perturbations, periastron advance and time dilation, were detected [12]. In Einstein's general theory of relativity (GR), these depend on the masses of the two stars, together with the well-known Keplerian parameters. Consequently, the masses could be determined and both were close to 1.4 $M_\odot$, confiming that the system consisted of two neutron stars in orbit around one another. Most importantly, measurement of the masses enabled prediction of a third relativistic term, orbital decay due to the emission of GWs from the system. This was also measured in 1979 [12] and was in accordance with the predictions of GR. Later measurements [13] show that the observed orbit decay $\dot{P}_b$ is within 0.2% of the GR prediction. These observations are therefore, not only a confirmation of the accuracy of GR as a theory of relativistic gravitation, but also the first observational evidence for the existence of GWs. These important results were recognised by the award of the 1993 Nobel Prize in Physics to Joseph Taylor and Russell Hulse.

An even more amazing binary system, the Double Pulsar, PSR J0737−3039A/B, was discovered at Parkes in 2003 [14, 15]. This system has a orbital period of 2.4 h, about one third that of PSR B1913+16 and is a DNS system. However, unlike any other DNS system, both neutron stars are detectable as pulsars. The A pulsar, the first-born neutron star of the two, was spun up to its short period of 22 ms by accretion of mass and angular momentum from its evolving companion which subsequently collapsed to form the second neutron star. This is now detected as a relatively young pulsar (B) with a pulse period of 2.77 s.[4] With a predicted periastron advance of 16.9° yr^{-1}, relativistic effects are even larger than for PSR B1913+16, and hence more stringent tests can be performed [17]. Six relativistic effects have now been detected for this system, including two describing the Shapiro delay, geodetic precession of the spin axis of pulsar B [18], and a much more precise measurement of the orbital decay (Kramer et al., in preparation). These results show that GR accurately describes the system at the 0.02% level, an order of magnitude better than the PSR B1913+16 test and the most stringent test so far of GR in the strong-field regime.

[4] The B pulsar became undetectable in 2008, most likely because of spin-axis precession, but it should reappear sometime within the next decade or so [16].

Despite the success of GR in accounting for the details of ToA orbital modulations in the DNS systems, it remains possible that departures from the predictions of GR will be found in the future. There are many alternative theories of gravity and pulsars provide several mechanisms for testing the parameters of such theories. Some of these theories violate the Strong Equivalence Principle (SEP), for example, allowing differences between the gravitational and inertial masses. As pointed out by [19], this would result in bodies with different gravitational self-energy falling at different rates in an external gravitational field. For binary pulsars with white dwarf or other low-mass companions, an orbital eccentricity would be induced by the gravitational field of the Galaxy. A limit on this, expressed in terms of the parameter $|\Delta| = |m_g/m - 1|$ of $\sim 4.6 \times 10^{-3}$, where m_g and m are the gravitational and interial masses respectively, effectively of the neutron star, has been obtained from an analysis of the orbital eccentricities of 27 pulsar – white dwarf systems with low intrinsic eccentricity [20].

Other gravitational theories predict the existence of dipolar gravitational radiation. A strong limit on this has been placed by Freire et al. [21] from an analysis of the various contributions to $\dot{P}_b$ in the pulsar – white dwarf system PSR J1738+0333.

The recent discovery of the very interesting triple system, PSR J0337+1715, by Ransom et al. [22] has opened up the possibility of an even more stringent test of the SEP. This system consists of a 2.73-ms pulsar in a 1.7-day orbit around a 0.19 $M_\odot$ white dwarf. In a co-planar orbit of period 327 days around both these stars is a somewhat more massive 0.41 $M_\odot$ white dwarf. Aside from interesting questions about the orbital dynamics and origin and evolution of this system, it is possible that the gravitational field of the outer white dwarf will induce an eccentricity in the orbit of the inner white dwarf. Since the gravitational field of the outer white dwarf at the position of the inner white dwarf is about 10^6 times as strong as the Galactic gravitational field, this test is potentially much more sensitive than the SEP test based on the pulsar – white-dwarf binary systems [20].

4. Pulsar timing arrays

Although MSPs have extremely stable periods, one cannot rule out the possibilty of intrinsic variations in the observed timing parameters of a given pulsar. Uncorrected variations in interstellar propagation delays are also possible. Therefore, in order to search for extrinsic modulations due to (for example) low-frequency GWs propagating through the Galaxy, it is necessary to observe many MSPs and to search for correlated timing variations among the sample. Such a set of MSPs, widely distributed on the sky and with frequent high-precision timing observations over a long data span is known as a "pulsar timing array" (PTA). PTAs are sensitive to low-frequency (nanoHertz) gravitational waves that could be generated by orbiting super-massive black holes in the cores of distant galaxies. Although individual binary systems could be detectable, the most likely GW source to be detected by PTAs is a stochastic background generated by many such binary systems in galaxies at redshifts of between one and two [23]. Predicted amplitudes of such a background suggest that observations of about 20 MSPs over 10 years are necessary for a detection.

There are three main PTA projects around the world: the European Pulsar Timing Array (EPTA) which uses data from several large radio telescopes in Europe, the North American NANOGrav array, which uses data from Arecibo and the Green Bank Telescope, and the Parkes Pulsar Timing Array (PPTA) which uses data from the Parkes 64-m telescope in Australia. Each of these has high-precision timing data on at least 20 MSPs with data spans ranging between a few years and nearly 30 years. Typically the pulsars in each PTA are observed at intervals of 2 – 3 weeks in two or three different radio-frequency bands.

Detection of GWs by PTAs relies on the correlations between the timing signals for pulsars in different directions on the sky induced by GWs passing over the Earth. Hellings & Downs [24] showed that, for a stochastic background, the correlations depend only on the angular

IOP Conf. Series: Journal of Physics: Conf. Series **932** (2017) 012002 doi:10.1088/1742-6596/932/1/012002

separation of the pulsar pairs, not their sky location. Pulsars close together on the sky have positively correlated signals but, because of the quadrupolar nature of gravitational radiation, pulsars that are in perpendicular directions have negatively correlated signals. The correlation returns to a positive value for pulsars that are diametrically opposed on the sky. GWs passing over the pulsars produce signals that are uncorrelated between the pulsars in the array, just adding a noise component to the GW signal.

Other correlated signals may exist in the data. For example, irregularities in the reference timescale (typically International Atomic Time, TAI), will have the same effect on timing residuals for all pulsars in the array. In terms of the sky distribution, this is a monopolar effect. Errors in the solar-system ephemeris used to refer the ToAs to the solar-system barycentre are equivalent to errors in the computed position of the Earth and so will have a dipolar signature on the sky. These different dependencies on sky position can be used to separate the different effects.

Up to now, there has been no positive detection of nanoHertz GWs by pulsar timing arrays. However, just by using the best few pulsars in an array, it is possible to set a limit on the strength of the GW background in the Galaxy. The best limit so far is from the PPTA as illustrated in Figure 3. This P15 limit, based on Parkes observations in the 10cm band ($\sim$ 3 GHz) of the four best-performing MSPs in the PPTA sample, effectively rules out all "standard" models for generation of a low-frequency GW background. This means that one or more of the assumptions that went into the construction of these models, for example, the merger rate of galaxies or the mass function for super-massive black holes in the cores of galaxies, is in error. Shannon et al. [25] consider that the most likely cause for the lower-than-expected GW background level is that, at a late stage in their evolution, super-massive black-hole binary systems lose energy to surrounding gas rather than to GWs. It should be remarked that recently uncovered differences in solar-system ephemerides may affect the derived GW background limits.

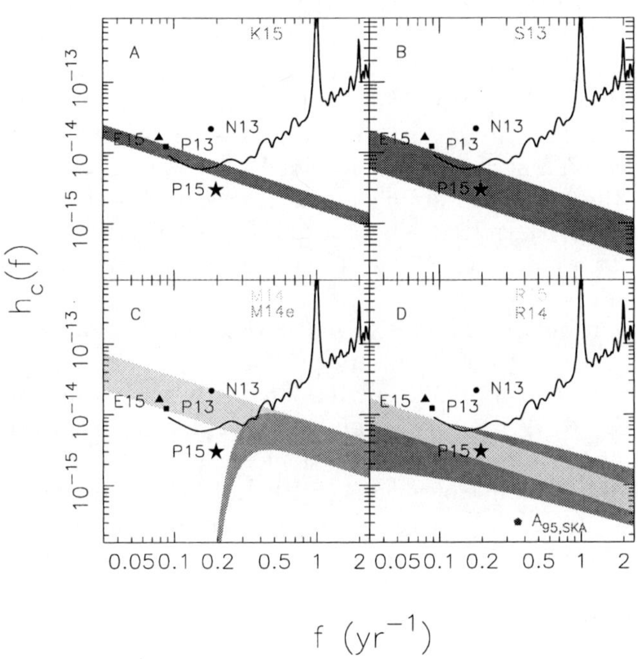

Figure 3. Limits on the characteristic strain of a GW background in the Galaxy from PTA observations. Marked points are 95% confidence limits from NANOGrav (N13, [26]), the EPTA (E15, [27], the 2013 PPTA limit (P13, [28]), and the most recent PPTA limit (P15, [25]). Different predictions for the spectrum of the GW background are shown in the different panels. The black line is the nominal spectral sensitivity of the PPTA observations used to derive the P15 limit and the point marked $A_{95,\mathrm{SKA}}$ gives the expected sensitivity of a PTA based on the Square Kilometre Array. (From Shannon et al., Science, 2015, [25]. Reprinted with permission from AAAS.)

An analysis of the GW senstivity of a PTA as a function of the PTA parameters [29] shows that, when the self-noise from GWs passing over the pulsars of a PTA is significant, the most efficient way to increase the sensitivity of the array is to increase the number of pulsars in it. To

IOP Conf. Series: Journal of Physics: Conf. Series **932** (2017) 012002 doi:10.1088/1742-6596/932/1/012002

this end, the three regional PTAs have combined to form the International Pulsar Timing Array (IPTA). Combined IPTA data sets for 49 pulsars are now available [30] and are starting to be used to further PTA science objectives, e.g., [31]. Other ways in which the number of pulsars in PTAs can be increased is through finding previously unknown MSPs with good timing properties in pulsar searches, e.g. [32], and through the use of new large radio telescopes such as the just-commissioned Five hundred metre Aperture Spherical Telescope (FAST) in China [33] and the Square Kilometer Array (SKA), Phase 1 of which should be operational around 2023 [34].

5. Pulsar timescales and navigation

As mentioned in the previous section, PTAs are sensitive to irregularities in the timescale to which the observations are referenced, normally either TAI or the post-corrected version of TAI, BIPMxx, where xx represents the year of creation [35]. One can invert the normal process of pulsar timing and use the pulsars to "time the timescale", in effect establishing a "pulsar timescale". Such a pulsar timescale differs from terrestrial timescales based on atomic frequency standards in a number of important ways:

- it is independent of terrestrial timescales
- it is based on entirely different physics – the rotation of macroscopic bodies rather than quantum processes
- it will be continuous for billions of years.

Because of the nature of pulsar timing, pulsar timescales are only competitive to (or of greater stability than) atomic timescales over long intervals, months to years and decades. With their very different basis, pulsar timescales can provide a valuable independent check on the long-term stability of terrestrial atomic timescales [36].

Figure 4 shows the results of analyses of the IPTA DR1 dataset [30] for the "common-mode" or clock term relative to TT(TAI) and to TT(BIPM15)[5] using a frequentist (TEMPO2) method and a Bayesian method (Guo et al., in preparation). The pulsar offsets define a timescale which we label TT(IPTA16). The left panels of Figure 4 show that TT(IPTA16) has significant deviations from TT(TAI) and, within the uncertainties, is consistent with TT(BIPM15). Directly referencing the pulsar ToAs to TT(BIPM15), as shown in the right panels, confirms the agreement of the two timescales. The figure also shows that consistent results are obtained with the two analysis methods. These results demonstrate that the long-term stability of a timescale based on the best available pulsar datasets is comparable to that of the best available atomic timescales. Furthermore, we have shown that the post-corrected timescale TT(BIPM15) is indeed more stable than TT(TAI).

Another interesting application of precision pulsar timing is to navigation of spacecraft that are distant from the Earth, even outside the solar system. Pulsars can form a "celestial GPS" system – instead of assuming a fixed observatory position and solving for the pulsar parameters as in normal pulsar timing, one can analyse the ToAs from a set of pulsars with known parameters and solve for the observatory position. For spacecraft navigation, it is most practical to use an X-ray telescope on the spacecraft and to observe a sample of pulsars with significant X-ray pulsed emission. Analysis of a realistic simulation [37] has shown that position location with an accuracy of better than 20 km is possible using observations of just four MSPs. Autonomous operation of the system is possible, but accuracy is improved with updates of the pulsar parameters from Earth-based observations. It is interesting to note that in 2016 the Chinese launched a satellite, XPNAV, dedicated to exploring X-ray pulsar navigation, and that NASA's recently launched NICER mission on the International Space Station has a project (SEXTANT) devoted to this topic as well.

[5] Terrestrial Time (TT) is a uniform theoretical timescale based on the value of the SI second on the Earth's geoid. TT(TAI) and TT(BIPM15) are realisations of TT based on TAI and BIPM15 respectively.

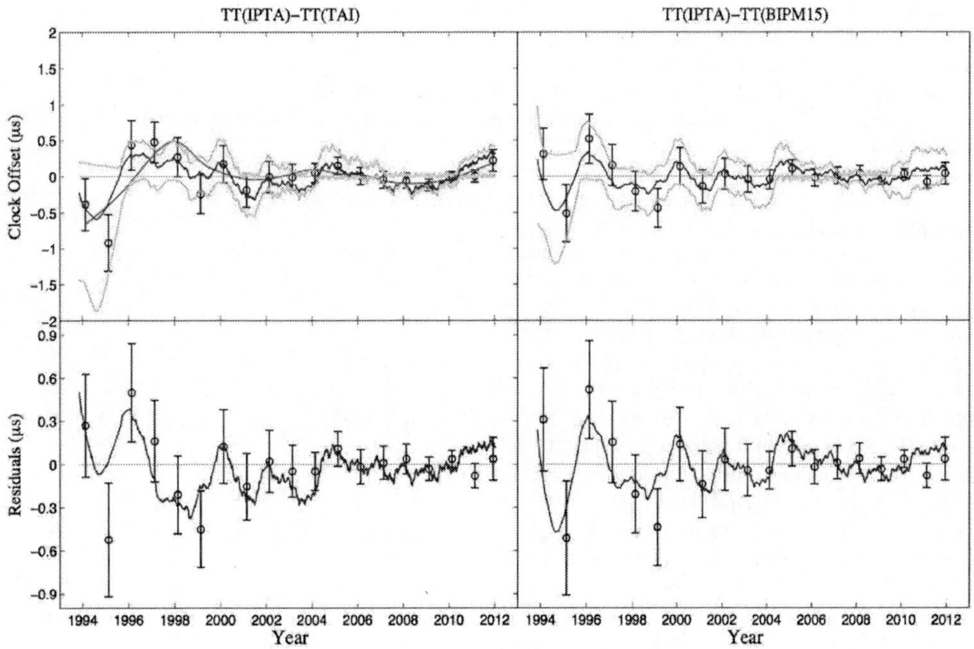

Figure 4. Offsets of the pulsar timescale TT(IPTA16) with respect to TT(TAI) (left panels) and TT(BIPM15) (right panels) derived from a frequentist analysis (black circles with error bars) and a Bayesian analysis (blue line with 1-σ error ranges in green) of the IPTA DR1 dataset. For the TT(TAI) reference, the red line gives the offset of TT(BIPM15) from TT(TAI) after subtracting quadratic terms. The lower panels show the residuals from TT(BIPM15). (Guo et al., in preparation)

6. Summary

Pulsar timing has many possible applications in physics and astrophysics. Their remarkable intrinsic period stability, especially for MSPs, their distribution throughout the Galaxy and the fact that many pulsars, including more than half of all MSPs, are in binary orbits around another star, allow explorations of theories of gravitation, properties of the interstellar medium, studies of binary and stellar evolution and many other topics. Pulsar timing arrays (PTAs) can be used as detectors of low-frequency gravitational radiation and to establish a "pulsar timescale" that is independent of terrestrial atomic timescales. In most of these areas, pulsars provide unique results and insights into the relevant physical processes. With new and more sensitive radio telescopes coming soon, the sample of pulsars that can contribute to these studies will grow, giving higher precision for existing applications and opening up new fields of research.

Acknowledgments
I thank my colleagues for their many contributions to the topics discussed and the referee for helpful suggestions.

References

[1] Manchester R N, Hobbs G B, Teoh A and Hobbs M 2005 *AJ* **129** 1993–2006
[2] Bhattacharya D and van den Heuvel E P J 1991 *Phys. Rep.* **203** 1–124
[3] Freire P C C *et al* 2017 *MNRAS* **471** 857–76

[4] Manchester R N and Taylor J H 1977 *Pulsars* (San Francisco: Freeman)
[5] Karako-Argaman C *et al* 2015 *ApJ* **809** 67
[6] Haberl F 2007 *Astrophys. Space Sci.* **308** 181–90
[7] Espinoza C M, Lyne A G and Stappers B W 2017 *MNRAS* **466** 147–62
[8] Lyne A G, Jordan C A, Graham-Smith F, Espinoza C M, Stappers B W and Weltevrede P 2015 *MNRAS* **446** 857–64
[9] Lyne A G, Pritchard R S, Graham-Smith F and Camilo F 1996 *Nature* **381** 497–8
[10] Akbal O, Alpar M A, Buchner S and Pines D 2017 *MNRAS* **469** 4183–92
[11] Hulse R A and Taylor J H 1975 *ApJ* **195** L51–3
[12] Taylor J H, Fowler L A and McCulloch P M 1979 *Nature* **277** 437
[13] Weisberg J M, Nice D J and Taylor J H 2010 *ApJ* **722** 1030–4
[14] Burgay M *et al* 2003 *Nature* **426** 531–3
[15] Lyne A G *et al* 2004 *Science* **303** 1153–7
[16] Perera B B P *et al* 2010 *ApJ* **721** 1193–205
[17] Kramer M *et al* 2006 *Science* **314** 97–102
[18] Breton R P, Kaspi V M, Kramer M, McLaughlin M A, Lyutikov M, Ransom S M, Stairs I H, Ferdman R D, Camilo F and Possenti A 2008 *Science* **321** 104–7
[19] Damour T and Schäfer G 1991 *Phys. Rev. Lett.* **66** 2549
[20] Gonzalez M E *et al* 2011 *ApJ* **743** 102
[21] Freire P C C, Wex N, Esposito-Farèse G, Verbiest J P W, Bailes M, Jacoby B A, Kramer M, Stairs I H, Antoniadis J and Janssen G H 2012 *MNRAS* **423** 3328–43
[22] Ransom S M *et al* 2014 *Nature* **505** 520–4
[23] Sesana A, Vecchio A and Colacino C N 2008 *MNRAS* **390** 192–209
[24] Hellings R W and Downs G S 1983 *ApJ* **265** L39
[25] Shannon R M *et al* 2015 *Science* **349** 1522–5
[26] Demorest P B *et al* 2013 *ApJ* **762** 94
[27] Gerosa D and Sesana A 2015 *MNRAS* **446** 38–55
[28] Shannon R M *et al* 2013 *Science* **342** 334–7
[29] Siemens X, Ellis J, Jenet F and Romano J D 2013 *Class. Quant. Grav.* **30** 224015
[30] Verbiest J P W *et al* 2016 *MNRAS* **458** 1267–88
[31] Lentati L *et al* 2016 *MNRAS* **458** 2161–87
[32] Stovall K *et al* 2016 *ApJ* **833** 192
[33] Nan R, Li D, Jin C, Wang Q, Zhu L, Zhu W, Zhang H, Yue Y and Qian L 2011 *Intnl J. of Mod. Phys.* D **20** 989–1024
[34] Carilli C and Rawlings S 2004 *New Astronomy Reviews*, **48** 11–2
[35] Petrova S A 2003 *A&A* **408** 1057–63
[36] Hobbs G *et al* 2012 *MNRAS* **427** 2780–7
[37] Deng X P *et al* 2013 *Adv. Space Res.* **52** 1602–21

IOP Conf. Series: Journal of Physics: Conf. Series **932** (2017) 012003 doi:10.1088/1742-6596/932/1/012003

Conquering systematics in the timing of the pulsar triple system J0337+1715: Towards a unique and robust test of the strong equivalence principle

N V Gusinskaia[1], A M Archibald[1], J W T Hessels[1,2], D R Lorimer[3,4,5], S M Ransom[6], I H Stairs[7] and R S Lynch[5,4]

[1] Anton Pannekoek Institute for Astronomy, University of Amsterdam, Science Park 904, 1098 XH Amsterdam, The Netherlands.
[2] ASTRON, the Netherlands Institute for Radio Astronomy, Postbus 2, 7990 AA, Dwingeloo, The Netherlands
[3] Department of Physics and Astronomy, West Virginia University, PO Box 6315, Morgantown, WV 26506, USA
[4] Center for Gravitational Waves and Cosmology, West Virginia University, Chestnut Ridge Research Building, Morgantown, WV 26505, USA
[5] Green Bank Observatory, Green Bank, WV 24944, USA
[6] National Radio Astronomy Observatory, Charlottesville, VA 22903, USA
[7] Department of Physics and Astronomy, University of British Columbia, Vancouver, BC V6T 1Z1, Canada

E-mail: N.Gusinskaia@uva.nl

Abstract. PSR J0337+1715 is a millisecond radio pulsar in a hierarchical stellar triple system containing two white dwarfs. The pulsar orbits the inner white dwarf every 1.6 days. In turn, this inner binary system orbits the outer white dwarf every 327 days. The gravitational influence of the outer white dwarf strongly accelerates the inner binary, making this system an excellent laboratory in which to test the strong equivalence principle (SEP) of general relativity – especially because the neutron star has significant gravitational self-binding energy. This system has been intensively monitored using three radio telescopes: Arecibo, Green Bank and Westerbork. Using the more than 25000 pulse times of arrival (TOAs) collected to date, we have modeled the system using direct 3-body numerical integration. Here we present our efforts to quantify the effects of systematics in the TOAs and timing residuals, which can limit the precision to which we can test the SEP in this system. In this work we describe Fourier-based techniques that we apply to the residuals in order to isolate the effects of systematics that could masquerade as an SEP violation. We also demonstrate that tidal effects are insignificant in the modeling.

1. The pulsar triple system PSR J0337+1715

PSR J0337+1715 is a millisecond radio pulsar that was discovered in 2012 using the Green Bank Telescope (GBT) during a 350-MHz drift-scan survey [1]. It was initially identified as a member of a 1.6-day binary system with an optically detected white dwarf (WD) companion (hereafter WD_i). Soon it became clear that this system has another gravitationally significant component, which orbits the pulsar-WD_i binary every 327 days. This component (hereafter WD_o) has mass 0.4 $M_\odot$, and is an old, cold white dwarf that is invisible in optical observations. As such, PSR J0337+1715 is a member of a hierarchical triple system with two WD components. It is the only known pulsar in a *stellar* triple system.

Due to the gravitational interactions between the components, it is possible to completely characterize the orbits of this system using direct numerical integration of the three-body equations of motion with perturbations between companions. Fitting this orbital model using pulsar times of arrival (TOAs) gives precise values for the orbital parameters and the masses of all stellar components. Owing to the hierarchical structure of this system, the gravitational influence of the outer white dwarf strongly accelerates the inner binary. This gives an opportunity to test the strong equivalence principle (SEP) of general relativity: i.e. whether the strongly self-gravitating neutron star (PSR J0337+1715) and weakly self-gravitating WD_i fall in the same way in the external gravitational potential of WD_o. Precision timing of the millisecond pulsar PSR J0337+1715 (sub-μs TOAs), the hierarchical structure of the system and information about the precise masses and orbital parameters of the companions make this system an excellent laboratory in which to test the SEP.

In order to perform the best possible timing of the millisecond pulsar PSR J0337+1715, it was intensively monitored using three world-class radio telescopes: Arecibo, GBT and Westerbork. It was monitored almost daily with Westerbork during the first ~ 1.5 years after discovery and it has continued to be monitored regularly with GBT and Arecibo for the past 6 years (figure 1). These observations make PSR J0337+1715 one of the most intensively monitored pulsars and provide a unique data set for testing gravitational theories.

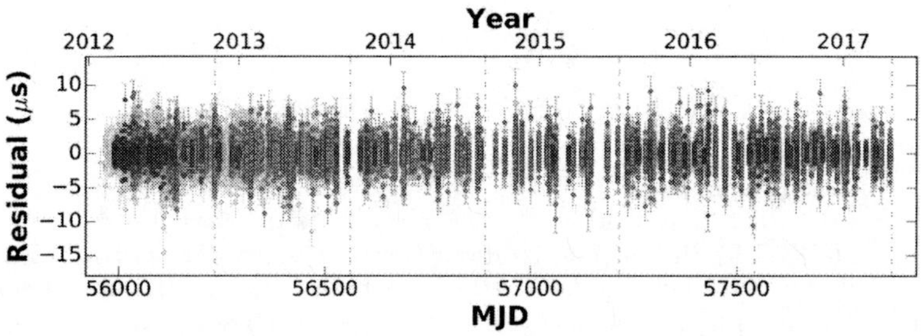

Figure 1. Residuals of the full numerical fit. Each color represents a different telescope: orange – Westerbork, green – GBT and blue – Arecibo. Dashed vertical lines represent T_{asc}=0 of the outer orbit.

2. Einstein's strong equivalence principle

The Einstein strong equivalence principle is a fundamental tenet of general relativity (GR). It postulates that gravitational mass equals inertial mass, even for objects with strong gravity. In other words, the gravitational interaction does not depend on the gravitational binding energy of the self-gravitating bodies. If the gravitational mass (M_G) and the inertial mass (M_i) are related according to equation:

$$M_G = (1 + \Delta)M_i, \tag{1}$$

where Δ is a dimensionless parameter, then the SEP states that $\Delta = 0$. This principle holds only in GR and all plausible alternatives to GR violate the SEP at some level ($\Delta \neq 0$). Thus, testing the SEP is an important way to test the validity of GR and to put constraints on its alternatives.

A straightforward method of testing the SEP is to compare how two bodies that have different gravitational binding energies fall in the presence of an external gravitational potential. Thus, the ideal test-beds for the SEP are hierarchical triple systems. If the SEP is violated ($\Delta \neq 0$), then the fractional difference in acceleration of the two bodies gives rise to excess eccentricity of their orbit in the direction of the external acceleration. In the case of the pulsar triple system, the pulsar inner orbit (pulsar - WD_i binary) would acquire an excess eccentricity towards the outer WD_o. As such, the pulsar inner orbit will be polarized in the direction of the WD_o, keeping this orientation as it moves in its outer orbit. This effect would lead to periodic changes of the orbital parameters with an exact frequency of $(-2\,\varphi_i + 1\,\varphi_o)$, where (φ_i) and (φ_o) are the inner and outer orbital frequencies ($\varphi = 1/P_{orb}$).

3. SEP in the pulsar triple system

In order to model orbital motion of the pulsar triple system we performed direct numerical integration of the three-body equations of motion in the first post-Newtonian approximation. We used this numerical orbital model and standard pulsar timing techniques (see e.g. [2]) to model the observed TOAs of PSR J0337+1715 (see Section 1), fitting for the pulsar and orbital parameters, including a possible difference between gravitational and inertial masses ($\Delta \neq 0$).

As a product of the linear part of the fit, we additionally produced derivatives, which represent the change of residuals corresponding to a *linear* change of each parameter of the numerical linear fit. The derivative of Δ represents the response of the timing residuals to a non-zero value of Δ, i.e. to the SEP violation effect. The signature of $\Delta \neq 0$ produces structure in the residuals, here folded at the inner orbital period (figure 2). Structure in the values of the residuals (diagonal stripes) illustrate the inner binary phase shift due to its eccentricity excess towards WD_o. The separation and orientation of the stripes confirms that this effect is periodic with frequency: $(-2\,\varphi_i + 1\,\varphi_o)$.

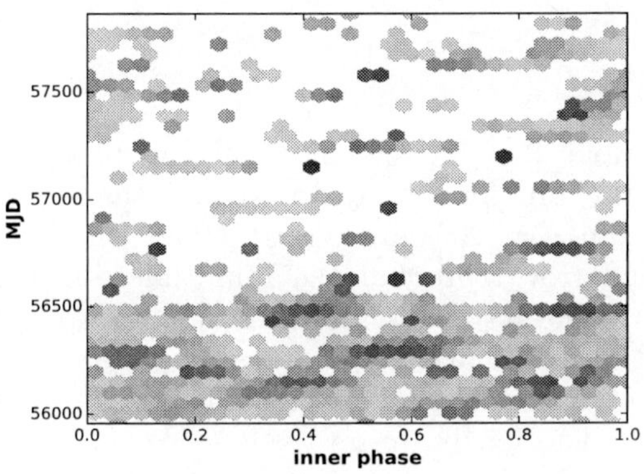

Figure 2. Change of the residuals of the fit corresponding to the effect of $\Delta \neq 0$ (derivatives of Δ), folded at the inner orbital period. Each hexagonal bin represents the average of the residuals for a particular range of inner phase and MJD. Diagonal stripes represent the phase shift due to the eccentricity excess of the pulsar inner orbit towards WD_o. Gaps represent the actual observing cadence.

4. Quasi-Fourier search

Ideally, the fitted value of Δ and its uncertainty would determine how well we constrain an SEP violation and whether GR is violated. However, fit uncertainties are only correct once we account for systematics in the residuals. Most importantly, such systematics could potentially masquerade as an SEP violation, especially if they have similar frequency structure. To quantify this, we used a Fourier-based technique to search for systematics whose frequency is harmonically related to the inner and outer orbital frequencies, assuming that these systematics have the same origin and, thus, roughly the same amplitude as the one related to Δ with $(-2\,\varphi_i + 1\,\varphi_o)$ frequency.

We perform a linear least-squares fit of the periodic signal to the residuals of the pulsar timing fit, in the form:

$$f(t) = A\,[\cos(k\varphi_i(t - T_{asc\,i}) + j\varphi_o(t - T_{asc\,o}));\ \sin(k\varphi_i(t - T_{asc\,i}) + j\varphi_o(t - T_{asc\,o})] \quad (2)$$

Here, t represents the TOAs (in MJD), $\mathrm{T}_{asc\,i}$ and $\mathrm{T}_{asc\,o}$ are times when the ascending nodes of the inner and outer orbits, respectively, equal zero, k and j are integer number of inner and outer orbital frequencies, and A is the amplitude of the harmonic systematic. As parameters of the fit, we use two vector components of the amplitude A: the amplitudes of sinusoid and cosinusoid with the same frequency. We fit these amplitudes to the residuals (and their uncertainties) provided by the pulsar timing fit. As a result, we measure the amplitude and phase of the signal at each frequency where $k \in [-3;3]$ and $j \in [0;3]$. Superposition of these components

IOP Conf. Series: Journal of Physics: Conf. Series **932** (2017) 012003 doi:10.1088/1742-6596/932/1/012003

is analogous to the discrete two-dimensional Fourier-transform. Note that the sinusoids with components $[k; j]$ are precisely the same as those with components $[-k; -j]$, thus we ignore the half-plane with negative j components.

We present the results of the quasi-Fourier search in vector plot form (figure 3): each arrow represent a frequency component. The position of the arrow represents the frequency of the systematics as a combination of k and j (horizontal and vertical axes values). The length of the arrow represents the amplitude (A) of this systematic, and the direction of this arrow represents its phase offset: i.e. sine and cosine components of A as vertical and horizontal projections of the arrow. Circles represent uncertainties on the linear least-squares fit parameters (projections of the amplitude A). We derived these uncertainties using the covariance of $[\mathrm{A}_{\cos}; \mathrm{A}_{\sin}]$ parameters for each systematic in the fit.

We performed our quasi-Fourier fit to all PSR J0337+1715 timing fit residuals (see blue arrows in figure 3). Our results show that there are still some periodic structures in our pulsar timing fit residuals that are not yet accounted for. The largest systematic that we found in our data has amplitude ≈ 50 ns, which is a factor of a few larger than our formal uncertainties.

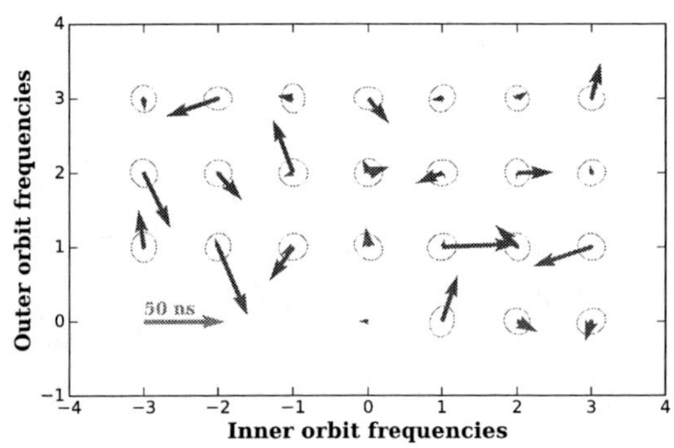

Figure 3. "Arrow plot" of the residuals of the full fit (blue arrows) and the signal of non-zero Δ (red arrows). The length of each arrow represents the amplitude A of the harmonic systematic in the data, whose phase is the combination of integer inner and outer orbital frequencies represented by values on the x and y axes, respectively (see eq 2). The direction of each arrow is the phase offset. Circles represent uncertainties on the linear least-squares fit.

With the red arrows in figure 3 we show the result of the quasi-Fourier search performed on the 'derivative' of the parameter Δ. These arrows represent the response of the timing residuals to the linear change of Δ, and this illustrates that a significant part of the $\Delta \neq 0$ signal is concentrated, indeed, in the $(-2\,\varphi_i + 1\,\varphi_o)$ component. The lengths of these arrows are arbitrary, and correspond to a value of Δ chosen for a convenient illustration.

The blue arrow that corresponds to the $(-2\,\varphi_i + 1\,\varphi_o)$ frequency is smaller than the fit uncertainty, which means that the timing fit provided the correct amplitude of the harmonic signal with this frequency. However, this signal does not necessarily correspond to the effect of $\Delta \neq 0$, since there are many other arrows that have amplitude significantly larger than the statistical uncertainty. Thus, the amplitude of the signal that the pulsar timing fit returns as a Δ value can actually be the result of an unknown systematic, similar to what is present in other frequency components. That means that we can not use the fitted value of Δ and its uncertainty provided by the fit to constrain an SEP violation, but instead our constraint should be based on a measure of the arrow length in the quasi-Fourier fit. In this way, we can use our quasi-Fourier search to put a reliable constraint on the parameters of the numerical fit and, consequently, on the SEP test results.

5. Tidal effects

One of the effects that could produce systematics in the data is secular changes of the pulsar orbit caused by tidal deformation of the inner WD companion. The neutron star raises tides on the inner WD, and this quadrupole moment introduces a non $1/r^2$ dependence of the gravitational pull. We do not model tidal effects in our timing fit. Thus, if they are present, they might appear as systematics in the timing residuals. However, even the inner binary of the triple system has a relatively wide orbit (16 lt-s), which suggests that tidal effects should be small.

In this section we provide calculations of the changes of longitude of periastron ($\dot{\omega}$) and shrinking of the orbit ($\dot{P}_{orb}$) of the pulsar – WD inner binary due to tidal effects. In our calculations we followed formulas provided by [3]. Since all the tidal effects strongly depend on the distance between objects, we do not provide the values of the tidal effects of the outer WD, whose orbit is a factor of 200 larger.

Secular periastron advance: The change of the longitude of periastron of the pulsar orbit caused by tidal deformation of its inner WD companion can be calculated using eq 3.7 of [3]:

$$\Delta\chi_{tidal} = k_2 \left(\frac{m_p}{m_c}\right) \left(\frac{R_c}{a}\right)^5 F(e) = 0.00036 \text{ deg/year} \quad (3)$$

Here, $R_c = 0.091\,R_\odot$ is the radius of WD_i, derived from optical observations [4], $m_p = 1.4478\,M_\odot$ is the mass of the pulsar, $m_c = 0.19751\,M_\odot$ is the mass of the WD_i, $a = 16$ lt-s is the semi-major axis of the pulsar – WD_i binary and $F(e) = 1$ is a function of its eccentricity. The orbital parameters and masses are derived from the numerical fit of the three-body interactions [1]. We used eq 27 from [5] to calculate the Love number: k_2 =0.102, which is consistent with other results for WDs (see e.g. [6]).

In summary, the change of χ_{tidal} of the pulsar inner orbit is about three orders of magnitude smaller than relativistic periastron advance and about one order of magnitude smaller than what can be detected with the current precision of the 6 years of PSR J0337+1715 timing. However this effect may be significant after $\sim$ 10 more years of observations.

Dissipative effect: The tidal deformation of WD_i can cause a loss of energy from the system (tidal lag) and thus shrinking of the orbit. We used eq 3.19 from [3] to calculate the characteristic timescale of the orbital period (P_{orb}) decrease:

$$\tau_p \equiv \frac{P_{orb}}{\dot{P}_{orb}} \approx \frac{3}{2}(1-e^2)^6 \left(\frac{m_c^2}{m_p(m_p+m_c)}\right) \left(\frac{n}{\omega_c}\right) \left(\frac{a}{R_c}\right)^8 \left(\frac{m_c}{<\mu> R_c}\right) = 1.16 \times 10^{14}\,\text{yr} \quad (4)$$

Following [3], here we make a conservative estimate, assuming that the inner WD_i has a high rotation rate ($\omega_c = 0.1$ rad/s) and a standard viscosity $<\mu>= 0.02 \times 10^{13}$ g/cm/s.

The characteristic timescale of the PSR J0337+1715 inner binary orbital decay due to tidal lag τ_p is $\sim$ 2 orders of magnitude longer than the characteristic timescale of the orbital decay due to gravitational wave emission τ_{gr}, which is $\approx 2 \times 10^{12}$ yr. (see e.g. [7] for formula for τ_{gr}). The current precision with which we can measure P_{orb} of the inner binary is $\approx 10^{-8}$ days. This means that we need about 10^4 years of observations in order to be able to detect dissipative tidal effects in the inner binary of the pulsar triple system.

References

[1] Ransom S M 2014 *Nature* **505** 520–4
[2] Lorimer D R and Kramer M (eds) 2012 *Handbook of Pulsar Astronomy* (Cambridge, UK: Cambridge University Press)
[3] Smarr L L and Blandford R 1976 *Astrophysical Journal* **207** 574–88
[4] Kaplan D L 2014 *The Astrophysical Journal Letters* **783** L23
[5] Hinderer T 2008 *The Astrophysical Journal* **677** 1216–20
[6] DallOsso S and Rossi E 2013 *MNRAS* **428** 518–31
[7] Wagoner R V 1975 *The Astrophysical Journal Letters* **196** L63–5

IOP Conf. Series: Journal of Physics: Conf. Series **932** (2017) 012004 doi:10.1088/1742-6596/932/1/012004

Proper motion of the radio pulsar B1727−47 and its association with the supernova remnant RCW 114

P S Shternin[1], M Yu[2], A Yu Kirichenko[3,1], Yu A Shibanov[1,4], A A Danilenko[1], M A Voronkov[5] and D A Zyuzin[1]

[1] Ioffe Institute, St. Petersburg, Russia
[2] National Astronomical Observatory, Chinese Academy of Science, Beijing, China
[3] Universidad Nacional Autonomia de Mexico, Ensenada, Baja California, Mexico
[4] Peter the Great St. Petersburg Polytechnic University, St. Petersburg, Russia
[5] CSIRO Astronomy & Space Science, Epping, New South Wales, Australia

E-mail: pshternin@gmail.com

Abstract. We report preliminary results of the analysis of the proper motion of the bright radio pulsar B1727−47. Using archival Parkes timing data, as well as original and archival ATCA interferometry observations, we, for the first time, constrain the pulsar proper motion at the level of 148±11 mas yr^{-1}. The backward extrapolation of the proper motion vector to the pulsar birth epoch points at the center of the Galactic supernova remnant RCW 114 suggesting the genuine association between the two objects. We discuss the implications of the association and argue that the distance to the system is less than 1 kpc. This value is at least two times lower than the dispersion measure distance estimates. This suggests that the existing Galaxy electron density models are incomplete in the direction to the pulsar.

1. Introduction

Identifying a pulsar (PSR) – supernova remnant (SNR) association leads to valuable conclusions on the properties of both components, such as distances, ages, evolution stages etc. Although there is a little doubt that each PSR was once born in a supernova explosion, only a few associations have been identified so far with high confidence (e.g., [1]). Even if the pulsar position projects on the remnant image, the probability is high that this is due to a chance coincidence [2]. Therefore each possible association should be considered in detail separately, and one of the strongest arguments can come from the measurement of the pulsar proper motion (p.m.) [3]. Here we accomplish this goal for PSR B1727−47 and SNR RCW 114.

2. PSR B1727−47 and its proper motion

PSR B1727−47 was among the first pulsars observed since the beginning of the Pulsar Era. It was discovered in 1968 with the Molonglo observatory [4] and is one of the brightest among the young (< 100 kyr) radio pulsars known. It has the period $P = 0.83$ s, the characteristic age $\tau \equiv P/(2\dot{P}) = 80$ kyr and the dispersion measure (DM) of 123 pc cm^{-3}. According to the widely-used NE2001 model of the Galactic electron density distribution [5], the DM-based distance to the pulsar is $D = 2.7$ kpc. However, the more recent model [1] places PSR B1727−47 as far as at 5.5 kpc.

Table 1. PSR B1727−47 p.m. obtained by various methods discussed in the text. CP – by comparing published positions; I – interferometry; T – from interglitch solutions; TN – TEMPONEST results; F – full combined solution. Cartesian components of the p.m. vector in RA (μ_α) and Dec. (μ_δ) directions and the total p.m. value μ are given.

	CP	I	T	TN	F
μ_α, mas yr^{-1}	63 ± 4	120 ± 9	65 ± 13	47 ± 14	73 ± 5
μ_δ, mas yr^{-1}	-92 ± 21	-162 ± 34	-224 ± 28	-132 ± 37	-127 ± 13
μ, mas yr^{-1}	111 ± 18	202 ± 28	232 ± 27	141 ± 36	148 ± 11

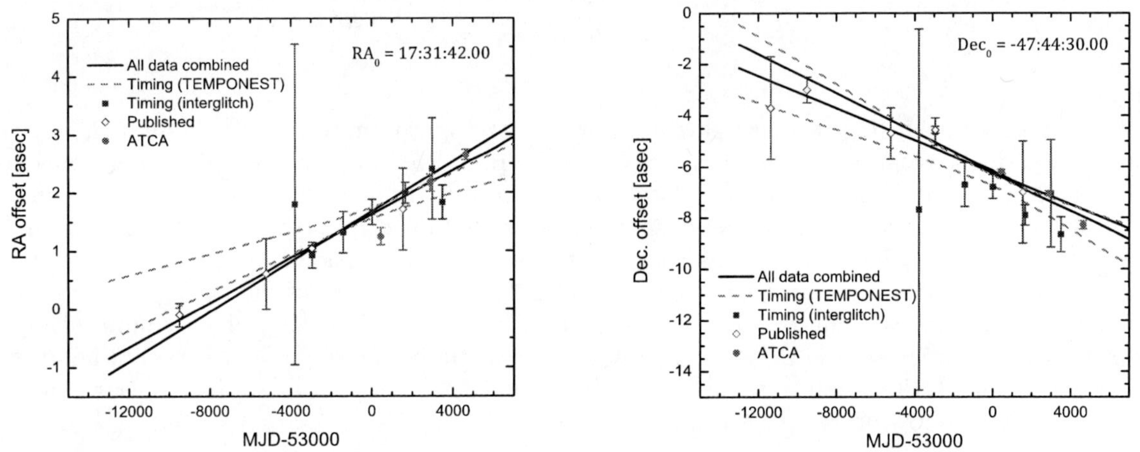

Figure 1. PSR B1727−47 positions discussed in this work. We show relative offsets in RA (*left*) and Dec. (*right*) directions from an auxiliary reference point, whose J2000 coordinates are indicated. Open diamonds correspond to previously published positions [6, 7]. Other positions are obtained in this work. Blue squares: Parkes timing solutions for interglitch epoches; red points: ATCA interferometry positions. Dashed lines bracket the 68 per cent TEMPONEST credible region and black lines are the same for all the data are combined, see text for details.

To the best of our knowledge, despite a long observational history, the p.m. of PSR B1727−47 has never been reported. However, a comparison of the published positions (blue diamonds in figure 1) suggests that its p.m. is significant. The result of a simple linear fit to these data is given in the first column in table 1. At $D = 2.7$ kpc, the estimated p.m. value $\mu^{\rm CP} = 111 \pm 18$ mas yr^{-1} leads to a suspiciously large transverse velocity $v_\perp = 1420 \pm 230$ km s^{-1} (note that a typical 2D pulsar velocity is about 250 km s^{-1} [8]). Thus the p.m. of PSR B1727−47 deserves a more detailed study. We have done this using Parkes timing and ATCA interferometry imaging data.

2.1. Parkes timing analysis

We used the processed timing data from the Parkes telescope archive[1]. We selected the data obtained from Feb 1993 to Mar 2014. Detailed information on the observations and filterbank systems is described in [7]. For analysis we used 222 times of arrival (TOAs) spanning 21 yr in total (MJD range 49043–56740). For the timing analysis we used the TEMPO2 package [9]. The pulsar is known to be glitching. Four glitches have been already reported [7]. In more

[1] http://data.csiro.au

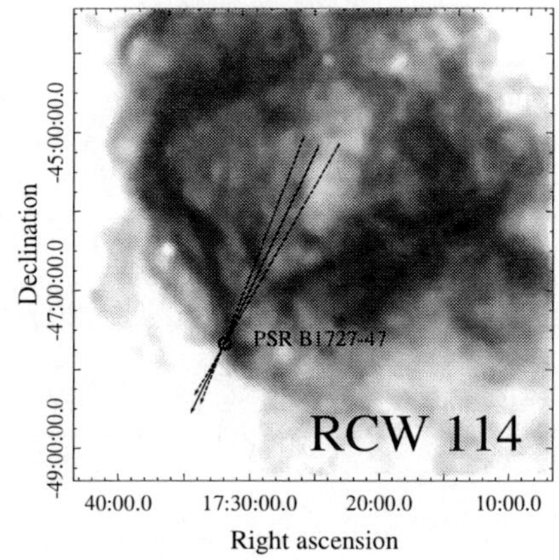

Figure 2. Hα image image of the RCW 114 field. The apparent radius of RCW 114 is about 2°. PSR B1727−47 position at the present epoch is marked by the circle, and the 68 per cent credibility region for the pulsar track extrapolated backwards is shown by dashed lines.

recent data we identified two new glitches, however the space limitations preclude us from the discussion of their parameters here. To measure the p.m., we first constructed timing solutions for seven inter-glitch periods in our data. The positions obtained in this way are shown in figure 1 with blue squares and the corresponding p.m. obtained from these positions is given in the 'T' column in table 1. The construction of the coherent solution for the full data span is complicated not only due to the presence of glitches but also because of the strong timing noise. To get rid of this, we employed the Bayesian TEMPONEST utility [10] which simultaneously optimizes the timing solution, glitches' parameters and timing noise model. The obtained p.m. is given in the 'TN' column of table 1, and the resulting 68 per cent credibility pulsar track is shown in figure 1 within red dashed lines.

2.2. ATCA radio interferometry analysis

PSR B1727−47 was observed with ATCA in 2004−2005 (project C1323) and in 2011 November (project C2566). For the C1323 observations, we used the data set obtained on 2005 March 26, as this session provides sufficient uv-coverage for accurate position measurements. The details of the array configurations and correlator modes can be found in the ATCA archive. The data were processed and analyzed in a standard way using the MIRIAD package [11]. The measured pulsar positions revealed a significant shift between the epochs in a direction compatible with the timing measurements (figure 1). To verify that the shift is not caused by some systematic effects, we performed observations (project CX367) of the pulsar with ATCA on 2016 September 15 using Director's time. All ATCA positions are shown in figure 1 with red dots, and the corresponding p.m. values are given in the 'I' column in table 1.

The interferometric and timing observations give qualitatively similar results. Taking the weighted mean of all the data (namely, ATCA positions, older published positions and TEMPONEST results) we obtain our final result given in the last column in table 1. The corresponding pulsar track uncertainty is constrained in figure 1 within black lines at 68 per cent credibility level.

3. Supernova remnant RCW 114

The backward extrapolation of the measured p.m. vector to the PSR B1727−47 birth epoch (using $\tau = 80$ kyr; the trajectory distortion due to effect of Galactic potential is negligible for this age) points roughly at the center of the nearby Hα nebula RCW 114, also known as SNR G343.0–6.0 [12] (figure 2). This strongly suggests a genuine association between the two objects and independently constrains the system age at $\sim 60 - 80$ kyr. The SNR nature of RCW 114 is supported by a recent discovery of C[IV] emission from the nebula [13]. This detection also allowed to place an upper limit on the distance to SNR, $D_{\rm SNR} \lesssim 1$ kpc [13]. On the other hand, detailed spectroscopic analysis of the stars projected on RCW 114 suggests $D_{\rm SNR} > 0.5$ kpc [14]. The distance range of $0.5 - 1$ kpc is plausible if the SNR is at the snowplough stage [15]. Using expressions from [15] for typical explosion energy of 10^{51} erg, ambient density of ~ 1 cm^{-3} and $D_{\rm SNR} = 0.7$ kpc, we estimated the expansion velocity of 96 km s^{-1}, which is in accord with the velocity of the faint outer filament of -80 km s^{-1} reported in [16].

4. Conclusions

- Based on radio timing and interferometric observations, we find the proper motion of PSR B1727−47 of 148±11 mas yr^{-1}. For the DM-based distance of 2.7 kpc (5.5 kpc) it leads to extremely high $v_\perp \approx 1900$ km s^{-1} (3900 km s^{-1}).
- The proper motion vector points at the center of the SNR RCW 114, suggesting SNR–PSR association. In this case the most plausible distance to the system is ~ 0.7 kpc and the p.m.-based age is $60 - 80$ kyr, compatible with the pulsar characteristic age. The relatively low distance value can be checked independently by the parallax measurements with very long baseline interferometry.
- Assuming $D = 0.7$ kpc, we find $v_\perp \approx 500$ km s^{-1}, which is well in line with the 2D pulsar velocity distribution [8]. This result suggests that the existing models of the electron density in the Galaxy require some tuning in the direction to PSR B1727−47. For instance, inclusion of a nearby 'clump' of enhanced electron density [5] in the model can reconcile the high DM value and the low distance inferred here.

Acknowledgments

We thank G B Hobbs, L Lentati and R N Manchester for discussions and the referee S Zane and the editor G G Pavlov for useful comments. The work was supported by the Russian Science Foundation, grant 14-12-00316. The Australia Telescope Compact Array is part of the Australia Telescope National Facility which is funded by the Australian Government for operation as a National Facility managed by CSIRO. This paper includes archived data obtained through the Australia Telescope Online Archive (http://atoa.atnf.csiro.au).

References

[1] Yao J M, Manchester R N and Wang N 2017 *ApJ* **835** 29
[2] Gaensler B M and Johnston S 1995 *MNRAS* **277** 1243–53
[3] Kaspi V M 1998 *Advances in Space Research* **21** 167–76
[4] Large M I, Vaughan A E and Wielebinski R 1968 *Nature* **220** 753–6
[5] Cordes J M and Lazio T J W 2002 NE2001.I. A new model for the Galactic distribution of free electrons and its fluctuations *Preprint* astro-ph/0207156
[6] Manchester R N, Newton L M, Goss W M and Hamilton P A 1983 *MNRAS* **202** 269–85
D'Alessandro F, McCulloch P M, King E A, Hamilton P A and McConnell D 1993 *MNRAS* **261** 883–94
Wang N, Manchester R N, Pace R T, Bailes M, Kaspi V M, Stappers B W and Lyne A G 2000 *MNRAS* **317** 843–60
[7] Yu M *et al* 2013 *MNRAS* **429** 688–724
[8] Hobbs G, Lorimer D R, Lyne A G and Kramer M 2005 *MNRAS* **360** 974–92
[9] Hobbs G B, Edwards R T and Manchester R N 2006 *MNRAS* **369** 655–72
[10] Lentati L, Alexander P, Hobson M P, Feroz F, van Haasteren R, Lee K J and Shannon R M 2014 *MNRAS* **437** 3004–23
[11] Sault R J, Teuben P J and Wright M C H 1995 A retrospective view of MIRIAD *Astronomical Data Analysis Software and Systems IV* (*Astronomical Society of the Pacific Conference Series* vol 77) ed Shaw R A, Payne H E *et al* p 433
[12] Green D A 2014 *Bulletin of the Astronomical Society of India* **42** 47–58
[13] Kim I J, Min K W, Seon K I, Han W and Edelstein J 2010 *ApJ* **709** 823–31
[14] Welsh B Y, Sallmen S, Jelinsky S and Lallement R 2003 *A&A* **403** 605–11
[15] Cioffi D F, McKee C F and Bertschinger E 1988 *ApJ* **334** 252–65
[16] Meaburn J, Goudis C, Solomos N and Laspias V 1991 *A&A* **252** 291–8

IOP Conf. Series: Journal of Physics: Conf. Series **932** (2017) 012005 doi:10.1088/1742-6596/932/1/012005

Neutron stars: Observational diversity and evolution

S Safi-Harb

Department of Physics and Astronomy, University of Manitoba, Winnipeg MB R3T 2N2, Canada

E-mail: samar.safi-harb@umanitoba.ca

Abstract. Ever since the discovery of the Crab and Vela pulsars in their respective Supernova Remnants, our understanding of how neutron stars manifest themselves observationally has been dramatically shaped by the surge of discoveries and dedicated studies across the electromagnetic spectrum, particularly in the high-energy band. The growing diversity of neutron stars includes the highly magnetized neutron stars (magnetars) and the Central Compact Objects shining in X-rays and mostly lacking pulsar wind nebulae. These two subclasses of high-energy objects, however, seem to be characterized by anomalously high or anomalously low surface magnetic fields (thus dubbed as 'magnetars' and 'anti-magnetars', respectively), and have pulsar characteristic ages that are often much offset from their associated SNRs' ages. In addition, some neutron stars act 'schizophrenic' in that they occasionally display properties that seem common to more than one of the defined subclasses.

I review the growing diversity of neutron stars from an observational perspective, then highlight recent and on-going theoretical and observational work attempting to address this diversity, particularly in light of their magnetic field evolution, energy loss mechanisms, and supernova progenitors' studies.

1. Introduction: Brief history of the neutron stars zoo

This conference celebrates 50 years of neutron stars discovery. Neutron Stars were discovered as Pulsars (PSRs) by Jocelyn Bell and Antony Hewish in 1967 but predicted to exist back in 1934 by Baade and Zwicky, just two years following the discovery of the neutron by Chadwick. It is with the Crab and Vela pulsars discovery in their respective supernova remnants (SNRs) that Baade and Zwicky's 1934 prediction that supernovae (SNe) make neutron stars was confirmed. The launch of imaging X-ray telescopes, such as *Einstein* (1980's), followed by *ROSAT* and *ASCA* (1990's) then by *Chandra*, *XMM-Newton* and *Suzaku* (2000's), showed us how these rotation-powered pulsars (RPPs) power synchrotron-dominated pulsar wind nebulae (PWNe).

One of the neutron star subclasses introduced in the 1990's is the 'anomalous' X-ray pulsars (AXPs). It was recognized first by Mereghetti & Stella [1] as a new class of X-ray pulsars, since their spectra were much softer than those of accretion-powered pulsars and had no binary companions; furthermore their X-ray luminosity exceeds the energy available from spin-down power, so they can not be powered by rotation. Duncan & Thompson [2] suggested that these are 'magnetars', i.e. highly magnetized neutron stars with a magnetic field exceeding the so-called 'quantum critical field' value of $B_{\rm QED} = 2\pi\ m_e^2c^3/eh = 4.4{\times}10^{13}$ G. AXPs are now merged with the Soft Gamma-ray Repeaters (SGRs), as bursting high-energy sources whose powerful outbursts are believed to be powered by their magnetic field decay. Dubbed under the

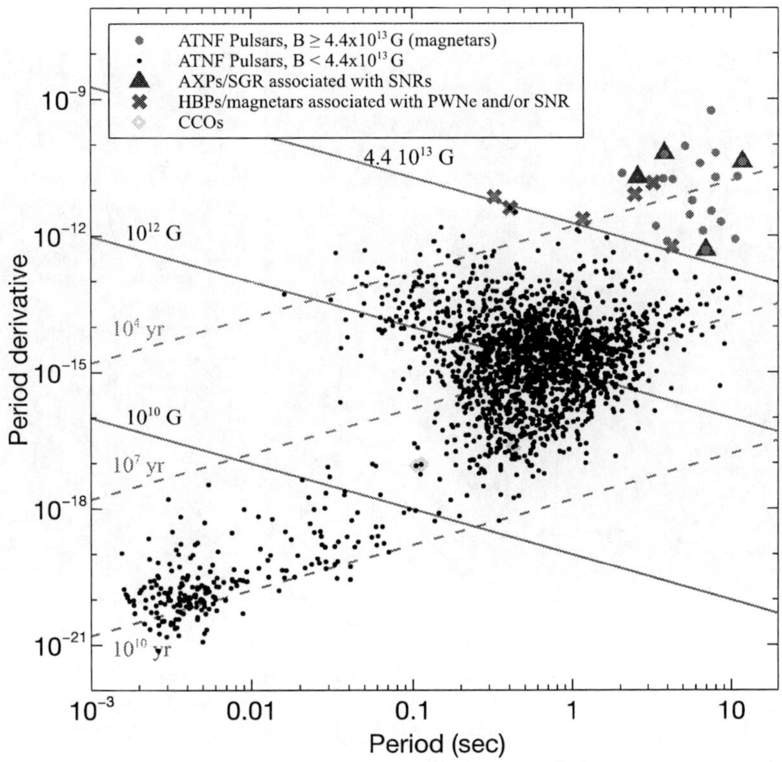

Figure 1. P–$\dot{P}$ diagram showing the diversity of neutron stars. Lines of constant dipole magnetic field (solid black) and characteristic age (dashed grey) are shown. The $4.4{\times}10^{13}$ G line corresponds to the QED value of the magnetic field that is traditionally used to separate magnetars from the rotation-powered pulsars. The 3 pulsars shown in cyan (diamond symbol) correspond to the 3 CCOs for which we have a measured P and $\dot{P}$. The 6 starred objects correspond to the 5 HBPs and 1 magnetar associated with a PWN and/or SNR, with the youngest and fastest two being PSR J1846–0258 in Kes 75 and J1119–6127 in G292.2–0.5, both having now displayed magnetar-like activity. The blue triangles correspond to the 3 AXPs + 1 SGR in our Galaxy known to be *securely* associated with an SNR *with known age.*

'Magnetars' family, these objects are among the most extreme objects known in the universe (see, e.g., [3]).

In the past decade, other sub-classes of neutron stars have emerged, thanks to sensitive radio and X-ray observations. These subclasses include the X-ray Dim Isolated Neutron Stars (XDINSs) (or the Magnificent Seven [4]), the Rotating Ratio Transients (RRATs [5]), the high-B radio pulsars (HBPs) with $B{\geq}B_{\rm QED}$ [6], and the Central Compact Objects (CCOs), typified by the Cas A CCO [7,8]. While we used to treat them as different classes, we are now into the era of unifying these sources thanks to a growing body of observational and theoretical works. This review will highlight these works, focusing on neutron stars with 'extreme' magnetic fields (i.e. much smaller or higher than the canonical value of $\sim 10^{12}$ G for the 'classical' rotation-powered pulsars) and stressing the fact that their hosting SNRs provide an independent means to address their unification.

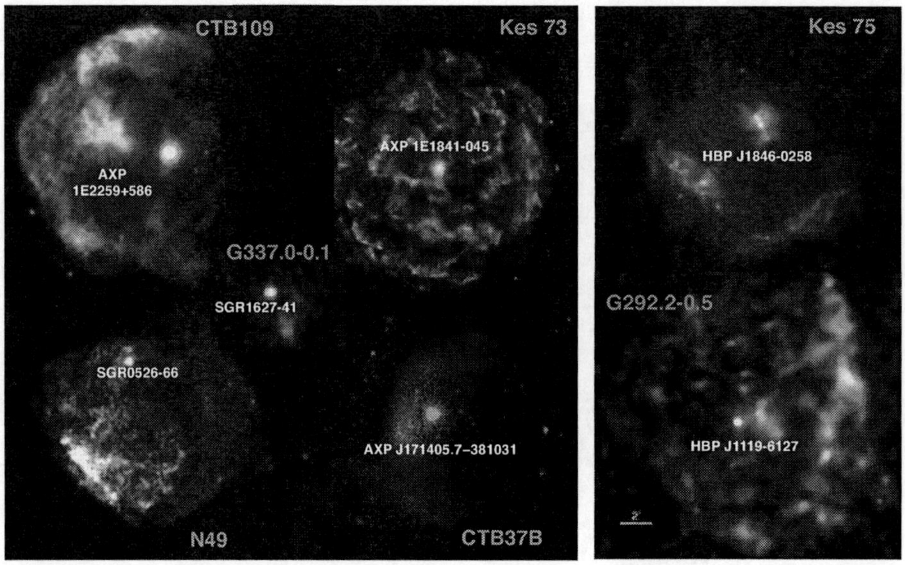

Figure 2. Left: AXPs and SGRs securely associated with SNRs. Right: HBPs associated with SNRs and having displayed a magnetar-like behaviour. Figure parts © AAS. Reproduced with permission: CTB 109 (ESA/*XMM-Newton*/M Sasaki *et al* [9]), Kes 73 (NASA/CXC/U. of Manitoba/H Kumar *et al* [10]), N49 (NASA/CXC/Caltech/S Kulkarni *et al* [11]), CTB 37B (MOST/ESA/*XMM-Newton*/H Kumar & S Safi-Harb), G337.0–0.1 (ESA/*XMM-Newton*/P Esposito *et al* [12]), Kes 75 (NASA/CXC/H Kumar & S Safi-Harb [13]), G292.2–0.5 (ESA/*XMM-Newton*/H Kumar *et al* [14]).

2. Pulsars Characteristics and diversity

The diversity of pulsars is connected with their positions on the so-called P–$\dot{P}$ diagram. The period P and period derivative $\dot{P}$ determine their spin-down energy $\dot{E}{=}I\Omega\dot{\Omega}$ (where I is their moment of inertia and $\Omega{=}2\pi/P$), their surface dipole field strength (at the magnetic equator) $B{=}3.2{\times}10^{19}\left(P\dot{P}\right)^{1/2}$ G, and their characteristic age $\tau_c{=}P/2\dot{P}$. For some pulsars, the braking index $n{=}\nu\ddot{\nu}/\dot{\nu}^2$ (where $\nu{=}1/P$) is measured (e.g, [15]).

Figure 1 shows the P-$\dot{P}$ diagram for the 2,613 currently known pulsars[1], revealing the growing diversity of neutron stars, and table 1 summarizes the secure PSR-SNR associations[2] for the pulsars with 'extreme' magnetic fields. Below we briefly comment on these growing subclasses of neutron stars, and then address the blurring diversity and magnetic field evolution.

2.1. Magnetars and HBPs

There are currently 29 known magnetars (including 6 candidates)[3]. Magnetars are traditionally discovered as high-energy sources, with no PWNe around them. Only 3 AXPs and 2 SGRs (one of which is an extragalactic pulsar in the LMC SNR N49) are securely associated with SNRs (see figure 2). The High-B Pulsars (HBPs) are a growing class of radio-detected pulsars with an inferred magnetic field close to, or just above, the QED value. We know of approximately a dozen such objects, and they are believed to be powered primarily by rotational energy. The youngest

[1] `http://www.atnf.csiro.au/research/pulsar/psrcat` (v1.56)
[2] `http://www.physics.umanitoba.ca/snr/SNRcat` (SNRcat)
[3] `http://www.physics.mcgill.ca/~pulsar/magnetar/main.html`

IOP Conf. Series: Journal of Physics: Conf. Series **932** (2017) 012005 doi:10.1088/1742-6596/932/1/012005

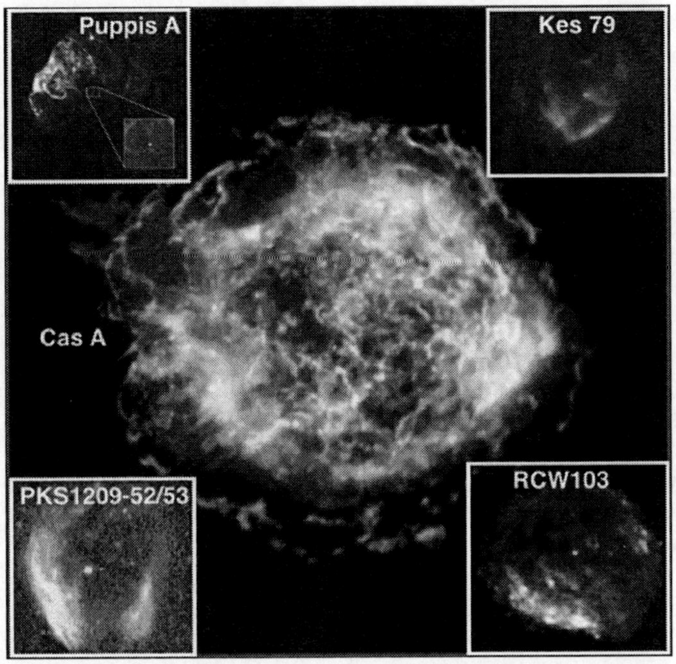

Figure 3. The CCOs with known spin properties, in addition to Cas A whose CCO does not yet show pulsations (see table 1). Figure parts © AAS. Reproduced with permission. Credits: Puppis A: *ROSAT*/R Petre *et al* [16] Kes 79: NASA/CXC//P Zhou *et al* [17]; PKS 1209–52/53: *ROSAT*/Max-Planck-Institut für extraterrestrische Physik; RCW 103: NASA/CXC/C Braun & S Safi-Harb; Cas A: NASA/CXC/GSFC/U Hwang *et al* [18].

HBPs, PSR J1846–0258 and J1119–6127 in the SNRs Kes 75 and G292.2–0.5, respectively (figures 1 and 2), have recently displayed magnetar-like activity providing conclusive evidence for the connection between HBPs and magnetars (see §2.3).

2.2. CCOs

CCOs are X-ray emitting neutron stars found near SNR centres and typified by the compact object in the SNR Cas A (see figure 3). The term CCO was coined by Pavlov *et al* [7] following the first light *Chandra* discovery of the object.[4]

There are currently 14 CCOs[5] (including 6 candidates) known in our Galaxy ([8, 19], SNRcat). These objects are X-ray emitters with no optical or radio counterparts, and with no evidence of PWNe surrounding them. X-ray pulsations have been discovered from 3 CCOs (see table 1). Their timing properties imply magnetic fields $\sim 10^{10}$–10^{11} G, much lower than those of the traditional RPPs and magnetars, which led Gotthelf & Halpern [20] to dub them 'anti-magnetars'. Spectroscopic studies support this interpretation through the discovery of spectral line features interpreted as cyclotron lines from a low B (e.g., [21–24]) or from modelling their thermal X-ray emission (e.g., [25, 26]).

These objects are also known to be quiet X-ray emitters (except for the 'CCO' in RCW 103) with their X-ray emission described by two blackbodies and their X-ray luminosity (ranging from $\sim$ a few$\times 10^{32}$–10^{34} erg s^{-1}) exceeding their spin-down energy. It is believed that their thermal spectra are derived from residual cooling and partly from accretion of supernova debris (e.g., [20]). One of their intriguing properties is their characteristic ages being orders of magnitude greater than their hosting SNRs' ages (see table 1 and figure 3).

[4] In a footnote referring to the Cas A compact object, Pavlov *et al* [7] state: 'According to the convention recommended by the Chandra Science Center, this source should be named CXO J232327.9+584842. We use the abbreviation CCO for brevity.'

[5] the total would be 15 CCOs if we include the variable X-ray source 1E 161348–5055 in RCW 103.

2.3. Diversity blurred

A growing body of observational and theoretical work has emerged blurring this apparent diversity: (1) The discovery of magnetar-like outbursts accompanied by spectral changes from the rotation-powered HBP J1846–0258 in the SNR Kes 75 [13, 27, 28] and more recently from the older HBP J1119–6127 in the SNR G292.2–0.5 [29–31] (2) The discovery of transient radio emission from magnetars (e.g., [32]) commonly discovered as high-energy objects. (3) The discovery of low-B magnetars, i.e. with $B \leq B_{\rm QED}$, through their bursting activity [33–35] (4) The discovery of magnetar-like activity from 1E 161348–5055 [36, 37], the variable central X-ray source in the SNR RCW 103, classified by some authors as a CCO. (5) The discovery of PWNe (which are characteristic of RPPs) around the magnetar, Swift J1834.9–0846 [38] and the HBP J1119–6127 [39–41] which, as mentioned above, recently revealed itself as a magnetar – see figure 1.

The thermal X-ray luminosities of some of the HBPs and magnetars are systematically higher than those of the traditional radio pulsars, suggesting that magnetic fields affect their X-ray emission. Using 2D simulations of the fully-coupled evolution of both temperature and magnetic field in neutron stars, Viganò *et al* [42] unified the phenomenological diversity of magnetars, HBPs and isolated nearby neutron stars by varying their initial magnetic field, mass and envelope composition. Furthermore, the magnetic field topology was argued to play a key role in the observed properties of the bursting activity of neutron stars. Perna & Pons [43] showed that the toroidal component, particularly its strength with respect to the poloidal component, plays a significant role in the frequency of the bursting activity and its dependence on the age of the system. In addition, the role of fallback disks around young isolated neutron stars has been highlighted by Alpar *et al* [44] to unify different classes of neutron stars.

3. SNRs associations shedding light on magnetic field evolution

Generally it is assumed that neutron stars lose energy by spinning down due to the emission of magneto-dipole radiation. However, this simple model does not describe the neutron star population for the following reasons. First, it predicts a braking index $n = 3$, which has not been observed in young neutron stars; most of the measured indices are smaller than 3 (e.g., [15]). Second, the pulsars securely associated with SNRs show a remarkable disparity between their characteristic age and the SNR age, in some cases differing by several orders of magnitude (especially for the CCOs). Under the standard assumption of constant magnetic field, we have: $P_0 = P\left[1 - (n-1)\tau\frac{\dot{P}}{P}\right]^{\frac{1}{(n-1)}}$. Replacing τ with the SNR age (table 1), we are generally unable to explain the observed braking indices and enforce the SNR's age to be equal to the PSR's age with a constant n (see, e.g., figure 2 in [45]). Rogers & Safi-Harb [45, 46] addressed magnetic field evolution focusing on the diverse population of neutron stars with 'anomalous' magnetic fields (i.e. much higher or lower than the canonical value of 10^{12} Gauss). These objects include the AXPs, SGRs, HBPs and CCOs securely associated with SNRs of known ages and listed in table 1.

3.1. Magnetic Field Decay

Magnetic field decay has been invoked to describe the evolution of AXPs and SGRs (e.g., [47–49]). Magnetic field decay channels depend on a variety of effects including Ohmic dissipation, ambipolar diffusion and the Hall drift. While for the low-B pulsars with ages $\sim$10–100 kyr the Ohmic dissipation is expected to dominate the magnetic field evolution, for magnetar-strength-B pulsars, the Hall effect provides the dominant mechanism for the field evolution on timescales comparable to their associated SNR ages.

Pulsar (SNR)	P	B ($\times10^{13}$ G)	τ (kyr)	$\tau_{\rm SNR}$ (kyr)
AXP 1E 1841–045 (Kes 73)	11.79	70.3	4.57	0.75–2.10
AXP 1E 2259+586 (CTB 109)	6.98	5.9	230	10.0–16.0
CXOU J171405.7–381031 (CTB 37B)	3.83	50	0.95	0.35–3.15
SGR 0526–66 (N49)	8.05	56	3.36	<4.80
SGR 1627–41 (G337.0–0.1)	2.59	22	2.2	<5.0
HBP J1846–0258 (Kes 75)	0.3265	4.9	0.73	0.90–4.30
HBP J1119–6127 (G292.2–0.5)	0.408	4.1	1.61	4.20–7.10
RX J0822.0–4300 (Puppis A)	0.113	3.27×10^{-3}	193	3.70–5.20
CXOU J185238.6+004020 (Kes 79)	0.105	3.05×10^{-3}	1.92×10^{5}	5.40–7.50
1E 1207.4–5209 (PKS 1209–51/52)	0.424	9.83×10^{-3}	3.02×10^{5}	2.00–20.0

Table 1. Pulsars with very high or very low magnetic field securely associated with SNRs with known ages.

3.2. Magnetic field growth

The hypothesis of fallback accretion goes back decades ago [50–52]. The submergence of the magnetic field due to fallback accretion has been explored by a number of authors (e.g., [53,54]), and revived recently [45,46,55–61] for the following reasons: (1) $n<3$ (see Section 3.3); (2) timing of CCOs giving low surface $B\sim10^{10}$–10^{11} G; (3) spectroscopic evidence for low surface magnetic field, as found, e.g., for the CCOs in Cas A, PKS 1209–51/52, and Puppis A; (4) the highly modulated pulsed signal in Kes 79's CCO implying a much higher internal magnetic field. This all supports growing evidence for the submerged field scenario due to fallback accretion.

3.3. Empirical model for magnetic field evolution

In Rogers & Safi-Harb [45, 46], both magnetic field growth and decay are described within the same basic framework, using empirical models for magnetic field evolution developed for various classes of neutron stars. In this approach the braking index is time-dependent: $n = 3 - 4\tau\dot{f}_j/f_j$, where the function $f_j(t)$ carries the time-dependence of the field. Thus, field decay gives $n > 3$ since $\dot{f}_D < 0$, and field growth gives $n < 3$ since $\dot{f}_G > 0$. The sample shown in table 1 (plus other RPPs with a measured braking index) gives solutions to the field evolution in neutron stars, some of which are shown in figure 4 (see [45] and figure 5 therein for a more detailed and complete description of the sample and fits considered). Magnetic field growth has been used to fit the 3 CCOs (cyan), the 2 SGRs (red) and the HBP J1846–0258 in Kes 75 (green). AXPs have been fit with magnetic field decay models highlighted by the grey area in figure 4 [47]. It is worth noting that the time evolution of the CCOs' characteristic age explains the apparent large discrepancy between the pulsars' ages (appearing very old) and the ages of the associated young SNRs. In particular, for the three systems shown, the PSR and SNR ages match at times $\geq10^{4.5}$ yr, by which time the SNR would have mostly dissipated. Therefore, the characteristic age for these pulsars does not reflect their true age as long as they are within their SNRs. This property, along with their inferred low asymptotic field strength, lead to the suggestion that CCOs could be ancestors of old isolated radio pulsars as long as they overcome the accretion phase (which would explain their X-ray dominant emission) and their surface field grows to the critical limit required for radio emission. The late time evolution of the CCOs may also link them to the class of objects known as XDINS; radio-quiet X-ray pulsars with long periods and no apparent SNR associations, with some of these objects having magnetic fields $\geq10^{13}$ G, similar to the HBPs.

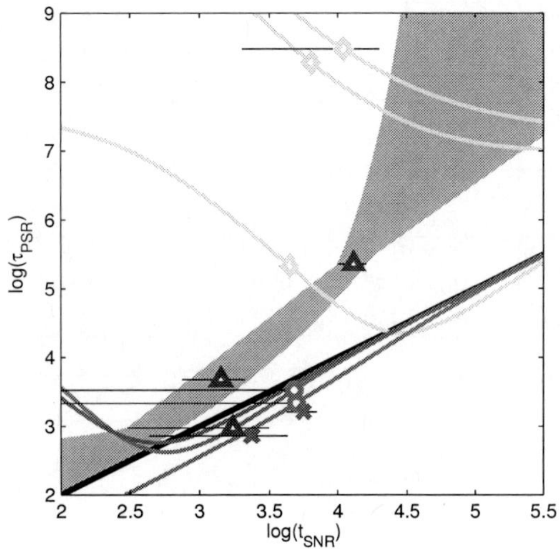

Figure 4. Pulsar age versus SNR age shown with evolutionary tracks for evolving magnetic fields in PSR–SNR pairs. The solid black diagonal line corresponds to equal PSR and SNR ages. The colours/symbols match those used in figure 1, with the cyan, red and green curves showing field growth fits to 3 CCOs, 2 SGRs and 1 HBP, respectively. See §3.3 for details.

4. SN progenitors

The diversity of neutron stars can be related to the diversity in SNe types resulting from core-collapse SNe (e.g., [62–64]).

4.1. Magnetar progenitors

Which stars make magnetars? Two scenarios of interest have been proposed in the literature. (1) The proto-neutron star model where the neutron star is born with a few milliseconds period [2], which predicts that the initial kinetic energy of the supernova would be reflected in a super-energetic hosting SNR. However, studies of magnetar SNRs and HBP SNRs (albeit limited to a very small sample and subject to low-resolution, CCD-type, X-ray spectroscopy) yield a 'typical' kinetic energy of the order of a few$\times 10^{50}$–10^{51} ergs (e.g., [14, 65]). Furthermore, the inferred ratio of initial to current spin-period (P_0/P) for the magnetars and HBPs is not much smaller than 1, unlike what is predicted by the proto-neutron star model. (2) The fossil field hypothesis where magnetars are born from the most massive, most magnetic, main-sequence stars [66]. Studies of magnetar SNRs point towards massive progenitors (see, e.g., [67]), supporting the fossil field hypothesis. However, this is not conclusive given that a few other studies point to lower mass progenitors [68, 69].

4.2. CCOs' progenitors

Progenitor studies have been performed on the CCOs-hosting SNRs: Cas A, Puppis A, Kes 79, RCW 103 and RX J1713.7–3946. It is interesting to note that these studies point to progenitor masses around 20 $M_\odot$ (although inconclusively). Below we summarize these studies.

- Cas A, the youngest historical type SNR in our Galaxy, harbours the CCO CXOU J232327.9+584842. It is commonly believed that Cas A results from a SN IIb event with a progenitor mass of $\sim$15–25 $M_\odot$ and may have lost its H envelope to binary interaction (e.g., [70]), although a slightly higher mass (up to $\sim$30 $M_\odot$) progenitor has been also discussed in the literature (e.g., [71]).
- Puppis A is a $\sim$4.5 kyr-old SNR harbouring the 112 ms CCO RX J0822–4300. Spatially resolved spectroscopic studies of SNR ejecta using *Suzaku*, *Chandra* and *XMM-Newton* point to a 15–25 $M_\odot$ progenitor [72, 73].

- Kes 79 is a ~5 kyr-old SNR harbouring the CCO CXOU J185238.6+004020 (PSR J1852+0040). A multi-wavelength study, together with an X-ray spectroscopic study of the ejecta compared to nucleosynthesis model yields, suggest a 15–20 $M_\odot$ progenitor [17]).
- RX J1713.7–3946 (G347.7–0.5) is a ~1–2 kyr-old SNR harbouring the CCO 1WGA J1713.4–3949. The measured metal abundance ratios suggest that the progenitor star was a relatively low-mass star ($\leq$20 $M_\odot$), However, based on the inferred blast wave velocity of ~6,000 km/s which is considered fast for such a core-collapse SNR, Katsuda *et al* [74] propose that RX J1713.7–3946 results from a SN Ib/c and that its progenitor is a member of an interacting binary.
- RCW 103 is a ~2 kyr-old SNR hosting near its centre the unusual CCO 1E 161348–5055 which had been recently proposed to be a magnetar [36, 37]. Comparison of the ejecta abundances inferred from X-ray spectra with supernova nucleosynthesis yields models suggests a progenitor mass of ~18–20 $M_\odot$ [75].

4.3. The Crab
Going back to the 'poster' outcome of a core-collapse SN, the Crab's progenitor is known to be an ~8–10 $M_\odot$ progenitor star. However, it has a low visible mass of ~5 $M_\odot$ and a small kinetic energy of $<10^{50}$ erg for a young core-collapse SNR. Two scenarios have been discussed: (1) a massive undetected shell beyond the visible PWN [76], (2) an electron-capture SN, an endothermic reaction of electrons captured in an O-Ne-Mg core of a super AGB star with a low energy ($\sim10^{50}$ erg) explosion [77].

This puzzle has been most recently addressed with the *Hitomi* X-ray satellite [78]. One of the science goals for *Hitomi* was to search for thermal X-ray emission from the synchrotron-dominated SNRs, thanks to its unprecedented spectral resolution and sensitivity to the thermal X-ray emission with the Soft X-ray Spectrometer. A brief observation took place just before the end of the *Hitomi* mission [79]. The gate valve was still closed, which hampered the sensitivity below 2 keV. While no thermal X-rays were detected, as in previous dedicated searches (e.g., with *Chandra* [80]), a more constrained upper limit on the X-ray emitting plasma has been established with *Hitomi* and past *Chandra* and *XMM-Newton* studies. Furthermore, while a low-energy supernova explosion has been favoured [81, 82], a higher energy Fe core-collapse explosion could not be ruled out but implies, depending on the environment, a very stringent upper limit on the ambient density or mass loss rate [79].

5. Future prospects
Despite significant advances in neutron stars physics in the past 50 years, many questions remain to be answered. We hope the answers will be obtained with the planned X-ray missions.

- The sample of PSR-SNR associations has been limited by sensitivity and resolution. Furthermore, the X-ray emission from CCO descendants and old radio pulsars would benefit from deep X-ray observations and/or a sensitive X-ray survey. The upcoming *eRosita* X-ray satellite (Germany/Russia), expected to be launched in 2018, will allow us to expand our sample and probe the X-ray emission from the different classes of neutron stars.
- To date, we do not have braking index measurement of magnetars (SGRs and AXPs) and other neutron stars subclasses, and most measurements done to-date point to an index $n<3$. Timing studies of the different classes of neutron stars is needed to shed light on their braking indices, which in turn addresses magnetic field evolution. The recently launched *ASTROSAT* (India/Canada), *NICER* (NASA), and the *HXMT* (China) X-ray satellites will provide a new window into timing studies of pulsars.

- X-ray polarimetry is still lacking in the field, but is crucial as it provides a direct measurement of the neutron star's magnetic field and sheds light on its topology (particularly for the magnetars and PWNe). *XIPE* (ESA), *IXPE* (NASA) and *XTP* (China) are currently being planned for the near future.
- High-resolution X-ray spectroscopy is needed to (a) provide an accurate measurement of SNR ages, (b) probe SN progenitors and (c) provide a direct measurement of the neutron star's magnetic field. Powerfully demonstrated by the glimpse of the *Hitomi* satellite on a few targets before its loss, high-resolution X-ray spectroscopy is now planned for the X-ray recovery mission (*XARM*, JAXA/NASA; 2021 timescale) and in the more distant future for *Athena* (ESA; 2028 timescale) and *Lynx* (NASA; beyond 2030).

Acknowledgments
This review made use of NASA's Astrophysics Data System, the U. of Manitoba's high-energy catalogue of SNRs (SNRcat [83]) and the ATNF PSR Catalogue [84]. The model described in Section 3.3 is based on work done with Adam Rogers. The author is supported by NSERC through the Canada Research Chairs and the Discovery Grants Programs, and by MITACS and the Canadian Space Agency. The author thanks the editors, particularly George Pavlov and Peter Shternin, and an anonymous referee for helpful comments, and is grateful to the organizers for their invitation and organization of an excellent conference.

References

[1] Mereghetti S and Stella L 1995 *ApJ (Letters)* **442** L17
[2] Duncan R C and Thompson C 1992 *ApJ (Letters)* **392** L9
[3] Turolla R, Zane S and Watts A L 2015 *Reports on Progress in Physics* **78** 116901 *Preprint* astro-ph/1507.02924
[4] Haberl F 2004 *Memorie della Societá Astronomica Italiana* **75** 454 *Preprint* astro-ph/0401075
[5] McLaughlin M *et al* 2006 *Nature* **439** 817
[6] Ng C-Y and Kaspi V M 2011 *Astrophysics of Neutron Stars: A Conference in Honor of M Ali Alpar* vol 1379, ed Göğüş *et al* (AIP Conf. Proc.) p 60 *Preprint* astro-ph/1010.4592
[7] Pavlov G G, Zavlin V E, Aschenbach B, Trümper J and Sanwal D 2000 *ApJ (Letters)* **531** L53
[8] Gotthelf E V, Halpern J P and Alford J 2013 *ApJ* **765** 58
[9] Sasaki M, Plucinsky P P, Gaetz T J, Smith R K, Edgar R J and Slane P O 2004 *ApJ* **617** 322
[10] Kumar H, Safi-Harb S, Slane P O and Gotthelf E V 2014 *ApJ* **781** 41
[11] Kulkarni S *et al* 2003 *ApJ* **585** 948
[12] Esposito P *et al* 2009 *ApJ (Letters)* **690** L105
[13] Kumar H and Safi-Harb S 2008 *ApJ (Letters)* **678** L43
[14] Kumar H, Safi-Harb S and Gonzalez M E 2012 *ApJ* **754** 96
[15] Espinoza C M 2013 *Neutron Stars and Pulsars: Challenges and Opportunities after 80 years* vol 291, ed J van Leeuwen (IAU Symp. Proc.) p 195 *Preprint* astro-ph/1211.5276
[16] Petre R, Becker C M and Winkler P F 1996 *ApJ (Letters)* **465** L43
[17] Zhou P, Chen Y, Safi-Harb S, Zhou X, Sun M, Zhang Z-Y and Zhang G-Y 2016 *ApJ* **831** 192
[18] Hwang U *et al* 2004 *ApJ (Letters)* **615** L117
[19] Pavlov G G, Sanwal D and Teter M A 2004 *Young Neutron Stars and Their Environments* vol 218, ed F Camilo and B M Gaensler (IAU Symp. Proc.) p 239 *Preprint* astro-ph/0311526
[20] Gotthelf E V and Halpern J P 2008 *40 Years of Pulsars: Millisecond Pulsars, Magnetars and More* vol 983, ed C Bassa *et al* (AIP Conf. Series) p 320 *Preprint* astro-ph/0711.1554
[21] Sanwal D, Pavlov G G, Zavlin V E and Teter, M A 2002 *ApJ (Letters)* **574** L61
[22] Bignami G F, Caraveo P A, De Luca A and Mereghetti S 2003 *Nature* **423** 725
[23] DeLuca A *et al* 2012 *MNRAS (Letters)* **421** L72
[24] Gotthelf E V and Halpern J P 2009 *ApJ* **695 (Letters)** L35
[25] Ho W C G and Heinke C O 2009 *Nature* **462** 71
[26] Suleimanov V S, Klochkov D, Pavlov G G and Werner K 2014, *ApJ Supplement Series* **210** 13
[27] Gavriil F P, Gonzalez M E, Gotthelf E V *et al* 2008 *Science* **319** 1802
[28] Ng C-Y, Slane P O, Gaensler B M and Hughes J P 2008 *ApJ* **686** 508
[29] Younes G, Kouveliotou C and Roberts P 2016 *GCN* **19736**
[30] Kennea J A *et al* 2016 *ATel* **9274**

[31] Göğüş E *et al* 2016 *ApJ* **829** L25
[32] Camilo F *et al* 2006 *Nature* **442** 892
[33] Rea N *et al* 2010 *Science* **330** 944
[34] Scholz P, Ng C-Y, Livingstone M A, Kaspi V M, Cumming A and Archibald R F 2012 *ApJ* **761** 66
[35] Zhou P, Chen Y, Li X-D, Safi-Harb S, Mendez M, Terada Y, Sun W and Ge M-Y 2014 *ApJ (Letters)* **781** L16
[36] D'Aì A *et al* 2016 *MNRAS* **463** 2394
[37] Rea N, Borghese A, Esposito P, *et al* 2016 *ApJ (Letters)* **828** L13
[38] Younes G *et al* 2016 *ApJ* **824** 138
[39] Gonzalez M E and Safi-Harb S 2003 *ApJ (Letters)* **591** L143
[40] Safi-Harb S and Kumar H 2008 *ApJ* **684** 532
[41] Blumer H, Safi-Harb S and McLaughlin M 2017 *ApJ (Letters)* **850** L18 *Preprint* astro-ph/1711.01600
[42] Viganò D, Rea N, Pons J A, Perna R, Aguilera D N and Miralles J A 2013 *MNRAS* **434** 123
[43] Perna R and Pons J A 2011 *ApJ (Letters)* **727** L51
[44] Alpar M A, Çalışkan, Ş and Ertan Ü 2013 *Feeding Compact Objects: Accretion on All Scales* vol 290, ed C M Zhang *et al* (IAU Symp. Proc.) p 93 *Preprint* astro-ph/1211.4721
[45] Rogers A and Safi-Harb S 2017 *MNRAS* **465** 383
[46] Rogers A and Safi-Harb S 2016 *MNRAS* **457** 1180
[47] Nakano T *et al* 2015 *PASJ* **67** 9
[48] Dall'Osso S *et al* 2012 *MNRAS* **422** 2878
[49] Colpi M, Geppert U and Page D 2000 *ApJ (Letters)* **529** L29
[50] Colgate S A 1971 *ApJ* **163** 221
[51] Zel'dovich Y B, Ivanova L N and Nadezhin D K 1972 *AZh* **49** 253
[52] Chevalier R A 1989 *ApJ* **346** 847
[53] Muslimov A and Page D 1995 *ApJ (Letters)* **440** L77
[54] Geppert U, Page D and Zannias T 1999 *A&A* **345** 487
[55] Ho W C G 2011 *MNRAS* **414** 2567
[56] Ho W C G 2015 *MNRAS* **452** 845
[57] Popov S B and Turolla R 2012 *Electromagnetic Radiation from Pulsars and Magnetars* vol 466, ed W Lewandowski *et al* (San Francisco: Astron. Soc. Pacific) p 191 *Preprint* astro-ph/1206.2819
[58] Bernal C G, Lee W H and Page D 2010 *RMxAA* **46** 309
[59] Bernal C G, Page D and Lee W H 2013 *ApJ* **770** 106
[60] Igoshev A P, Elfritz J G and Popov S B 2016 *MNRAS* **462** 3698
[61] Eksi K Y 2017 *MNRAS* **469** 1974
[62] Chevalier R A 2005 *ApJ* **619** 839
[63] Milisavljevic D and Fesen R 2017 *Preprint* astro-ph/1701.00891
[64] Patnaude D and Badenes C 2017 *Preprint* astro-ph/1702.03228
[65] Vink J 2008 *AdSpRes* **41** 503
[66] Ferrario L and Wickramasinghe D 2008 *MNRAS (Letters)* **389** L66
[67] Safi-Harb S and Kumar H 2013 *Neutron Stars and Pulsars: Challenges and Opportunities after 80 years* vol 291, ed J van Leeuwen (IAU Symp. Proc.) p 480 *Preprint* astro-ph/1210.5261
[68] Borkowski K and Reynolds S P 2017 *ApJ* **846** 13
[69] Davies *et al* 2009 *ApJ* **707** 844
[70] Young P A, Fryer C L, Hungerford A *et al* 2006 *ApJ* **640** 89
[71] Pérez-Rendón B, Garca-Segura G and Langer N 2009 *A&A* **506** 1249
[72] Katsuda S *et al* 2012 *ApJ* **757** 49
[73] Hwang U, Petre R and Flanagan K 2008 *ApJ* **676** 378
[74] Katsuda S *et al* 2015 *ApJ* **814** 29
[75] Frank K A, Burrows D N and Park, S 2015 *ApJ* **810** 113
[76] Chevalier R A 1977 *Supernovae, Astro. and Space Science Library* vol 66, ed D. Schramm (Dordrecht: Springer)
[77] Nomoto K, Sugimoto D, Sparks W M, Fesen, R A, Gull T R and Miyaji S 1982 *Nature* **299** 803
[78] Takahashi T *et al* 2014 *SPIE* **9144** 25 *Preprint* astro-ph/1412.1356
[79] Hitomi collaboration 2017 *PASJ* in press *Preprint* astro-ph/1707.00054
[80] Seward F D, Gorenstein P and Smith R K 2006 *ApJ* **636** 873
[81] Smith N 2013 *MNRAS* **434** 102
[82] Yang H and Chevalier R A 2015 *ApJ* **806** 153
[83] Ferrand G and Safi-Harb S 2012 *AdSpR* **49** 1313 (SNRcat)
[84] Manchester R N, Hobbs, G B, Teoh, A and Hobbs M 2006 *AJ*, **129**, 1993

Central compact objects in supernova remnants

A De Luca

INAF/IASF Milano, Via Bassini 15, I-20133 Milano, Italy

E-mail: deluca@iasf-milano.inaf.it

Abstract. Central Compact Objects (CCOs) are a handful of sources located close to the geometrical center of young supernova remnants. They only show thermal-like, soft X-ray emission and have no counterparts at any other wavelength. While the first observed CCO turned out to be a very peculiar magnetar, discovery that three members of the family are weakly magnetised Isolated Neutron Stars (INSs) set the basis for an interpretation of the class. However, the phenomeology of CCOs and their relationship with other classes of INSs, possibly ruled by supernova fall-back accretion, are still far from being well understood.

1. Introduction

A dozen of point-like, soft X-ray sources has been discovered over the last two decades, gently shining very close to the geometrical center of young (0.3–7 kyr) supernova remnants (SNR). Their spectra are thermal-like, usually well described by the sum of two blackbodies with high temperatures (0.2–0.5 keV) and very small emitting radii (ranging from 0.1 to a few km). Their emission is generally steady, with a luminosity of the order of 10^{33} erg s^{-1}, and pulsations are undetected in most cases. They have no counterparts at any wavelength. No associated structures of diffuse, non-thermal emission are seen. Such sources are clearly different from both standard rotation-powered pulsars and magnetars. They have been dubbed, as a class, "Central Compact Objects" (CCOs)[1] – indeed, a designation suggesting our rather poor understanding of their nature and physics. Discovery of fast pulsations and measurement of a tiny, positive period derivative in three sources set the basis of a framework for CCOs as young, Isolated Neutron Stars (INS) with a weak dipole field. In this short paper, I will focus on recent results and directions in CCO studies, including discovery of the magnetar nature of the first observed CCO – a possibly unique source. I will show that an explanation of the properties of these sources would be highly relevant for our overall understanding of NS production rate, physics of core-collapse, INS diversity and evolution. Previous reviews on CCOs were given by [2, 3, 4].

2. 1E 161348–5055: a unique, slowly-rotating magnetar.

1E 161348–5055 (1E 1613) in the 2 kyr-old SNR RCW103 has been the prototype for CCOs. It was the first candidate radio-quiet INS discovered in a SNR [5]. However, more recent observations clearly separated 1E 1613 from CCOs and settled the case for a truly unique phenomenology: 1E 1613 displays a dramatic long-term variability with large outbursts and a puzzling periodicity at 6.67 hours, together with a young age and lack of an optical/infrared counterpart [6, 7]. The nature of the source, possibly the first low-mass X-ray binary system

[1] The name has been used for the first time by G. Pavlov, referring to the central source in the Cas A SNR [1].

observed inside a SNR, or a very peculiar isolated, young magnetar with an extremely slow spin period, possibly braked by interaction with a surrounding disk [6], remained debated for a decade – even more exotic pictures were proposed [8, 9, 10, 11, 12, 13].

A partial answer to this puzzle came only very recently. Indeed, on 2016 June 22, the Burst Alert Telescope onboard *Swift* detected a magnetar-like, short X-ray burst from the direction of 1E 1613, with a spectrum well described by a blackbody model ($kT \sim 9$ keV) and a luminosity of $\sim 2 \times 10^{39}$ erg s^{-1}. A strong X-ray outburst was simultaneously observed from 1E 1613 with the *Swift* X-Ray Telescope (XRT), with the 0.5–10 keV flux a factor of 100 brighter than the quiescent level observed up to one month before [14, 15]. Follow-up observations performed with *Chandra* and *NuSTAR* unveiled (i) a dramatic change in the shape of the 6.67 hour modulation, two broad peaks per period replacing the sinusoidal shape that had been observed since 2005 [14]; (ii) a non-thermal spectral component extending up to ~ 30 keV (modulated up to ~ 20 keV), never detected before, well described by a hard power law with $\Gamma \sim 1.2$, superimposed to the thermal continuum dominating at lower energies [14]. Based on a series of *Swift*/XRT observations, the overall energy emitted in the outburst (impulsive plus persistent) was computed to be $\sim 2.6 \times 10^{42}$ ergs [14]. A likely counterpart in the near infrared was detected by [16] with the *Hubble Space Telescope* with AB magnitudes of 26.3 and 24.2 in the F110W and F160W filters, respectively (roughly corresponding to the J and H bands). The same source was detected with the ESO/VLT with Ks=20.9 ± 0.1 (De Luca et al, in preparation). All results from such multiwavelength observations strongly point to a magnetar interpretation for 1E 1613 [14, 15, 16]. Indeed, 1E 1613 stands out as a unique source because of its rotation period, slower by three orders of magnitude with respect to any other known magnetar candidate.

Assuming that 1E 1613 was born with a spin period of a few hours is hardly conciliable with angular momentum conservation in the progenitor star's collapsing core and also clashes with the commonly assumed dynamo mechanism for the generation of the gigantic magnetic field in magnetars, requiring a fast initial period (~millisecond) [17]. Disregarding such a possibility, one should explain how a fast-spinning NS could be slowed-down to a period of 6.67 hours in ~ 2 kyr. Standard magneto-dipole braking, coupled to magneto-thermal evolution, cannot account for such a slow rotation for any reasonable assumption on the newborn NS properties [14]. Braking of a strongly-magnetized NS by material/magnetic interaction with a low-mass companion star was discussed by [9]. However, near-infrared observations rule out any companion with a mass larger than an M8 dwarf [16] – it seems unlikely that a binary system with such an extreme mass ratio survive a supernova explosion[2]. An alternative possibility is to invoke the role of supernova fallback material [18, 19]. The NS could have been braked by propeller interaction with a long-lived residual disk. Such a possibility had been considered by [6, 8] and, more recently, by [20, 21]. Although the formation and stability of fallback disks surrounding newborn NSs has been questioned by [22], it was shown that a very low-mass disk ($\sim 10^{-9}$ $M_\odot$) could brake the NS to the observed spin rate in $\sim 1-3$ kyr, provided it is endowed by a dipole magnetic field of 5×10^{15} G [21], the largest in the magnetar family. As a further possibility, the NS could have been slowed-down in an earlier phase – propeller interaction starting at the onset of fallback accretion (with no need for formation and long-term survival of a disk). Such a picture was mentioned by [14, 23], but no detailed calculations have been presented so far.

3. CCO pulsars: neutron stars with a weak dipole magnetic field.

A key discovery has been the detection of fast pulsations from three objects of the class: 1E 1207.4—5209 in G296.5+10.0 ($P \sim 424$ ms, [24]), CXOU J185238.6+004020 in Kes 79 ($P \sim 105$ ms, [25]) and RX J0852.0–4622 in Puppis A ($P \sim 112$ ms, [26]). This proved with no doubt that these sources are NSs. Even more enlightening has been the measurement of their period

[2] The same argument, together with the difficulty in explaining the dramatic long-term variability in pulse shape and pulsed fraction, argues against an interpretation of the 6.67 hr cycle as the orbital period in a binary.

derivatives [4, 27], which turned out to be very small (indeed, very long observation campaigns were required). Implications are extremely interesting: (i) the dipole magnetic field inferred from standard magneto-dipole braking $B_{\rm dip} = 3.2 \times 10^{19}(P\dot{P})^{1/2}$ is $10^{10} - 10^{11}$ G, remarkably smaller than the one of the bulk of the rotation-powered pulsar population; (ii) the characteristic age ($\tau_C = P/2\dot{P}$) is 4–5 orders of magnitude larger than the age of the host SNR (indeed, the CCO birth period is likely very close to the currently observed one); (iii) the spin-down energy loss is more than 10 times smaller than the X-ray luminosity. However, several properties of the CCO pulsars do not fit easily in the picture of weakly magnetised INSs.

CXOU J185238.6+004020 displays an extremely high modulation (pulsed fraction $\sim 64\%$, with very little dependence on energy) of its thermal emission, consisting of the sum of two hot blackbodies with tiny emitting areas [27]. Reproducing the observed pulse shape and fraction with thermal emission models is challenging [28] and points to a peculiar thermal map, with a high temperature contrast between a small emitting region (a hot spot surrounded by a warmer region) and the remaining cooler and unobservable surface of the star. A highly elongated shape in the longitudinal direction (with respect to the spin axis) is required for the emitting region; a more conventional polar cap geometry requires the radiation to be highly beamed [29].

RX J0852.0–4622 too has a thermal continuum well described by two blackbodies. Its pulse shape is sinusoidal, with a very abrupt, 180° phase reversal at ~ 1.2 keV, where the hot and warm blackbodies switch dominance [30]. This is consistent with an anti-podal hot spot model, with two small emitting regions with different size and temperature located at opposite sides of the NS [30]. A spectral feature is also seen in the low-energy portion of the spectrum. It can be described either as an emission line at ~ 0.75 keV, or as a couple of absorption lines at ~ 0.46 keV and ~ 0.92 keV. Significant variability in the feature is seen between two observations performed in 2001 and 2009: assuming the emission line model, the central energy decreases from ~ 0.79 to ~ 0.71 keV [4, 31]. No further variability is seen after 2009 [4].

1E 1207.4–5209 displays multiple absorption features at harmonically spaced energies (0.7, 1.4, 2.1 and possibly 2.8 keV) superimposed to a thermal spectrum [32, 33, 34, 35]. After a long debate (see [3]), measurement of P and $\dot{P}$ points to an interpretation of the features as due to electron cyclotron scattering close to the star's surface, the feature at 0.7 keV being the fundamental. The magnetic field strength in the region where the lines are formed would be $\sim 8 \times 10^{10}$ G (accounting for a gravitational redshift $z = 0.3$), in broad agreement with the value inferred by magneto-dipole braking ($\sim 9.8 \times 10^{10}$ G). The relative strength of the harmonics was explained by taking into account resonances in the photospheric free-free opacity in the presence of the magnetic field [36, 37, 38]. The X-ray pulsation is dominated by the complex modulation of the spectral features as a function of the rotational phase [35]; the continuum, well described by the sum of two blackbody curves, being almost unpulsed. No simple constraints on the geometry of the thermally emitting regions could be set.

4. Other CCOs: homogeneous group or mixed bag?

The family of CCOs includes about ten more sources[3]. Thermal emission properties are pretty homogeneous, with no pulsations (rather deep upper limits have been set in a few cases), nor long-term variability (although multi-epoch coverage is limited in most cases). It seems reasonable to assume that such sources are INSs similar to the three CCO pulsars, possibly with even smaller dipole fields.

The most famous CCO was discovered close to the center of the very young (~ 350 yr) Cas A SNR in the *Chandra* "first light" image [39]. An early claim of a magnetar nature for this CCO, based on detection of the possible infrared echo of a giant flare occurred around 1950 [40] was later retracted [41, 42]. Multi-epoch observations with *Chandra* point to long-term

[3] see `www.iasf-milano.inaf.it/~deluca/cco/main.htm` for an updated list

evolution of the spectral shape and luminosity of this source, consistent with a $\sim 4\%$ temperature decrease in about 10 yr [43]. Such a direct observation of the NS cooling would have very important implications for our understanding of the properties of ultra-dense matter in the NS interior [44, 45, 46, 47, 48]. However, more recent analysis of *Chandra* data focused on possible systematics affecting the measurement (e.g., instrument calibration issues) and concluded that the reality and rate of the possible temperature decrease are uncertain [49, 50].

It was shown [51] that a carbon atmosphere NS model with low magnetic field provides a good description of the spectrum of this source and implies an emitting region consistent with the entire NS surface. This possibly solves the puzzling lack of pulsations coupled to the large temperature anisotropy (with uncomfortably small emitting areas) as derived from other spectral models (both blackbody curves and weakly-magnetised NS hydrogen atmosphere models), pointing to the picture of a weakly magnetized NS, with uniform emission from its surface and with active nuclear burning in its surface layers.

The same picture was proposed for other CCOs, as soon as high-quality X-ray spectra became available: the source in G353.6–0.7 [52, 53] (possibly the hottest known INS), as well as the source in G15.9+0.2 [54]. It was stressed that the carbon atmosphere models makes it possible to reconcile thermal luminosities with the current distance estimates and also to constrain the equation of state of the NS [53, 55]. As a matter of fact, however, such models cannot explain the phenomenology of all CCOs. For instance, the carbon atmosphere model would describe the emission of the CCO pulsar in Kes 79 as coming from the whole surface of the NS, but this is clearly at odds with the observed very high pulsed fraction [29]. Indeed, the current upper limits on the CCO pulsations are not stringent enough to rule out highly anisotropic surface temperature distributions [56, 57] that would be consistent with different atmosphere chemical composition as well as with different properties of the NSs.

5. CCOs as a "class".

The picture of CCOs as weakly-magnetised, young NSs has to face two issues.

First issue: the highly-anisotropic, high contrast-ratio temperature surface distribution cannot be easily explained in this framework. The hot spots cannot be heated by rotation-powered particle bombardment (ruled out by the spin-down energetics); they cannot be accretion-powered (ruled out by timing [27] as well as by deep optical limits to any companion star, e.g., [58]). Hot spots could be due to localized crustal heating by magnetic field decay. This would require a magnetic field with strong non-dipolar components ($10^{14} - 10^{15}$ G to account for the observed luminosity [27]), coupled to a factor 10^4 less intense dipolar component. As an alternative possibility, hot spots could be powered by residual heat. This would require highly anisotropic heat transfer from the NS interior, pointing to a strong magnetic field in the crust. For instance, a toroidal field of order 10^{14} G could screen large part of the surface around the equator, channelling the heat flux towards the poles [28, 59]. A poloidal field as low as $10^{10} - 10^{11}$ G could be enough to make the magnetic field configuration stable [60].

Second issue: the region in the $P - \dot{P}$ parameter space where CCO pulsars are located is highly underpopulated, at odds with expectations [61]. Such region is well above the "death line" for radio pulsars. Indeed, rotation-powered radio emission is seen from several sources with similar spin parameters. Moreover, CCOs are a common outcome of core-collapse supernovae – they are found in SNR as frequently as other classes of NSs (more frequently than magnetars, see, e.g., [62]). Frequent formation, coupled with a very slow evolution of the spin period would lead to a large number of low B-field sources above the death line. As a possible explanation of this puzzle, CCOs could be intrinsically radio-quiet and become essentially invisible after their host SNR fade away. Alternatively, radio luminosity in rotation-powered emission could depend on spin-down power, as suggested by radio pulsar population synthesis (e.g., [63]). As a further possibility, CCOs could evolve and move to a different region of the $P - \dot{P}$ diagram. This would

IOP Conf. Series: Journal of Physics: Conf. Series **932** (2017) 012006 doi:10.1088/1742-6596/932/1/012006

ease a further related issue: the sum of the inferred birth rates for different classes of NSs exceeds the Galactic core-collapse supernova rate (see, e.g., [64], although these authors did not discuss the case of CCOs) – this points to evolutionary links among different NS families.

Different frameworks for CCOs as young INSs with weak dipole field have been discussed. A first possible scenario [65, 66] links the origin of the weak field to a slow rotation of the collapsing progenitor star core, which would result in an inefficient dynamo mechanism [67]. Such a picture does not seem very appealing: the birth period for CCOs is not "long", according to population synthesis models for radio pulsars (e.g., [63] point to a wide distribution of birth periods with a mean of 300 ms and a dispersion of 150 ms). Moreover, the origin of the peculiar thermal anisotropy would remain unexplained. As a second possibility, CCOs could be quiescent magnetars, with an extremely weak dipole field, but with a strong crustal magnetic field, emerging in local "sunspot" structures. Such a scenario, however, seems rather unlikely because of the general lack of variability in CCOs, at odds with the ubiquitous variability seen in magnetars (but we cannot exclude the picture to be correct for a fraction of the sample). A third picture is known as the "buried field" scenario: prompt accretion of supernova fallback material [18] could bury the magnetic field of the newborn NS beneath its surface; the field could then re-emerge by diffusion on a time scale depending on the amount of accreted matter [19, 68, 69]. Recent models suggest that spherical accretion of $\sim 10^{-4} - 10^{-2}$ $M_\odot$ is required to screen typical fields in the $10^{12} - 10^{14}$ G range for $\sim 10^3 - 10^5$ yr [70, 71, 72, 73, 74, 75]. According to these studies, (i) the strong, hidden crustal field could explain peculiarities in the thermal map; (ii) a low dipole field could be common in young NSs ($\leq$ few kyr); (iii) CCOs could turn into radio pulsars at an older age (see also [76]). It is difficult to test the buried field scenario on CCOs by X-ray timing: evidence for a re-emerging magnetic field could come from a measurement of their braking index $n = 2 - P\ddot{P}/\dot{P}^2$, expected to be equal to 3 in the case of pure magneto-dipole braking for a constant magnetic field, and smaller than 3 in case of a growing dipole field. However, the tiny $\dot{P}$ of CCOs make any measurement of their n extremely challenging. Some supporting evidence for this picture is provided by timing investigation of young ($\tau_c \sim 10^3 - 10^4$ yr) rotation-powered pulsars, yielding low values for the braking index that can be interpreted as a hint of a growing magnetic field [77, 78]; indirect support is also provided by the absence of evidence for weakly magnetised ($\leq 10^{11}$ G) NSs in High-mass X-ray binaries, consistent with field re-emergence in these sources on a time scale of $\sim 10^4$ yr [79].

A different way to attack the problem is to search for CCO descendants. Such investigations assume that at least some CCO may be a radio pulsar; CCOs would be much younger than "standard" pulsars in a given $P - \dot{P}$ parameter space region and thus they should be distinguishable because of a high thermal luminosity. A search for CCO descendants was performed among "mildly-recycled" pulsars (NSs that have accreted gas from a companion star in a binary system for a short time before a second supernova explosion halted their spin-up) – a class of pulsars proposed to explain the sparse population of sources in the same $P - \dot{P}$ region where CCOs have been recently located. However, none was found [80]. A second attempt was carried out, considering seemingly old radio pulsars spatially coincident with known SNRs, with negative results [81]. A third investigation did not identify any plausible CCO among radio pulsars with weak field ($B < 10^{11}$ G), but energetics larger than mildly recycled pulsars [82]. These studies suggest that CCO descendants, if not radio-quiet, should hide among pulsars with larger B-field and energetics. Interestingly, the peculiar neutron star dubbed *Calvera* [83] has been proposed as a possible example of an evolved CCO, whose SNR is no more detectable in X-rays. It is a radio-quiet pulsar, with $P \sim 59$ ms [84] and $\dot{E}_{\rm rot} \sim 6 \times 10^{35}$ erg s^{-1} [85], only showing soft X-ray thermal-like emission in spite of its high spin-down luminosity. Calvera lies in the $P - \dot{P}$ plane along the path of growing B-field between CCOs and young radio pulsars; however, a lack of any constraint on its distance (hence on its luminosity) as well as on its true age does not allow us to draw a firm conclusion [86].

6. Conclusions.

CCOs, a handful of candidate young INSs with elusive properties, proved to include extraordinarily interesting objects. Explanation of their phenomenology challenges standard models and points to a very complex scenario, in which accretion of supernova fallback material in different regimes, coupled to a variety of initial conditions for the magnetic field strength and configuration, plays an important role in shaping the properties of newborn NSs. This has very important implications towards a physical understanding of the different phenomenological classes of INSs and of their evolution. There is still a large space for discoveries in CCO studies, especially in view of forecoming observing facilities such as *eROSITA*, *ATHENA*, SKA. Deep searches for periodicity and long-term flux monitoring in the X-rays will allow one to assess the nature of non-pulsating CCOs – proving them to be similar to the three CCO pulsars, or unveiling any (peculiar) magnetar among them – as a matter of fact, the spectrum of a quiescent magnetar and of a CCO are almost undistinguishable. Large surveys will possibly allow one to identify new candidate CCOs and/or candidate CCO descendants. (Targeted) radio pulsar searches will unveil if CCOs are intrinsically radio-quiet, or simply radio-faint. Using already available data, modeling the phase-resolved behaviour of CCO pulsar will allow one to constrain the physics of the spectral features, the surface thermal map and magnetic field topology, as well as to test expectations of the buried field scenario.

6.1. Acknowledgments

I warmly thank the SOC and the LOC for organizing a very interesting meeting and for the invitation to speak. I thank P Esposito for a critical reading of the manuscript and G F Bignami (who left us prematurely on May 24th, 2017), A Borghese, P A Caraveo, F Coti Zelati, G L Israel, S Mereghetti, N Rea, A Tiengo for many useful discussions.

References

[1] Pavlov G G, Zavlin V E, Aschenbach B, Trümper J and Sanwal D 2000 *ApJ* **531** L53-6
[2] Pavlov G G, Sanwal D and Teter M A 2004 *IAUS* **218** 239
[3] De Luca A 2008 *AIPC* **983** 311-9
[4] Gotthelf E V, Halpern J P and Alford J 2013 *ApJ* **765** 58
[5] Tuohy I and Garmire G 1980 *ApJ* **239** L107-10
[6] De Luca A, Caraveo P A, Mereghetti S, Tiengo A and Bignami G F 2006 *Science* **313** 814-7
[7] De Luca A *et al* 2008 *ApJ* **682** 1185-94
[8] Li X-D 2007 *ApJ* **666** L81-4
[9] Pizzolato F, Colpi M, De Luca A, Mereghetti S and Tiengo A 2008 *ApJ* **681** 530-42
[10] Bhadkamkar H and Ghosh P 2009 *A&A* **506** 1297-307
[11] Esposito P, Turolla R, De Luca A, Israel G L, Possenti A and Burrows D N 2011 *MNRAS* **418** 170-5
[12] Ikhsanov N R, Kim V Y, Beskrovnaya N G and Pustil'nik L A 2013 *Ap&SS* **346** 105-9
[13] Liu X W, Xu R X, van den Heuvel E P J, Qiao G J, Han J L, Han Z W and Li X D 2015 *ApJ* **799** 233
[14] Rea N, Borghese A, Esposito P, Coti Zelati F, Bachetti M, Israel G L and De Luca A 2016 *ApJ* **828** L13
[15] D'Aì A *et al* 2016 *MNRAS* **463** 2394-404
[16] Tendulkar S P, Kaspi V M, Archibald R F and Scholz P 2017 *ApJ* **841** 11
[17] Thompson C and Duncan R C 1993 *ApJ* **408** 194-217
[18] Colgate S A 1971 *ApJ* **163** 221
[19] Chevalier R A 1989 *ApJ* **346** 847-59
[20] Tong H, Wang W, Liu X W and Xu R X 2016 *ApJ* **833** 265
[21] Ho W C G and Andersson N 2017 *MNRAS* **464** L65-9
[22] Perna R, Duffell P, Cantiello M and MacFadyen A I 2014 *ApJ* **781** 119
[23] Popov S B, Kaurov A A and Kaminker A D 2015 *PASA* **32** e018
[24] Zavlin V E, Pavlov G G, Sanwal D and Trümper J 2000 *ApJ* **540** L25-8
[25] Gotthelf E V, Halpern J P and Seward F D 2005 *ApJ* **627** 390-6
[26] Gotthelf E V and Halpern J P 2009 *ApJ* **695** L35-9
[27] Halpern J P and Gotthelf E V 2010 *ApJ* **709** 436-46
[28] Shabaltas N and Lai D 2012 *ApJ* **748** 148

[29] Bogdanov S 2014 *ApJ* **790** 94
[30] Gotthelf E V, Perna R and Halpern J P 2010 *ApJ* **724** 1316-24
[31] De Luca A *et al* 2012 *MNRAS* **421** L72-6
[32] Sanwal D, Pavlov G G, Zavlin V E and Teter M A 2002 *ApJ* **574**, L61-4
[33] Mereghetti S, De Luca A, Caraveo P A, Becker W, Mignani R and Bignami G F 2002 *ApJ* **581** 1280-5
[34] Bignami G F, Caraveo P A, De Luca A and Mereghetti S 2003 *Nature* **423** 725-7
[35] De Luca A, Mereghetti S, Caraveo P A, Moroni M, Mignani R P and Bignami G F 2004 *A&A* **418** 625-37
[36] Suleimanov V F, Pavlov G G and Werner K 2010 *ApJ* **714** 630-5
[37] Potekhin A Y 2010 *A&A* **518** A24
[38] Suleimanov V F, Pavlov G G and Werner K 2012 *ApJ* **751** 15
[39] Tananbaum H 1999 *IAUC* **7246** 1
[40] Krause O *et al* 2005 *Sci* **308** 1604-6
[41] Kim Y, Rieke G H, Krause O, Misselt K, Indebetouw R and Johnson K E 2008 *ApJ* **678** 287-96
[42] Dwek E and Arendt R G 2008 *ApJ* **685** 976-87
[43] Heinke C O and Ho W C G 2010 *ApJ* **719** L167-71
[44] Shternin P S, Yakovlev D G, Heinke C O, Ho W C G and Patnaude D J 2011 *MNRAS* **412** L108-12
[45] Yakovlev D G, Ho W C G, Shternin P S, Heinke C O and Potekhin A Y 2011 *MNRAS* **411** 1977-88
[46] Page D, Prakash M, Lattimer J M and Steiner A W 2011 *PhRvL* **106** 081101
[47] Yang S-H, Pi C-M and Zheng X-P 2011 *ApJ* **735** L29
[48] Noda T, Hashimoto M, Yasutake N, Maruyama T, Tatsumi T and Fujimoto M 2013 *ApJ* **765** 1
[49] Elshamouty K G *et al* 2013 *ApJ* **777** 22
[50] Posselt B, Pavlov G G, Suleimanov V and Kargaltsev O 2013 *ApJ* **779** 186
[51] Ho W C G and Heinke C O 2009 *Nature* **462** 71-3
[52] Klochkov D, Puehlhofer G, Suleimanov V, Simon S, Werner K and Santangelo A 2013 *A&A* **556** A41
[53] Klochkov D, Suleimanov V, Puelhlhofer G, Yakovlev D G, Santangelo A and Werner K 2015 *A&A* **573** A53
[54] Klochkov D, Suleimanov V, Sasaki M and Santangelo A 2016 *A&A* **592** L12
[55] Ofengeim D D, Kaminker A D, Klochkov D, Suleimanov V and Yakovlev D G 2015 *MNRAS* **454** 2668-76
[56] Pavlov G G and Luna G J M 2009 *ApJ* **703** 910-21
[57] Suleimanov V F, Klochkov D, Poutanen J and Werner K 2017 *A&A* **600** A43
[58] De Luca A *et al* 2011 *A&A* **525** A106
[59] Pérez-Azorín J F, Miralles J A and Pons J A 2006 *A&A* **451** 1009-24
[60] Braithwaite J 2009 *MNRAS* **397** 763-74
[61] Kaspi V M 2010 *PNAS* **107** 7147-52
[62] Halpern J P and Gotthelf E V 2010 *ApJ* **710** 941-7
[63] Faucher-Giguère C-A and Kaspi V M 2006 *ApJ* **643** 332-55
[64] Keane E F and Kramer M 2008 *MNRAS* **391** 2009-16
[65] Gotthelf E V and Halpern J P 2007 *ApJ* **664** L35-8
[66] Halpern J P, Gotthelf E V, Camilo F and Seward F D 2007 *ApJ* **665** 1304-10
[67] Bonanno A, Urpin V and Belvedere G 2006 *A&A* **451** 1049-52
[68] Muslimov A and Page D 1995 *ApJ* **440** L77-80
[69] Geppert U, Page D and Zannias T 1999 *A&A* **345** 847-54
[70] Ho W C G 2011 *MNRAS* **414** 2567-75
[71] Viganò D and Pons J A 2012 *MNRAS* **425** 2487-92
[72] Bernal C G, Page D and Lee W H 2013 *ApJ* **770** 106
[73] Torres-Forné A, Cerdá-Durán P, Pons J A and Font J A 2016 *MNRAS* **456** 3813-26
[74] Bernal C G and Fraija N 2016 *MNRAS* **462** 3646-59
[75] Igoshev A P, Elfritz J G and Popov S B 2016 *MNRAS* **462** 3689-702
[76] Rogers A and Safi-Harb S 2016 *MNRAS* **457** 1180-9
[77] Ho W C G 2015 *MNRAS* **452** 845-51
[78] Marshall F E, Guillemot L, Harding A K, Martin P and Smith D A 2016 *ApJ* **827** L39
[79] Popov S B and Turolla R 2012 *ASPC* **466** 191
[80] Gotthelf E V, Halpern J P, Allen B and Knispel B 2013 *ApJ* **773** 141
[81] Bogdanov S, Ng C-Y and Kaspi V M 2014 *ApJ* **792** L36
[82] Luo J, Ng C-Y, Ho W C G, Bogdanov S, Kaspi V M and He C 2015 *ApJ* **808** 130
[83] Rutledge R E, Fox D B and Shevchuk A H 2008 *ApJ* **672** 1137-43
[84] Zane S *et al* 2011 *MNRAS* **410** 2428-45
[85] Halpern J P, Bogdanov S and Gotthelf E V 2013 *ApJ* **778** 120
[86] Halpern J P and Gotthelf E V 2015 *ApJ* **812** 61

International Conference Physics of Neutron Stars - 2017. 50 years after. IOP Publishing
IOP Conf. Series: Journal of Physics: Conf. Series **932** (2017) 012007 doi:10.1088/1742-6596/932/1/012007

Phase-dependent absorption features in X-ray spectra of X-ray Dim Isolated Neutron Stars

A Borghese[1], N Rea[1,2], F Coti Zelati[2], R Turolla[3,4], A Tiengo[5] and S Zane[4]

[1] Anton Pannekoek Institute for Astronomy, University of Amsterdam, Amsterdam, The Netherlands
[2] Institute of Space Sciences (IEEC-CSIC), Campus UAB, Barcelona, Spain
[3] Dipartimento di Fisica e Astronomia, Università di Padova, Padova, Italy
[4] Mullard Space Science Laboratory, University College London, UK
[5] Scuola Universitaria Superiore IUSS Pavia, piazza della Vittoria 15, I-27100 Pavia, Italy

E-mail: a.borghese@uva.nl

Abstract. A detailed phase-resolved spectroscopy of archival *XMM–Newton* observations of X-ray Dim Isolated Neutron Stars (XDINSs) led to the discovery of narrow and strongly phase-dependent absorption features in two of these sources. The first was discovered in the X-ray spectrum of RX J0720.4-3125, followed by a new possible candidate in RX J1308.6+2127. Both spectral lines have similar properties: they are detected for only $\sim 20\%$ of the rotational cycle and appear to be stable over the timespan covered by the observations. We performed Monte Carlo simulations to test the significance of these phase-variable features and in both cases the outcome has confirmed the detection with a confidence level $> 4.6\sigma$. Because of the narrow width and the strong dependence on the pulsar rotational phase, the most likely interpretation for these spectral features is in terms of resonant proton cyclotron absorption scattering in a confined high-B structure close to the stellar surface. Within the framework of this interpretation, our results provide evidence for deviations from a pure dipole magnetic field on small scales for highly magnetized neutron stars and support the proposed scenario of XDINSs being aged magnetars, with a strong non-dipolar crustal B-field component.

1. Introduction

The XDINSs form a group of seven thermally-emitting, radio-quiet [1], nearby ($\lesssim 500\,\mathrm{pc}$) X-ray pulsars, originally discovered by *ROSAT* (see [2] for a review). Timing studies have shown that XDINSs are slow rotators with periods in the range $3 - 12\,\mathrm{s}$ and spin down regularly with period derivatives of the order of $10^{-14} - 10^{-13}\,\mathrm{s\,s^{-1}}$ [3]. The inferred surface dipolar magnetic field is typically of the order of $10^{13}\,\mathrm{G}$, being close to the magnetar ones. Their X-ray luminosity, $\approx 10^{31-32}\,\mathrm{erg\,s^{-1}}$, exceeds the spin-down luminosity making XDINSs different from the rotation-powered pulsars.

XDINSs appear to be thermally cooling with emission in the soft X-ray band and faint optical and/or ultraviolet counterparts [4]. This thermal emission is thought to be due to residual heat and to come directly from the stellar surface. XDINSs show thermal spectra which can be well described by blackbody models with inferred temperature in the range $kT \sim 50 - 100\,\mathrm{eV}$. In the past decades *XMM–Newton* observations have detected deviations from a pure

IOP Conf. Series: Journal of Physics: Conf. Series **932** (2017) 012007 doi:10.1088/1742-6596/932/1/012007

blackbody model in six of the XDINSs, in the form of broad absorption features in the phase-averaged spectra. Their origin is still an open issue: they could be due to proton cyclotron resonances/atomic transitions in a magnetized atmosphere [5] or to an inhomogeneous surface temperature distribution [6].

Recently, phase-dependent features have been observed in two low-field magnetars, SGR 0418+5729 [7] and Swift J1822.3-1606 [8]. These findings have motivated our search for similar features in the XDINS sample. According to the most recent magneto-thermal evolution models [9], XDINSs are thought to be descendants of magnetars, therefore we expect to find such phase-dependent features in their X-ray spectra.

In this proceeding we summarise the discoveries published in Borghese et al. [10, 11]. Moreover, we present additional results for RX J0720.4-3125; in particular, we repeated our analysis assuming the new rotational period recently proposed by Hambaryan et al. [12].

2. Observations and data analysis

We re-analysed all the available *XMM–Newton* observations of the XDINSs, focusing on the data acquired by the EPIC pn camera which provides the spectra with the highest counting statistics owing to its larger effective area than the MOS cameras. The raw data were processed following the standard threads from the Science Analysis Software (see Borghese et al. [10, 11] for details about the data reduction, the observation logs and the timing solutions).

To study spectral variations as a function of the rotational phase without making assumptions about the spectral energy distribution, normalized energy versus phase images were created by binning the source counts in rotational phase and energy bins, and then normalizing to the phase-averaged spectrum and pulse profile. We found possible candidates in RX J0720.4-3125 and RX J1308.6+2127: the images (fig. 1) exhibit features in a narrow phase interval, which are produced by the lack of counts with respect to the nearby energy channels. These might correspond to absorption spectral lines present in a limited interval of the rotational phase. To further investigate the presence of these features we performed a detailed phase-resolved spectral analysis by dividing the rotational phase in bins and extracting the corresponding spectra.

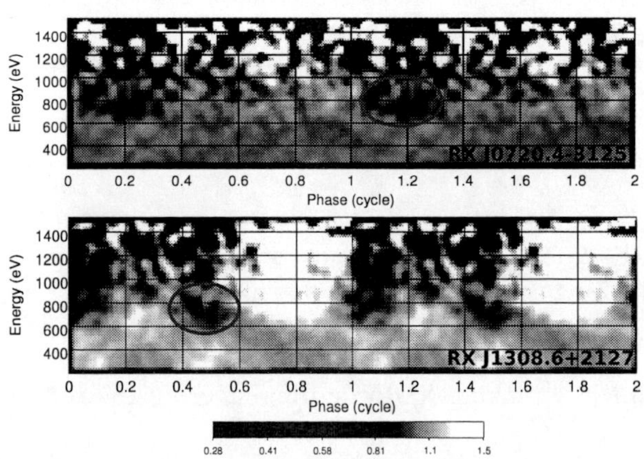

Figure 1. Normalized energy versus phase images for the longest *XMM–Newton* observations of RX J0720.4-3125 (top, 51 ks, folded on a period of 8.39 s) and RX J1308.6+2127 (bottom, 35.1 ks, folded on a period of 10.31 s) obtained by binning EPIC pn source counts into 100 phase bins and 25-eV-wide energy channels. The blue ellipses mark the phase-dependent features. Two cycles are shown for better visualization.

2.1. RX J0720.4-3125

RX J0720.4-3125 was discovered as an isolated, pulsating neutron star with a spin period of 8.39 s [13]. The phase-averaged spectrum is well modelled by a blackbody with temperature $kT_{BB} \sim 85$ eV and the inclusion of a broad absorption feature ($E_{line} \sim 270$ eV, $\sigma \sim 70$ eV) [14]. Hambaryan et al. (2017) claimed a new rotational period, which is twice that reported in

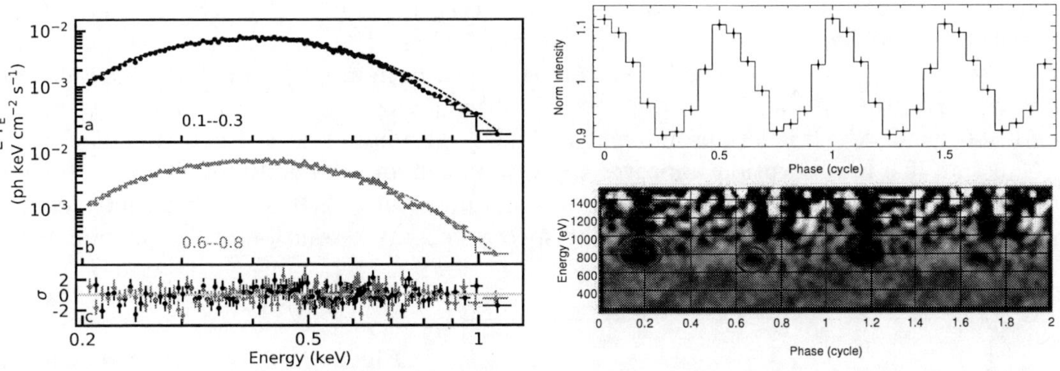

Figure 2. Results for RX J0720.4-3125 using the new period of $\sim$ 16 s, for the observation performed on 2nd May 2003 (51 ks). Left-hand, panel a: 0.1 – 0.3 phase-resolved spectrum fitted with an absorbed blackbody plus a Gaussian profile (solid line). The dashed line represents the blackbody component. Left-hand, panel b: 0.6 – 0.8 phase-resolved spectrum fitted with an absorbed blackbody plus a Gaussian profile (solid line). The dashed line represents the blackbody component. Left-hand, panel c: residuals with respect to the best-fitting models. Right-hand panel, top: light curve folded on the period 16.78 s and sampled in 16 phase bins. Right-hand panel, bottom: normalized energy versus phase image using the above-mentioned period; for the colour scale see fig. 1. The blue ellipses mark the narrow features. Two cycles are shown for better visualization.

literature. We performed our analysis assuming a period of 8.38 s (see fig. 1, top panel and [11]) and of 16.78 s, for which we report our results here (fig. 2).

XMM–Newton observed RX J0720.4-3125 twenty times between May 2000 and September 2012. We started our analysis from the longest observation (51 ks) performed on 2nd May 2003. We fitted the phase-averaged spectrum with an absorbed blackbody model in the energy range 0.3 – 1.2 keV (excluding the broad absorption feature in order to reduce the degrees of freedom). The corresponding phase versus energy image gives us a hint for a narrow spectral line in the phase bin 0.1 – 0.3 if we fold the light curve with a period of $\sim$ 8 s (fig. 1, top panel); using the new period, the feature appears in two phase intervals, 0.1 – 0.3 and 0.6 – 0.8 (fig. 2, right-hand panel). For the spectrum relative to these phases, the addition of a Gaussian line in absorption to the blackbody continuum leads to an improvement in the residuals, whereas a simple blackbody model gives an acceptable fit for all the other phase intervals. Considering the case of the new period, the spectral feature has consistent energies within the errors (90% confidence level), $\sim$ 740 eV, while it shows different widths σ and equivalent widths EqW. For the 0.1 – 0.3 phase-resolved spectrum the line is broader and deeper ($\sigma \sim 43$ eV and $EqW \sim$ 44 eV), while we obtained an upper limit for the width of $\sim$ 62 eV and an EqW of $\sim$ 19 eV for the feature in the 0.6 – 0.8 phase bin.

2.2. RX J1308.6+2127

RX J1308.6+2127 was identified as a possible nearby isolated neutron star rotating at 10.31 s in the *ROSAT* Bright Survey [15]. The phase-averaged spectrum is fitted with a combination of an absorbed blackbody ($kT_{BB} \sim 85$ eV) and a Gaussian absorption line ($E_{line} \sim 270$ eV, $\sigma \sim$ 155 eV) with the largest equivalent width ($\sim$ 180 eV) in the XDINS sample [16].

Thirteen observations were carried out by *XMM–Newton* between December 2001 and June 2007; we excluded observations with an exposure time less than 10 ks (because of insufficient statistics) and merged together those performed within a month, focusing on six data sets in total. For each of them the normalized images display an additional narrow feature in certain rotational phases, 0 – 0.2 and 0.4 – 0.6 (see fig. 1, bottom panel for the longest observation, and [10]). To fit the six phase-resolved spectra simultaneously, we used a combination of an absorbed blackbody with an absorption line for the phase-averaged feature. Because of a dependence of the phase-averaged feature energy on the rotational phase [16], we allowed this parameter to vary for the different phase-resolved spectra. As shown in figure 3 panel b, the residuals in the spectra relative to 0.4 – 0.6 phase bin reveal a discrepancy between the model and the data around ~ 0.7 keV, which is less prominent in the 0 – 0.2 phase-resolved spectra. The inclusion of a Gaussian line in absorption improves the residual shape especially for the 0.4 – 0.6 phase-resolved spectra (see panel c). The feature has an energy of ~ 740 eV and an equivalent width of ~ 15 eV, the width is compatible with the spectral energy resolution of the pn camera.

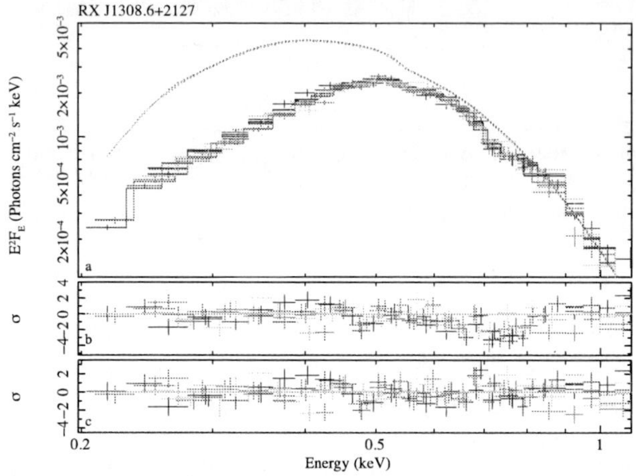

Figure 3. Results for RX J1308.6+2127. Panel a: the unfolded 0.4 – 0.6 phase-resolved spectra from six different observations considered in the analysis fitted simultaneously. The solid line represents the best-fitting model (an absorbed blackbody plus two absorption Gaussian lines), while the dashed line is the blackbody component. Panel b: residuals after setting the normalization of the phase-dependent line to zero. Panel c: residuals with respect to the best-fitting model.

2.3. Upper limits for other XDINSs
For the other sources in the sample (RX J1856.5-3754, RX J1605.3+3249, RX J2143.0+0654, RX J0806.4-4123, RX J0420.0-5022) no features have been observed in the normalized images. For the phase-averaged spectrum of the longest observation we derived 3σ upper limits on the equivalent width of a Gaussian line in absorption with a width of 0 eV (i.e. compatible with the energy resolution of the pn camera) and of 100 eV (for more details see [10]).

3. Significance of the features
To test the significance of the phase-dependent narrow spectral features detected in RX J0720.4-3125 and RX J1308.6+2127 we performed Monte Carlo simulations [17]. In both cases we simulated 10^5 spectra based on the model for the phase-averaged spectrum (the null model) and fitted them with an alternative model, i.e. including a narrow Gaussian line in absorption to account for the phase-dependent feature. In none of the simulated spectra the feature has an equivalent width higher than the observed value for both sources. Therefore the probability of the features being a fluctuation is $< 10^{-5}$. This value can be interpreted as a p-value. The outcome of the simulations relies on the assumption that the null model is true, so a p-value $< 10^{-5}$ means that the features are unlikely to have occurred by chance if the null model holds. A p-value $< 10^{-5}$ corresponds to a confidence level $> 4.6\sigma$, supporting thus our detections (for further details see [10]).

4. Discussion
A visual inspection of the normalized energy versus phase images gave us a hint for phase-dependent features in the spectra of two XDINSs, RX J0720.4-3125 and RX J1308.6+2127. Their presence was confirmed thanks to a detailed phase-resolved spectral analysis and Monte Carlo simulations. In both sources the spectral lines are detected in a limited phase interval ($\sim 1/5$ of the rotational cycle). They have an energy of ~ 750 eV and width consistent with the spectral resolution of the pn camera. Interestingly similar features have been detected in two

low-B magnetars, SGR 0418+5729 [7] and Swift J1822.3-1606 [8]; in these cases the line energy is higher (> 2 keV) and shows a stronger variation in phase. These discoveries strengthen the evolutionary connection between magnetars and XDINSs, considered to be aged magnetars.

The similarities between the features in the two different classes suggest that the same mechanism might be at work in both cases: the features might be explained invoking proton cyclotron resonant absorption scattering in a confined magnetic loop close to the star surface [7]. The proton cyclotron energy measured by a distant observer is $E_c = (eB\hbar/m_p)/(1+z) = 0.63B_{14}/(1+z)$ keV, where $B_{14} = B/10^{14}$ G, $z = 2GM_{NS}/R_{NS}c^2 \sim 0.4$ (assuming a neutron star mass $M_{NS} = 1.4M_{\odot}$ and a radius $R_{NS} = 10$ km) and $1+z$ is the gravitational redshift. The implied magnetic field in the loop is then $B_{loop} \sim 2 \times 10^{14}$ G, higher than the dipolar magnetic field at the equator ($B_{dip} \sim 2.5 \times 10^{13}$ G and $\sim 3.4 \times 10^{13}$ G for RX J0720.4-3125 [18] and RX J1308.6+2127, [19] respectively).

Atomic transitions in a magnetized atmosphere is an alternative explanation; however, current atmosphere models cannot successfully describe the spectra. Moreover the line energy is too high to be explained by transitions in light element atoms. The strong dependence on the rotational phase requires that the atmosphere is on top of a small part of the surface, while the primary emission comes from a larger region, but it is difficult to understand how only a small part of the surface can be covered by some absorbing material. Therefore, the first interpretation is preferred and, if confirmed, supports the scenario according to which the magnetic field of highly magnetized neutron stars is more complex than a pure dipole on small scale, with localized high B-field bundles.

5. Conclusions

A careful re-analysis of archival *XMM–Newton* EPIC pn data allowed us to discover phase-dependent features in the X-ray spectra of RX J0720.4-3125 and RX J1308.6+2127, and to derive upper limits for the other XDINSs. The properties of the features are naturally explained in the proton cyclotron scattering scenario if localized magnetic field bundles are present close to the surface of the star. Moreover, these discoveries provide important implications on the evolutionary link between magnetars and XDINSs.

References

[1] Kondratiev V I, McLaughlin M A, Lorimer D R, Burgay M, Possenti A, Turolla R, Popov S B and Zane S 2009 *Astrophys. J.* **702** 692–706
[2] Turolla R 2009 *Astrophysics and Space Science Library* vol 357 ed Becker W pp 141–163
[3] Pires A M, Haberl F, Zavlin V E, Motch C, Zane S and Hohle M M 2014 *Astron. Astrophys.* **563** A50–61
[4] Kaplan D L, Kamble A, van Kerkwijk M H and Ho W C G 2011 *Astrophys. J.* **736** 117–27
[5] van Kerkwijk M H and Kaplan D L 2007 *Astrophys. Space Sci.* **308** 191–201
[6] Viganò D, Perna R, Rea N and Pons J A 2014 *Mon. Not. R. Astron. Soc.* **443** 31–40
[7] Tiengo A *et al.* 2013 *Nature* **500** 312–14
[8] Rodríguez Castillo G A *et al.* 2016 *Mon. Not. R. Astron. Soc.* **456** 4145–55
[9] Viganò D, Rea N, Pons J A, Perna R, Aguilera D N and Miralles J A 2013 *Mon. Not. R. Astron. Soc.* **434** 123–41
[10] Borghese A, Rea N, Coti Zelati F, Tiengo A, Turolla R and Zane S 2017 *Mon. Not. R. Astron. Soc.* **468** 2975–83
[11] Borghese A, Rea N, Coti Zelati F, Tiengo A and Turolla R 2015 *Astrophys. J. Lett.* **807** L20–4
[12] Hambaryan V, Suleimanov V, Haberl F, Schwope A D, Neuhäuser R, Hohle M and Werner K 2017 *Astron. Astrophys.* **601** A108–25
[13] Haberl F, Motch C, Buckley D A H, Zickgraf F J and Pietsch W 1997 *Astron. Astrophys.* **326** 662–68
[14] Haberl F, Zavlin V E, Trümper J and Burwitz V 2004 *Astron. Astrophys.* **419** 1077–85
[15] Schwope A D, Hasinger G, Schwarz R, Haberl F and Schmidt M 1999 *Astron. Astrophys.* **341** L51–4
[16] Hambaryan V, Suleimanov V, Schwope A D, Neuhäuser R, Werner K and Potekhin A Y 2011 *Astron. Astrophys.* **534** A74–80
[17] Protassov R, van Dyk D A, Connors A, Kashyap V L and Siemiginowska A 2002 *Astrophys. J.* **571** 545–59
[18] Kaplan D L and van Kerkwijk M H 2005 *Astrophys. J. Lett.* **628** L45–8
[19] Kaplan D L and van Kerkwijk M H 2005 *Astrophys. J. Lett.* **635** L65–8

The missing links of neutron star evolution in the eROSITA all-sky X-ray survey

A M Pires

Leibniz-Institut für Astrophysik Potsdam (AIP), An der Sternwarte 16, 14482 Potsdam, Germany

E-mail: apires@aip.de

Abstract. The observational manifestation of a neutron star is strongly connected with the properties of its magnetic field. During the star's lifetime, the field strength and its changes dominate the thermo-rotational evolution and the source phenomenology across the electromagnetic spectrum. Signatures of magnetic field evolution are best traced among elusive groups of X-ray emitting isolated neutron stars (INSs), which are mostly quiet in the radio and γ-ray wavelengths. It is thus important to investigate and survey INSs in X-rays in the hope of discovering peculiar sources and the long-sought missing links that will help us to advance our understanding of neutron star evolution. The Extended Röntgen Survey with an Imaging Telescope Array (eROSITA), the primary instrument on the forthcoming Spectrum-RG mission, will scan the X-ray sky with unprecedented sensitivity and resolution. The survey has thus the unique potential to unveil the X-ray faint end of the neutron star population and probe sources that cannot be assessed by standard pulsar surveys.

1. Introduction

An isolated neutron star (INS) radiates at the expense of its rotational, thermal, and magnetic energy. The emission is produced through various mechanisms, with photon energies covering the entire electromagnetic spectrum. Fundamental properties inherited at birth (mass, composition, spin rate, magnetic field, etc), as well as their subsequent temporal evolution, determine the dominant emission mechanism at a given age. These, however, are subject to many theoretical uncertainties and remain poorly constrained by observations.

The population of over 2,600 neutron stars observed in our Galaxy is dominated by radio pulsars [1]. Of those, about 50 peculiar X-ray emitting sources are not detected by radio and γ-ray pulsar surveys: they are the young and energetic magnetars, usually identified by their remarkable spectral properties and bursts of high energy emission [2]; the local group of thermally emitting middle-aged INSs discovered by ROSAT and dubbed the magnificent seven (M7) [3]; and the young and seemingly weakly-magnetised central compact objects (CCOs), which are thermal X-ray sources located near the centre of supernova remnants [4]. These groups are important as they provide a privileged view of a variety of emission processes and evolutionary channels that cannot be probed through the bulk of the pulsar population.

On the one hand, this is realised by the correlation of a neutron star's thermal luminosity and its inferred magnetic field [5], which holds true for sources with $B \gtrsim 10^{13}$ G (Figure 1). The standard cooling theory [6] falls short at describing the temperature of these INSs (Figure 2). Whereas for most neutron stars magneto-rotational and thermal evolution proceed almost

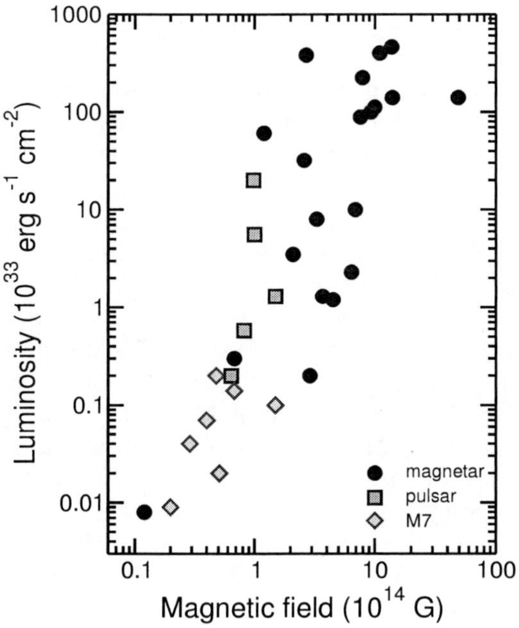

Figure 1. Correlation between thermal luminosity and magnetic field intensity for high magnetic field INSs (data extracted from the Magnetar Outburst Online Catalog, `http://magnetars.ice.csic.es`).

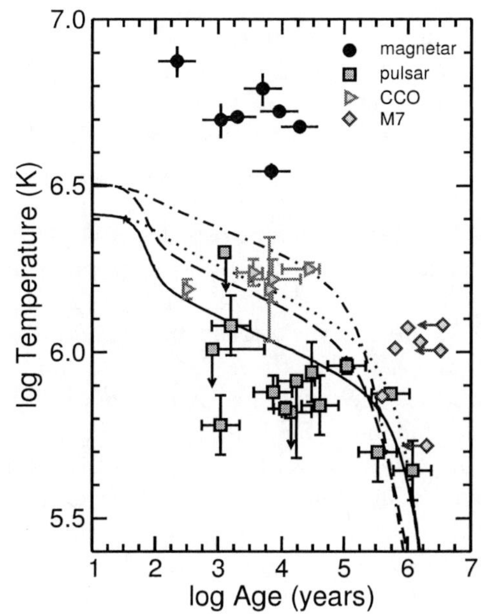

Figure 2. Cooling-age diagram for INSs with thermal emission, in relation to the cooling scenario with no magnetic field decay (solid, dashed, dotted, and dotted-dahsed lines).

independently, the cooling of strongly magnetised sources is affected by field decay and magneto-rotational energy dissipation [7, 8]. Evolutionary models that take these effects into account [9] show that strong fields at birth can brake the INS spin to an asymptotic value over a relatively short timescale; moreover, field dissipation keeps the stellar crust hot for a longer time than expected from the standard cooling theory. While accounting for the spectral and timing properties of individual sources, the model implies an evolutionary link between magnetars and the M7.

At the opposite end of the magnetic field distribution, CCOs challenge the stereotype of a young INS. The very low measured spin-down values of three CCOs (dubbed *anti-magnetars*) imply very low magnetic field strengths and characteristic ages much older than expected from their association with a supernova remnant. Moreover, the lack of detected pulsations in most members indicates that the atmosphere is weakly magnetised [10]. This contradicts the evidence that some sources conceal strong magnetic fields [11]. The fact that no old (> 10 kyr) anti-magnetar with similar timing properties is recognised in either radio or X-ray surveys [12, 13] implies that their magnetic field is not constant at secular timescales.

CCOs might represent a different outcome of neutron star evolution, where they are recovering from an early episode of field burial by hypercritical accretion [14]. In this scenario, if a large amount of debris falls back onto the neutron star after the supernova explosion, accretion is likely to occur at such high rates that any field can be hidden [15]. This may delay the onset of radio pulsar activity for several $1 - 100$ kyr, until the field can diffuse back to the surface [16].

2. Evolutionary scenarios: issues and open questions

The role played by the decay of the magnetic field in heating the neutron star crust cannot be overlooked, as it partially explains the observed neutron star diversity. However, even state-of-the-art models of field decay are built over uncertain assumptions, concerning, for instance, the initial field configuration and the level of impurity of the crust [9]. At present, the observed radio and X-ray pulsar population is not sufficient to constrain different hypotheses [17]. While the phenomenology of CCOs has triggered a lively debate over their formation and fate, to our knowledge no attempt has been made in population synthesis to include anti-magnetars as a possible outcome of neutron star evolution.

The paucity of these INS groups has hindered their use in population syntheses on the Galactic scale. While particular aspects of neutron star evolution and emissivity have been investigated through an extensive number of population syntheses of radio pulsars[1], overall progress can only be hoped for when the various groups are unified in a 'grand scheme' of neutron star evolution [19]. There are the works which attempt to join the standard radio pulsar population with those of magnetars and the M7, through models of field decay [20, 17]. A crucial result is that, to reproduce the number of X-ray thermally emitting INSs we observe to date, the mean strength of the magnetic field distribution at birth has to be significantly higher (and the distribution wider) than one would expect from radio pulsar studies alone. Moreover, the models usually predict a large number of very long spin period neutron stars ($P > 20\,\mathrm{s}$), which are not observed.

Also important is to assess the relative contributions of the various INS groups to the total number of neutron stars populating the Milky Way. If each INS group is treated independently, a likely consequence is that the Galactic core-collapse supernova rate cannot account for the entire estimated population of neutron stars [21]. The neutron star birthrate derived from population synthesis is even more discrepant from the expected value when field decay is taken into account [22, 17]. Moreover, while active magnetars are not expected to contribute significantly to the absolute number of observable neutron stars in the Galaxy, the existence of transient sources, which are believed to possess low quiescent X-ray luminosities, also has implications for both the total magnetar birthrate as for the ensemble of the population. The birthrate of CCOs is similarly unclear, either as anti-magnetars (that is, as neutron stars born with intrinsically weak magnetic fields), or within the 'hidden magnetic field' scenario. The fraction of neutron stars that undergo a phase of fallback accretion and field submergence is unknown, although these events could commonly take place in the early life of a neutron star (see e.g. [15], for a discussion).

3. A forecast for the eROSITA survey

As long as only the X-ray bright end of the radio-quiet INS population is known, the observational and theoretical advances seen in recent years may reach a halt. The Extended Röntgen Survey with an Imaging Telescope Array (eROSITA) [23] is the primary instrument on the forthcoming Spectrum-RG mission. The four-year eROSITA all-sky survey (eRASS) is expected to detect a large sample of clusters of galaxies, active galactic nuclei, transient phenomena, and Galactic compact objects, among other interesting case studies. eRASS offers a timely opportunity for a better sampling of a considerable number of neutron stars that cannot be assessed by radio and γ-ray surveys. Eventually, the identification and investigation of eROSITA sources at faint X-ray fluxes will help us to test alternative neutron star evolutionary scenarios and constrain the rate of core-collapse supernovae in our Galaxy.

To estimate the number of INSs to be detected by eRASS through their thermal X-ray emission, we performed Monte Carlo simulations of a population synthesis model [24]. Briefly, neutron stars are created from a progenitor population of massive stars distributed in the spiral arms of the Galactic disk; after the supernova explosion and the imparted 'kick' velocity, their evolution in the Galactic potential is followed while they cool down, isotropically emitting soft

[1] Interestingly, recent results considering only the radio pulsar population show that pulsar evolutionary paths in $P - \dot{P}$ space can signficantly deviate from those expected along lines of constant field, when the decay of the angle between spin and magnetic axes is included in the population synthesis [18].

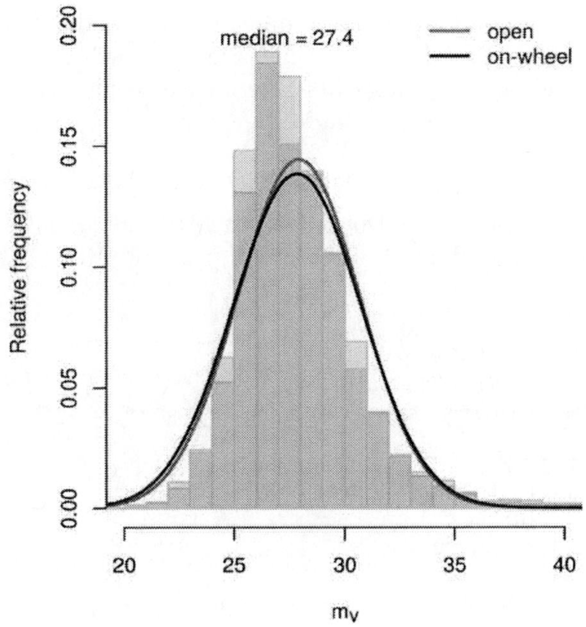

Figure 3. Histogram of the optical V magnitudes required for ruling out other classes of X-ray emitter among INS candidates, based on X-ray-to-optical flux ratios. We considered in the simulations two possible configurations of optical blocking filters and the resulting total effective area of the seven telescopes, averaged over the field of view [24]. A minimum X-ray-to-optical flux ratio of $10^{3.5}$ is assumed for the eROSITA-detected sample of neutron stars. Distances to the detected sources are within 400 pc and 8 kpc and the accumulated survey exposure ranges from 1.6 to 8 ks, with a median around 1.9 ks. The limiting flux is around 10^{-14} erg s^{-1} cm^{-2}, which is 50 times deeper than that of the ROSAT Bright Star Catalogue.

X-rays. The expected source count rates and total flux are then computed in the eROSITA detectors, taking into account the absorbing material in the line-of-sight, the detector and survey properties, and the celestial exposure after four years of all-sky observation. We apply an analytical description of the interstellar medium, based on layers of hydrogen in both atomic and molecular form, and commonly adopted cross-sections. All neutron stars in our simulations cool down at the same fiducial rate.

Our study indicates an expected number of up to $\sim 85 - 95$ thermally emitting INSs[2] to be detected in the eRASS with more than 30 counts (0.2−2 keV). In Figure 3 we show the histogram of V magnitudes required to select promising INS candidates for deep follow-up investigations (assuming a logarithmic X-ray-to-optical flux ratio of $\log(F_X/F_V) = 3.5$, which is sufficient to exclude other classes of X-ray emitters). Although optical follow-up will require very deep observations – in particular, the identification of the faintest candidates will have to wait for the next generation of extremely large telescopes – sources at intermediate fluxes can be selected for follow-up investigations using current observing facilities. We anticipate ~ 25 discoveries in the first years after the completion of the all-sky survey (see [24], for details).

4. Summary and conclusions

The observed neutron star phenomenology is arguably governed by the properties of the magnetic field: specifically, by its magnitude at birth, whether it decays or grows during the star's lifetime, and how these characteristics affect the rotational and thermal history of the neutron star. Overall, crucial aspects of neutron star evolution and emissivity cannot be probed through the radio and γ-ray pulsar population alone, and are not yet described by theory. It is thus important to investigate and survey INSs in X-rays in the hope of discovering missing links in their evolution, and to gain insight into neutron star physics and phenomenology. The forthcoming all-sky survey of eROSITA, expected for launch in September 2018, is a unique oportunity to unveil the X-ray faint end of the neutron star population. Both coordination with photometric and spectroscopic surveys at other wavelengths and catalogue cross-correlation procedures yielding

[2] Depending on the exact configuration of optical blocking filters [24].

actual probabilities of source identification are required to pinpoint promising INS candidates among the millions of eROSITA-detected sources. Dedidated follow-up investigations will tackle the evolutionary state of individual targets to evaluate the impact of alternative scenarios on the evolution and observability of the neutron star population in the Milky Way.

Acknowledgments

The collaboration of Axel Schwope and Christian Motch is acknowledged. AMP is financially supported by the Deutsches Zentrum für Luft- und Raumfahrt (DLR) under grant 50 OR 1511.

References

[1] Manchester R N, Hobbs G B, Teoh A and Hobbs M 2005 *AJ* **129** 1993–2006
[2] Kaspi V M and Beloborodov A 2017 *ArXiv e-prints* (*Preprint* 1703.00068)
[3] Turolla R 2009 *Astrophysics and Space Science Library* (*Astrophysics and Space Science Library* vol 357) ed Becker W pp 141–63
[4] Gotthelf E V, Halpern J P and Alford J 2013 *ApJ* **765** 58
[5] Pons J A, Pérez-Azorín J F, Miralles J A and Miniutti G 2007 *Ap&SS* **308** 247–57
[6] Yakovlev D G and Pethick C J 2004 *ARA&A* **42** 169–210
[7] Aguilera D N, Pons J A and Miralles J A 2008 *ApJ* **673** L167–70
[8] Pons J A, Miralles J A and Geppert U 2009 *A&A* **496** 207–16
[9] Viganò D, Rea N, Pons J A, Perna R, Aguilera D N and Miralles J A 2013 *MNRAS* **434** 123–41
[10] Ho W C G and Heinke C O 2009 *Nature* **462** 71–3
[11] Rea N, Borghese A, Esposito P, Coti Zelati F, Bachetti M, Israel G L and De Luca A 2016 *ApJ* **828** L13
[12] Gotthelf E V, Halpern J P, Allen B and Knispel B 2013 *ApJ* **773** 141
[13] Luo J, Ng C Y, Ho W C G, Bogdanov S, Kaspi V M and He C 2015 *ApJ* **808** 130
[14] Ho W C G 2011 *MNRAS* **414** 2567–75
[15] Geppert U, Page D and Zannias T 1999 *A&A* **345** 847–54
[16] Muslimov A and Page D 1995 *ApJ* **440** L77–80
[17] Gullón M, Pons J A, Miralles J A, Viganò D, Rea N and Perna R 2015 *MNRAS* **454** 615–25
[18] Johnston S and Karastergiou A 2017 *MNRAS* **467** 3493–9
[19] Kaspi V M 2010 *Proceedings of the National Academy of Science* **107** 7147–52
[20] Popov S B, Pons J A, Miralles J A, Boldin P A and Posselt B 2010 *MNRAS* **401** 2675–86
[21] Keane E F and Kramer M 2008 *MNRAS* **391** 2009–16
[22] Gullón M, Miralles J A, Viganò D and Pons J A 2014 *MNRAS* **443** 1891–9
[23] Predehl P 2017 *Astronomische Nachrichten* **338** 159–64
[24] Pires A M, Schwope A D and Motch C 2017 *Astronomische Nachrichten* **338** 213–9

IOP Conf. Series: Journal of Physics: Conf. Series **932** (2017) 012009 doi:10.1088/1742-6596/932/1/012009

X-ray properties of the mode-switching pulsar PSR B0943+10

S Mereghetti[1], L Kuiper[2], A Tiengo[3,1,4], J Hessels[5,9], W Hermsen[2,9], K Stovall[6], A Possenti[7], J Rankin[8], P Esposito[9], R Turolla[10,11], D Mitra[8,12,13], G Wright[14], B Stappers[14], A Horneffer[15], S Oslowski[15,16,17], M Serylak[18,19], J-M Griessmeier[20,19] and M Rigoselli[1,21]

[1] INAF-IASF Milano, Italy
[2] SRON, Utrecht, The Netherlands
[3] Scuola Universitaria Superiore IUSS Pavia, Italy
[4] INFN, Sezione di Pavia, Italy
[5] ASTRON, Dwingeloo, The Netherlands
[6] Department of Physics and Astronomy, University of New Mexico, Albuquerque, NM, USA
[7] INAF - Osservatorio Astronomico di Cagliari, Selargius, Italy
[8] Physics Department, University of Vermont, Burlington, VT 05405, USA
[9] Anton Pannekoek Institute for Astronomy, Univ. of Amsterdam, The Netherlands
[10] Dipart. di Fisica e Astronomia, Università di Padova, Italy
[11] MSSL-UCL, Holmbury St. Mary, Dorking, UK
[12] National Centre for Radio Astrophysics, Ganeshkhind, Pune , India
[13] Janusz Gil Institute of Astronomy, Univ. of Zielona Góra, Poland
[14] Jodrell Bank Centre for Astrophysics, Univ. of Manchester, UK
[15] Max-Planck-Institut für Radioastronomie, Bonn, Germany
[16] Fakultät für Physik, Universität Bielefeld, Germany
[17] currently at Swinburne Univ. of Technology, Australia
[18] Dept. of Physics & Astronomy, Univ. of the Western Cape, Bellville, South Africa
[19] Station de Radioastronomie de Nançay, Observatoire de Paris, CNRS, Nançay, France
[20] LPC2E - Université d'Orléans, France
[21] Università di Milano Bicocca, Milano, Italy

E-mail: `sandro@iasf-milano.inaf.it`

Abstract.
The mode-switching pulsar PSR B0943+10 has been extensively studied in the radio band for many years and, more recently, it has been found to vary also in X-rays, with a flux anticorrelated with the radio emission. Here we review the results of long observations of PSR B0943+10 carried out with *XMM-Newton* and the LOFAR, LWA and Arecibo radio telescopes in 2014. These results support a scenario in which both unpulsed non-thermal emission, likely of magnetospheric origin, and pulsed thermal emission from a small polar cap ($\sim$1500 m^2) with a strong non-dipolar magnetic field ($\sim 10^{14}$ G), are present during both radio modes and vary in intensity in a correlated way.

1. Introduction

PSR B0943+10 can be considered the prototypical mode-switching radio pulsar. It was soon discovered that its radio emission alternates between two different states: when the pulsar is in B (bursting or bright) mode, its radio emission displays a regular pattern of drifting subpulses, while it is chaotic, and on average fainter, when the pulsar is in the Q (quiescent) mode [1, 2].

Its spin period P=1.1 s and period derivative $\dot{P}$=3.5×10^{-15} s s^{-1} imply a characteristic age τ=5 Myr, a dipolar surface magnetic field B=4×10^{12} G, and a spin-down power $\dot{E}_{rot}$=10^{32} erg s^{-1}. The distance of PSR B0943+10, based on its dispersion measure (DM=15.32 pc cm^{-3} [3]) and the Galactic electron density distribution of [4], is ~630 pc. According to the more recent electron density model of [5] the same DM gives instead a distance d=890 pc.

The faint (a few 10^{-15} erg cm^{-2} s^{-1}) X-ray emission of PSR B0943+10 was first detected in 2003 [6], and studied in much more detail thanks to a set of *XMM-Newton* observations carried out in 2011. These observations, providing a useful exposure of about 100 ks, were complemented by simultaneous radio monitoring with LOFAR and the GMRT. This made it possible to analyse separately the X-ray data corresponding to the two radio modes, thus leading to the discovery of X-ray flux variations anticorrelated with the radio modes [7]. In fact, it was found that in the Q-mode, when the radio emission is fainter, the X-ray flux is larger by more than a factor two than in the B-mode.

X-ray pulsations at the rotation period of 1.1 s were detected only in the Q-mode. The X-ray emission during the Q-mode was well described by a blackbody plus power law spectrum, while that of the X-ray fainter B-mode was less constrained: both a single power law and a single blackbody could fit it. The pulsed spectrum (in Q-mode) was found to be a blackbody, consistent in flux and temperature with the blackbody component seen in the total spectrum. Furthermore, the power-law component of the total Q-mode spectrum was consistent, in slope and normalization, with the best fit power-law spectrum of the B-mode. These results led to the suggestion that PSR B0943+10 emits only an unpulsed non-thermal component when it is in B-mode and its higher luminosity in Q-mode is caused by the addition of a 100%-pulsed thermal component [7].

However, since there is evidence from radio data that PSR B0943+10 is a nearly aligned rotator (i.e. small angle between rotation and magnetic axis) and that the line of sight is also close to the rotation axis [8, 9, 10], one would not expect such a large modulation from a hot spot in the polar region. Furthermore, it is unclear what could cause the rapid appearance/disappearance of the pulsed thermal emission when the pulsar switches radio mode. A new longer campaign of simultaneous X-ray and radio observations was therefore carried out in November 2014 in order to clarify these issues. The X-ray observations, performed with the EPIC instrument of the *XMM-Newton* satellite, consisted of seven long pointings (~16–19 hr each), which, after screening to remove periods of high background, provided a net exposure of ~120 ks during the pulsar Q-mode and ~175 ks during B-mode (for comparison, the previous data consisted of ~50 ks in each mode). Simultaneous radio coverage was provided by using three radio telescopes at different longitudes: the Low-Frequency Array (LOFAR) at 110–190 MHz, the Arecibo telescope at 302–352 MHz, and the Long Wavelength Array (LWA) at 40–80 MHz. Here, we briefly report on the main results obtained in these observations, referring to [11] for more details.

2. Results

The larger X-ray count statistics obtained in the 2014 observations, compared to that of the previous data (in particular for the B-mode), made it possible to reveal that the periodic modulation caused by the star's rotation is present also during the B-mode, when the X-ray flux is lower (see figure 1). By comparing the folded light curves obtained before and after T$_{QB}$+ 3 hr (where T$_{QB}$ is the time of the Q-to-B mode transitions), we found some evidence

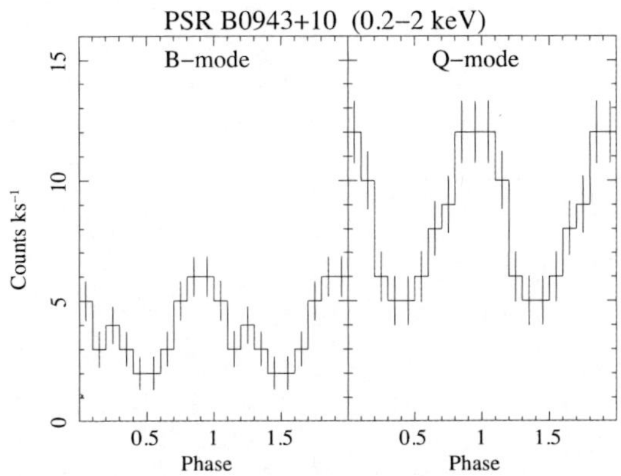

Figure 1. Background-subtracted pulse profiles of PSR B0943+10 in the 0.2-2 keV energy range obtained in the 2014 *XMM-Newton* observations.

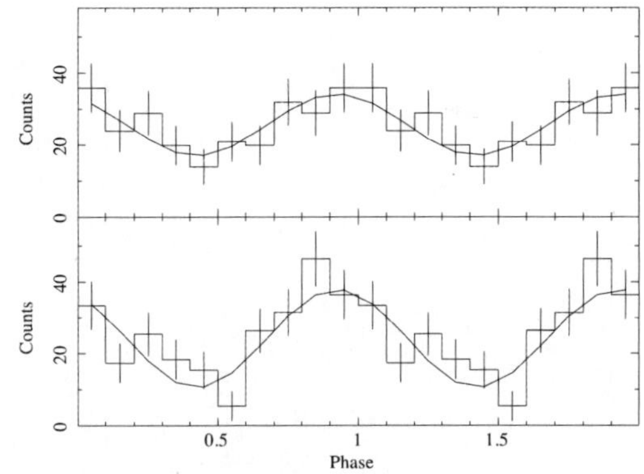

Figure 2. B-mode pulse profiles in the 0.5-2 keV range (background subtracted). Top: initial 3 hr after the Q-to-B transitions. The pulsed fraction is (27±8)%. Bottom: data obtained later than 3 hr after the Q-to-B transitions. The pulsed fraction is (42±8)%.

that the pulsed fraction increases along the B-mode (see figure 2). This possible variation, although only at the $\lesssim 2\sigma$ level, is potentially interesting and might be related to the fact that several properties of the PSR B0943+10 radio emission show an evolution along the B-mode [2, 12, 13, 3]. The possible presence of an increase in the X-ray pulsed fraction is also supported by the failure to detect the B-mode pulsations in the 2011 observations, which covered mostly the first few hours after the Q-to-B mode transitions. Simulations show that, with the count statistics of the 2011 data, the pulsations in B-mode should have been detected at a significance $> 3\sigma$ with a probability of ~50%, if they had the same pulsed fraction measured in 2014.

The spectral analysis confirmed that the Q-mode emission cannot be fit by single component models, as found in [7]. A good fit could be obtained with a blackbody with temperature kT=0.27 keV plus a power law with photon index Γ=2.4, or with the sum of two blackbodies with temperatures kT_1=0.11 keV and kT_2=0.34 keV.

On the other hand, the new data allowed a much better characterization of the B-mode spectrum, which, at variance with the findings of the 2011 observations, could not be fit by a single power law. In fact, a good fit could be obtained either with a single blackbody with temperature kT=0.24 keV, or with the sum of a blackbody and a power law (kT=0.24 keV, Γ=2.3).

Very interesting results were obtained by applying a 3-D maximum likelihood analysis (ML) [14] to *simultaneously* derive the spectra of the pulsed and unpulsed emission. The findings of [7] for the Q-mode were confirmed: the pulsed flux is thermal, being well fit by a blackbody (kT=0.29 keV) but not by a power-law. The opposite is true for the unpulsed flux, which can be fit by a power-law (Γ=2.5) and not by a blackbody. The blackbody describing the pulsed emission and the power law describing the unpulsed emission are consistent with the two components used to fit the total spectrum of the Q-mode. The situation is not so well constrained for the B-mode, where, both for the pulsed and unpulsed emissions, it was impossible to distinguish between a power law and a blackbody, which gave equally good fits.

It is reasonable to assume that the spectral components in the B-mode are the same as in the

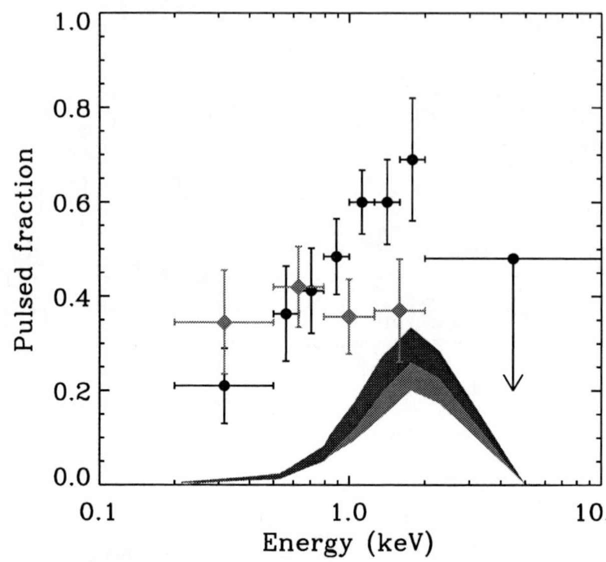

Figure 3. X-ray pulsed fraction as function of energy for the Q-mode (black dots) and for the B-mode (red squares). The blue regions show the values expected for the polar cap emission computed with model atmospheres of different temperatures (light blue kT=0.25 keV, dark blue kT=0.14 keV, see [11] for details) for the most likely ranges [10] of the angles between the magnetic and rotation axis ($10° < \xi < 15°$) and between the rotation axis and the line of sight ($5° < \chi < 10°$).

Q-mode, i.e. the thermal emission is pulsed and the non-thermal is non-pulsed. The opposite situation (thermal unpulsed plus non-thermal pulsed emission), although statistically consistent with the data, is more difficult to interpret physically. In fact, a significant and finely-tuned change in both spectral components would be required when the pulsar switches between the different radio modes.

These results are illustrated in figure 4, where the spectra obtained with the 3-D ML for the pulsed (red) and unpulsed (blue) components are compared with those derived for the total emission (black).

3. Discussion

The above results strongly suggest that PSR B0943+10 emits a combination of thermal and non-thermal X-rays in both radio modes. The X-ray flux variation, anticorrelated with that in the radio band, involves both spectral components and, quite surprisingly, the pulsations are due to the thermal component.

The best-fit temperature of the thermal component in the Q-mode is ~30% higher than that of the B-mode. Thus the flux variation can, in principle, be explained only by the temperature change of a fixed emitting area. However, the temperatures and emitting areas derived by the

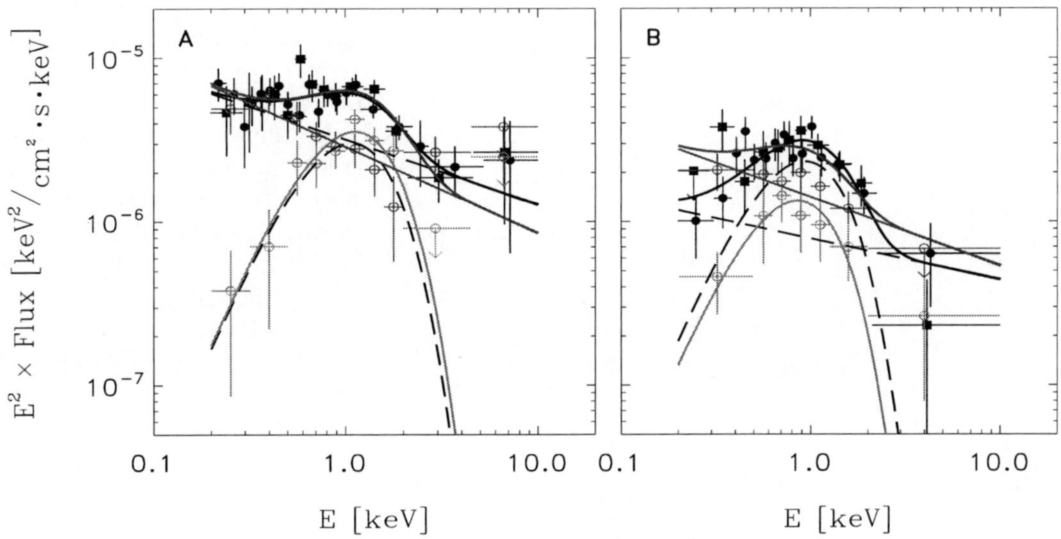

Figure 4. Spectra of the Q-mode (left panel) and B-mode (right panel) obtained in the 2014 observations of PSR B0943+10. All the spectra have been derived with a Maximum Likelihood analysis, which gives the best sensitivity for faint X-ray sources [7]. Black lines and points refer to the spectra of the total (i.e. pulsed plus unpulsed) emission (circles = pn, squares = sum of the two MOS; the solid lines are the best-fit blackbody plus power law models, while the dashed lines indicate the individual components). The red and blue points and lines refer to the spectra simultaneously derived for the pulsed and non-pulsed components (see [11, 14] for details). Blue indicates the unpulsed component (best-fit with a single power-law) and red indicates the pulsed component (best-fit with a single blackbody). Note that, in both modes, the sum of the blue and red lines (indicated in purple) is fully consistent with the solid black line.

fits are strongly correlated and, as shown in figure 5, it is also possible that an area variation is entirely (or partially) responsible for the difference in flux between the two modes.

In any case, it is clear that the thermally-emitting area is a small fraction of the star surface, as expected for an old pulsar like PSR B0943+10, in which only the magnetic polar caps are heated by backward-accelerated magnetospheric particles, while the bulk of the surface has cooled below detectability in the X-ray band. Even with the upward revised distance of 890 pc, the projected emitting area derived from the blackbody fits is smaller than that expected for the polar cap defined by the last open field lines of a dipolar field ($A \sim \pi R_{pc}^2 = \frac{2\pi^2 R^3}{cP} = 1.3 \times 10^5$ m^2, for a star radius $R = 13$ km).

The flux of the unpulsed power-law increases by a factor ~2 from the B- to the Q-mode, while that of the pulsed blackbody increases by a factor ~3. The flux correlation between the thermal and non-thermal X-rays is not surprising in the framework of the Vacuum Gap models [15, 16, 17, 18] in which the pairs are produced relatively close to the star surface and the accelerated particles are responsible both for the non-thermal emission in the magnetosphere and for the heating of the polar regions through return currents. On the other hand, Space Charge Limited Flow models with particles leaving freely the star surface (e.g. [19, 20]), predict the pair-creation front much higher in the magnetosphere. Since in this case only a few percent

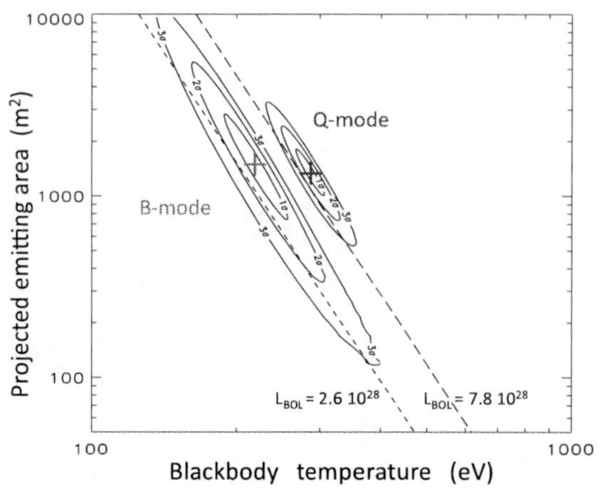

Figure 5. Error regions of the parameters derived for the thermal emission of PSR B0943+10 in the B- and Q-modes. The crosses indicate the best fit values of blackbody temperature and projected emitting area, while the contours show the corresponding uncertainty regions (1, 2, and 3σ c.l.). The dashed lines indicate values of the bolometric luminosities (in erg s^{-1}) of the polar cap, corrected for the (small) projection effect and assuming d=630 pc.

of the return particles might reach the surface, one would not naturally expect a correlation beteween the thermal and non-thermal emission.

The high pulsed fraction of the thermal X-ray component (figure 3) is rather puzzling considering the geometry of PSR B0943+10 derived from radio observations [8, 10, 3]. Although one should remember that such estimates are subject to model dependent uncertainties, there is a broad consensus that PSR B0943+10 is a nearly aligned rotator seen approximately pole-on. Therefore, the polar cap hot spot should not produce strongly pulsed X-rays, unless the emission is significantly beamed, e.g. due to magnetic effects. Another possibility that has been proposed to solve this puzzle is that of an offset magnetic dipole [21].

4. Conclusions

Extensive radio observations of the prototypical mode-switching pulsar PSR B0943+10, carried out for more than 40 years, have provided a wealth of information and a detailed knowledge of its complex phenomenology. However, the cause of the mode-switching in radio pulsars remains unknown. Thanks to the high sensitivity of the *XMM-Newton* satellite, it has now become possible to supplement the radio studies with a high-energy view, which might help to understand better this phenomenon. In this respect it is useful to obtain X-ray observations of similar objects. So far, this has been done for PSR B1822–09 and PSR B0823+26.

PSR B1822–09 is younger ($\tau = 0.2$ Myr) and more energetic ($\dot{E}_{rot} = 4.6 \times 10^{33}$ erg s^{-1}) than PSR B0943+10 and is believed to be a nearly orthogonal rotator [22, 23]. Differently from PSR B0943+10, its mode changes occur very frequently, with average mode durations of only a few minutes. t was extensively observed with *XMM-Newton* (200 ks of exposure) in 2013–2014, but no evidence for X-ray variations was found [14]. Its spectrum was well fit by two blackbodies with temperatures $kT_1 = 0.08$ keV, $kT_2 = 0.19$ keV and emitting radii $R_1 = 2$ km, $R_2 = 100$ m. Both the hotter and the cooler blackbody components seem to contribute to the X-ray pulsations, which are present in both radio modes. The apparent lack of X-ray changes related to the radio-modes could be due to the short duration of the modes in PSR B1822–09. This could arise, e.g., if the timescale for the X-ray variation is longer than the mode duration.

However, this seems to be in contradiction with the recent X-ray observations of PSR B0823+26 [24], in which a high X-ray state lasting only 6 minutes has been seen. This pulsar has timing parameters very similar to those of PSR B0943+10 (P=0.5 s, $\dot{P}$=1.7×10^{-15} s s^{-1}) and it also shows a Bright and a Quiet radio mode [25], but, contrary to PSR B0943+10, the X-ray flux is higher during the B-mode [24].

It thus appears that the multiwavelength picture that is emerging from the observations of these three pulsars is rather heterogeneous and further X-ray observations of these and other mode-switching pulsars are required to advance. Unfortunately, most mode-switching radio pulsars are relatively old and not very energetic. This means that they are too faint for detailed studies in the X-ray range with the current facilities. Future X-ray satellites with a large effective area will be essential to provide high quality data, provided they have adequate timing capabilities and good sensitivity also in the soft range below ∼1–2 keV, where the emission from old pulsars dominates.

Acknowledgments
This work received support through PRIN INAF 2014.

References

[1] Suleimanova S A and Izvekova V A 1984 *Soviet Astronomy* **28** 32–5
[2] Rankin J M and Suleymanova S A 2006 *Astron. Astrophys.* **453** 679–86
[3] Bilous A V et al. 2014 *Astron. Astrophys.* **572** A52
[4] Cordes J M and Lazio T J W 2002 *ArXiv Astrophysics e-prints* (*Preprint* astro-ph/0207156)
[5] Yao J M, Manchester R N and Wang N 2017 *Astrophys. J.* **835** 29
[6] Zhang B, Sanwal D and Pavlov G G 2005 *Astrophys. J. Lett.* **624** L109–12
[7] Hermsen W *et al.* 2013 *Science* **339** 436
[8] Lyne A G and Manchester R N 1988 *MNRAS* **234** 477–508
[9] Rankin J M 1993 *Astrophys. J. Suppl. Ser.* **85** 145–61
[10] Deshpande A A and Rankin J M 2001 *MNRAS* **322** 438–60
[11] Mereghetti S et al. 2016 *Astrophys. J.* **831** 21
[12] Suleymanova S A and Rankin J M 2009 *MNRAS* **396** 870–7
[13] Backus I, Mitra D and Rankin J M 2011 *MNRAS* **418** 1736–45
[14] Hermsen W, Kuiper L, Hessels J W T, Mitra D, Rankin J M, Stappers B W, Wright G A E, Basu R, Szary A and van Leeuwen J 2017 *MNRAS* **466** 1688–708
[15] Ruderman M A and Sutherland P G 1975 *Astrophys. J.* **196** 51–72
[16] Zhang B, Qiao G J, Lin W P and Han J L 1997 *Astrophys. J.* **478** 313–21
[17] Gil J, Melikidze G and Zhang B 2007 *MNRAS* **376** L67–71
[18] Szary A, Melikidze G I and Gil J 2015 *MNRAS* **447** 2295–306
[19] Arons J and Scharlemann E T 1979 *Astrophys. J.* **231** 854–79
[20] Zhang B and Harding A K 2000 *Astrophys. J.* **532** 1150–71
[21] Storch N I, Ho W C G, Lai D, Bogdanov S and Heinke C O 2014 *Astrophys. J. Lett.* **789** L27
[22] Backus I, Mitra D and Rankin J M 2010 *MNRAS* **404** 30–41
[23] Latham C, Mitra D and Rankin J 2012 *MNRAS* **427** 180–9
[24] Hermsen W *et al.* 2018 *in preparation for MNRAS* –
[25] Sobey C *et al.* 2015 *MNRAS* **451** 2493–506

IOP Conf. Series: Journal of Physics: Conf. Series **932** (2017) 012010 doi:10.1088/1742-6596/932/1/012010

XMM-Newton observations of the γ-ray pulsar J0633+0632

A V Karpova[1], **D A Zyuzin**[1], **A A Danilenko**[1] **and Yu A Shibanov**[1,2]
[1]Ioffe Institute, Politekhnicheskaya 26, St. Petersburg, 194021, Russia
[2]Peter the Great St. Petersburg Polytechnic University, Politekhnicheskaya 29, St. Petersburg, 195251, Russia

E-mail: annakarpova1989@gmail.com

Abstract.
We briefly report the results of *XMM-Newton* X-ray observations of the γ-ray pulsar J0633+0632. We detected, for the first time, X-ray pulsations from J0633+0632 with the pulsar period. The pulse profile is typical for spin-modulated thermal emission from the surface of a neutron star. The pulsed fraction in the 0.3–2 keV band is $23 \pm 6\%$. We confirmed previous results that the X-ray spectrum of J0633+0632 contains a non-thermal power law and a thermal (described either by the blackbody or the neutron star atmosphere models) components. However, *XMM-Newton* observations do not confirm the absorption line in the pulsar spectrum at $\approx$ 0.8 keV found in the *Chandra* data (Danilenko *et al* 2015 *PASA* **32** e038).Using the interstellar absorption–distance relations, we constrained the distance to the pulsar in the range of 0.7–2.2 kpc.

1. Introduction

The radio-quiet *Fermi* pulsar J0633+0632 (hereafter J0633) has a period P = 297 ms, a characteristic age τ = 59 kyr, a spin-down luminosity $\dot{E} = 1.2 \times 10^{35}$ erg s^{-1}, and a magnetic field $B = 4.9 \times 10^{12}$ G [1]. *Chandra* observations revealed the pulsar and its wind nebula (PWN) [2] (see figure 1, left). The J0633 X-ray spectrum contains a non-thermal power-law (PL) and a thermal component. Reanalysing *Chandra* observations, Danilenko et al [3] found an absorption feature in the J0633 spectrum at $\approx$ 0.8 keV with equivalent width of $\approx$ 60 eV. The interstellar absorption–distance relation suggests that the distance to J0633 is 1–4 kpc [3]. The PWN morphology and extent indicate a likely high pulsar proper motion of 80 mas yr^{-1} = $380 D_{1\rm kpc}$ km s^{-1}, where $D_{1\rm kpc}$ is the distance in units of 1 kpc [3].

Besides J0633, only one rotation-powered pulsar, J1740+1000, has absorption lines in its spectrum [4]. Short *Chandra* observations of J0633 did not allow to constrain the shape and nature of the spectral line (e.g., cyclotron resonance scattering, atomic transitions in the interstellar medium (ISM) or the neutron star (NS) atmosphere, an observational artefact). Here we briefly report the results of our longer (93 ks) *XMM-Newton* observations of J0633 which were carried out on 2016 March 31 (ObsID 0764020101, PI Danilenko).

2. *XMM-Newton* images

XMM-Newton data were filtered from the periods of high background that resulted in exposures of 51.8, 63.6 and 33.0 ks for EPIC-MOS1, -MOS2 and -pn detectors, respectively. The image

IOP Conf. Series: Journal of Physics: Conf. Series **932** (2017) 012010 doi:10.1088/1742-6596/932/1/012010

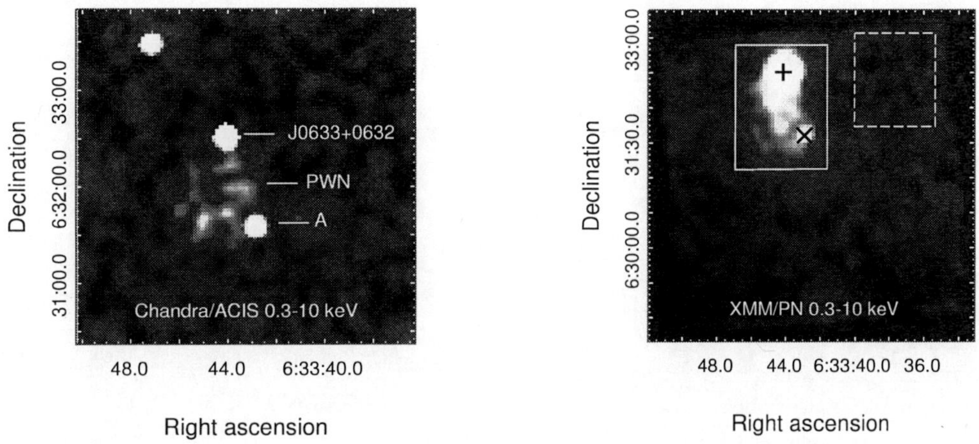

Figure 1. *Left*: 0.3–10 keV *Chandra*/ACIS image of the J0633 field. J0633, its tail-like PWN and unrelated source 'A' projected onto the PWN are marked. *Right*: Field of view of the *XMM-Newton* EPIC-pn camera in the 0.3–10 keV band. J0633 and source 'A' positions obtained from *Chandra* data are marked by the '+' and 'x' symbols, respectively. The solid and dashed boxes show regions used for extraction of the PWN and background spectra, respectively.

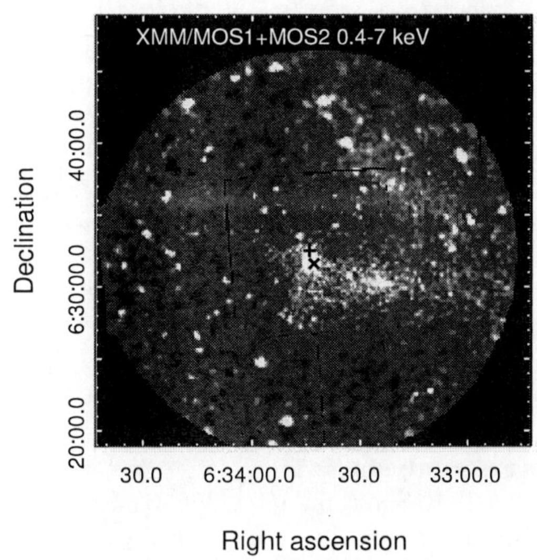

Figure 2. Combined *XMM-Newton* MOS1+MOS2 exposure-corrected image of the J0633 field in 0.4–7 keV. J0633 and source 'A' positions are marked by the '+' and 'x' symbols, respectively.

obtained with the pn detector, operated in Small Window mode, is presented in figure 1, right. J0633 and unrelated source 'A' positions derived from *Chandra* data are marked. We also constructed the combined image from MOS detectors operated in the Full Frame mode following the recipe from *XMM-Newton* Extended Source Analysis Software Cookbook [5]. The resulting image in 0.4–7.0 keV energy range is shown in figure 2. Besides the bright PWN seen in *Chandra* and *XMM-Newton*/pn images, this image also revealed a fainter large-scale emission south, west and north-west to J0633. Detailed analysis of this emission is beyond the scope of this Paper.

IOP Conf. Series: Journal of Physics: Conf. Series **932** (2017) 012010 doi:10.1088/1742-6596/932/1/012010

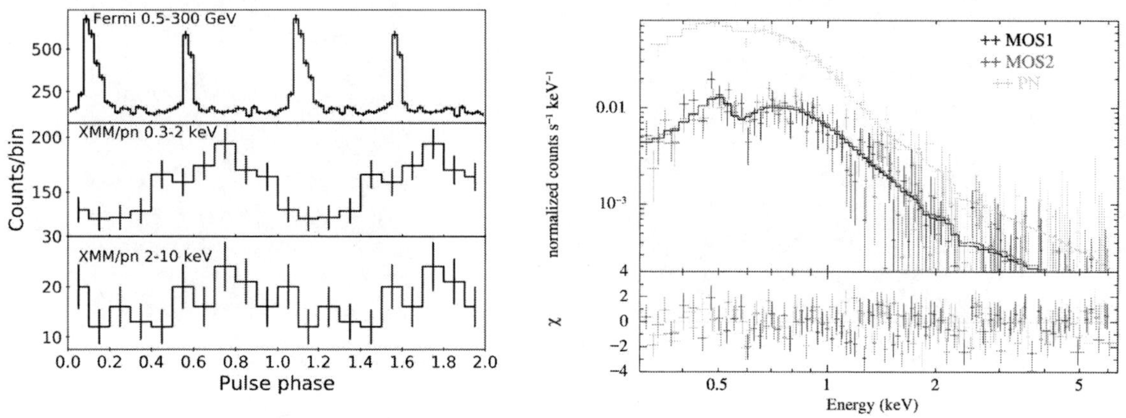

Figure 3. *Left*: J0633 pulse profiles in different energy bands indicated in the panels. *Right*: Observed phase-averaged J0633 spectrum, best-fit BB+PL model and fit residuals. The data from different instruments are shown by different colours, as indicated in the panel.

3. Timing

To search for pulsations from J0633 we used the pn data with the temporal resolution of $\approx$6 ms and Z_1^2-test [6]. Event times extracted from the 15″-radius aperture were corrected to the solar system barycentre using J0633 *Chandra* coordinates (R.A. = 06:33:44.142, Dec = +06:32:30.40; [2]). We ran Z_1^2-test in the 3.35–3.37 Hz frequency range. It yielded a peak $Z_{1,\max}^2 = 38.7$ at the frequency 3.3623316(14) Hz in the 0.3–2 keV band[1]. This corresponds to a pulsation detection at $\approx 4.5\sigma$. The pulsation frequency is consistent with the pulsar rotation frequency extrapolated from the epoch of the *Fermi* timing solution[2] [8]. Thus, we used the *Fermi* ephemeris to assign a pulse phase to X-ray photons. In γ-rays, the J0633 pulse profile shows two distinct peaks representing the magnetospheric emission while in X-rays the profile is broad, as typical for the spin-modulated thermal emission from a NS surface (see figure 3, left). The X-ray pulsed fraction PF $= (I_{\max} - I_{\min})/(I_{\max} + I_{\min})$, where $I_{\max}$ and $I_{\min}$ are maximum and minimum intensities of the pulse profile, corrected for the background is $23 \pm 6\%$ in the 0.3–2 keV band.

4. Phase-averaged spectral analysis

For the spectral analysis we used MOS and pn data. J0633 and PWN spectra were extracted from the 15″-radius aperture and the solid box shown in the right panel of figure 1, respectively (J0633 and source 'A' were excluded from the PWN region). The background spectrum was extracted from the dashed box (figure 1, right panel). J0633 and PWN spectra were fitted simultaneously with a common absorbing column density in the 0.3–10 keV band using C-statistic [9] and the model for a Poisson background, where each bin in the background spectrum is described by its own parameter. For the J0633 spectrum, we used models consisting of a non-thermal (PL) component, describing the pulsar magnetosphere emission, and a thermal component, representing NS surface emission. For the latter, we applied the blackbody (BB) model and the NS atmosphere model NSMAX [10]. The PWN spectrum was fitted with a PL model. The best-fit parameters are presented in table 1 while the pulsar spectrum and the best-fit BB+PL model are shown in figure 3, left. No absorption line at 0.8 keV is seen in the phase-averaged spectra. The derived spectral parameters within uncertainties are in agreement with values from [3].

[1] We used the formula from [7] to calculate 1σ frequency uncertainty
[2] http://www.slac.stanford.edu/~kerrm/fermi_pulsar_timing/

Table 1. Best-fit spectral parameters for J0633 and its PWN. All uncertainties correspond to 90% confidence. $N_\mathrm{H}[10^{21}\ \mathrm{cm}^{-2}]$ is the absorption column density, Γ are photon indices, $K[10^{-5}$ ph s^{-1} cm^{-2} keV$^{-1}]$ are normalizations of PL components. For the NSMAX model, the redshift parameter $1+z_g = 1.21$ and magnetic field $B = 4 \times 10^{12}$ G. Temperatures T[eV] and emitting area radii R[km] at the distance of 1 kpc are given as measured by a distant observer. d.o.f. = degrees of freedom.

Model	N_H	Γ_psr	K_psr	T	R	Γ_pwn	K_pwn	C/d.o.f.
BB+PL	$1.5^{+0.4}_{-0.3}$	$1.85^{+0.29}_{-0.30}$	$1.0^{+0.3}_{-0.3}$	120^{+9}_{-8}	$1.1^{+0.5}_{-0.3}$	$1.58^{+0.13}_{-0.12}$	$2.4^{+0.3}_{-0.3}$	769/765
NSMAX+PL	$1.9^{+0.4}_{-0.4}$	$1.36^{+0.30}_{-0.33}$	$0.5^{+0.3}_{-0.2}$	54^{+6}_{-5}	$9.2^{+5.3}_{-3.6}$	$1.67^{+0.13}_{-0.13}$	$2.7^{+0.3}_{-0.4}$	772/765

5. Distance to J0633

We estimated the distance to J0633 using the column densities N_H obtained from the X-ray fits (table 1) and different correlations between the distance and the reddening $E(B-V)$ [11, 12, 13]. The $E(B-V)$ values were obtained using an empirical correlation $N_\mathrm{H} = (0.7 \pm 0.1) \times E(B-V) \times 10^{22}\ \mathrm{cm}^{-2}$ [14]. This results in the distance range of 0.7–2.2 kpc.

6. Summary

We detected, for the first time, the X-ray pulsations from J0633 with the pulsar period. The pulse profile is broad as expected for spin-modulated thermal emission. The pulsed fraction in the 0.3–2 keV band is $23 \pm 6\%$. The X-ray phase-averaged spectrum of J0633 can be well described either by the BB+PL or NSMAX+PL model. Resulting parameters are in agreement with [3] though *XMM-Newton* observations do not confirm the absorption line at 0.8 keV in the pulsar spectrum detected in *Chandra* data. The distance to J0633 estimated using interstellar extinction−distance relations is 0.7–2.2 kpc.

Acknowledgments

The work of AK, DZ and AD was supported by RF Presidential Programme MK−2566.2017.2.

References

[1] Abdo A A *et al.* 2013 *ApJS* **208** 17
[2] Ray P S *et al.* 2011 *ApJS* **194** 17
[3] Danilenko A, Shternin P, Karpova A, Zyuzin D and Shibanov Y 2015 *PASA* **32** e038
[4] Kargaltsev O, Durant M, Misanovic Z and Pavlov G G 2012 *Science* **337** 946
[5] Snowden S L and Kuntz K D 2014 *Cookbook for analysis procedures for XMM-Newton EPIC observations of extended objects and the diffuse background* URL `http://heasarc.gsfc.nasa.gov/docs/xmm/esas/cookbook`
[6] Buccheri R *et al.* 1983 *A&A* **128** 245–51
[7] Chang C, Pavlov G G, Kargaltsev O and Shibanov Y A 2012 *ApJ* **744** 81
[8] Kerr M, Ray P S, Johnston S, Shannon R M and Camilo F 2015 *ApJ* **814** 128
[9] Cash W 1979 *ApJ* **228** 939–947
[10] Ho W C G, Potekhin A Y and Chabrier G 2008 *ApJS* **178** 102-9
[11] Drimmel R, Cabrera-Lavers A and López-Corredoira M 2003 *A&A* **409** 205–15
[12] Green G M *et al.* 2015 *ApJ* **810** 25
[13] Sale S E, Drew J E, Barentsen G, Farnhill H J, Raddi R, Barlow M J, Eislöffel J, Vink J S, Rodríguez-Gil P and Wright N J 2014 *MNRAS* **443** 2907–22
[14] Watson D 2011 *A&A* **533** A16

IOP Conf. Series: Journal of Physics: Conf. Series **932** (2017) 012011 doi:10.1088/1742-6596/932/1/012011

Pulsar radio emission mechanism: Why no consensus?

D B Melrose and M Z Rafat

SIfA, School of Physics, University of Sydney, Sydney, NSW 2006, Australia

E-mail: donald.melrose@sydney.edu.au

Abstract. Pulsar radio emission involves a coherent emission mechanism involving relativistic particles, but there is no consensus on what the specific emission mechanism is. There are arguments for and against several suggested mechanisms of long-standing, including two, relativistic plasma emission and anomalous Doppler emission, that depend explicitly on the dispersive properties of waves in a pulsar plasma. Models that do not include dispersion due to the relativistic spread in energy of the plasma particles can be misleading, and including the effects of this spread on the wave dispersion leads to severe constraints on these two suggested mechanisms. It is argued that all suggested mechanisms are implausible. A variant of relativistic plasma emission involving large-amplitude oscillations needs to be explored as a possible alternative.

1. Introduction

Five decades after the discovery of radio pulsars there is no consensus on what the pulsar radio emission mechanism is. There is wide agreement on several general features of the emission mechanism: the very high brightness temperature implies some form of coherent emission; the emission involves relativistic electrons (and positrons) [1] generated in pair cascades [2]; and the source region is on "open" field lines that define the "polar-cap" regions of the magnetosphere [3]. There are two well-established coherent emission mechanisms in astrophysical plasmas [4]: plasma emission and electron cyclotron emission. Pulsar radio emission is different from either of these, e.g. because it involves relativistic particles, and is a third form of coherent emission. Several emission mechanisms were suggested within the first decade or two after the discovery of pulsars and continue to be of interest: coherent curvature emission (CCE), relativistic plasma emission (RPE), anomalous Doppler emission (ADE), and linear acceleration emission (LAE) or free-electron maser emission (FEM). There are plausible arguments in favor of each of these mechanisms, but there are also arguments against each of them. The various arguments against the suggested mechanisms are so strong that one could conclude that all of them are untenable as the generic pulsar radio emission mechanism.

The suggested emission mechanisms are discussed critically in §2. Wave dispersion in a "pulsar plasma" is discussed in §3 and its negative implications for "beam-driven" RPE are discussed in §4. In §5 we speculate on non-beam-driven forms of RPE and summarize our conclusions in §6.

2. Critique of existing mechanisms

Historically, CCE was the first suggested emission mechanism [1, 5, 6, 7], and there are ongoing arguments in favor of it [8]. CCE is due to one-dimensional (1D) motion of a highly relativistic particle along a curved magnetic field line. The frequency for incoherent curvature emission peaks at about $(c/R_c)\gamma^3$, where R_c is the radius of curvature of the magnetic field line, and high γ is required for the frequency to be in the observed radio range. However, there are strong arguments against the postulated coherence due to bunches of highly relativistic particles [3, 9, 10], and although coherence due to maser is possible in principle [11, 12, 13], it does not provide a realistic alternative. CCE is not discussed further here.

LAE and FEM involve large-amplitude structures oscillating in time (frequency ω_0) and space (wavevector $\mathbf{k}_0$), respectively. Incoherent emission occurs at $\omega \lesssim \omega_0\gamma^2, k_0c\gamma^2$, respectively, and maser emission is possible for a distribution that is an increasing function of γ. Because LAE and FEM require a separate mechanism to generate the postulated large-amplitude structures, they may be interpreted as mechanisms for generating escaping radiation in the presence of such structures, and in this sense may be regarded as the second stage of a RPE mechanism.

All suggested mechanisms involve highly relativistic electrons, and three characteristic Lorentz factors are invoked in different models: a very large γ_p of order 10^7 for "primary" particles (which are not essential [14]), and two characteristic Lorentz factors associated with the "secondary" pair plasma, bulk outflow at γ_s and an intrinsic spread $\langle\gamma\rangle$ in the rest frame of the outflowing plasma. Models for the pair cascade [15, 16] suggest values for the latter two of order $\gamma_s \sim 10^3$ and $\langle\gamma\rangle \sim 10$–$10^2$. RPE and ADE depend intrinsically on the dispersive properties of the "pulsar plasma", whereas CCE and LAE/FEM exist in vacuo, albeit modified by the dispersive properties of the plasma.

Recently it has been argued [22] that the so-called nanoshots from the Crab pulsar [23] favor a specific form of beam-driven RPE [24, 25]. RPE may be regarded as a pulsar counterpart of conventional plasma emission which involves two stages, with the first stage being an instability that generates turbulence in Langmuir-like waves, and a subsequent stage involving partial conversion of energy in this turbulence into escaping radio waves. Difficulties with both stages of RPE have long been recognized: the growth rates for various suggested instabilities in the first stage are too slow to be effective, and ways in which this might be overcome have been explored [26, 27, 28, 29]; the difficulty with the second stage has been referred to as a "bottle-neck" [30]. In §3 a further difficulty is pointed out with beam-driven RPE: there are effectively no "Langmuir-like waves" that can grow due to a beam instability. This provides a strong argument against beam-driven RPE. A similar difficulty arises with suggested beam-driven growth of Alfvén waves [31, 32, 33, 34].

3. Wave dispersion in pulsar plasma

There is an extensive literature on wave dispersion in a pulsar plasma, defined here as a highly relativistic, strongly-magnetized, electron-positron pair plasma. The relativistic bulk streaming of the plasma may be included in the wave dispersion by treating it either in the plasma rest frame and Lorentz transforming to the pulsar frame [17, 18] or in the pulsar frame directly [19]. We adopt the former approach. Although these two approaches are equivalent in principle, they lead to different results in practice due to different approximations. With the former approach, there are only three relevant wave modes, two of which depend strongly on a relativistic plasma dispersion function (RPDF), due to an assumed spread $\langle\gamma\rangle \gg 1$ in the rest frame. With the latter approach, the RPDF is approximated by a moment of the distribution and four wave modes are identified [20, 21]. We note that the inconsistency involving the fourth mode is one of several unresolved problems relating to wave dispersion relevant to beam-driven RPE [17], even when inhomogeneity (field-line curvature [20]) and plasma turbulence [32], are neglected.

It is helpful to consider the cold and nonrelativistic cases as a prelude to discussion of the

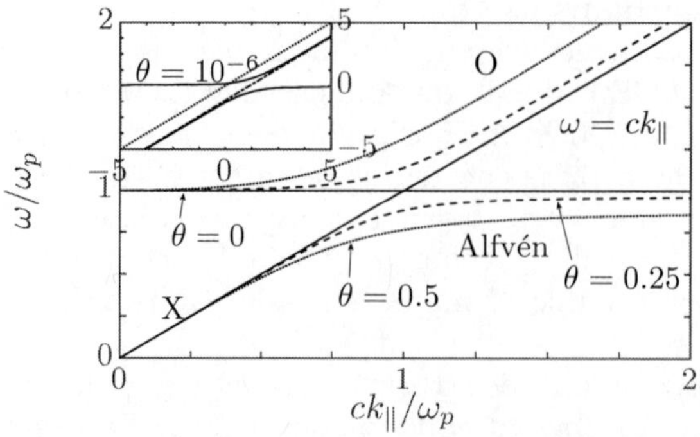

Figure 1. Dispersion relations for cold pulsar plasma. For $\theta = 0$ the L-mode is the horizontal line at $\omega/\omega_p = 1$, and there are two degenerate modes (A and X) at $k_\parallel c/\omega = z_A \approx 1$. For $\theta \neq 0$ the dispersion curves L and A reconnect to form the O mode and the Alfvén mode which move increasingly apart with increasing θ. Insert: $(\omega/\omega_p - 1) \times 10^6$ vs $(k_\parallel c/\omega_p - 1) \times 10^6$ for $\theta = 10^{-6}$.

relativistic case $\langle\gamma\rangle \gg 1$. A cold-plasma model for a pulsar plasma follows by assuming that in a cold plasma model [35] ions are replaced by positrons, and with electron cyclotron frequency, Ω_e, assumed much greater than both the plasma frequency, ω_p, and the wave frequency, ω. A conventional labeling [36] of the modes in a pulsar plasma is based on this cold-plasma, or magnetoionic approximation: the extraordinary, or X mode, has refractive index $kc/\omega = 1/z_A$ with $1/z_A^2 = 1 + \omega_p^2/\Omega_e^2$, and the ordinary mode separates into a higher-frequency branch, called the O mode, and a lower-frequency branch identified as the Alfvén mode. For parallel propagation the dispersion curves for the latter two modes correspond to straight lines, cf. Figure 1: $\omega = \omega_p$ for the parallel longitudinal (L) mode and $z = \omega/k_\parallel c = z_A$ for parallel Alfvén (A) mode. The two dispersion curves cross, and for nonzero angle of propagation, θ, they reconnect to form the O mode and the (oblique) Alfvén mode. In the limit $\theta \to 0$, the O mode coincides with the L mode from the cutoff at $\omega/\omega_p = 1$ and for $k_\parallel c/\omega_p \in [0, 1/z_A]$ and to the A mode at higher frequencies. The Alfvén mode coincides with the A mode from $\omega = 0$ to $\omega = \omega_p$ and to the L mode from $k_\parallel c/\omega = 1/z_A$ to the resonance $k_\parallel \to \infty$. The (reconnected) O mode and Alfvén separate from each other with increasing θ, the O mode to higher ω and larger z, and the Alfvén mode to lower ω and smaller z. There is a stop band between the resonance in the A mode, at $\omega \approx \omega_p|\cos\theta|$ for $\Omega_e \gg \omega_p$, and the cutoff in the O mode at $\omega = \omega_p$.

An important difference between these wave properties and those in the more familiar case $\Omega_e < \omega_p$ concerns the resonance. For $\Omega_e \ll \omega_p$ there is a resonance in the (upper-frequency branch of the magnetoionic) extraordinary mode at $\omega > \omega_p > \Omega_e$, and Langmuir waves correspond to a thermally modified form of these resonant waves. For $\Omega_e > \omega_p$ the resonance is in the (lower frequency branch of the magnetoionic) ordinary mode, at $\omega < \omega_p < \Omega_e$, interpreted as a resonance is in the Alfvén mode. As a consequence the concept of a "Langmuir-like" wave mode becomes ill-defined in a pulsar plasma with $\Omega_e \gg \omega_p$.

The inclusion of a nonrelativistic velocity spread modifies the wave properties and these modifications become extreme when the spread is relativistic. For parallel propagation, the dispersion relation for the A mode is modified by the inclusion of relativistic effects in β_A, with $\beta_A^2 \approx \Omega_e^2/\omega_p^2\langle\gamma\rangle$. The dispersion relation for longitudinal waves becomes [17, 18] $\omega = \omega_L(z)$ with $\omega_L^2(z) = \omega_p^2 z^2 W(z)$ where the RPDF $z^2W(z)$ is plotted in Figure 2 for a 1D thermal distribution for temperature $5 \times 10^9/\rho$ K with $\rho = 50, 10, 1$. The cold-plasma limit corresponds to $z^2W(z) \to 1$. For nonrelativistic temperatures, $\rho \gg 1$, $z^2W(z)$ is negative for very small z, crosses zero and rises to a maximum and then asymptotes to $1 + 3/\rho z^2$ for $\rho z^2 \gg 1$. The dispersion relation for the L mode is then approximated by $\omega_L(z) \approx \omega_p^2 + 3k_\parallel^2 V_e^2$, with

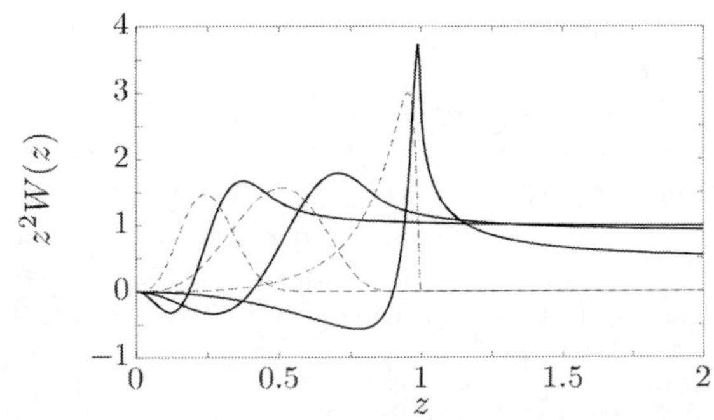

Figure 2. The real (solid) and negative imaginary (dashed) parts of the RPDF $z^2W(z)$ are plotted as a function of z for three 1D Jüttner distributions: leftmost $\rho = 50$, center $\rho = 10$ and rightmost $\rho = 1$. The imaginary part is identically zero for $z \geq 1$.

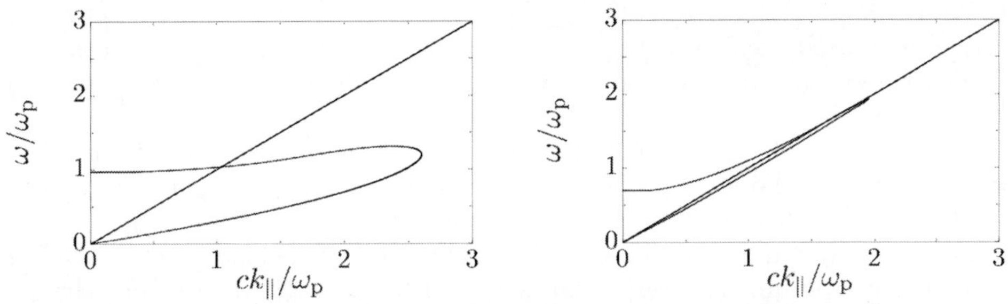

Figure 3. Dispersion curves for parallel propagation: left $\rho = 20$; right $\rho = 1$.

$V_e^2/c^2 = 1/\rho \ll 1$.

In the highly relativistic limit, $\langle\gamma\rangle \approx 1/\rho$, the peak in $z^2W(z)$ becomes very large, with its maximum $\approx 2.7/\rho$ at $z = z_m \approx 1 - 0.013\rho^2$ very close to the light line, $z = 1$. The imaginary part of $z^2W(z)$ is strictly zero for $z > 1$ and has a maximum at z slightly less than z_m, where Landau damping is strong.

The dispersion curves for the two parallel modes intersect at a crossover frequency $\omega_{\rm co} = \omega_L(z_A)$, as illustrated in Figure 3 for a nonrelativistic value (left: $\rho = 20$) and a relativistic value (right: $\rho = 1$). The maximum in the curve for the L mode is at the maximum of the RPDF, cf. Figure 2. In a highly relativistic plasma, the L mode dispersion curve starts at the cutoff frequency, ω_x, given by $\omega_x^2 = \omega_L^2(\infty) = \omega_p^2\langle 1/\gamma^3\rangle$, and crosses the light line at ω_1, determined by $\omega_1^2 = \omega_L^2(1) = \omega_p^2\langle\gamma\beta^2\rangle$. The line $z = z_m$ through the maximum in $\omega_L(z)$ moves closer to the light line $z = 1$ with increasing $\langle\gamma\rangle$. Reconnection between the L and A modes occurs for $z_A > z_0$, with the A mode strongly damped for $z_0 < z_A \lesssim z_m$ and weakly damped for $z \gtrsim z_m$. The condition $z_A > z_m$ is equivalent to $\beta_A^2 > \gamma_m^2 = 1/(1 - z_m^2)$, with $\gamma_m \approx 6\langle\gamma\rangle$. Provided this condition is satisfied, the reconnected O and Alfvén modes move away from each other with increasing θ; the maximum is then in the (reconnected) Alfvén mode.

For $\theta = 0$, the L mode is superluminal over the frequency range, $\omega_x \leq \omega \leq \omega_1 \approx \omega_x\langle\gamma\rangle$. For $\theta \neq 0$ the reconnected O mode moves to the upper left, as in the cold-plasma limit Figure 1. The cutoff frequency is independent of θ, and the frequency at which the dispersion curve crosses the light line increases to infinity over $0 < \theta \lesssim 1/\beta_A$. There are subluminal (L or O) waves only in this tiny range of angles.

The reconnected Alfvén mode is dominated by the peak in the RPDF, at $z = z_m$. As θ increases the maximum frequency of the Alfvén mode decreases, remaining at $z \approx z_m$. Landau damping becomes strong at $z < z_m$, and Alfvén waves are weakly damped only on the portion of the dispersion curve at $z \gtrsim z_m$.

The application to pulsars in general and to any specific pulsar in particular requires estimates of Ω_e, ω_p and β_A as functions of height r in the magnetosphere. For a given pulsar, the observed period P and period derivative $\dot{P}$ imply the surface magnetic field $\propto (P\dot{P})^{1/2}$. A dipolar magnetic field falls off $\propto 1/r^3$ for $r \ll r_L$ with $r_L = Pc/2\pi$. One finds $\Omega_e/2\pi$ of order $3\times10^7(\dot{P}/P^3)^{1/2}(r/r_L)^{-3}$ Hz. An estimate of the plasma frequency is given by assuming that the electron number density is greater than the corotation value by a multiplicity κ, implying $\omega_p^2 \approx \kappa\Omega_*\Omega_e \propto (r/r_L)^{-3}$, with $\Omega_* = 2\pi/P$. These estimates also imply $\beta_A^2 = \Omega_e^2/\omega_p^2\langle\gamma\rangle \propto (r/r_L)^{-3}$. The fiducial values adopted here are $P = 1\,\mathrm{s}$, $\dot{P} = 10^{-15}$, $\kappa = 10^5$ [37] and $\langle\gamma\rangle = 10$. Evaluating the parameters at $r = 0.1r_L$ gives $\Omega_e/2\pi = 30\,\mathrm{GHz}$, $\omega_p/2\pi = 20\,\mathrm{kHz}$ and $\beta_A^2 = 3 \times 10^4$. At $r = 0.1r_L$ the conditions $\omega \ll \Omega_e$ and $\beta_A > \gamma_m$ are well satisfied and plausibly satisfied, respectively, and they are more strongly satisfied at all $r < 0.1r_L$.

4. Beam-driven RPE

In a beam-driven instability waves are assumed to grow due to negative absorption resulting from resonance between the wave and a distribution function $g(u)$ with $dg(u)/du > 0$ below a peak at $u = u_b$ say, with $u_b = \gamma_b\beta_b$. The (Cerenkov) resonance condition $\omega - k_\parallel v_\parallel = 0$ corresponds to $z - \beta = 0$. Beam-driven waves have $z = \beta < \beta_b$. All velocities of relevance to the discussion here are very close to c, and it is convenient to refer to the corresponding Lorentz factors: $\gamma_\phi^2 = 1/(1 - z^2)$, $\gamma_m^2 = 1/(1 - z_m^2)$, $\beta_A^2 = 1/(1 - z_A^2)$, $\gamma_b^2 = 1/(1 - \beta_b^2)$. The resonance condition becomes $\gamma_\phi = \gamma$ with $\gamma < \gamma_b$ for the beam. The value of γ_b depends on the assumptions made about the beam. Assuming extremely large $\gamma_b \sim 10^7$, corresponding to primary particles, gives a growth rate that is too small. To overcome this difficulty an inhomogeneous model was proposed [26, 27] in which bursts of pair creation result in localized bunches, with faster particles in a following bunch overtaking slower particles in a preceding bunch. Waves grow in such a model only if there is an identifiable beam, requiring that γ_b be significantly greater than $\langle\gamma\rangle$. Although a detailed model is required to discuss this point quantitatively, it seems that the model is viable only if the following bunch has $\langle\gamma\rangle$ greater than that of the preceding bunch, with γ_b of order the larger value.

A beam-driven instability is relevant only to either the O mode or the Alfvén mode: the X mode has polarization perpendicular to **B** and does not couple to a field-aligned beam. It follows that there are two possible versions of beam-driven RPE, involving growth of subluminal ($z < 1$) O mode waves and Alfvén-mode waves, respectively. In most treatment of these instabilities unrealistic assumptions are (implicitly) made about the dispersive properties of these waves: it is essential that the wave dispersion is calculated assuming $\langle\gamma\rangle \gg 1$ in the rest frame of the plasma.

For $\beta_A > \gamma_m$, so that the modes reconnect as discussed above, the O mode is subluminal only for $\omega > \omega_1$ and in a tiny range of angles, $\theta < 1/\beta_A$. For resonance to be possible between the beam and such O mode waves requires $\gamma_b > \beta_A$. The extremely large value of β_A for $r/r_L \ll 1$ requires an even larger γ_b, which is possible only for primary particles and for these the growth rate is known to be unacceptably small. The condition $\gamma_b > \beta_A$ could be satisfied for $r/r_L \gtrsim 0.1$, but even in this case it seems implausible that the conditions for effective growth could be satisfied. One difficulty is that the polarization of O mode waves has a large component along **B** only for $\theta \ll 1/\beta_A$ and ω just above ω_1, so that although growth at higher frequencies is possible in principle for sufficiently small θ, the growth rate is very small. Another difficulty is that the expected frequency, $\omega \approx \omega_1$, of the resulting waves at $r/r_L \gtrsim 0.1$ is too low (even after Lorentz boosting to the pulsar frame) to provide a plausible explanation for the highest frequencies in pulsar radio emission [10]. On the other hand, a favorable aspect of beam-driven O mode waves is that the waves can escape directly: no second stage in the emission process is required.

The other possibility for beam-driven wave growth involves Alfvén waves [31, 32, 33, 34]. Consider resonance between a beam and Alfvén waves in a pulsar plasma with $\langle\gamma\rangle \gg 1$ in the limit of $\theta \to 0$. The dispersion curve is then $z = z_A$ for $\omega < \omega_L(z_A)$ and $\omega = \omega_L(z)$ for $z < z_A$. In the limit $\theta \to 0$, the transverse polarization precludes growth along the former branch, and only waves on the latter branch can grow, and for small $\theta \neq 0$ this continues to be approximately the case. As θ increases waves in the range $z_A > z > z_m$ move to lower frequency, and it is only these waves that are of interest here. Resonance and hence growth is possible for $z_A > \beta_b > z_m$ or $\beta_A > \gamma_b > \gamma_m$. With $\gamma_m \approx 6\langle\gamma\rangle$, this requires $\gamma_b > 6\langle\gamma\rangle$. Thus the conditions under which these waves would be expected to grow are less extreme than for the O mode. However, unlike the O mode case, these waves cannot escape directly and a second stage in the RPE process is required.

ADE also depends intrinsically on the dispersive properties of the plasma. The anomalous Doppler resonance, $\omega - k_\parallel v_\parallel = -\Omega_e/\gamma$, may be rewritten as $\omega = z\Omega_e/(\beta - z)\gamma$ and approximated by $\omega = \Omega_e \gamma\gamma_\phi^2/(\gamma^2 - \gamma_\phi^2)$. The frequency of ADE must satisfy both the resonance condition and the dispersion relation for a specific mode. At the expected high frequency the only relevant wave modes are the O mode and X mode. For small θ both modes have γ_ϕ of order β_A, and resonance requires $\gamma > \beta_A$. An order of magnitude estimate gives $\omega \sim \beta_A\Omega_e$ for $\gamma \sim \beta_A$. This estimate applies in the rest frame of the plasma, and the frequency in the pulsar frame is boosted by a Lorentz factor of order γ_s. This estimate of ω is higher than an earlier estimate based on a calculation of wave dispersion in the pulsar frame [19]. This higher frequency severely exacerbates the difficulty of explaining the observed radio frequency range with the source of ADE anywhere within the light cylinder. Due to this difficulty ADE is also at least implausible and arguably untenable.

5. Alternative forms of RPE

The argument in favor of RPE [22] seems strong. The foregoing criticism of RPE applies to the assumption that it requires a"beam-driven" instability to produce the assumed "coherent charge bunching and strong electromagnetic turbulence" [22]. The favorable aspects of RPE can be retained if an alternative way of generating the relevant "turbulence" can be shown to be effective. An interesting possibility is the large amplitude longitudinal oscillations (LAOs) that are driven directly by the electrodynamics [38, 39, 40]. An interpretation is that the inductive electric field associated with an obliquely rotating magnetic dipole has a component $E_\parallel$ along the magnetic field; charges are accelerated along field lines by $E_\parallel$ tending to set up a potential electric field that opposes $E_\parallel$. Such screening of $E_\parallel$ is oscillatory setting up the LAOs and triggering a pair cascade that increases the number of pairs and facilitates the screening. The source of the energy in the LAOs is the rotational energy, through the inductive $E_\parallel$. (The idea that rotational energy may be partly converted directly into energy in LAOs has also been proposed in a different model involving a parametric instability [41, 42].) A model for LAOs [43] needs to be based on the actual wave dispersion of longitudinal waves in a pulsar plasma, that is, of the L mode waves discussed in §3. A version of RPE based on such waves has the attractive feature that the waves escape directly, by evolving into O mode waves as they propagate outwards. Alternatively the escaping radiation may result from a maser version of LAE associated with LAOs [44, 45, 46].

6. Conclusions

Why no consensus? One reason is that all suggested pulsar radio emission mechanisms encounter major difficulties and are arguably untenable, allowing a diversity of opinions on which is the least implausible. This diversity of opinions underlies the lack of consensus. Our opinion is that the least implausible is a form of RPE based on LAOs driven directly by the electrodynamics. A detailed model needs to be developed for such RPE.

Acknowledgments

We thank Alpha Mastrano for helpful comments on the manuscript.

References

[1] Radhakrishnan V 1969 *Proc. Astron. Soc. Australia* **1** 254
[2] Sturrock P A 1971 *Astrophys. J.* **164** 529
[3] Melrose D B 1981 Maser pulse emission mechanisms *Pulsars: 13 Years of Research on Neutron Stars* (*IAU Symposium* vol 95) ed Sieber W and Wielebinski R pp 133–9
[4] Melrose D B 2017 *ArXiv e-prints*(*Preprint* 1707.02009)
[5] Komesaroff M M 1970 *Nature* **225** 612–4
[6] Ruderman M A and Sutherland P G 1975 *Astrophys. J.* **196** 51–72
[7] Benford G and Buschauer R 1977 *Mon. Not. Roy. Astron. Soc.* **179** 189–207
[8] Mitra D, Gil J and Melikidze G I 2009 *Astrophys. J. Lett.* **696** L141–5
[9] Melrose D B 1995 *J. Astrophys. Astron.* **16** 137
[10] Lesch H, Jessner A, Kramer M and Kunzl T 1998 *Astron. Astrophys.* **332** L21–4
[11] Zheleznyakov V V and Shaposhnikov V E 1979 *Aust. J. Phys.* **32** 49–59
[12] Luo Q and Melrose D B 1992 *Mon. Not. Roy. Astron. Soc.* **258** 616–20
[13] Luo Q and Melrose D B 1995 *Mon. Not. Roy. Astron. Soc.* **276** 372–82
[14] Timokhin A N 2010 *Mon. Not. Roy. Astron. Soc.* **408** 2092–114
[15] Hibschman J A and Arons J 2001 *Astrophys. J.* **560** 871–84
[16] Arendt Jr P N and Eilek J A 2002 *Astrophys. J.* **581** 451–69
[17] Melrose D B and Gedalin M E 1999 *Astrophys. J.* **521** 351–61
[18] Melrose D B, Gedalin M E, Kennett M P and Fletcher C S 1999 *J. Plasma Phys.* **62** 233–48
[19] Lyutikov M, Blandford R D and Machabeli G 1999 *Mon. Not. Roy. Astron. Soc.* **305** 338–52
[20] Beskin V, Gurevich A and Istomin Y 1993 *Physics of the Pulsar Magnetosphere* (Cambridge: Cambridge Univ. Press)
[21] Lyne A G and Graham-Smith F 2006 *Pulsar Astronomy* (Cambridge: Cambridge Univ. Press)
[22] Eilek J A and Hankins T H 2016 *J. Plasma Phys.* **82** 635820302
[23] Hankins T H and Eilek J A 2007 *Astrophys. J.* **670** 693–701
[24] Weatherall J C 1997 *Astrophys. J.* **483** 402–13
[25] Weatherall J C 1998 *Astrophys. J.* **506** 341–6
[26] Usov V V 1987 *Astrophys. J.* **320** 333–5
[27] Ursov V N and Usov V V 1988 *Astrophys. Space Sci.* **140** 325–36
[28] Asseo E and Melikidze G I 1998 *Mon. Not. Roy. Astron. Soc.* **301** 59–71
[29] Gedalin M, Gruman E and Melrose D B 2002 *Phys. Rev. Lett.* **88** 121101
[30] Usov V V 2000 Radiating Regions in Pulsar Magnetospheres: From Theory to Observations and Back *IAU Colloq. 177: Pulsar Astronomy - 2000 and Beyond* (*Astron. Soc. Pacific Conf. Series* vol 202) ed Kramer M *et al* p 417
[31] Tsytovich V N and Kaplan S A 1972 *Astrofizika* **8** 441–60
[32] Lominadze J G, Stenflo L, Tsytovich V N and Wilhelmsson H 1982 *Physica Scripta* **26** 455–8
[33] Lyutikov M 1999 *J. Plasma Phys.* **62** 65–86
[34] Lyutikov M 2000 *Mon. Not. Roy. Astron. Soc.* **315** 31–6
[35] Stix T H 1962 *The Theory of Plasma Waves* (New York: McGraw-Hill)
[36] Arons J and Barnard J J 1986 *Astrophys. J.* **302** 120–37
[37] Timokhin A N and Harding A K 2015 *Astrophys. J.* **810** 144
[38] Levinson A, Melrose D, Judge A and Luo Q 2005 *Astrophys. J.* **631** 456–65
[39] Beloborodov A M and Thompson C 2007 *Astrophys. J.* **657** 967–93
[40] Beloborodov A M 2008 *Astrophys. J. Lett.* **683** L41
[41] Machabeli G Z and Rogava A D 1994 *Phys. Rev.* A **50** 98–103
[42] Machabeli G Z, Rogava A, Chkheidze N, Osmanov Z and Shapakidze D 2016 *J. Plasma Phys.* **82** 635820305
[43] Luo Q and Melrose D 2008 *Mon. Not. Roy. Astron. Soc.* **387** 1291–302
[44] Melrose D B, Rafat M Z and Luo Q 2009 *Astrophys. J.* **698** 115–23
[45] Melrose D B and Luo Q 2009 *Astrophys. J.* **698** 124–30
[46] Reville B and Kirk J G 2010 *Astrophys. J.* **715** 186–93

IOP Conf. Series: Journal of Physics: Conf. Series **932** (2017) 012012 doi:10.1088/1742-6596/932/1/012012

So how do radio pulsars slow-down?

V S Beskin[1,2], A K Galishnikova[2], E M Novoselov[2], A A Philippov[3,0] and M M Rashkovetskyi[2]

[1] P.N.Lebedev Physical Institute, Leninsky prosp., 53, Moscow, 119991, Russia
[2] Moscow Institute of Physics and Technology, Dolgoprudny, Institutsky per., 9, Moscow Region, 141700, Russia
[3] Department of Astronomy, University of California Berkeley, Berkeley, CA 94720-3411

E-mail: beskin@lpi.ru

Abstract. We show that the increasing of energy losses of radio pulsars $W_{\rm tot}$ with the inclination angle χ obtained numerically by many authors can be explained by the separatrix currents circulating in the pulsar magnetosphere which do not outflow into pulsar wind.

1. Introduction

From the very beginning magneto-dipole radiation was considered as a main mechanism of the pulsar braking [1, 2]. But later it was shown that if the pulsar magnetospheric fills fully with plasma so that $E_{\parallel} = 0$, for zero longitudinal electric currents circulating in the pulsar magnetosphere the energy losses $W_{\rm tot}$ vanish for any inclination angle χ [3]. This effect confirmed later in [4] results from full screening of the magneto-dipole radiation of neutron star by magnetospheric plasma. This implies that the pulsar braking results fully from the impact of the torque $\mathbf{K}$ due to longitudinal currents flowing in the magnetosphere.

Nevertheless, magneto-dipole losses are still often mentioned in connection with energy release mechanism. On the other hand, MHD simulations obtained first in [5] and later confirmed in many papers [6, 7, 8, 9] give the following dependence $W_{\rm tot}$ on the inclination angle χ

$$W_{\rm tot}^{\rm MHD} \approx \frac{1}{4}\frac{B_0^2\Omega^4R^6}{c^3}(1+\sin^2\chi). \quad (1)$$

As we see, in addition to unity this expression has the same $\sin^2\chi$ dependence as for magneto-dipole losses. Recent numerical simulations have not clarified this question as well. Indeed, as was shown in [9, 10], in the pulsar wind there is a noticeable component depending on time. In particular, for orthogonal rotator one can write-down [9]

$$B_r \approx B_0\frac{R^2}{r^2}\sin\theta\cos(\varphi-\Omega t+\Omega r/c); \qquad B_{\varphi} = E_{\theta} \approx -B_0\frac{\Omega R^2}{cr}\sin^2\theta\cos(\varphi-\Omega t+\Omega r/c). \quad (2)$$

Below we analyze the torque acting on a rotating star and the angular distribution of the energy flux inside the light cylinder. It is shown that MHD solution (1) does not contain magneto-dipole wave. Inside the light cylinder almost 100% of the energy flux outflows through very narrow angular domain of open magnetic field region, not quasi-isotropical, as it should have been for magneto-dipole radiation (e.g., in Deutsch solution [11]).

[0] Einstein Fellow

IOP Conf. Series: Journal of Physics: Conf. Series **932** (2017) 012012 doi:10.1088/1742-6596/932/1/012012

2. Current losses

As is well-known, the key role in particle motion in the pulsar magnetosphere should be played by the electric drift associated with a strong electric field produced by the rotation of a neutron star. As a result, the transverse electric current $\mathbf{j}_{\perp}$ can be written in the hydrodynamical form $\mathbf{j}_{\perp} = \rho_{\rm e}\mathbf{U}_{\rm dr}$, where $\mathbf{U}_{\rm dr} = c[\mathbf{E}\times\mathbf{B}]/B^2$ is the electric drift velocity. On the other hand, for fast enough rotation when the potential drop near magnetic poles is much smaller than the maximum possible value one can write down $\mathbf{E} = -\beta_{\rm R}\times\mathbf{B}$, where $\beta_{\rm R} = [\mathbf{\Omega}\times\mathbf{r}]/c$. Using this relation we finally obtain

$$\mathbf{j} = \rho_{\rm e}[\mathbf{\Omega}\times\mathbf{r}] + i_{\parallel}\mathbf{B}, \tag{3}$$

where $i_{\parallel}$ a scalar function. In particular, for orthogonal wind solution (2) we have

$$i_{\parallel} = -3\,\frac{\Omega}{c}\cos\theta. \tag{4}$$

The convenience of expression (3) is due to the fact that the scalar function $i_{\parallel}$ must be constant along magnetic field lines [12].

Further, general expression for energy losses of a rotating magnetized sphere looks like [13]

$$W_{\rm tot} = \frac{c}{4\pi}\oint[\mathbf{E}\times\mathbf{B}]{\rm d}s = -\mathbf{\Omega}\cdot\mathbf{K}; \quad \mathbf{K} = \frac{R}{4\pi}\oint[\mathbf{n}\times\mathbf{B}]\,(\mathbf{B}\cdot\mathbf{n})\,{\rm d}s. \tag{5}$$

Here the first bracket in the relation for the torque $\mathbf{K}$ corresponds to the surface current $\mathbf{J}_{\rm s}$, and the second one to magnetic field $\mathbf{B}$ in the Ampere force $\mathbf{F} = [\mathbf{J}_{\rm s}\times\mathbf{B}]/c$.

It is convenient to introduce two components of the torque $\mathbf{K}$ parallel and perpendicular to the magnetic dipole $\mathbf{m}$ and connect them with dimensionless current in the polar cap zone $i = j_{\parallel}/j_{\rm GJ}$ separating it into symmetric and anti-symmetric contributions, $i_{\rm s}$ and $i_{\rm a}$, depending upon whether the direction of the current is the same in the north and south parts of the polar cap, or opposite. As one can easily check, $K_{\parallel} \propto i_{\rm s}$, and $K_{\perp} \propto i_{\rm a}$. Here and below we apply normalization to the 'local' Goldreich-Julian current density, $j_{\rm GJ} = |\mathbf{\Omega}\cdot\mathbf{B}|/2\pi$ (with scalar product). As a result, we obtain

$$W_{\rm tot} = -\Omega\,\mathcal{K}_{\parallel}\cos^2\chi - \Omega\,\mathcal{K}_{\perp}\sin^2\chi, \tag{6}$$

where we put $K_{\parallel} = \mathcal{K}_{\parallel}\cos\chi$ and $K_{\perp} = \mathcal{K}_{\perp}\sin\chi$.

As was shown in [14], the direct action of the Ampère force on the star $\mathbf{K} = \int[\mathbf{r}\times[\mathbf{J}_{\rm s}\times\mathbf{B}]/c]\,{\rm d}s$ by surface currents $\mathbf{J}_{\rm s}$ which close volume longitudinal electric currents circulating in the pulsar magnetosphere can be written as

$$K_{\parallel}^{\rm sur} \approx -\frac{B_0^2\Omega^3R^6}{c^3}i_{\rm s}, \quad K_{\perp}^{\rm sur} \approx -\frac{B_0^2\Omega^3R^6}{c^3}\left(\frac{\Omega R}{c}\right)i_{\rm a}. \tag{7}$$

Thus, to satisfy MHD energy losses $W_{\rm tot}$ (1) for $\chi = 90^{\circ}$ the anti-symmetric current $i_{\rm a}$ is to be large enough: $i_{\rm a} \approx (\Omega R/c)^{-1}$. On the other hand, equating the total current $I = \int i_{\parallel}\mathbf{B}{\rm d}\mathbf{s}$ with $i_{\parallel}$ from (4) flowing through the upper hemisphere of the orthogonal wind (2) and the total current flowing through the northern part of the polar cap on the star surface (and remembering that $i_{\parallel}$ is constant along magnetic field lines), one can obtain

$$i_{\rm a} \approx \left(\frac{\Omega R}{c}\right)^{-1/2}. \tag{8}$$

Thus, antisymmetric current itself is too small to explain MHD energy losses.

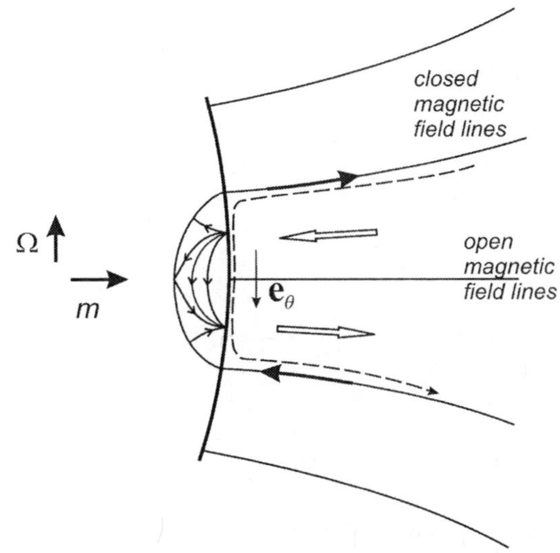

Figure 1. The structure of the volume (contour arrows), separatrix (fat arrows) and surface (thin arrows) currents near the polar cap of the orthogonal rotator. The additional separatrix current is shown by a dashed curve.

3. Additional torques

To find another possible braking mechanisms responsible for MHD solution, let us consider arbitrarily magnetized sphere rotating in a vacuum. According to (7), braking torque $\mathbf{K}$ must correspond to the third power of the expansion with respect to the small parameter $\varepsilon = \Omega R/c$. On the other hand, one can show that the first order terms $B^{(1)}$ turn out to be zero (see, e.g., [15]). As a result, general expression for the braking torque can be written down as

$$\mathbf{K} = \frac{R}{4\pi} \oint \{[\mathbf{n} \times \mathbf{B}^{(3)}](\mathbf{B}^{(0)} \cdot \mathbf{n}) + [\mathbf{n} \times \mathbf{B}^{(0)}](\mathbf{B}^{(3)} \cdot \mathbf{n})\} \mathrm{d}s. \quad (9)$$

In particular, in Deutsch [11] solution ($B_n^{(3)} = 0$) the only contribution is given by the first term. On the other hand, for the point orthogonal dipole [13] only two thirds of the losses will still be determined by the first term, while one third with the second one. Certainly, the total energy losses and the inclination angle evolution do not depend on the choice of the solution.

Thus, in addition to current losses (7) mentioned above, pulsar braking can be connected with the disturbance of normal component of the magnetic field B_n. It is important that this disturbance covers the entire surface of the neutron star. Such additional contribution could be related to the violation of the exact mutual compensation between the magneto-dipole radiation of a central star and the radiation of the magnetosphere itself resulting from longitudinal currents circulating in the magnetosphere (e.g., due to the changing of a shape of closed magnetic field lines region). For zero longitudinal currents exact mutual compensation takes place.

To discuss another possible additional torque, let us consider in more detail the braking of an orthogonal rotator. It is convenient to rewrite the energy losses $W_{\rm tot}$ (5) in the form (do not forget that the radio pulsar has two poles!)

$$W_{\rm tot} = 2\,\frac{\Omega R}{c} \int J_\theta B_n \mathrm{d}s. \quad (10)$$

An accurate analysis shows [14] that surface currents should in fact be arranged in such a way that the current $< J_\theta >$ averaged over the polar cap surface would be zero. In this case, as

IOP Conf. Series: Journal of Physics: Conf. Series **932** (2017) 012012 doi:10.1088/1742-6596/932/1/012012

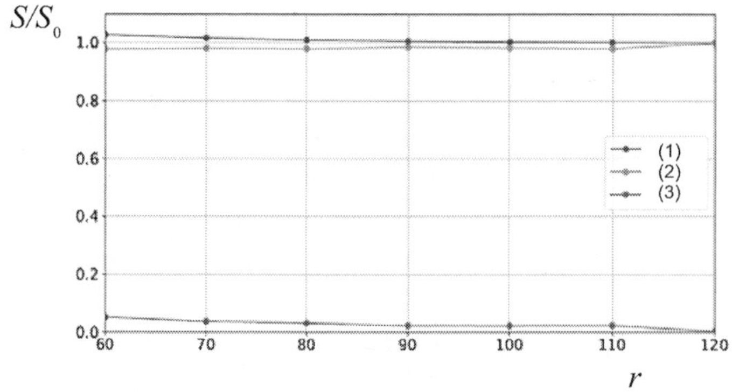

Figure 2. Electromagnetic energy flux from the region of closed (lower curve) and open (middle curve) magnetic field lines. The upper curve corresponds to the total energy loss. The radius of the neutron star (40 cells) is 10 times smaller than the radius of the light cylinder.

shown in Figure 1, along the separatrix separating open and closed magnetic field line domains must flow surface current $I_{\rm sep}$ comparable to the total volume current $I_{\rm vol}$ circulating in the magnetosphere, but flowing in the opposite direction. E.g., for a round polar cap and for a local GJ current the reverse current must be 3/4 from the volume current ($I_{\rm sep}/I_{\rm vol} \approx 0.75$) [16]. Such a reverse current was later confirmed in numerical simulations, but its value turned out to be smaller: $I_{\rm sep}/I_{\rm vol} \approx 0.2$ [17].

Here, however, one very important remark should be made. The above conclusion was based on the assumption that there is no additional current circulating in the magnetosphere, but not outgoing into the pulsar wind region. Indeed, as shown in Figure 1, additional separatrix currents must inevitably lead to a nonzero average surface current $< J_\theta > \neq 0$ on the polar cap, and, consequently, to additional energy losses.

4. Results

Thus, there are two possibilities to explain large enough energy losses of the orthogonal rotator. Previously not only to determine analytically, but even evaluate corresponding contributions was not possible. Now we can get an answer by analyzing directly the results of numerical simulation. Figure 2 shows the dependence of the flux of electromagnetic energy on the radius r through the region of closed (lower curve) and open (middle curve) field lines for the inclination angle 60°. The size of the star is 40 cells, and the light cylinder is 400 cells ($\varepsilon = 0.1$).

As we see, practically all the energy flux is concentrated within open magnetic field region. Hence, the energy losses of orthogonal rotator are associated with additional separatrix currents. There are no energy flux from the closed region which should have occurred if the second tern in (9) was important. This result does not leave any room for magneto-dipole losses as well.

Now using relation (10), one can obtain an expression for the surface current averaged over the polar cap $< J_\theta >$, which provides the MHD loss $W_{\rm tot}$ (1) for the orthogonal rotator

$$< J_\theta > = \frac{c}{4\pi f_*} B_0 \left(\frac{\Omega R}{c}\right)^2. \tag{11}$$

Accordingly, the averaged toroidal magnetic field $< B_\varphi >$ within the open field lines at the distance r is to equal to

$$< B_\varphi > = \frac{1}{f_*} B_0 \left(\frac{\Omega R}{c}\right)^2 \frac{R}{r}. \tag{12}$$

It can be easily obtained by determining the flux of the Poynting vector through the corresponding area $s(r) = f_* \pi(\Omega/c)^{1/2} r^{3/2}$ (f_* is a dimensionless polar cap area). As we see, toroidal magnetic field B_φ behaves in the same way as in a magneto-dipole wave. Finally, we note that the total additional separatrix current must be $(\Omega R/c)^{1/2}$ times smaller than the total current circulating in the magnetosphere.

Finally, let us write down the additional braking torque in general form as

$$K_{\perp}^{\rm add} = -A\,\frac{B_0^2 \Omega^3 R^6}{c^3}\, i_{\rm a}, \tag{13}$$

and try to estimate the dimensionless constant A from the results of numerical modeling. Since, as was shown, in this case the dimensionless longitudinal current is estimated as $i_{\rm a} \sim (\Omega R/c)^{-1/2}$, then the coefficient A turns out to be equal to

$$A \sim (\Omega R/c)^{1/2}. \tag{14}$$

For such a small value $A \ll 1$, one can neglect the magnetospheric contribution $K_{\perp}^{\rm add}$ (13) for the local GJ current $i_{\rm a}^{\rm A} \sim 1$, which, in fact, was done within the framework of BGI model [14].

5. Conclusion

Thus, direct analysis of the energy flux angular distribution inside the light cylinder unambiguously indicate the absence of magneto-dipole losses. This implies that pulsar wind should be considered as an example of a relativistic magneto-hydrodynamical wave. For example, the angular distribution of the energy flux varies from $\sin^2\theta$ for axisymmetrical case to $\sin^4\theta$ for the orthogonal rotator. Already in this point there is a significant difference from magneto-dipole losses for which the losses are proportional to $(1 + \cos^2\theta)$ [13].

Besides, we recognized the importance of additional separatrix current providing the most energy losses for orthogonal rotator. The amplitude of this current is to be much smaller than the total current circulating in the magnetosphere. For this reason the structure of this current near the light cylinder is unclear. Numerical simulations give us only the qualitative indication on its existence; quantitative consideration was not produced yet.

Acknowledgments

This research was partially supported by the government of the Russian Federation (agreement No. 05.Y09.21.0018) and by Russian Foundation for Basic Research (Grant no. 17-02-00788).

References

[1] Pacini F 1967 *Nature* **216** 567–8
[2] Ostriker J P and Gunn J E 1969 *Nature* **221** 454–6
[3] Beskin V S, Gurevich A V and Istomin I N 1983 *Sov. Phys. JETP* **58** 235–53
[4] Mestel L, Panagi P and Shibata S 1999 *Mon. Not. Royal Astron. Soc.* **309** 388–94
[5] Spitkovsky A 2006 *Astrophys. J. Lett.* **648** L51–4
[6] Kalapotharakos C and Contopoulos I 2009 *Astron. Astrophys.* **496** 495–502
[7] Pétri J 2012 *Mon. Not. Royal Astron. Soc.* **424** 605–19
[8] Tchekhovskoy A, Spitkovsky A and Li J G 2013 *Mon. Not. Royal Astron. Soc.* **435** L1–5
[9] Tchekhovskoy A, Philippov A and Spitkovsky A 2016 *Mon. Not. Royal Astron. Soc.* **457** 3384–95
[10] Kalapotharakos C, Contopoulos I and Kazanas D 2012 *Mon. Not. Royal Astron. Soc.* **420** 2793–8
[11] Deutsch A 1955 *Ann. d'Astrophys.* **18** 1–10
[12] Mestel L 1973 *Astrophis. Space Sci.* **24** 289–97
[13] Landau L D and Lifshitz E M 1971 *The Classical Theory of Fields.* (Oxford: Pergamon Press)
[14] Beskin V, Gurevich A and Istomin Y 1993 *Physics of the Pulsar Magnetosphere* (Cambridge: Cambridge University Press)
[15] Beskin V S and Zheltoukhov A A 2014 *Phys. Uspekhi* **57** 799–806
[16] Beskin V S 2010 *MHD flows in Compact Astrophysical Objects* (Berlin: Springer)
[17] Bai X N and Spitkovsky A 2010 *Astrophys. J.* **715** 1282–301

IOP Conf. Series: Journal of Physics: Conf. Series **932** (2017) 012013 doi:10.1088/1742-6596/932/1/012013

Pulsed emission from a rotating off-centred magnetic dipole in vacuum

A Kundu and J Pétri

Observatoire astronomique de Strasbourg, Université de Strasbourg, CNRS, UMR 7550, 11 rue de l'université, F-67000 Strasbourg, France

E-mail: anu.kundu@astro.unistra.fr

Abstract. Studying electromagnetic field around neutron stars is one of the vital methods to understand the physics of pulsars. From the very beginning of the efforts made to understand these objects, most of the works have been based on the assumption of a standard centred dipolar electromagnetic field. However, lately some studies have been focussing on explaining that including higher multipolar field components could modify our current ideas regarding these objects. Also, quite recently a more generalized picture has been put forward for pulsars in which the magnetic dipole moment is shifted off from the geometrical centre of the star showing how a rotating off-centred dipole can be expanded into multipolar components. We discuss the effects of such off-centred rotating dipole on various characteristic features of pulsars in vacuum like shape of the polar caps, radio and high energy emission phase plots and light curves along with a comparison with the standard centred case.

1. Introduction

Pulsars are compact objects with strong magnetic fields and short spin periods. Despite of being discovered five decades ago, our understanding of these objects involves many open questions. The associated physical processes are not completely known and studying magnetic field of pulsars assist in the understanding of the underlying physics since the physical processes occurring in the magnetosphere are closely linked to the topology of the field lines.

Almost all the works done in the field begin from a basic assumption which considers a centred dipolar field. There is no specific reason to assume that except the associated simplification which follows from it but complicated objects like pulsars, indeed, need to be visualized from an unconventional perspective. Some recent works did go beyond the dipolar assumption and have shown how the inclusion of multipolar components severely affects our understanding of these objects [1]. Also, a more general picture has been introduced which presents the magnetic dipole moment as shifted off from the geometrical centre of the star [2, 3].

The offset behaviour is not entirely alien and has been established for stars and even for planets [5, 6]. Several new studies have applied the idea to show how this approach is a potential candidate to justify the characteristics of these objects (refer [7, 9]). Also, recent observation of a high braking index pulsar pointed towards the possibility of an off-centred geometry [8].

The exact analytical solutions for the electromagnetic field of a rotating off-centred dipole were presented in [2] in form of an infinite series. The terms of the series represent individual multipolar field components and we use the equations only up to a first order correction here

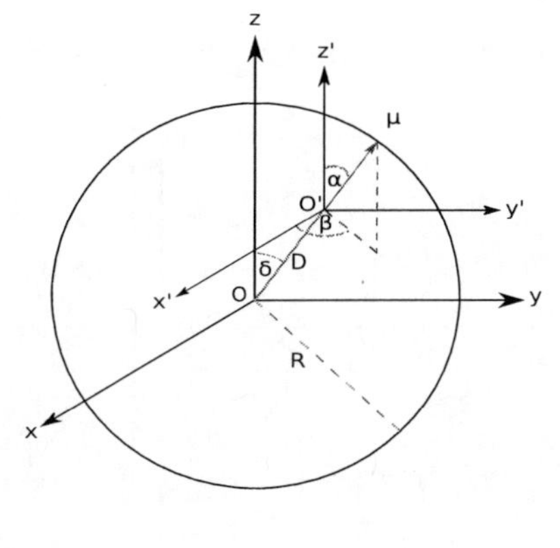

Figure 1: Geometry of an off-centred pulsar (after [3])

i.e. quadrupolar term. We derive the consequences of this field topology on radiation properties by focusing on the polar cap geometry, emission sites and mechanism [4].

2. Off-centred geometry

This section gives a brief introduction to the parameters involved and the geometry of an off-centred pulsar. A section of pulsar of radius R is shown in figure 1.There are two reference frames O and O' with the former attached to the geometrical centre of the star and the latter shifted off from O by a distance D. The rotation axis coincides with the $z - axis$ of the centred frame O. The magnetic moment is represented by blue vector and angles (α, β, δ) define its orientation as $\vec{\mu} = m(\sin\alpha\cos\beta, \sin\alpha\sin\beta, \cos\alpha)$ where m represents the scalar magnetic moment. The location of the magnetic moment with respect to the centred frame is given by $\vec{d} = D(\sin\delta, 0, \cos\delta)$. As it is physically impossible for the magnetic moment to lie outside pulsar we need to constraint the shift. The parameter $\epsilon = D/R$ is defined to consider the constraint with condition $\epsilon \ll 1$ i.e. weak off-centring.

Throughout the paper, the pulsar period is set to 2 ms. $\epsilon = 0$ and $\epsilon = 0.2$ are considered for the centred and the off-centred calculations respectively. The angles α, β and δ are shuffled to achieve different orientations.

3. Results and Discussions

3.1. Polar cap geometry

Polar caps are considered to be the locus of the feet of the last closed magnetic field lines. The idea is central to the pulsed emission as theoretical models use the location of the polar caps to differentiate between several emission regions. We use the polar cap model [10] to study the radio emission and slot gap model [11] to study the high energy emission.

The results for the polar cap geometry are shown in figure 2 where θ and ϕ represent polar angle and azimuth for frame O attached to the rotation axis of the pulsar. The $x-axis$ represent the normalized ϕ and $y - axis$ represent θ ranging from 0° to 180°. The centred case ($\epsilon = 0$) and the off-centred case ($\epsilon = 0.2$) are shown in red and blue respectively. The crosses represent the geometrical poles of the pulsar.

As apparent, the two geometries show significant differences. A prominent shift is observed in two cases but most significant point is the difference in size of the polar caps (see cases with $\delta = 0°$). The size of the polar cap is directly related to the pulse width and hence, such differences imply that the width could be larger or smaller depending on which pole we are looking at. This

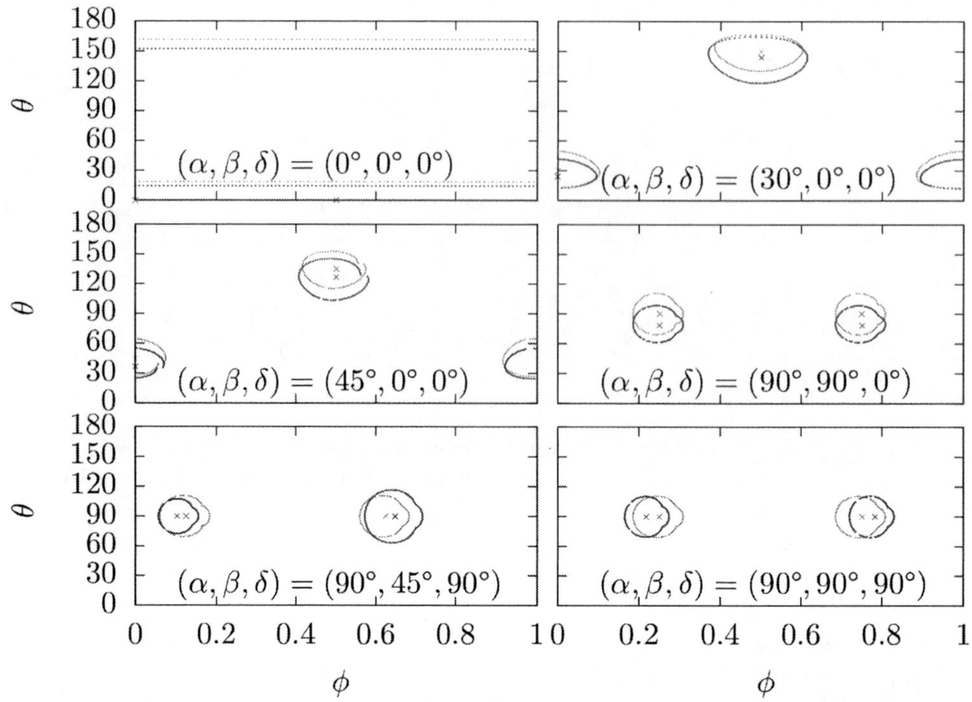

Figure 2: Polar cap geometry comparison for centred case with $\epsilon = 0$ (in red) and off-centred case with $\epsilon = 0.2$ (in blue) for various cases. The x-axis and y-axis represent the azimuth ϕ and the polar angle θ respectively. Crosses are the location of the poles.

could be one of the potential contributing factors to justify observations showing pulse width discrepancies [12]. Also, interesting to note are the off-centred cases with $\delta = 90°$ that show a shift in ϕ in opposite direction as compared to the centred case i.e. the shift in ϕ is greater than 180° for the former but equal for the latter.

3.2. Radio and high energy emission

We study the phase plots to better understand the emission features of pulsars. To derive them, we use the two pole caustic model explained by [13] and revised later by [14] which considers the relativistic effects on the photon propagation. The aberration effect is corrected by transforming the photon propagation in the corotating frame to the lab frame. The rotational phase delay which arises because of the time delay in photon propagation as a result of the finite speed of light is also taken in account. The differences between the centred and the off-centred cases are too subtle to conclude from the phase plots, hence, we show only the derived light curves to highlight the differences.

The light curves for orientation $(\alpha, \beta, \delta) = (30°, 0°, 0°)$ are plotted in figure 3 where some specific line of sights (ζ) are chosen. The light curves shown are phase zero curves where we introduce a phase shift to consider the zero phase as the point where radio emission is at its peak. From the plots the complementing behaviour of the two regimes is clearly visible, meaning that the range where high energy emission is at its maximum is also the range for the radio emission to be at its lowest which is expected from the models we consider. The dip in the high energy emission corresponds to the size of the polar cap as high energy emission is at its lowest in vicinity of the polar caps. As the emission is related to pulse width, the differences seen in figure represent pulse width differences in a way. We showed that already in Section 3.1

IOP Conf. Series: Journal of Physics: Conf. Series **932** (2017) 012013 doi:10.1088/1742-6596/932/1/012013

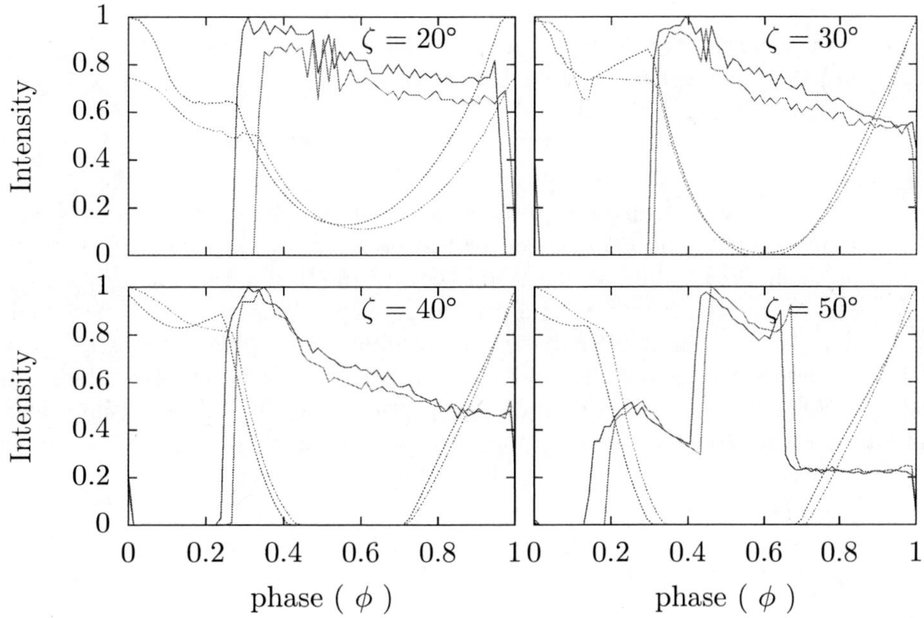

(a) $\zeta = 20°$, $30°$, $40°$, $50°$

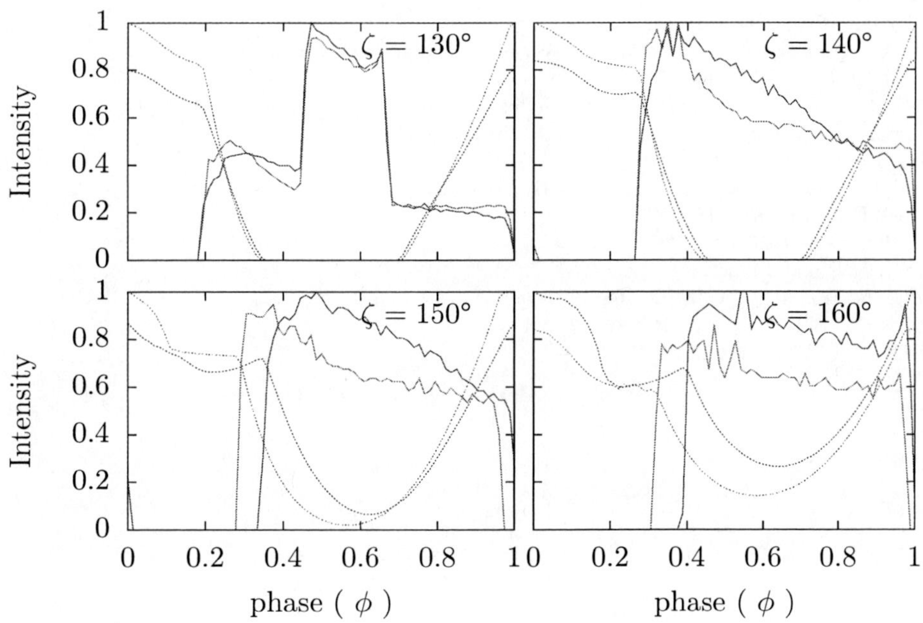

(b) $\zeta = 130°$, $140°$, $150°$, $160°$

Figure 3: High energy and radio emission light curves with zero phase considered to be the phase of the peak of radio emission for $(\alpha, \beta, \delta) = (30°, 0°, 0°)$ for centred case ($\epsilon = 0$, in red) and off-centred case ($\epsilon = 0.2$, in blue) for various line of sights (ζ). In (a) we have ζ ranging from 20° to 50° and (b) has ζ from 130° to 160° in steps of 10° each. The high energy emission curves are represented in solid lines and the radio emission curves in dashed lines. (Note that ϕ here is similar to the one in figure 2 but here it is not just a geometrical idea and is a characteristic of emission process representing phase of photons.)

but it was interpreted purely from a geometrical point of view there, while here, we see it as a characteristic of emission process. Also, if we compare the phase between the radio and the high energy emission peaks for the corresponding cases we notice that there are phase delays as compared to the centred geometry. The second Fermi catalgoue [15] mentions observations of time lags and the apparent phase delays discussed above might contribute to justification of such observed lags.

4. Conclusions

The off-centred topology is a reliable approach which makes an attempt to better explain polar caps and fit light curves. The comparison of the polar cap geometry for the centred and the off-centred approach shows shift in the polar caps highlighting the difference in size and can be considered as one of the reasons to justify the observed pulse width discrepancies. Study of phase plots gives us an insight into the site of production of pulsed radiation so as to better understand the emission mechanism and the comparison of the emission light curves shows phase delay between the radio and high-energy profiles which could potentially help to explain the observational time lag signatures between the two regimes.

Acknowledgments

This work has been supported by the French National Research Agency (ANR) through the grant No. ANR-13-JS05-0003-01 (project EMPERE) and by the University of Strasbourg IdEx programme.

References

[1] Pétri J 2015 *MNRAS* **450** 714–42
[2] Pétri J 2016 *MNRAS* **463** 1240–68
[3] Pétri J 2017 *MNRAS* **466** L73–7
[4] Kundu A and Pétri J 2017 *MNRAS* `https://doi.org/10.1093/mnras/stx1711`
[5] Stift M J 1974 *MNRAS* **169** 471–6
[6] Komesaroff M M 1976 *Proceedings of the Astronomical Society of Australia* **3** 51–3
[7] Harding A K and Muslimov A G 2011 *ApJ* **726** L10
[8] Archibald R F et al 2016 *ApJ* **819** L16
[9] Barnard M, Venter C and Harding A K 2016 *ApJ* **832** 107
[10] Sturrock P A 1971 *ApJ* **164** 529–56
[11] Arons J 1983 *ApJ* **266** 215–41
[12] Weltevrede P and Johnston S 2008 *MNRAS* **391** 1210–26
[13] Dyks J and Rudak B 2003 *ApJ* **598** 1201–06
[14] Bai X N and Spitkovsky A 2010 *ApJ* **715** 1270
[15] Abdo A A et al 2013 *ApJ* **208** 17

Abrupt changes in pulsar pulse profile through multiple magnetospheric state switching

R Yuen[1,2,3] **and D B Melrose**[3]

[1]Xinjiang Astronomical Observatory, Chinese Academy of Sciences, 150 Science 1-Street, Urumqi, Xinjiang, 830011, China
[2]Key Laboratory of Radio Astronomy, Chinese Academy of Sciences, Urumqi, Xinjiang, 830011, China
[3]SIfA, School of Physics, University of Sydney, Sydney, NSW 2006, Australia

E-mail: `ryuen@xao.ac.cn`, `donald.melrose@sydney.edu.au`

Abstract. A purely magnetospheric model is introduced for observed abrupt changes in pulsar radio profile. Motion of magnetospheric plasma is described by a drift frequency, $\omega_{\rm dr}$, that depends on a parameter $0 \leq y \leq 1$, and a change in the magnetospheric state corresponds to a change in y. Emission is assumed to arise from m spots distributed uniformly around the magnetic axis, so that spots drift by at the rate $m\omega_{\rm dr}$. Observable features, such as subpulses, appear to rotate as $\omega_{\rm R} = m\omega_{\rm dr} - \omega_{\rm V}$. The motion of the visible point, $\omega_{\rm V}$, is ignored in a "standard" version of the viewing geometry that assumes a fixed line of sight (rather than a fixed line-of-sight direction), implying $\omega_{\rm V} = 0$. With $\omega_{\rm V} \neq 0$, the apparent motion of subpulses is not constant. An abrupt (or more gradual) change in y implies a change in $\omega_{\rm R}$, which affects the observed pulse structure and the average profile. We apply the model for profile shifts observed with PSR B0919+06.

1. Introduction

Discernible drifting subpulses was detected from more than 50% of the radio pulsars sampled [1] implying difference in the rotation frequency of the subpulses, $\omega_{\rm R}$, from that of the star, ω_*. Abrupt changes in the drift pattern was observed from some pulsars [2], which implies that $\omega_{\rm R}$ can change abruptly and then return to its initial value. Furthermore, instantaneous switches between different subpulse drift rates are also known to correlate with changes in the average pulse profile [3–6], implying that a change in $\omega_{\rm R}$ may be responsible for a change in the pulse profile (we omit 'average' hereafter where no confusion is resulted). Changes are not necessarily abrupt as evident from the nulling PSR B0809+74 whose drift rate recovers to its normal value over tens of pulse periods after a null [7]. Another example is the "swooshing" in PSR B0919+06 [8, 9], which shows quasi-periodic shifts in the pulse profile that also last for tens of periods [10].

An implication of $\omega_{\rm R} \neq \omega_*$ is that the magnetospheric plasma does not corotate with the star, and the correlated changes in $\omega_{\rm R}$ and the emission characteristics imply that the configuration of the magnetosphere changes when $\omega_{\rm R}$ changes which affects the radio emission. The suggested interpretation of the abrupt changes of emission mode in PSR B1931+24 [11] involves switching between two limiting magnetospheric states, with one corresponding to the state of charge starvation in which screening of the vacuum electric field is inefficient due to insufficient plasma (and slowing down is due to magnetic dipole radiation as there is no wind), and the other to

IOP Conf. Series: Journal of Physics: Conf. Series **932** (2017) 012014 doi:10.1088/1742-6596/932/1/012014

the corotation or a "force-free" state [12, 13]. Switching of magnetospheric states then leads to changes to the global current, and the associated spin-down torque on the star, and this evidently also affects the pulse profile. The two limits are unrealistic for a real pulsar magnetosphere, and the switches must be between magnetospheric states intermediate between these limiting states. In this paper, we present and explore a model for $\omega_{\rm R}$ in a magnetosphere of multiple magnetospheric states as an alternative for interpreting the observed profile changes.

2. Apparent subpulse drift

We outline the three ingredients in our model for $\omega_{\rm R}$:

Plasma flow ($\omega_{\rm dr}$): We identify a variable drift frequency, $\omega_{\rm dr}$, of the magnetospheric plasma in the source region for the radio emission as given by the electric drift, $\mathbf{E}\times\mathbf{B}/B^2$. In non-corotating magnetosphere, it has the form $\omega_{\rm dr} = y\omega_{\rm ind} + (1-y)\omega_{\rm cor\perp}$ with values intermediate between drift due, respectively, to the inductive (ind) electric field for a rotating dipole in vacuo and the electric fields corresponding to (oblique) corotation (cor) [14]. In addition to depending on the viewing angle, ζ, between the rotation axis of the pulsar and the line of sight, the obliquity angle, α, between the rotation and the magnetic axes, and the pulsar phase, $\psi = \omega_\star t$, the drift frequency also depends on y and a change in magnetospheric state corresponds to a change in y.

Emission structure (m): A purely magnetospheric structure $\propto \cos(m\phi_b)$ is assumed, where m is an integer and ϕ_b is azimuthal angle around the magnetic pole. The structure may arise from an instability in the magnetosphere such as diocotron instability [15, 16]. Subpulses are then identified as emission spots (regions of emission) from a periodic wave-like disturbance with a specific spherical harmonic, given by l and m, being favored implying m nodes (underdense) and antinodes (overdense) as a function of ϕ_b. The observer would then see individual emission spots drifting by at the rate $m\omega_{\rm dr}$.

Motion of the visible point ($\omega_{\rm V}$): For radiation from near the stellar surface where field-line distortions may be ignored, pulsar visibility requires that visible radiation be emitted tangent to the magnetic field line at the source point and must be in the direction of the line of sight [17]. An implicit assumption in the "standard viewing geometry" is that the line of sight is a fixed line through the center of the star. In the exact geometry, the line of sight is treated as a fixed direction and not a fixed line, and this tangent line (in the line-of-sight direction) changes continuously as the pulsar rotates tracing out the trajectory of the visible point. The visible point moves along the trajectory with an angular frequency, denoted by $\omega_{\rm V}$, where it sub-rotates at some phases and super-rotates at other phases.

The implications of $\omega_{\rm V} \neq 0$ on the trajectory through an open region is shown in figure 1 for the exact model as compared with the standard model in which the motion of the visible point is ignored. The two versions coincide only at one point ($\psi = 0°$), and deviate as $|\psi|$ increases. An equal time interval is indicated as separation of any two consecutive dots, and it is clear that the traversal time through the open region, assumed at $0.1r_{\rm L}$, where $r_{\rm L}$ is the light-cylinder radius, is also different in the two versions. The difference is a result of a relative motion between the visible point and the pulsar rotation, both moving in the same direction, as opposed to the standard version in which such motion is absent due to the line of sight being assumed fixed. The result is that more dots are predicted in the exact version implying that the trajectory traverses the region over a longer duration, and enters and leaves the region at earlier and later ψ, respectively, than that implied in the standard version.

Combining the above three ingredients to drifting subpulses in dipolar magnetic field structure, the observer sees emission spots flowing past at angular speed, ω_R, given by

$$\omega_{\rm R}(y;\psi) = m\omega_{\rm dr}(\theta_{\rm V},\phi_{\rm V};y) - \omega_{\rm V}(\zeta,\alpha;\psi), \tag{1}$$

where the dependence on y and ψ are indicated, and $\theta_{\rm V}$ and $\phi_{\rm V}$ represent coordinates of the

IOP Conf. Series: Journal of Physics: Conf. Series **932** (2017) 012014 doi:10.1088/1742-6596/932/1/012014

trajectory of the visible point [17]. In the remainder of this paper, we (arbitrarily) fix $m = 20$ [18, 19].

3. Effects on pulse profiles

We consider observable changes in pulse profiles as implied by the model described in the previous section, and show how these can be related to y. We assume a pulse profile composed of several different Gaussian components formed from radiation of their respective emission spots within the profile. Fig. 2 illustrates the effect of state switching on observed profiles initially placed at $\psi = 0°$, $\pm 25°$, and $\pm 50°$ (red) all with unit amplitude and width of two units for $y = 0$ and $\omega_V = 0$. When variation of ω_V is included, $\omega_R/\omega_\star$ depends on phase ψ and the visible point crosses emission spots at a non-uniform rate. The effect is that the profiles appear to be displaced from their original locations, as shown in black, with the displacement increasing as $|\psi|$ increases with symmetry about $\psi = 0°$. As $m\omega_{dr}$ approaches ω_V, the visible point takes longer time to pass an emission spot giving the illusion of a broader profile. The broadening for profiles in black relative to that of the corresponding profiles in red at 10% amplitude in the order of increasing $|\psi|$ are 27%, 32% and 54%, respectively, as shown in Fig. 2.

In our model, a change in the magnetospheric state corresponds to a change in y and manifests as a change in ω_R. Fig. 3 shows a pulse profile with its peak initially placed at $\psi = -10°$ (black) for $y = 0$. As y increases, $\omega_R/\omega_\star$ decreases implying that the visible point meets the emission spot at an earlier phase and traverses it in a longer elapsed time as measured by an observer. The profile peak then shifts to increasingly negative ψ at $-11.0°$ for $y = 0.3$ (dot-dashed), $-12.2°$ for $y = 0.6$ (gray) and $-14.0°$ for $y = 0.9$ (dashed), with broadened width of 1.1 ($y = 0.3$), 1.2 ($y = 0.6$) and 1.4 ($y = 0.9$) times the original at 10% amplitude. Hence, the peak longitudinal phase and width of the observed profile changes as a result of a change in y, with the variations being functions of y.

Table 1. Values for the amplitude (A, normalized at the peak of the profile), width (σ) and peak phase of the emission components for simulating the 'stable' profiles of PSR B0919+06 at 1425 MHz.

Comp.	1	2	3	4	5	6	7
A	0.12	0.09	0.09	0.28	0.46	0.16	0.19
σ	3.70	0.60	0.33	2.40	0.87	2.01	1.51
$-\psi°$	17.5	15.4	14.6	14.4	13.2	12.8	11.1

The observations of PSR B0919+06 reveal shifting of the profile peak toward earlier phases where it remains for tens of pulsar periods before returning to its usual location [8]. During the event, known as swooshing, the profile peak shifts by $\sim -5°$ together with an increase in the profile width. Assuming $\zeta = 58.1°$, $\alpha = 53°$ and a profile width of 20° [8], we estimate seven emission spots (see Table 1) intersected by the trajectory of the visible point. The 'stable' profile at 1425 MHz is given in black in Fig. 4. As y varies, the emission spots shift, with the shift in ψ due to a change in ω_R implying that the emission spots are intersected by the trajectory differently. The combined effect from each emission spots is that (i) the profile peak is shifted to increasingly negative phases and (ii) the profile width is broadened revealing some of the previously unseen emission components. The changes to the profiles during the event exhibit as gradually moving away from its stable position as y increases, reaching $y = 0.42$, then the pattern reverses as ω_R returns to its usual value as y decreases and the profile appears to return to its initial phase. This results in an overall shift in the profile peak of $\sim -4°$ consistent with observations [8].

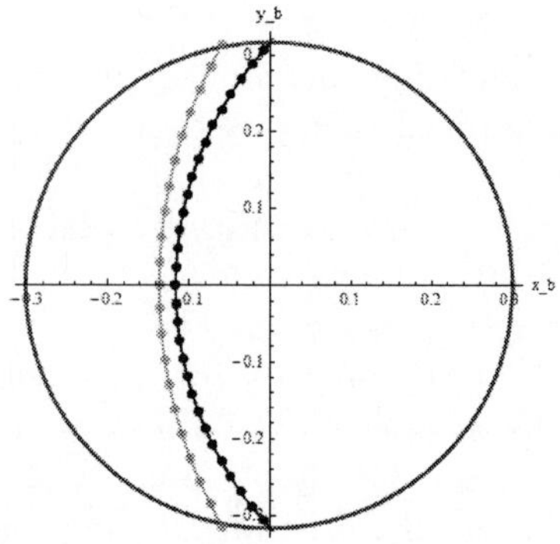

Figure 1. Differences in the trajectory through an open-field region (brown circle), in the magnetic frame, as predicted in the standard (gray) and the exact (black) models for $\zeta = 40°$, $\alpha = 30°$. The gray path is shifted by -0.02 along the x_b axis for clarity.

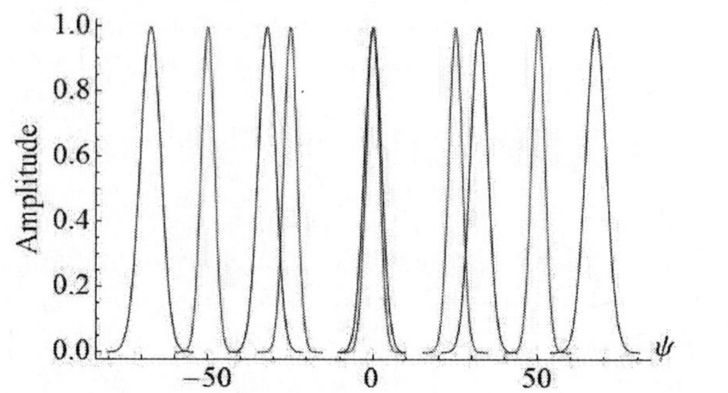

Figure 2. Effects of $\omega_V \neq 0$ on profiles originally placed at $\psi = 0°$, $\pm 25°$ and $\pm 50°$ for $y = 0$ and $\zeta = 10°, \alpha = 20°$ but assuming $\omega_V = 0$ (red). When ω_V is taken into account, the profiles are displaced to an earlier or later ψ (black) with increased width at larger ψ.

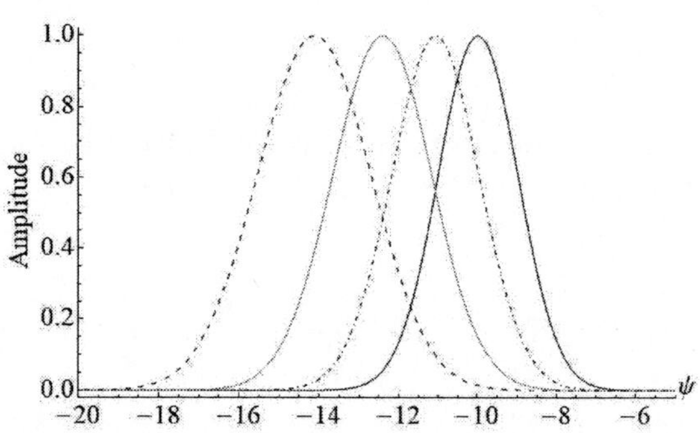

Figure 3. Simulation showing gradual shifts in profile peak, to more negative ψ, and increase in the pulse width due to state switching from $y = 0$ (solid black) to 0.3 (dot-dashed), 0.6 (gray) and 0.9 (dashed) for $\zeta = 10°$, $\alpha = 20°$ incorporating $\omega_V \neq 0$.

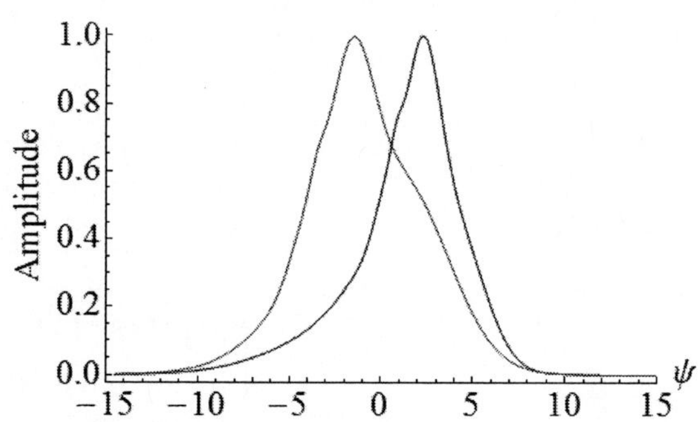

Figure 4. Simulation for 'stable' profile of PSR B0919+06 (black) using the parameters in Table 1 and the 'shifted' profile (blue) at 1425 MHz. Profile shapes and locations of the pulse peak are changed as ω_R changes due to switching from y =0 to 0.42.

4. Discussion and conclusions

We present a model for multiple magnetospheric states of a pulsar that exhibits time-variations in subpulse phase. We outline a quantitative definition for continuous set of "magnetospheric states" parameterized by $0 \leq y \leq 1$ where the inferred changes in magnetospheric states correspond to changes in y. We investigate the implications of the model for the interpretation of observed profile changes and show how the changes may be used to infer the change in y. Our model includes a traditionally neglected effect of the motion of the visible point and we show that it has observable consequences in the emission region and the shape of pulse profiles. We attribute the profile shifts to a different phase to changes in magnetospheric state (y) through its effect on $\omega_{\rm R}$ in a non-corotating pulsar magnetosphere. We show, using a Gaussian profile, that the peak phase of the profile and its width are unique functions of y for given ζ, α in our model. We consider PSR B0919+06 as a test case and argue that our model can reproduce the shift in the phase, the broadened profile and several related features. Other implications include:

- The observed subpulse drift rate results from the motion of the visible point at $\omega_{\rm V}$ imposed on the 'true' flow rate of the emitting plasma. An equally-spaced distribution of subpulses that flow uniformly around the magnetic axis does not imply a constant subpulse drift rate. Even if the actual motion of subpulses around the magnetic pole is uniform, the observed drift rate across a pulse window is a function of ψ, and depends on ζ, α and y.
- Pulsar profile shape and peak phase variations due to multiple state switching imply variations in the time of arrival (ToA) of pulses thus limiting the timing accuracy. Jitter noise may result from small random changes in y.
- In our model, switching in the magnetospheric state causes a change in the plasma flow ($\omega_{\rm R}$) and hence the profile. A change in y also causes a change in the ratio of the particle current to the displacement current and this evidently affects the global current and consequently $\dot{\nu}$. This implies a causal link between radio emission and the torque on the star as revealed by the observed correlated changes in the $\dot{\nu}$ and the pulse profile from some pulsars [20]. A quantitative treatment of such connections would require further exploration of the model to determine the current flow along open-field lines and how this current changes when y changes, and it is beyond the scope of this paper.

Acknowledgements

RY acknowledges supports from NSFC Project no. 11573059; the Joint Funds of NSFC Grant No. U1531137; the Technology Foundation for Selected Overseas Chinese Scholar, Ministry of Personnel of China; and the Strategic Priority Research Program "The Emergence of Cosmological Structures" of the Chinese Academy of Sciences, Grant No. XDB09000000. DBM acknowledges support from the Australian Research Council through grant DP160102932.

References

[1] Weltevrede P, Edwards R T and Stappers B 2006 *A&A* **445** 243
[2] Smits J M, Mitra D and Kuijpers J 2005 *A&A* **440** 683
[3] Wright G A E and Fowler L A 1981 *Astron. Astrophys.* **101** 356
[4] Redman S L, Wright G A E and Rankin J M 2005 *MNRAS* **357** 859
[5] Joshi B C 2012 *Neutron Stars and Pulsars: Challenges and Opportunities after 80 years* ed van Leeuwen J Proc. IAU Symposium 291 (CUP)
[6] Rankin J M, Wright G A E and Brown A M 2013 *MNRAS* **433** 445
[7] van Leeuwen A G J, Stappers B W, Ramachandran R and Rankin J M 2003 *A&A* **399** 223
[8] Rankin J M, Rodriguez C and Wright G A W 2006 *MNRAS* **370** 673
[9] Wahl H M, Orfeo D J, Rankin J M and Weisberg J M 2016 *MNRAS* **461** 3740
[10] Perera B B P, Stappers B W, Weltevrede P, Lyne A G and Bassa C G 2015 *MNRAS* **446** 1380
[11] Kramer M, Lyne A G, O'Brien J T, Jordan C A and Lorimer D R 2006 *Science* **312** 549
[12] Li J, Spitkovsky A and Tchekhovskoy A 2012 *ApJL* **746** L24
[13] Kalapotharakos C, Harding A K, Kazanas D and Contopoulos I 2012 *ApJ* **754** L1
[14] Melrose D B and Yuen R 2014 *MNRAS* **437** 262
[15] Fung P K, Khechinashvili D and Kuijpers J 2006 *A&A* **445** 779
[16] Pétri J 2007 *A&A* **464** 135
[17] Yuen R and Melrose D B 2014 *PASA* **31** e039
[18] Godoberidze G, Machabeli G Z, Melrose D B and Luo Q 2005 *MNRAS* **360** 669
[19] Mitra D and Rankin J M 2008 *MNRAS* **385** 606
[20] Lyne A G, Hobbs G, Kramer M, Stairs I H and Stappers B 2010 *Science* **329** 408

IOP Conf. Series: Journal of Physics: Conf. Series **932** (2017) 012015 doi:10.1088/1742-6596/932/1/012015

Luminosity of synchrotron radiation from pulsar magnetospheres

S Kisaka[1] and S J Tanaka[2]

[1] Department of Physics and Mathematics, Aoyama Gakuin University, 5-10-1 Fuchinobe, Sagamihara 252-5258, Japan

[2] Department of Physics, Faculty of Science and Engineering, Konan University, 8-9-1 Okamoto, Kobe, Hyogo 658-8501, Japan

E-mail: kisaka@phys.aoyama.ac.jp

Abstract. We calculate the maximum luminosity of the synchrotron radiation in X-ray and optical bands, which is originated from the electromagnetic cascade in the magnetosphere of rotation-powered pulsars. We find that for pulsars with spin-down luminosity $L_{\rm sd} \leq 10^{35}$ erg s^{-1}, even if the full allowed energy range of the γ-ray photon emitted by the accelerated particles is taken into account, the observed non-thermal luminosities are higher than the maximum luminosity of synchrotron radiation from pairs created by two-photon collision. The synchrotron radiation from the inner region could explain the observed non-thermal luminosities, so that the γ-ray pulsars with $L_{\rm sd} \leq 10^{35}$ erg s^{-1} would have multiple particle accelerators in the magnetosphere.

1. Introduction

In the magnetosphere of rotation-powered pulsars, charged particles are accelerated and created, and emit pulsed emission in broad energy range. The accelerated particles with Lorentz factor $\gamma > 10^6$ emit γ-ray photons. A fraction of the γ-ray photons converts to electrons and positrons. The created particles emit synchrotron radiation in X-ray and optical bands (e.g. [1]). Therefore, the high-energy emission provides useful tool to prove the particle acceleration and creation in the magnetosphere.

Observed γ-ray pulsed emission comes from the outer magnetosphere [2]. At the outer magnetosphere, the particles are created via two-photon pair creation process. Synchrotron radiation from the created particles could produce the observed non-thermal emission. On the other hand, particles could be created near the stellar surface via magnetic pair creation process (e.g., [3]). These particles could also contribute to the observed non-thermal luminosities. In this proceeding, we calculate the maximum luminosity of synchrotron radiation based on the model of Kisaka & Tanaka [4], which considers synchrotron radiation from particles created by both two-photon collision and magnetic pair creation processes. In section 2, we briefly review the model of Kisaka & Tanaka [4]. The difference of Kisaka & Tanaka [4] is that we take into account the full allowed range of the characteristic energy of the γ-ray photon $E_{\rm cur}$. In section 3, we show the maximum luminosity of synchrotron radiation in X-ray and optical bands, and compare with the observations. We also discuss the effect of the particle trajectory on the luminosity of synchrotron radiation.

2. Model

We adopt the model described in Kisaka & Tanaka [4]. The basic picture is as follows. A fraction of the spin-down luminosity $L_{\rm sd}$ (a fraction η) converts to the energy flux $\gamma_{\rm p} m_{\rm e} c^2 \dot{N}_{\rm p}$ of the accelerated particles with Lorentz factor $\gamma_{\rm p}$ at a given radius r from the centre of a neutron star, where $\dot{N}_{\rm p}, m_{\rm e}$ and c are the number flux of the accelerated particles, the electron rest mass, and the speed of light, respectively. The accelerated particles emit γ-ray photons with the characteristic energy $E_{\rm cur}$ via curvature radiation. The ratio of the γ-ray luminosity to the energy flux of the accelerated particles is determined by comparing the advection timescale $t_{\rm ad} \sim r/c$ with the radiative cooling timescale $t_{\rm cool,cur} \sim \gamma_{\rm p} m_{\rm e} c^2 / P_{\rm cur}$, where $P_{\rm cur}$ is the power of curvature radiation for a particle. A part of γ-ray photons converts to electron-positron pairs with Lorentz factor $\gamma_{\rm s,pair}$. For pair creation processes, two-photon collision and magnetic pair creation work in the magnetosphere. In the two-photon collision process, the optical depth $\tau_{\gamma\gamma}$ for γ-ray photons and the luminosity of curvature radiation determine the energy flux of the created particles. The thermal emission from the heated polar cap is considered as seed photons to create the pairs. In the electromagnetic cascade via magnetic pair creation, higher generation pairs could be created due to high pair conversion efficiency. The maximum number of the created particles at a given r is estimated that the γ-ray photons with the energy $E_{\rm cur}$ are split to the photons with the maximum escapable energy $E_{\rm esc}$, and all the photons with $E_{\rm esc}$ convert to pairs. Note that in this case, almost all γ-ray luminosity converts to the energy flux of the created particles. The created pairs have non-zero pitch angle in general, so that the particles emit synchrotron radiation in X-ray and optical bands. The ratio of the luminosity of synchrotron radiation for a given frequency $\nu_{\rm obs}$ to the energy flux of the created particles is determined by the momentum perpendicular to the magnetic field, and the comparison between the advection timescale $t_{\rm ad} \sim r/c$ and the radiative cooling timescale $t_{\rm cool,syn} \sim \gamma_{\rm s,syn} m_{\rm e} c^2 / P_{\rm syn}$, where $P_{\rm syn}$ is the power of synchrotron radiation for a particle, and $\gamma_{\rm s,syn}$ is the Lorentz factor of particles whose characteristic frequency of synchrotron radiation equals to $\nu_{\rm obs}$.

Some constraints are imposed on the model. First, in order to emit synchrotron radiation at a frequency $\nu_{\rm obs}$, the Lorentz factor $\gamma_{\rm s,syn}$ has to be lower than the Lorentz factor $\gamma_{\rm s,pair}$ determined by the energy of the parent γ-ray photons, $E_{\rm cur}$ or $E_{\rm esc}$. Second, the frequency $\nu_{\rm obs}$ has to be higher than the turnover frequency for the synchrotron radiation, $\nu_{\rm obs} > \nu_{\rm g}/\alpha$ [5], where $\nu_{\rm g} = eB/(2\pi m_{\rm e} c)$, e is the charge of an electron, B is the magnetic field at the emission region, and α is the pitch angle. The constraint also gives the upper limit on the magnetic field $B \sim 8.6 \times 10^{10} \alpha (h\nu_{\rm obs}/1{\rm keV})$ G at the emission region, where h is the Planck constant. Then, we describe a free electron in a magnetic field in classical regime. Third and forth conditions are the thresholds of the pair creation via two-photon collision and magnetic pair creation, respectively.

In the model, the parameter is only η in the dipole magnetic field case. For the pitch angle, we use the approximated value $\alpha \sim \sqrt{r R_{\rm lc}}$, where $R_{\rm lc}$ is the radius of the light cylinder. In the non-dipole field case, we take the maximum value of the pitch angle $\alpha = 1$ at any r. The strength of the non-dipole magnetic field does not depend on the maximum luminosity of synchrotron radiation [4]. For the radius to the emission region r, we take the value which gives the maximum value of the luminosity of synchrotron radiation for given observables such as the surface dipole field $B_{\rm s}$ and the spin-down luminosity $L_{\rm sd}$.

Different from Kisaka & Tanaka [4], we take into account the full allowed range of the energy $E_{\rm cur}$. The observed characteristic energy could depend on the inclination and viewing angles (e.g., [6, 7]). Then, typical energy of parent photons $E_{\rm cur}$, which determines the energy of created particles, could be different from the observed γ-ray photon energy. In the model, there is a lower limit on the energy $E_{\rm cur}$ to create the pairs via two-photon collision for a given temperature of the thermal emission from the polar cap surface $T_{\rm pc}$,

$$E_{\rm cur} \geq 1.4 \left(\frac{T_{\rm pc}}{10^{6.5}\ {\rm K}} \right)^{-1} \ {\rm GeV}. \quad (1)$$

The observed temperature is $T_{\rm pc} \sim 10^6$-$10^{6.5}$ K (e.g., [8]). On the other hand, the full potential drop across the polar cap $\Delta V_{\rm pc}$ gives the upper limit on the energy,

$$E_{\rm cur} \leq 0.29 \frac{3}{4\pi} \frac{hc}{R_{\rm cur}} \gamma_{\rm max}^3 \tag{2}$$

where $R_{\rm cur}$ is the curvature radius of the particle trajectory, and $\gamma_{\rm max}$ is the maximum Lorentz factor,

$$\gamma_{\rm max} = \frac{e\Delta V_{\rm pc}}{m_{\rm e}c^2}. \tag{3}$$

3. Results and Discussion

Figure 1 shows the maximum luminosities of synchrotron radiation $L_{\rm syn}$ (left) and the ratios $L_{\rm cur}/L_{\rm syn}$ (right) in the two-photon collision ($\gamma\gamma$) and the magnetic pair creation cases (Bγ), where $L_{\rm cur} = P_{\rm cur}\dot{N}_{\rm p} \min\{t_{\rm cool,cur}, t_{\rm ad}\}$ is the luminosity of curvature radiation. For example, the maximum luminosity of synchrotron radiation for conditions $t_{\rm cool,syn} < t_{\rm ad}$ and $t_{\rm cool,cur} < t_{\rm ad}$ at the emission region are described by

$$L_{\rm syn} \sim \begin{cases} 1.5\times10^{31}\eta \left(\frac{\nu_{\rm obs}}{1\ {\rm keV}}\right)^{1/2} \left(\frac{E_{\rm cur}}{1\ {\rm GeV}}\right)^{-1} \left(\frac{L_{\rm pc}}{10^{-3}L_{\rm sd}}\right) \left(\frac{T_{\rm pc}}{10^{6.5}\ {\rm K}}\right)^{-1} \\ \quad \times \left(\frac{B_{\rm s}}{10^{12}\ {\rm G}}\right)^{-1/4} \left(\frac{L_{\rm sd}}{10^{35}\ {\rm erg\ s^{-1}}}\right)^{15/8} {\rm erg\ s^{-1}}\ (\gamma\gamma), \\ 4.6\times10^{31}\eta \left(\frac{\nu_{\rm obs}}{1\ {\rm keV}}\right)^{5/7} \left(\frac{B_{\rm s}}{10^{12}\ {\rm G}}\right)^{-1/7} \left(\frac{L_{\rm sd}}{10^{35}\ {\rm erg\ s^{-1}}}\right)^{17/14} {\rm erg\ s^{-1}}\ ({\rm B}\gamma, {\rm dipole}), \\ 4.5\times10^{33}\eta \left(\frac{\nu_{\rm obs}}{1\ {\rm keV}}\right) \left(\frac{L_{\rm sd}}{10^{35}\ {\rm erg\ s^{-1}}}\right) {\rm erg\ s^{-1}}\ ({\rm B}\gamma, {\rm non-dipole}), \end{cases} \tag{4}$$

where $L_{\rm pc}$ is the thermal luminosity from the heaped polar cap. The plotted synchrotron luminosity in the two-photon collision case is the maximum value for the full allowed range of $E_{\rm cur}$ and the range of the surface magnetic field $B_{\rm s} > 10^{11}$G. We also plot the observed non-thermal luminosities $L_{\rm X}$ and $L_{\rm opt}$ and the flux ratios $F_\gamma/F_{\rm X}$ and $F_\gamma/F_{\rm opt}$. For the pulsars with $L_{\rm sd} \leq 10^{35}$ erg s^{-1}, the observed luminosities and flux ratios are not explained by the two-photon collision case. The maximum luminosities for the magnetic pair creation case with non-dipole field are higher than the all observed values, so that the γ-ray pulsars would have the accelerator at the inner region of the magnetosphere.

The dependence on $E_{\rm cur}$ is described as follows. Once the energy flux of the accelerated particles is given by the efficiency parameter η, the energy $E_{\rm cur}$ determines the number flux of the γ-ray photons in the model. The number flux of the γ-ray photon and the optical depth give the number flux of created particles. The higher number flux of the created particles gives the higher synchrotron luminosity. The lowest value is $E_{\rm cur} \sim 1.4$ GeV from inequality (1), so that the luminosity $L_{\rm syn}$ is about twice higher than that in Kisaka & Tanaka (2017) [4] which fixes $E_{\rm cur} = 3$ GeV. On the other hand, for pulsars with $L_{\rm sd} \leq 10^{34} - 10^{36}$ erg s^{-1} and the lower value of $E_{\rm cur}$, the upper limit on the radius of the emission region given by the energy constraint $\gamma_{\rm s,pair} > \gamma_{\rm s,syn}$ becomes smaller than the light cylinder radius. Then, the pitch angle α at the emission region $r < R_{\rm lc}$ becomes smaller than unity, so that the higher $E_{\rm cur}$ gives the higher synchrotron luminosity. Note that the energy of the accelerated particle cannot exceed the full potential drop across the polar cap. This gives the upper limit on the energy $E_{\rm cur}$ and the resulting synchrotron luminosity. As a result, the maximum luminosity does not significantly change from the case of Kisaka & Tanaka [4].

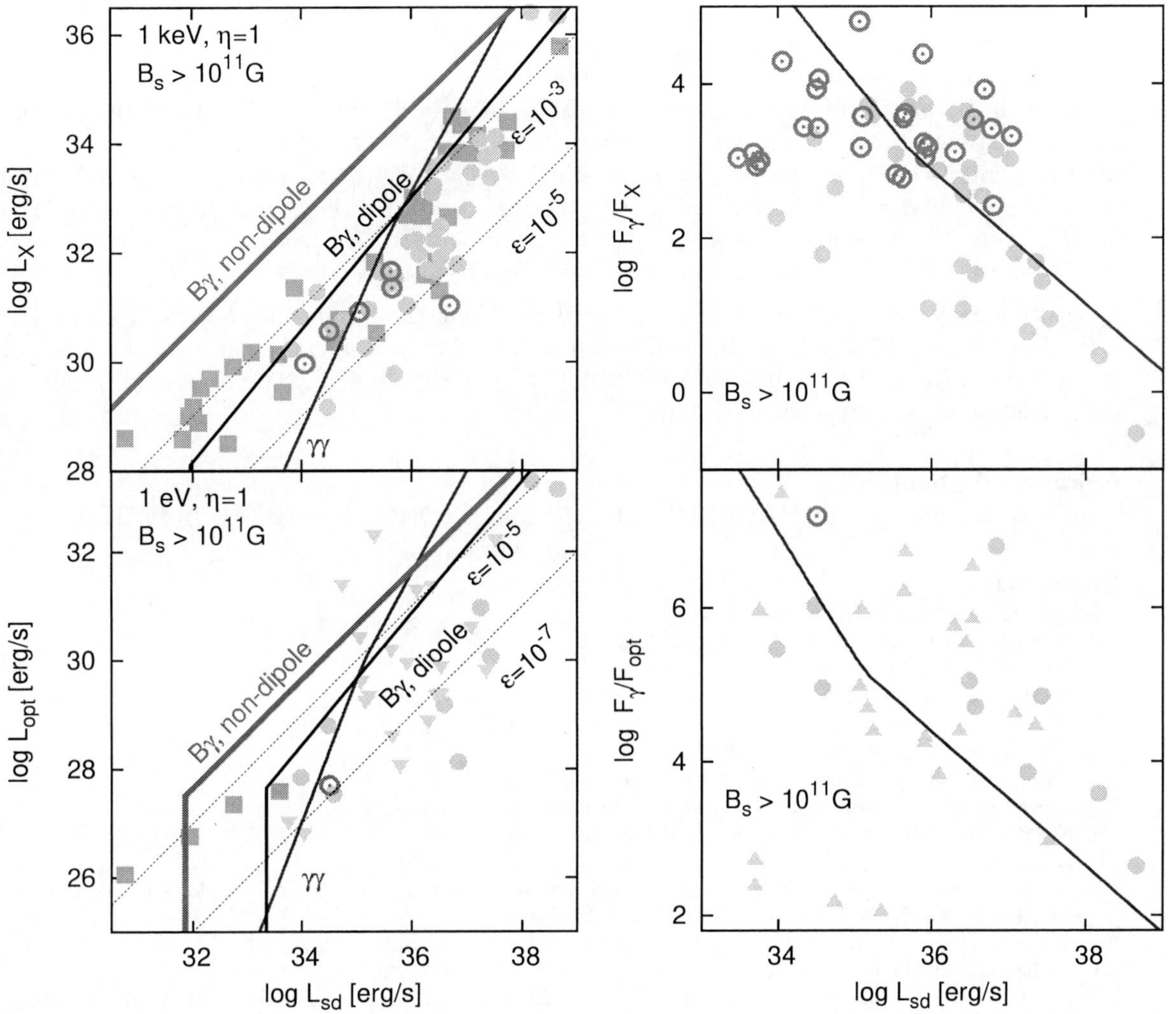

Figure 1. Left: Maximum luminosities of synchrotron radiation in 1 keV (upper) and 1 eV (lower) for two-photon collision case ($\gamma\gamma$, blue), magnetic pair creation cases (Bγ) with dipole field (black) and non-dipole fields (red) as a function of the spin-down luminosity L_{sd}. Right: Flux ratios F_γ/F_X (upper) and F_γ/F_{opt} (lower) as a function of L_{sd} for γ-ray pulsars. Blue curves are the flux ratio of curvature to synchrotron radiation in two-photon collision case. Observed data are taken from [8, 9, 10, 11, 12, 13, 14, 15, 16, 17, 18, 19, 20, 21, 22, 23, 24, 25, 26, 27]. Filled green and open magenta circles are radio-loud and radio-quiet γ-ray pulsars, respectively. Orange squares denote the non-γ-ray pulsars. Upper and lower limits on the optical luminosity L_{opt} and the flux ratio F_γ/F_{opt}, respectively, are also plotted as cyan triangles.

For the curvature radius of the trajectory of the accelerated particles, we use the approximation formula, $R_{cur} \sim \sqrt{rR_{lc}}$ [29]. At the outer magnetosphere, the curvature radius of the particle trajectory could be significantly different from that of the magnetic field line (e.g., [6]), which may change curvature radiation from the accelerated particles. However, the curvature radius does not significantly change the resulting synchrotron luminosity from the created particles. As long as the cooling timescale of curvature radiation is shorter than the advection timescale, $t_{cool,cur} < t_{ad}$, the total luminosity of the curvature radiation is $L_{cur} = \eta L_{sd}$, which does not depend on the curvature radius. Most pulsars satisfied the condition $t_{cool,cur} < t_{ad}$

[4], so that the synchrotron luminosity does not depend on the curvature radius.

In Figure 1, we consider the range of the surface magnetic field $B_s > 10^{11}$ G. Here, we briefly discuss the pulsars with a low surface magnetic field $B_s \sim 10^8$ G such as millisecond pulsars. In the two-photon collision case, since the conditions $\gamma_{s,pair} > \gamma_{s,syn}$ and $t_{cool,syn} > t_{ad}$ are satisfied at the light cylinder R_{lc}, the dependence of the luminosity on the magnetic field is described by $L_{syn} \propto B_s^{-1/4}$ [4]. Then, the maximum luminosity could be ~ 10 times higher than that of the plotted blue curves in Figure 1. On the other hand, in the magnetic pair creation case, the X-ray frequency is higher than ν_g/α at the surface. Then, the luminosity, $L_{syn} \propto B_s^{1/8}$, becomes lower for the lower dipole field. If the non-dipole field dominates at the surface, the luminosity could become higher for the higher strength of the non-dipole field. The maximum luminosity is the same as in the case with $B_s > 10^{11}$ G.

Acknowledgments

This work is supported by KAKENHI 16J06773 (S.K.), 15H05440, 17K18270 (S.J.T.).

Reference

[1] Romani R W 1996, *Astrophys. J.*, **470** 469-78
[2] Abdo A A et al 2009, *Astrophys. J.*, **696** 1084-93
[3] Sturrock P A 1971, *Astrophys. J.*, **164** 529-56
[4] Kisaka S and Tanaka S J 2017 *Astrophys. J.* **837** 76
[5] O'Dell S L and Sartori L 1970 *Astrophys. J. Lett.* **161** L63-4
[6] Hirotani K 2013 *Astrophys. J.* **766** 98
[7] Philippov A A and Spitkovsky A 2017 *Preprint* arXiv:1707.0423
[8] Becker W 2009 *Neutron Stars and Pulsars* (Berlin: Springer) p 91-140
[9] Kargaltsev O and Pavlov G G 2008 *AIP Conf. Ser.* **983** 171-85
[10] Kargaltsev O, Durant M, Pavlov G G and Garmire G P 2012 *Astrophys. J. Suppl.* **201** 37
[11] Posselt B, Pavlov G G Manchester R N, Kargaltsev O and Garmire G P 2012 *Astrophys. J.* **749** 146
[12] Abdo A A et al 2013 *Astrophys. J. Suppl.* **208** 17
[13] Acero F, Gallant Y, Ballet J, Renaud M and Terrier R 2013 *Astron. Astrophys.* **551** A7
[14] Zarikov S and Mignani R P 2013 *Mon. Not. R. Astron. Soc.* **435** 2227-33
[15] Beronya D M, Shibanov Y A, Zyuzin D A and Komarova V N 2015 *J. Phys.: Conf. Series* 661 012001
[16] Kirichenko A, Danilenko A, Shternin P, Shibanov Y, Ryspaeva E, Zyuzin D, Durant M, Kargaltsev O, Pavlov G and Cabrera-Lavers A 2015 *Astrophys. J.* **802** 17
[17] Kuiper L and Hermsen W 2015 *Mon. Not. R. Astron. Soc.* **449** 3827-66
[18] Marelli M, Mignani R P, De Luca A, Saz Parkinson P M, Salvetti D, Den Hartog P R and Wolff M T 2015 *Astrophys. J.* **802** 78
[19] Prinz T and Becker W 2015 *Preprint* arXiv:1511.07713
[20] Klingler N, Kargaltsev O, Rangelov B, Pavlov G G, Posselt B and Ng C-Y 2016 *Astrophys. J.* **828** 70
[21] Mignani R P, Rea N, Testa V, Marelli M, De Luca A, Pierbattista M, Shearer A, Torres D F and De Oa Wilhelmi E 2016 *Mon. Not. R. Astron. Soc.* **461** 4317-28
[22] Mignani R P, Testa V, Marelli M, De Luca A, Pierbattista M, Razzano M, Salvetti D, Belfiore A, Shearer A and Moran P 2016 *Mon. Not. R. Astron. Soc.* **463** 2932-38
[23] Mignani R P, Testa V, Marelli M, De Luca A, Salvetti D, Belfiore A, Pierbattista M, Razzano M, Shearer A and Moran P 2016 *Astrophys. J.* **825** 151
[24] Shibanov Y, Danilenko A, Zharikov S, Shternin P and Zyuzin D 2016 *Astrophys. J.* **831** 112
[25] Karpova A, Shternin P, Zyuzin D, Danilenko A and Shibanov Y 2017 *Mon. Not. R. Astron. Soc.* **466** 1757-63
[26] Hermsen W, Kuiper L, Hessels J W T, Mitra D, Rankin J M, Stappers B W, Wright G A E, Basu R, Szary A and van Leeuwen J 2017 *Mon. Not. R. Astron. Soc.* **466** 1688-08
[27] Szary A, Gil J, Zhang B, Haberl F, Melikidze G I, Geppert U, Mitra D and Xu R-X 2017 *Astrophys. J.* **835** 178
[28] Hirotani K 2011 *Astrophys. J. Lett.* **733** L49
[29] Kisaka S and Tanaka S J 2014 *Mon. Not. R. Astron. Soc.* **443** 2063-76

IOP Conf. Series: Journal of Physics: Conf. Series **932** (2017) 012016 doi:10.1088/1742-6596/932/1/012016

Magnetic energy stored in relativistic force-free magnetosphere

Y Kojima

Department of Physics, Hiroshima University, Higashi-Hiroshima 739-8526, Japan

E-mail: ykojima-phys@hiroshima-u.ac.jp

Abstract. A magnetar is one of variable sources, whose activity is supplied by huge magnetic energy. Magnetar magnetosphere is gradually twisted due to shearing motion at the surface. At the same time, the energy is stored there. When it exceeds a threshold, the energy is abruptly released on a dynamical timescale as observed in energetic flares. Here, static and axially symmetric force-free magnetospheres are calculated in the exterior of non-rotating relativistic star. The magnetic energy increases along a sequence characterized by the twist degree. In a highly twisted case, a magnetic flux rope, in which large amount of toroidal field is confined, is detached in vicinity of the star. It is found that larger amount of energy is capable to be stored in the relativistic magnetosphere. In an extreme case, the magnetic energy in presence of current flow increases up to a few times larger than that of current-free dipole. There is an upper limit in any model, and a catastrophic event like a giant flare would occur, when the field is further twisted.

1. Introduction

A diversity of neutron stars has been gradually disclosed in 50 years since the first discovery of a radio pulsar. Magnetars belong to a peculiar class. Their strong magnetic fields cause a remarkable difference, although there are some exceptional sources such as 'low-dipole-field' magnetars. So, the problem is not simple. The activity of magnetars is widely believed to be supplied by the strong magnetic field of $B_0 = 10^{14}$-10^{15}G. Enormous energy ($\sim 10^{44}$-10^{46} erg) instantaneously released in giant flares is a part of magnetic energy $10^{48}(B_0/10^{15}\mathrm{G})^2(R/10\mathrm{km})^3$ erg. Such strong field may be amplified by dynamo action at the birth, or the progenitor itself may be different. See also [1]. Stellar properties of magnetars are not well-known. For example, suppose that their masses have a bias toward massive one ($M \approx 2M_\odot$), then such information would be a key to their origin. Information of stellar radius R or a ratio M/R is also equivalently valuable, since their relation is inferred by nuclear equation of state. However, direct observational evidences will be rarely found. Theoretical study, which predicts a particular property through different observed events, might be helpful.

A giant flare model on magnetars is discussed in an analogy of solar flares. In the model, the energy is stored in the magnetospheres by twisting the magnetic field lines. Here, the magnetospheres are calculated by taking into account general relativistic effect ($M/R = 0.25$ in geometrical units of $c = G_\mathrm{N} = 1$). Results are compared with those obtained in a flat spacetime. It is found that the general relativistic effect can not be neglected, and is therefore necessary for an appreciate model construction.

IOP Conf. Series: Journal of Physics: Conf. Series **932** (2017) 012016 doi:10.1088/1742-6596/932/1/012016

2. Relativistic force-free magnetosphere

We consider the static magnetic configuration in curved spacetime outside a non-rotating compact object with mass M. The metric is given by

$$ds^2 = -\alpha^2 dt^2 + \alpha^{-2} dr^2 + r^2 d\theta^2 + \varpi^2 d\phi^2, \tag{1}$$

where

$$\alpha = \left(1 - \frac{2M}{r}\right)^{1/2}, \quad \varpi = r \sin\theta. \tag{2}$$

The magnetic field for the axially symmetric case is generally given by

$$\vec{B} = \vec{\nabla} \times \left(\frac{G}{\varpi} \vec{e}_{\hat{\phi}}\right) + \frac{S}{\alpha\varpi} \vec{e}_{\hat{\phi}} = \frac{\vec{\nabla} G \times \vec{e}_{\hat{\phi}}}{\varpi} + \frac{S}{\alpha\varpi} \vec{e}_{\hat{\phi}}, \tag{3}$$

where $\vec{e}_{\hat{\phi}} \equiv \varpi^{-1} \partial_\phi$ is a unit vector in the azimuthal direction. The force-free magnetic field $\vec{j} \times \vec{B} = 0$ means that current flows along the magnetic field lines. Poloidal component $(\vec{j} \times \vec{B})_\phi = 0$ is reduced to $S = S(G)$, and $4\pi\alpha\vec{j} = S'\vec{B}$. Azimuthal component of the Biot-Savart law is expressed as

$$\alpha^2 \frac{\partial}{\partial r}\left(\alpha^2 \frac{\partial G}{\partial r}\right) + \frac{\alpha^2 \sin\theta}{r^2} \frac{\partial}{\partial\theta}\left(\frac{1}{\sin\theta} \frac{\partial G}{\partial\theta}\right) = -SS'. \tag{4}$$

We adopt a power-law form for the current model with strength parameter γ and index n, so that the right hand side is $SS' = \gamma G^n$. When S is a linear function of G, i.e., $n = 1$, eq.(4) is a wave equation with angular frequency $\gamma^{1/2}$. General relativistic effect appears by a factor α^2 of eq.(4). When $\alpha^2 = 1$, we have models in flat spacetime. The correction factor α^2 on typical neutron star surface is $\alpha^2 \approx 0.5$ at most, so that the effect might be expected to be small. However, actual results given in next section show that the relativistic correction is not small, since we consider a non-linear system. Qualitative difference due to α^2 becomes small with decreasing index n. In this paper, we concentrate on $n = 7$ model. See [2] for details of other models.

Magnetic energy is given by integrating over 3-dimensional volume

$$E_{\rm EM} = \int \frac{\alpha B^2}{8\pi} \sqrt{g_3} d^3x = \frac{1}{4} \int \left[\left(\alpha \frac{\partial G}{\partial r}\right)^2 + \left(\frac{1}{r}\frac{\partial G}{\partial\theta}\right)^2 + \left(\frac{S}{\alpha}\right)^2\right] \frac{dr d\theta}{\sin\theta}, \tag{5}$$

where $\sqrt{g_3} (= \alpha^{-1} r^2 \sin\theta)$ is a determinant of 3-dimensional space metrics. Total relative helicity $H_{\rm R}$ defined by the integration range $r \geq R$ is

$$H_{\rm R} = \int \vec{A} \cdot \vec{B} \sqrt{g_3} d^3x = 4\pi \int \frac{GS}{\alpha^2} \frac{dr d\theta}{\sin\theta}, \tag{6}$$

where $\vec{A}$ is a vector potential of $\vec{B}$.

3. Results

We first construct a magnetosphere model with a dipole in vacuum, and gradually increase the toroidal component for a given boundary condition. It is known from previous calculations[3] in flat spacetime that a free parameter γ is not suitable for specifying the sequence of models, because there are multiple solutions concerning γ. Instead, the azimuthal magnetic flux or helicity is used as the degree of twist, and the constant γ is posteriorly determined. We follow the same method, and calculate the energy and the relative helicity along the sequence of solutions.

Results are shown in Figure 1. The left panel shows a model in flat spacetime $M/R = 0$, while the right a relativistic model with $M/R = 0.25$. Increment of magnetic energy ΔE from the potential dipole field, which is discriminated by poloidal ΔE_p or toroidal components ΔE_t, is given as a function of γ. Relative helicity $H_{\rm R}$ is also given. Each symbols in the figure represents one force-free solution. There is a maximum in $H_{\rm R}$, and it is impossible to calculate equilibrium model beyond it. A possible evolutionary sequence is constructed by connecting models in order of $H_{\rm R}$. That is, magnetosphere is gradually distorted by helicity injection from initial potential vacuum field. The structure of magnetic field lines also changes, and two examples are displayed: potential dipole ($\Delta E = 0$ and $\gamma = 0$) at the bottom, and highly twisted state at the top. Contour lines of magnetic function G are quite different. In the presence of toroidal component, magnetic field lines are outwardly stretched. In an extreme case of relativistic models, a loop structure is formed in the vicinity of the surface. The maximum of magnetic function G is located at $(r/R, \theta) \approx (1.5, \pi/2)$. There-dimensional structure of it is axially symmetric torus or rope. Both magnetic field components are strong there, since $B_{\hat{\phi}} = (\alpha\varpi)^{-1}S \propto G^4$. Strong currents ($\vec{j} \propto \alpha^{-1}S'\vec{B}$) also flow there. Thus, flux rope is highly non-linear equilibrium structure.

We consider how magnetic energy stored in the magnetosphere increases by twisting fields. The dipolar potential field energy E_0 is expressed as $E_0 = 0.33B_0^2R^3$ in a flat spacetime, and $E_0 = 0.71B_0^2R^3$ in a curved spacetime with $M/R = 0.25$[2]. Here we used typical strength $B_0(\equiv \mu R^{-3})$, which is determined by magnetic dipole moment μ asymptotically at infinity. The difference in E_0 between relativistic and non-relativistic models comes from the normalization. For example, the surface field strength in relativistic model is a few times larger. It is therefore

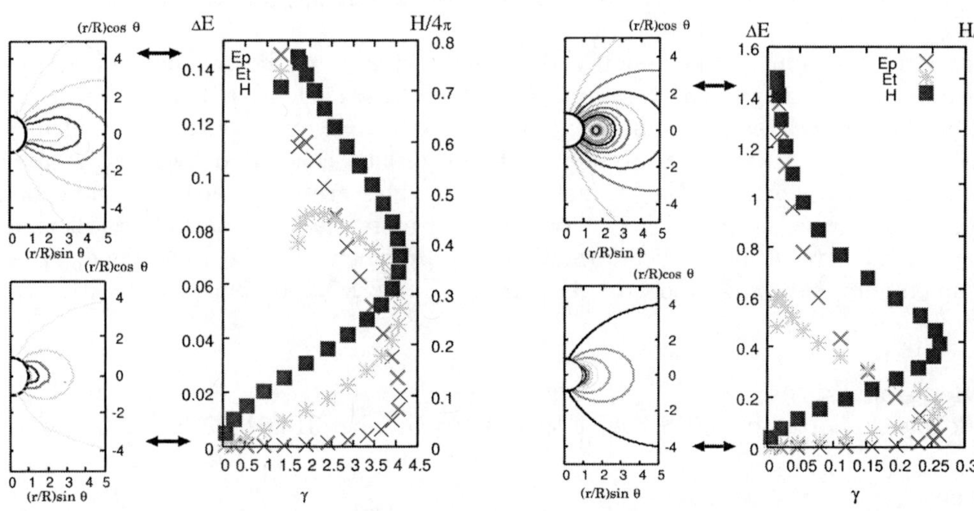

Figure 1. Comparison of models in flat (left panel) and curved spacetime (right panel). Increase in magnetic energy $\Delta E/(B_0^2R^3)$ from the potential dipole field is shown in the left scale and total relative helicity $H_{\rm R}/(4\pi B_0^2R^4)$ is shown in the right scale, where typical field strength B_0 is defined by magnetic dipole moment μ and the surface radius R, $B_0 \equiv \mu R^{-3}$. The poloidal component of the energy is denoted by crosses, the toroidal component by asterisks, and helicity by squares. The horizontal axis denotes the dimensionless value $\gamma(B_0R^2)^7$. Contour lines of magnetic function G in the r-θ plane are also displayed nearby: the potential dipolar field at the bottom and, the highly twisted state at the top.

better to compare maximum increment ΔE, in a ratio to E_0: $\Delta E/E_0 \approx 0.2$ for a model in a flat spacetime, $\Delta E/E_0 \approx 2$ for a relativistic model. There is still a big difference, which comes from the formation of the flux rope, a braid of poloidal and toroidal magnetic fields. A larger amount of energy is capable of being stored there.

4. Discussion

A family of axially symmetric static force-free magnetosphere models is calculated. Total magnetic energy increases with increasing helicity. Numerical calculations suggest that there is a maximum of the helicity deposited in the magnetosphere. Beyond this, we could not find any static solutions.

We have also compared models in flat and curved spacetimes. General relativistic effects are significant. A larger amount of energy is capable of being stored in the relativistic model. In extreme cases, the magnetic energy increases up to a few times larger than that of a current-free dipole. This marks a contrast with the models in flat spacetime, in which the increment of the magnetic energy is less than 20 percent. This big difference is related to the formation of an axially symmetric magnetic rope in the vicinity of the surface. The rope contains a large amount of magnetic energy constructed by poloidal and toroidal components. The flux rope is not inherent in relativistic magnetosphere, but possibly originates from some nonlinearities. Actually, this kind of structure is also realized in flat spacetime, but the condition to the power-law index n is severe. Namely, the structure is evident only in the model with stronger nonlinearity, $n > 7$[3]. Curvature due to general relativity helps the confinement. See also numerical models[4] obtained by solving relativistic MHD equations with a low-density plasma, in which an evident flux rope can be seen in their magnetosphere. Thus, relativistic model is preferable for larger energy events. At moment, it is impossible to determine the relativistic factor M/R of magnetar only by magnetosphere model. Further improvement of the model is necessary.

The existence of a maximum energy or helicity stored in the magnetosphere seems to be reasonable, and is important for a catastrophic event such as a flare. When the total helicity injected into the magnetosphere exceeds a certain capacity, a transient event may occur subsequently. The transition of the magnetosphere can be calculated by resistive dynamical simulation e.g., [5, 6]. After a burst, magnetic field lines are open with the same boundary condition at surface, after some energy transfers to mass ejection, particle acceleration and the radiation. The transition is therefore allowed, only when open field energy $E_{\rm open}$ is larger than the maximum energy $E_{\rm max}$ before the burst. The work to compare $E_{\rm max}$ with $E_{\rm open}$ is in preparation.

Acknowledgments

This work was supported by JSPS KAKENHI Grant Number JP26400276.

References

[1] Obergaulinger M Magnetic field amplification during core collapse *this volume*
[2] Kojima Y 2017 *Mon. Not. R. Astron. Soc.* **468** 2011
[3] Flyer N, Fornberg B, Thomas S and Low B C 2004 *ApJ* **606** 1210
[4] Pili A G, Bucciantini N and Del Zanna L 2015 *Mon. Not. R. Astron. Soc.* **447** 2821
[5] Parfrey K, Beloborodov A M and Hui L 2013 *ApJ* **774** 92
[6] Kojima Y and Kato Y E 2014 *Progress of Theoretical and Experimental Physics,* **2** 023E01

IOP Conf. Series: Journal of Physics: Conf. Series **932** (2017) 012017 doi:10.1088/1742-6596/932/1/012017

The influence of the positronium photoionization rate on the polar cap X-ray luminosity of radio pulsars

D P Barsukov[1,2] and M V Vorontsov[2]

[1] Ioffe Institute, Saint Petersburg, Russia

[2] Peter the Great St. Petersburg Polytechnic University, Saint Petersburg, Russia

E-mail: bars.astro@mail.ioffe.ru

Abstract. The influence of the positronium photoionization rate on the polar cap X-ray luminosity of old radio pulsars is considered. It is assumed that the polar cap is heated only by reverse positrons accelerated in the pulsar diode. It is supposed that the pulsar diode is in a stationary state with the lower plate located near the star surface (polar cap model) occupies all the pulsar tube cross section and operates in the regime of steady space charge by the limited electron flow. The influence of a small-scale magnetic field on the electric field inside the pulsar diode is taken into account. The reverse positron current is calculated in the framework of two models: rapid and gradual screening. To calculate the production rate of electron-positron pairs we take into account only the curvature radiation of primary electrons and its absorption in the magnetic field. It is assumed that some fraction of electron-positron pairs is created in a bound state (positronium). Later, such positroniums are photoionized by thermal photons from the polar cap.

1. Introduction

The influence of the generation of electron-positron pairs in a bound state (positronium) on the polar cap X-ray luminosity of radio pulsars has been thoroughly analyzed in many papers, see for example [1, 2]. In this paper we consider the influence of the positronium generation and its photoionization on the polar cap heating in the case of the inner gap with the free electron emission from the neutron star surface taking into account the presence of a small-scale surface magnetic field. The calculations are performed in two models based on extreme assumptions about the value of the reverse positron current.

2. Model

Let the neutron star have a radius $r_{\rm ns}$ and a dipolar magnetic moment $\vec{m}$ (its field at the magnetic pole is $B_{\rm dip} = 2m/r_{\rm ns}^3$). We assume also that a small-scale magnetic field with the strength $B_{\rm sc}$ and the characteristic scale $\ell \sim r_{\rm ns}/10$ is present on the polar cap. For simplicity, we suppose that $\vec{B}_{\rm sc}$ is parallel to the surface and the angle ϕ_Ω determines its direction, $0 < \phi_\Omega < \pi/2$, see figure 1 and figure 2 [3]. We consider only the case of the inner gap [4] and assume that the inner gap occupies all the pulsar tube cross section and resides as low as possible. Let us denote the altitudes of the inner gap lower plate (cathode) and the upper plate (anode) by $z_{\rm lo}$ and is $z_{\rm hi}$ respectively, see figure 1. In most cases, the inner gap resides exactly at the neutron star surface ($z_{\rm lo} = 0$), see [5] for details. We suppose that the inner gap is stationary and operates in the regime of a charge limited steady flow. For simplicity, we take into account the generation

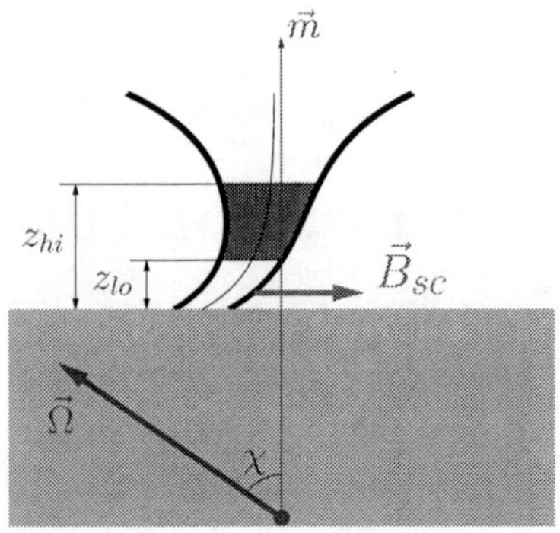

Figure 1. A sketch of the vicinity of the inner gap. Neutron star is shown by the gray color, boundaries of the pulsar tube are shown by black lines and the inner gap is shown by the brown color.

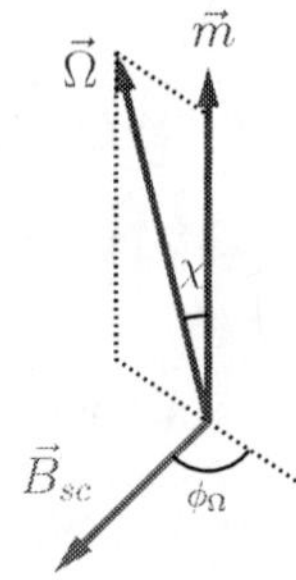

Figure 2. The definition of angles χ and ϕ_Ω.

of electron-positron pairs only by the curvature radiation of primary electrons in the magnetic field. We also take into account the generation of pairs in the bound state (positronium). For simplicity, we assume that the probability P_b that a pair is created in bound state is defined as follows: $P_b = 0$ if $B < B_{\rm low}$ (no positroniums are created), $P_b = (B - B_{\rm low})/(B_{\rm high} - B_{\rm low})$ if $B_{\rm low} \leq B \leq B_{\rm high}$ and $P_b = 1$ if $B > B_{\rm high}$ (all pairs are created in the bound state), where B is the magnetic field strength at the point of the pair creation, $B_{\rm low} = 0.04\,B_{\rm cr}$, $B_{\rm high} = 0.15\,B_{\rm cr}$, $B_{\rm cr} \approx 4.41 \times 10^{13}$ G [2]. In order to simplify the calculation, we assume that the pair generation and its properties do not depend on the photon polarization. The positronium decay is also neglected. However, we take into account the photoionization of positronium by thermal photons from the hot polar cap. The photoionization rate is estimated using the formula [2]

$$\frac{dN}{dt} = W_0 \left(\frac{10^2}{\Gamma}\right)^3 \left(\frac{T}{10^6 K}\right)^2 (1 - \cos\theta_{\rm cap}), \tag{1}$$

where Γ is the positronium Lorentz factor, T is the polar cap temperature, $\theta_{\rm cap}$ is the polar cap angular radius. The coefficient W_0 determines the photoionization rate. The question of its value is not completely solved yet [6, 7]: according to [2] it is equal to $W_0 = 6 \times 10^5$ s^{-1} and it is estimated as $W_0 \sim 10^8$ s^{-1} in [8]. In the latter case, the positronium is photoionized almost immediately and we can assume that all pairs are created in the unbound state [1]. Hence, in this paper we consider three cases: all pairs are created in the unbound state, $W_0 = 1.2\times10^6$ s^{-1} and $W_0 = 6\times10^5$ s^{-1}. The reverse positron current is calculated using two models: the model of rapid screening [9] according to which the electron-positron plasma screens the parallel electric field $E_{||} = (\vec{E}\cdot\vec{B})/B$ almost immediately and the model of gradual screening [10, 11], which allows the parallel electric field to penetrate deep into the electron-positron plasma, see details of calculations in [12].

3. Results

The dependence of the polar cap luminosity $L_{\rm pc}$ of the pulsar B0656+14 on the small-scale magnetic field strength $B_{\rm sc}$ under the assumption that all pairs are generated in the unbound state (no positroniums) is shown in figure 3, $B_{\rm dip} = 9.3\times10^{12}$ G, the pulsar period is $P = 0.385$ s,

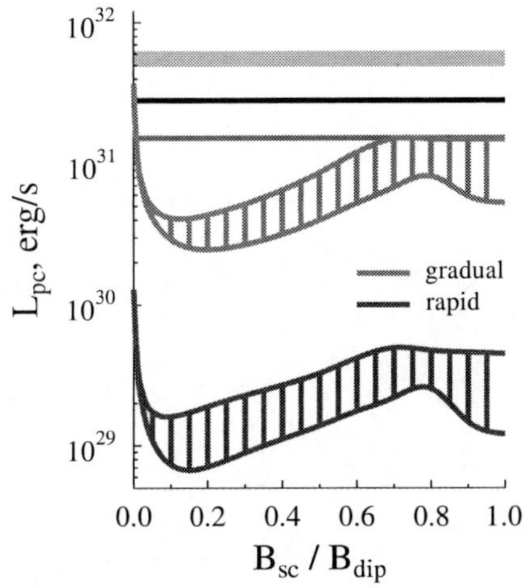

Figure 3. The polar cap luminosity $L_{\rm pc}$ of the pulsar B0656+14 under the assumption that all pairs are created in the unbound state (no positroniums). Blue dashed area corresponds to the reverse positron current calculated with the rapid screening model, red dashed area corresponds to the gradual screening model. Upper boundaries of both areas correspond to $\phi_\Omega = 0$, lower boundaries correspond to $\phi_\Omega = \pi/2$. Various observed values of $L_{\rm pc}$ are shown by the yellow color [15] or by black [16] and green [17] lines.

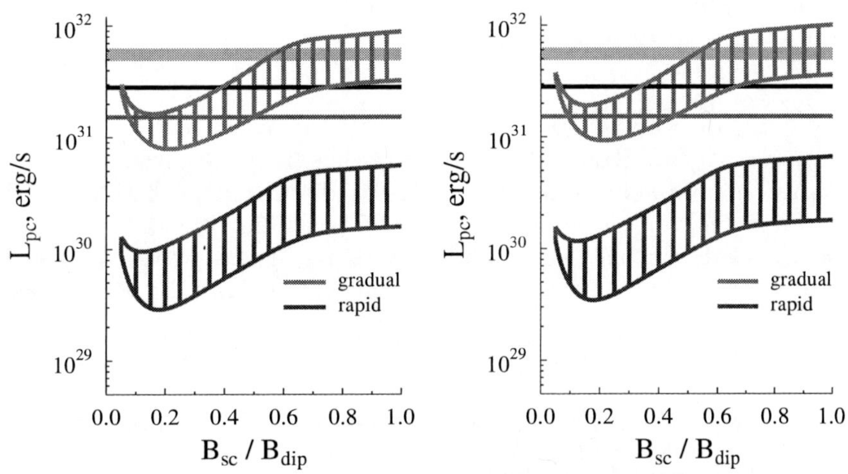

Figure 4. The same as in figure 3, but with positronium generation and photoionization taking into account. The case of $W_0 = 1.2 \times 10^6$ s^{-1} is shown on the left panel and the case of $W_0 = 6 \times 10^5$ s^{-1} is shown on the right panel.

its age is $\tau = 1.1 \times 10^5$ years, $\dot{E} = 3.8 \times 10^{34}$ erg/s [13] and the inclination angle is $\chi = 22°$ [14]. The polar cap luminosity $L_{\rm pc}$ obtained assuming that some fraction of pairs is created in the bound state is shown in figure 4. It is easy to see that graphs in figure 4 are almost the same and differ only slightly from figure 3. The dependence of $z_{\rm hi}$ and $z_{\rm lo}$ on the strength of the small-scale magnetic field $B_{\rm sc}$ for two orientations of the vector $\vec{B}_{\rm sc}$ is shown in figure 5. According to the used model, the altitude $z_{\rm lo}$ does not depend on the pair production at all [5]. However, the altitude of the upper plate $z_{\rm hi}$ has almost doubled after the positronium generation was taken into account. But it depends only slightly on the photoionization rate.

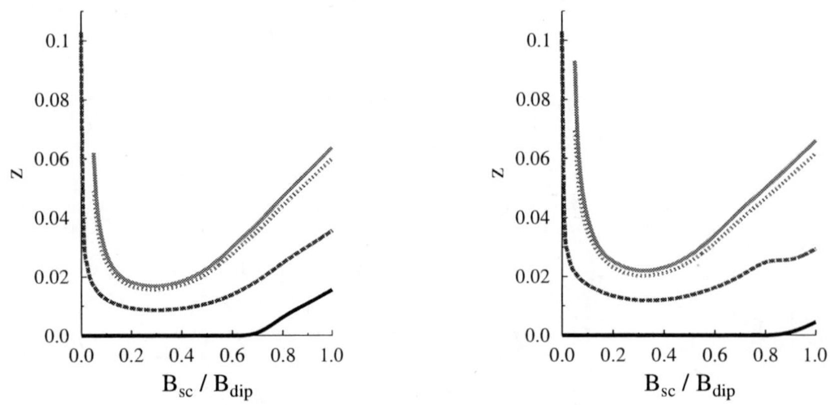

Figure 5. The altitudes z (in units of r_{ns}) of the pulsar diode plates at $\phi_\Omega = 0$ are shown on the left panel and at $\phi_\Omega = \pi/2$ on the right panel. Black line shows the altitude of the lower plate of the diode while blue line corresponds to the upper plate altitude if all pairs are created in the unbound state (no positroniums). Green and red lines correspond to the cases where the positronium generation is taken into account: the green one corresponds to $W_0 = 1.2 \times 10^6$ s^{-1} photoionization rate and the red one corresponds to $W_0 = 6 \times 10^5$ s^{-1}.

Acknowledgments
We sincerely thank A.I. Tsygan, K.Yu. Kraav, V.M. Kontorovich, D.A. Rumyantsev, D.N. Sobyanin, I.F. Malov and V.A. Urpin for help, comments and useful discussions, I.F. Malov, E.B. Nikitina for providing values of inclination angles, G.G. Pavlov for providing observational data, A.I. Chugunov, M.E. Gusakov, E.M. Kantor, Yu.A. Shibanov, D.A. Zyuzin, A.A. Danilenko, A.Yu. Kirichenko, M.A. Garasyov and D.M. Sedrakian for useful discussions and D.A. Baiko for help.

References
[1] Gil J, Melikidze G and Zhang B 2007 *Astrophysics and Space Science* **308** 325-33
[2] Usov V V and Melrose D B 1995 *Australian Journal of Physics* **48** 571-612
[3] Kantor E M and Tsygan A I 2003 *Astronomy Reports* **47** 615-22
[4] Shibata S 1991 *ApJ* **378** 239-54
[5] Barsukov D P, Polyakova P I and Tsygan A I 2009 *Astronomy Reports* **53** 1146-54
[6] Baring M G and Harding A K 2001 *ApJ* **547** 929-48
[7] Baring M G 2008 Photon Splitting and Pair Conversion in Strong Magnetic Fields *Computing Anticipatory Systems: CASYS'07-Eighth International Conference. AIP Conference Proceedings* **1051** 53-64
[8] Herold H, Ruder H and Wunner G 1985 *Phys. Rev. Letters* **54** 1452-55
[9] Arons J and Scharlemann E T 1979 *ApJ* **231** 854-79
[10] Harding A K and Muslimov A G 2001 *ApJ* **556** 987-1001
[11] Lyubarskii Yu E 1992 *A&A* **261** 544-50
[12] Barsukov D P, Goglichidze O A and Tsygan A I 2016 *Astronomy Reports* **60** 586-97
[13] Manchester R N, Hobbs G B, Teoh A and Hobbs M 2005 *Astron. J.* **129** 1993-2006 http://www.atnf.csiro.au/research/pulsar/psrcat, catalogue version 1.56
[14] Malov I F and Nikitina E B 2011 *Astronomy Reports* **55** 19-30
[15] Gil J, Haberl F, Melikidze G, Geppert U, Zhang B and Melikidze G Jr 2008 *ApJ* **686** 497-507
[16] Szary A 2013 Non-dipolar magnetic field at the polar cap of neutron stars and the physics of pulsar radiation *Preprint* astro-ph/1304.4203
[17] Pavlov G, Kargaltsev O, Posselt B and Durant M 2013 X-ray observations of rotation-powered pulsars *The Fast and the Furious: Energetic Phenomena in Isolated Neutron Stars, Pulsar Wind Nebulae and Supernova Remnants, held 22-24 May, 2013 in Madrid, Spain.* id 9

IOP Conf. Series: Journal of Physics: Conf. Series **932** (2017) 012018 doi:10.1088/1742-6596/932/1/012018

On the light-curve anomalies of radio pulsars

H L Hakobyan[1], A A Philippov[2,0], V S Beskin[3,4], A K Galishnikova[4], E M Novoselov[4] and M M Rashkovetskyi[4]

[1] Department of Astrophysical Sciences, Peyton Hall, Princeton University, Princeton, NJ 08544, USA

[2] Department of Astronomy, University of California Berkeley, Berkeley, CA 94720-3411

[3] P.N.Lebedev Physical Institute, Leninsky prosp., 53, Moscow, 119991, Russia

[4] Moscow Institute of Physics and Technology, Dolgoprudny, Institutsky per., 9, Moscow Region, 141700, Russia

E-mail: beskin@lpi.ru

Abstract. The wave propagation theory (Beskin and Philippov 2012 *MNRAS* **425** 814; Hakobyan *et al* 2017 *MNRAS* **469** 2704) allowed us to understand the main polarization properties of mean profiles. However, some radio pulsars indicate clear deviation from the theory predictions. Below we show how such anomalies can be explained within our theory without additional assumptions about orthogonal modes mixing.

1. Introduction

Quantitative theory of the radio waves propagation [1, 2] based on Kravtsov & Orlov approach allows us to describe general properties of mean profiles such as the position angle of the linear polarization *p.a.* and the circular polarization for the realistic structure of the magnetic field in the pulsar magnetosphere. The main theoretical prediction is the correlation of signs of the circular polarization, V, and derivative of the position angle with respect to pulsar phase, $\mathrm{d}p.a./\mathrm{d}\phi$: for extraordinary X-mode the signs should be the same, and for ordinary O-mode the signs should be opposite. In most cases it gives us the possibility to recognize the orthogonal mode, ordinary or extraordinary, playing the main role in the formation of the mean profile. On the other hand, there are some pulsars for which observations are in disagreement with above prediction. Below we demonstrate that complex behaviour of pulsar light curves and polarization profiles can be explained by including aberration effects into consideration.

2. Changeover effect

According to [1], the theory predicts O-X-O sequence where two orthogonal modes are detected in the three-component mean profile. However, some pulsars show polarization profiles that poorly fit into this simplified model. E.g., *p.a.* curve of pulsar PSR J2048−1616 shows one orthogonal mode whereas Stokes parameter V changes sign along the mean profile.

To understand this discrepancy, remember that according to [1] the sign of the circular polarization V is determined by the sign of the derivative $\mathrm{d}(\beta_B + \delta)/\mathrm{d}r$ along the ray r in the vicinity of escape radius $r_{\rm esc}$

$$r_{\rm esc} \sim 10^3 R \cdot \lambda_4^{2/5} \nu_{\rm GHz}^{-2/5}, \tag{1}$$

[0] Einstein Fellow

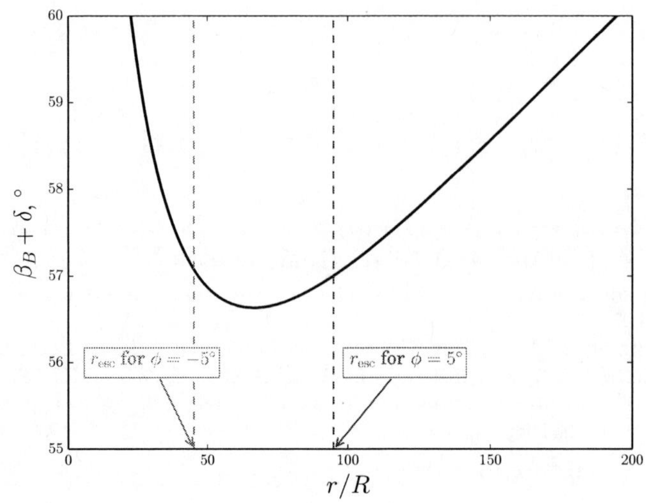

Figure 1. The plot of $\beta_B + \delta$ along the propagation ray. If the polarization is formed below the extremum, the sign is governed by the derivative of β_B, and when it's formed above — the sign is determined by the derivative of δ.

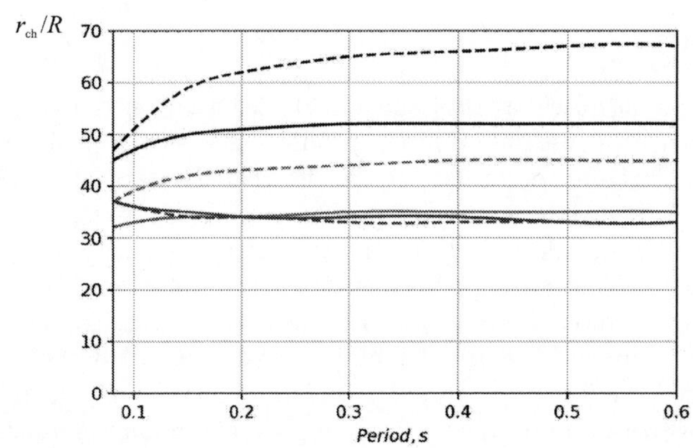

Figure 2. Changeover radius $r_{\rm ch}$ as a function of period P for different parameters. Black line corresponds to start parameters, red line corresponds to smaller radiation radius $r_{\rm rad}$, blue line corresponds to another inclination angle χ. Dashed lines correspond to another phase ϕ.

where the polarization characteristics cease to depend on the distance r from the neutron star. Here $\lambda = n_{\rm e}/n_{\rm GJ}$ (in units 10^4) is the multiplicity parameter and the angle β_R determines the orientation of the external magnetic field in the picture plane. Finally, aberration angle δ results from electric drift motion of particles in the pulsar magnetosphere:

$$\tan\delta = \frac{\cos\theta\, U_y/c}{U_x/c - \sin\theta}, \tag{2}$$

where U_x and U_y are two components of the $\mathbf{E} \times \mathbf{B}$ drift velocity, and θ is the angle between wave vector $\mathbf{k}$ and external magnetic field $\mathbf{B}$.

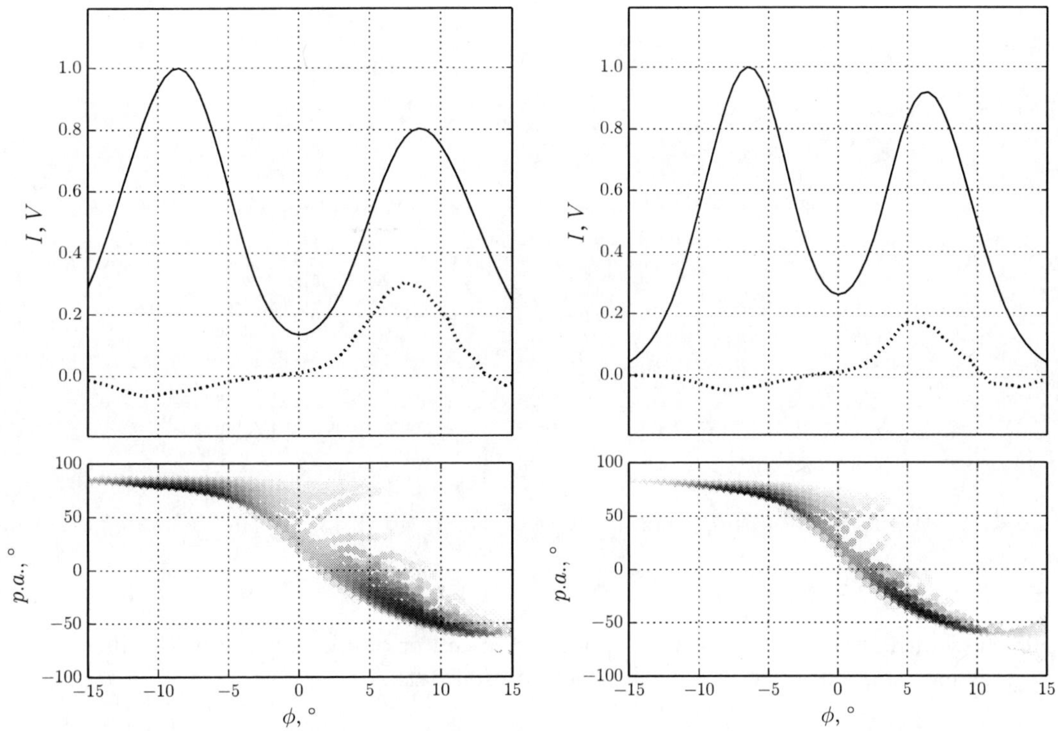

Figure 3. Simulated profiles for O-mode pulsar at two distinct frequencies 430 MHz (left) and 1414 MHz (right).

It is important to stress that the mode classification presented above corresponds to the case when it is the aberration angle δ that plays the leading role in the derivative $\mathrm{d}(\beta_B + \delta)/\mathrm{d}l$. As is shown on figure 1, it takes place at large enough distances $r > r_{\rm ch}$ where $r_{\rm ch}$ is so-called changeover radius. As is shown on figure 2, it weakly depends on pulsar's parameters:

$$r_{\rm ch} \approx (30 - 60)R. \tag{3}$$

But as the escaping radius $r_{\rm esc}$ (1) depends on frequency ν and, via the multiplicity parameter λ, on the number density $n_{\rm e}$, at high frequencies and at that part of the directivity pattern where in the vicinity of the escaping radius the number density of the outflowing plasma is low, the Stokes parameter V can change sign at some parts of the mean profile although radio emission corresponds to the same orthogonal mode (see figure 3). Thus, in reality the circular polarization of a given mode can switch its sign, without supposing the existence of another radiation mode or other effects.

3. Width of the emitting region

Analyzing the width of the *p.a.* curve obtained in [3] one can formulate a method which allows us to determine the radius-to-frequency mapping. Indeed, assuming that the width of this curve results from different radiation radii for the given frequency, we can evaluate the height and characteristic depth of the radiation region. To do that, we compare the results of our simulation with the corresponding observational data obtained in [3].

In figure 3 we show simulated mean profile corresponding to O-mode double profile. As one

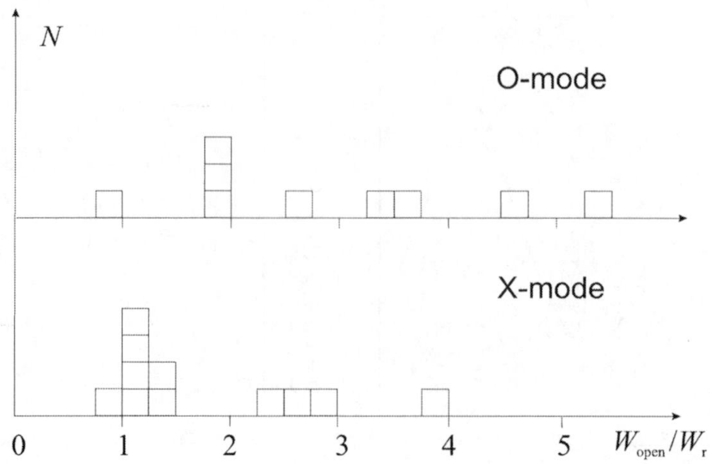

Figure 4. Distribution of pulsars taken from [5] on the ratio $W_{\rm open}/W_{\rm r}$ for X-mode and O-mode separately.

can see, the width of the $p.a.$ curve is slimmer in the center of a profile and wider near the pulse edges. This is in a good agreement with observational data.

4. 'Shift to the right' and O-mode directivity pattern

Finally, let us remember that aberration results in so-called 'shift to the right' effect, i.e., the $\delta\phi = 4\Omega r_{\rm rad}/c$ shift between the position of the maximum derivative $\mathrm{d}p.a./\mathrm{d}\phi$ and the center of the mean profile [4]. As was shown in [5], this effect can help us to determine the generation level $r_{\rm rad}$ which gives quite reasonable values.

Besides, knowing generation level $r_{\rm rad}$, one can evaluate the width of the mean profile as $W_{\rm r} \approx 2(\Omega r_{\rm rad}/c)^{1/2}$ and compare it with observable $W_{\rm obs}$. As is shown on figure 4, for X-mode pulsars the distribution of the ratio $W_{\rm obs}/W_{\rm cal}$ has a sharp maximum near unity, while for O-mode pulsars these ratios are much larger. This difference can be easily explained as the width of the O-mode pulsars is to be larger due to its refraction.

5. Concluding Remarks

In this work we demonstrate that complex behavior of pulsar light curves and polarization profiles can be explained by propagation theory developed in [1]. In particular, one can understand different signs V at different phases for the same polarization mode. In addition, our method provides direct information about the width of the radiating region.

Acknowledgments

This research was partially supported by the government of the Russian Federation (agreement No. 05.Y09.21.0018) and by Russian Foundation for Basic Research (Grant no. 17-02-00788).

References

[1] Beskin V S and Philippov A A 2012 *Mon. Not. Royal Astron. Soc.* **425** 814–40
[2] Hakobyan H L, Beskin V S and Philippov A A 2017 *Mon. Not. Royal Astron. Soc.* **469** 2704–19
[3] Hankins T H and Rankin J M 2010 *Astron. J.* **139** 168–75
[4] Blaskiewicz M, Cordes J M and Wasserman I 1991 *Astrophys. J.* **370** 643–69
[5] Rookyard S C, Weltevrede P and Johnston S 2015 *Mon. Not. Royal Astron. Soc.* **446** 3367–88

IOP Conf. Series: Journal of Physics: Conf. Series **932** (2017) 012019 doi:10.1088/1742-6596/932/1/012019

Analytical theory of neutral current sheets with a sheared magnetic field in collisionless relativistic plasma

V V Kocharovsky[1,2], **Vl V Kocharovsky**[2], **V Yu Martyanov**[3] **and A A Nechaev**[2]

[1] Dept. of Physics and Astronomy, Texas A&M Univ., College Station, TX 77843, USA
[2] Institute of Applied Physics of the RAS, Nizhny Novgorod, 603950, Russia
[3] Intel Corp., 5000 W Chandler Blvd, Chandler, AZ 85226, USA

E-mail: ant.a.nech@gmail.com

Abstract. We derive and describe analytically a new wide class of self-consistent magnetostatic structures with sheared field lines and arbitrary energy distributions of particles. To do so we analyze superpositions of two planar current sheets with orthogonal magnetic fields and cylindrically symmetric momentum distribution functions, such that the magnetic field of one of them is directed along the symmetry axis of the distribution function of the other. These superpositions satisfy the pressure balance equation and allow one to construct configurations with an almost arbitrarily sheared magnetic field. We show that most of previously known current sheet families with sheared magnetic field lines are included in this novel class.

1. Introduction

Quasistationary neutral current sheets are important elements of various structures in collisionless plasma [1, 2, 3, 4], including space plasma or the relativistic plasma of neutron stars' magnetosphere and wind. Most analytical studies of these structures are limited by planar configurations of magnetic field lines and ignore the effect of shear (see, e.g., [5, 6, 7]), although the latter appears naturally in the system 'rotating magnetosphere – outgoing wind' [3, 4].

In this report we present a new wide class of neutral self-consistent magnetostatic structures with sheared field lines using superpositions of planar current sheets with orthogonal magnetic fields for widely tunable momentum and arbitrary energy distributions of particles. We consider particle distributions as functions of generalized particle momenta and restrict ourselves to the case when the total particle distribution function expands into the sum of two cylindrically symmetric distribution functions with orthogonal axes. In this simple, yet poorly explored situation the equations of magnetostatics break down into two independent nonlinear Grad–Shafranov equations, each corresponding to a current sheet with a planar magnetic field.

Following our initial suggestion in [8], we analyze various superpositions of two neutral current sheets with orthogonal magnetic fields, in which the field of one of them is parallel to the current in the other. They satisfy the 'plasma + field' pressure balance equation and allow us to construct configurations with an almost arbitrarily sheared magnetic field. It turns out that one-dimensional periodic helical structures and localized current sheets exist with either monotonous magnetic field vector rotation, or with several switches in it.

We show that the major part of previously known current sheets with sheared magnetic field lines, e.g. [9, 10, 11, 12], are particular cases of this novel wide class of current sheets.

2. Analytical description of self-consistent current sheets with the magnetic shear

The analysis of stationary current structures in a collisionless plasma is based on the recently developed method of particle motion invariants, details on which could be found in [8]. Let all quantities depend on a single spatial coordinate x and the magnetic field $\vec{B}$ be orthogonal to the x-axis, so that it can be described with two components of the vector potential $\vec{A}$: $B_{z,y} = \pm\, dA_{y,z}/dx$. The basic fact is that in the case of a neutral plasma, with zero electrostatic potential, any distribution function of the form $f_\alpha = f_\alpha(\varepsilon, P_y, P_z)$ for particles of the sort α automatically satisfies the stationary kinetic equation. Here $\varepsilon = \sqrt{p^2 + m_\alpha^2 c^4}$ is the particle energy, $\vec{p}$ the particle momentum, c the velocity of light, $P_{y,z} = p_{y,z} + (q_\alpha/c)A_{y,z}$ are components of the generalized momentum, m_α and q_α are particle mass and electric charge.

For the distribution functions representable as the sum of two cylindrically symmetric ones,

$$f_\alpha = f_\alpha^{(y)}\left(\varepsilon, p_y + \frac{q_\alpha}{c}A_y\right) + f_\alpha^{(z)}\left(\varepsilon, p_z + \frac{q_\alpha}{c}A_z\right), \tag{1}$$

with orthogonal axes y and z, the equations of magnetostatics for the vector potential $\vec{A}$ break down into two independent equations of the Grad–Shafranov type:

$$\frac{d^2 A_{y,z}}{dx^2} = -\frac{dU^{(y,z)}}{dA_{y,z}}. \tag{2}$$

Here the functions $U^{(y,z)}(A_{y,z}) = 4\pi P_{xx}^{(y,z)} + \text{const}$ are called the Grad–Shafranov potentials and $P_{xx}^{(y,z)} = \sum_\alpha \iiint p_x v_x f_\alpha^{(y,z)}\, d^3\vec{p}$ is the xx-component of the pressure tensor, $v_x = p_x c^2/\varepsilon$.

Thus the use of simple superpositions (1) of two one-dimensional sheets, in which the magnetic field of one sheet is directed along the symmetry axis of the distribution function of the other, allows us to construct a wide class of self-consistent current sheets with a sheared magnetic field. Obviously, a planar external field may stand for the magnetic field of each sheet. The resultant solutions to (2), $\vec{B}\{0, B_y, B_z\}$, may have rather diverse shear spatial profiles, determined by the shape of the respective Grad–Shafranov potentials [8] (Fig. 1). In the following we describe only a few representative combinations (see Fig. 2–5).

The mechanical analogy for the Harris sheet [6], $B_{y,z} \propto \tanh((x - d_{1,2})/\lambda_{1,2})$, is the reflection of a point A' from the potential wall (Fig. 1b). Profiles of the magnetic field and its shear angle, $\psi(x)$, defined as the angle between the vector $\vec{B}$ and the positive z-axis, of the combinations of a pair of those sheets are shown in Fig. 2 for different relations of spatial scales, $\lambda_{1,2}$, and shift

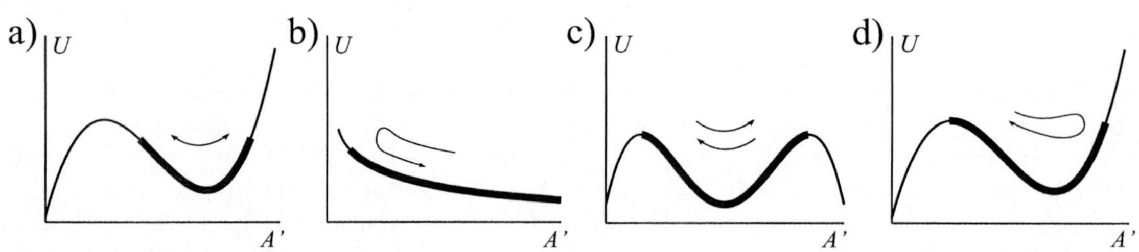

Figure 1. Various types of the Grad–Shafranov potential (see [8] for details) corresponding to a) periodic sheets; b) isolated sheets (as Harris sheet); c) double sheets (as Nicholson sheet); d) symmetrical shielded sheets. Arrows show the motion of a point A' in the potential wells.

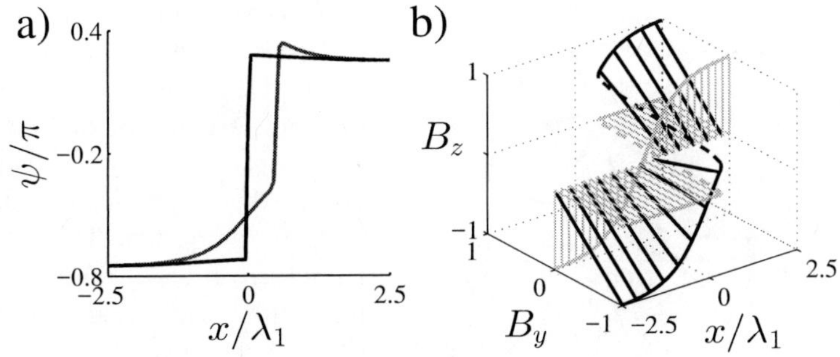

Figure 2. Current sheet of a type 'Harris + Harris', $B_{y,z} \propto \tanh((x - d_{1,2})/\lambda_{1,2})$:
a) the shear angle ψ/π: black curve — the ratio between the scales of two sheets, λ_2/λ_1, is 1.2, $d_1 = d_2$; blue — $\lambda_2/\lambda_1 = 20$, the sheets are shifted along each other by $d_1 - d_2 = 0.5\lambda_1$;
b) black — total magnetic field of a current sheet with parameters used for the blue curve in a), light blue and orange — fields of orthogonal sheets.

values, $d_{1,2}$. For the given parameters, with x ranging from $-\infty$ to ∞, vector $\vec{B}(x)$ switches its rotation direction once, gathering $\psi = \pi$ radians in total. Similar self-consistent structures with sheared field lines were found in [9, 10] in a different way and only for non-relativistic Maxwell distribution function. Note that in our solutions particle energy distributions may be arbitrary, and similar sheared structures may be obtained for spatial profiles different from the Harris one and defined by other Grad–Shafranov potential profiles $U^{(y,z)}(A_{y,z})$.

The 'motion' type of the vector potential in Fig. 1c constitutes a self-consistent structure with a localized field as in Nicholson sheet [7]: $B_z \propto \cosh^{-1}((x - d_2)/\lambda_2)$. Its combination with a Harris-type one is shown in Fig. 3 (cf. [11]), with another localized sheet — in Fig. 5a.

Combining two symmetrical shielded current sheets with $B_{y,z} \propto d/dx \cosh^{-2}((x-d_{1,2})/\lambda_{1,2})$, corresponding to Fig. 1d, one can obtain a localized structure with several switches in the rotation direction of $\vec{B}$ (Fig. 4). Finally, a periodic sheet is presented in Fig. 5b. It is force-free, i.e., $\vec{B}^2(x) \equiv \mathrm{const}$, and the pressure balance equation, following from (2), $P_{xx} + \vec{B}^2/8\pi = \mathrm{const}$, leads to $P_{xx} \equiv \mathrm{const}$. Other examples of force-free sheets, resembling Figs. 4a, 5b, can be found in [12]. A general classification of the considered sheared magnetostatic structures may be easily given on the basis of known classification [8] of the shearless current sheets.

3. Conclusions

We describe and analyze a new wide class of neutral current sheets with a widely tunable momentum and arbitrary energy distributions of particles as well as rather arbitrarily sheared

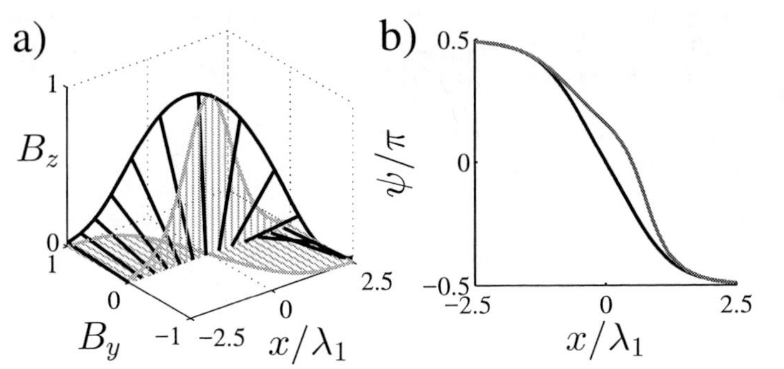

Figure 3. 'Harris + Nicholson' sheet, $B_y \propto \tanh((x - d_1)/\lambda_1)$, $B_z \propto \cosh^{-1}((x - d_2)/\lambda_2)$, $\lambda_2/\lambda_1 = 1.2$:
a) black curve — total magnetic field, light blue and orange — fields of orthogonal sheets with $d_1 = d_2$;
b) black — the shear angle for the parameters in a), red — for the shifted sheets, $d_1 - d_2 = 0.5\lambda_1$.

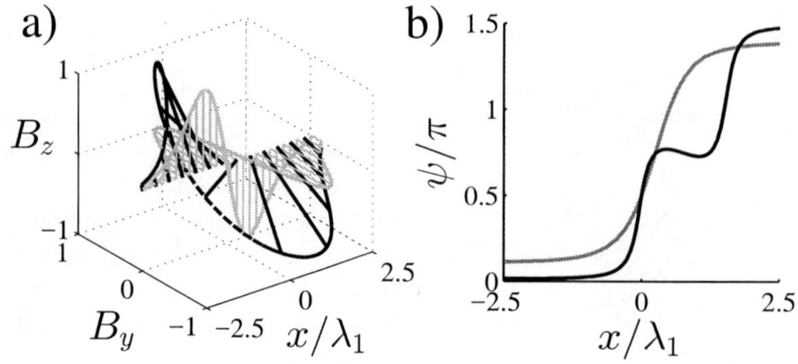

Figure 4. Superposition of two localized sheets with antisymmetrical magnetic field, $B_{y,z} \propto d/dx \cosh^{-2}((x - d_{1,2})/\lambda_1)$: a) black curve — total magnetic field, light blue and orange — fields of orthogonal sheets, $d_1 - d_2 = 0.5\lambda_1$; b) red curve — the shear angle for the parameters in a), black — for the increased shift, $d_1 - d_2 = 1.5\lambda_1$.

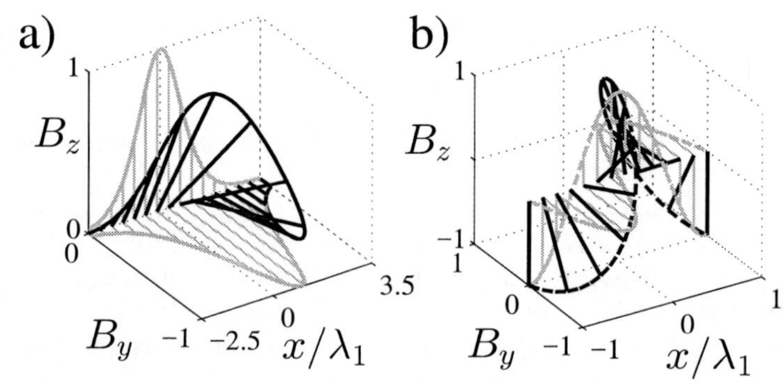

Figure 5. a) Two localized sheets, $B_{y,z} \propto \cosh^{-1}((x - d_{1,2})/\lambda_{1,2})$, $\lambda_2/\lambda_1 = 1.2$, $d_1 - d_2 = \lambda_1$; b) periodical force-free current sheet, $B_y \propto \sin(\pi x/\lambda_1)$, $B_z \propto \cos(\pi x/\lambda_1)$. Black curves — total magnetic fields; light blue and orange — fields of orthogonal sheets.

magnetic field. These sheets may have almost arbitrary spatial profile of the field vector's angle in the plane orthogonal to the axis of the sheet's inhomogeneity, including several switches in its derivative. We give analytical examples of sheared structures of this kind and demonstrate qualitatively that the suggested class of current sheets may be used to model quasistationary equatorial current structures in pulsar magnetospheres.

Acknowledgments
This work was supported by the Russian Science Foundation grant No. 16-12-10528.

References
[1] Komar C M, Fermo R L and Cassak P A 2015 *J. Geophys. Res. Space Physics* **120** 276
[2] Komar C M, Cassak P A 2016 *J. Geophys. Res. Space Physics* **121** 5105
[3] Tchekhovskoy A, Philippov A and Spitkovsky A 2016 *MNRAS* **457** 3384
[4] Cerutti B, Philippov A, Parfrey K and Spitkovsky A 2015 *MNRAS* **448** 606
[5] Komissarov S S, Barkov M and Lyutikov M 2007 *MNRAS* **374** 415
[6] Harris E G 1962 *Nuovo Cimento* **23** 115
[7] Nicholson R B 1963 *Phys. Fluids* **6** 1581
[8] Kocharovsky V V, Kocharovsky Vl V, Martyanov V Yu and Tarasov S V 2016 *Phys. Uspekhi* **59** 1165
[9] Alpers W 1969 *Astrophys. Space Sci.* **5** 425
[10] Channell Paul J 1976 *J. Phys. Fluids* **19** 1541
[11] Kan J R 1972 *J. Plasma Phys.* **7** 445
[12] Wilson F and Neukirch T 2011 *Phys. Plasmas* **18** 082108

IOP Conf. Series: Journal of Physics: Conf. Series **932** (2017) 012020 doi:10.1088/1742-6596/932/1/012020

Radiation reflection from star surface reveals the mystery of interpulse shift and appearance of high frequency components in the Crab pulsar

V M Kontorovich[1,3] **and S V Trofymenko**[2,3]

[1]Institute of Radio Astronomy NAS of Ukraine, Art st. 4, 61002 Kharkov, Ukraine
[2]NSC 'Kharkov Institute of Physics and Technology', Akademicheskaya st. 1, 61108 Kharkov, Ukraine
[3]Karazin Kharkov National University, Svoboda sq. 4, 61022 Kharkov, Ukraine

E-mail: vkont1001@gmail.com; trofymenko@kipt.kharkov.ua

Abstract. A new mechanism of radiation emission in the polar gap of a pulsar is discussed. It is based on the curvature radiation which is emitted by positrons moving towards the surface of neutron star along field lines of the inclined magnetic field and reflects from the surface. This mechanism explains the mystery of the interpulse shift and appearance of additional components in the emission of Crab pulsar at high frequencies discovered by Moffett and Hankins twenty years ago. We discuss coherence, energy flux and spectrum of the reflected radiation, appearance and disappearance of the interpulse position shift with the frequency increase. It is also possible that a nonlinear reflection (stimulated scattering) from the star surface is observed in the form of HF components. The frequency drift of these components, discovered by Hankins, Jones and Eilek, is discussed. The nonlinear reflection is associated with "Wood's anomaly" at the diffracted waves grazing along the star surface. Two components can arise due to slow and fast waves which are present in the magnetospheric plasma. The possible scheme of their appearance due to birefringence at the reflection is also proposed.

1. Introduction

Twenty years later after the discovery of interpulse shift and detection of the high frequency (HF) components in the radiation of the Crab pulsar by D Moffet and T Hankins [1], T Hankins, G Jones and J Eilek returned to observation of these phenomena at even higher frequencies [2]. As one of the results, a frequency drift of HF components has been discovered. The detailed analysis of physical results and problems is discussed in the authors' review [3]. The problems of interpulse shift and HF components appearance are marked there as unresolved.

The figure (figure 1) from the paper [1] by Moffett and Hankins, 1996 shows that at around 3 GHz there appears a phase shift δ of interpulse (IP) of about 7 degrees comparing to the initial IP positon at lower frequencies. Moreover, two more distinct pulses – high-frequency components [2] – appear at the same frequencies [1, 2]. In our preprint [4] (and subsequent papers [5, 6]) and in the current work we propose the explanation of these phenomena applying the idea of the radiation reflection from the neutron star surface by returning positrons moving in the polar gap region toward the surface of the star in the inclined magnetic field (figure 2).

Presently we consider in detail the curvature radiation (CR) reflected from the pulsar surface.

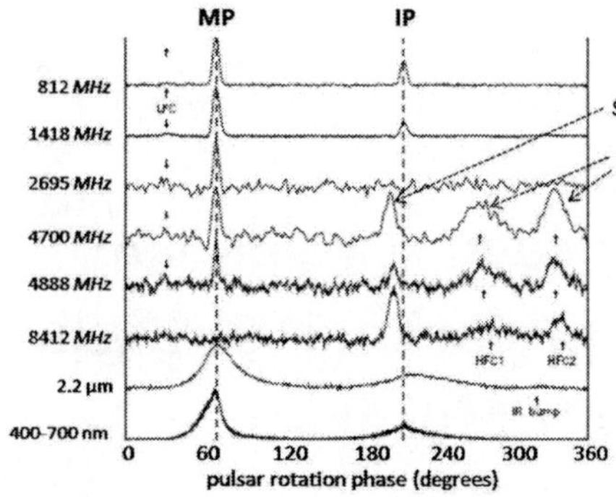

Figure 1. Fragment from the average light curves obtained at multi-frequency observations of Crab pulsar in [1]. The shifted (to about 7^0) interpulse and high-frequency components are marked. By the courtesy of the authors. ©AAS. Reproduced with permission.

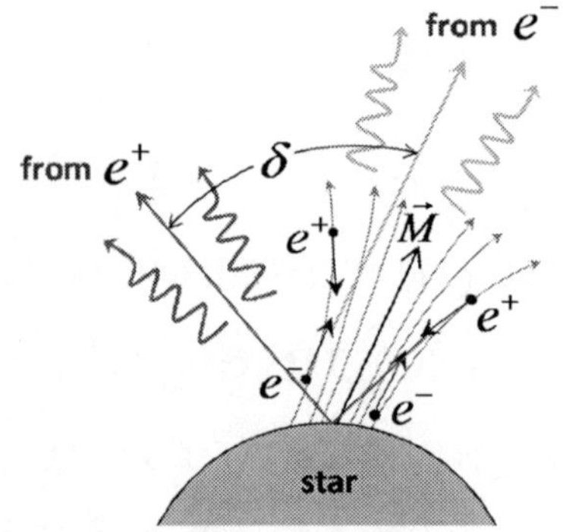

Figure 2. Schematic picture of motion and radiation by electrons and positrons in the polar gap of the pulsar in the case of a tilted magnetic axis. The directions of radiation by electrons and reflected radiation by positrons are shifted at the angle of mirror reflection

We show that such mechanism of (linear, mirror) reflection also provides the possibility of explaining the intensity of IP coherent radiation, its energy flux and spectrum.[1]

2. Estimation of the reflected radiation flux forming the shifted IP

The positrons moving towards the surface of the pulsar in the polar gap can be returned from the lower layers of the magnetospheric plasma by the same electric field which accelerates the electrons outwards the star. We estimate the flux of the reflected radiation above the entire area of the polar cap. The observations indicate the necessity of coherent character of radiation (see the recent review on the problem, [8]). We make estimation taking into account this fact which is discussed in detail in the Appendix of [6]. For the estimation of CR intensity and derivation of the radiation energy spectrum at first it is enough to accept the very fact of existence of inhomogeneous (clumped) positron flux in the gap. In this case the main contribution to radiation is made by coherently radiating volumes, which we estimate further. The number of such volumes can be rather roughly estimated through 'division' of the total radiating volume by the coherence volume. The dependence of such volume on the wavelength, Lorenz-factor and the curvature of the positrons trajectories is taken into account.

The radiation spectral density at frequency ω produced by a single positron, which moves along an arc of a circle of the (curvature) radius R per unit path can be estimated as $W_\omega \sim \left(\omega R/c\right)^{1/3} e^2/2\pi Rc$. It directly follows from the known expression for spectral density of radiative energy loss by a particle moving along a circular trajectory.

The size of a volume which encloses amount of particles radiating coherently we take as $V_{coh} = r_{||} r_\perp^2 \sim \gamma^2 \lambda^3/(4\pi^3)$, where the distances $r_{||} \sim \lambda/\pi$ and $r_\perp \sim \gamma\lambda/(2\pi)$ define the linear dimensions of such coherently radiating volume both in the direction of the particles motion and in the perpendicular one. It is the square of the average number $N_{coh} \sim \kappa \times n_{GJ} V_{coh}$ of positrons in such volume that radiation flux produced by these particles is proportional to. Here $n_{GJ} \sim 10^{11} cm^{-3}$ is the Goldreich-Julian density and $\kappa < 1$ is some coefficient defining the

[1] The idea of (nonlinear) reflection from the pulsar surface is applied in [5] for consideration of high frequency components (HFCs) since they appear at the same frequencies as the shifted IP [1, 2]. (For the independent treatment of the HFCs, not connected with the IP shift, see the article of S. Petrova [7]).

IOP Conf. Series: Journal of Physics: Conf. Series **932** (2017) 012020 doi:10.1088/1742-6596/932/1/012020

difference of n_{GJ} from the real particle density in the returning positron flux. The amount of the volumes V_{coh} which lie within the surface element $2\pi r dr$ of the polar cap at certain altitude is then $dN_\perp \sim 2\pi r dr/r_\perp^2 = (2\pi)^3 r dr/\gamma^2\lambda^2$. The upper estimation for the number of such volumes falling on the polar cap surface segment of the order of $r_\perp^2$ per unit of time is then $dN_{||}/dt \sim c/\lambda$. The effective solid angle which encloses the considered radiation emitted by positrons with the Lorentz-factor γ can be estimated as $\Omega_{eff} \sim \pi/\gamma^2$ that gives for the effective area on the distance d from the star: $S_{eff} \sim \pi d^2/\gamma^2$.

Summarizing the previous considerations we can present the expression for the flux of the reflected radiation by positrons as $J(\omega) \sim \int dz dN_\perp \times dN_{||}/dt \times W_\omega N_{coh}^2/S_{eff}$. One can express the positron coordinate through its Lorenz-factor as $dz = \sqrt{mc^2\bar{h}/2eE(r)}\, d\gamma/\sqrt{\gamma}$ at the linear growth of accelerating electric field strength in the actual region of the trajectories contributing to the coherent radiation flux. Here $\bar{h}$ is an averaged length of the positron trajectory interval of the gap size order, $z = R\alpha$ (see figure 4) is the positron coordinate along the magnetic field line and $E(r) = E_0(1 - r^2/R_{PC}^2)$ is the accelerating electric field strength, vanishing on the polar cap boundary R_{PC}.

The dependence of the curvature radius of the magnetic field lines on r in the vicinity of the star surface in the case of a dipole magnetic field is described by the expression: $R(r) = 4R_*^2/3r$, where $R_* \sim 10^6$ cm is the pulsar radius. Substitution of the explicit expressions for the quantities presented here gives the following:

$$J(\omega) \sim \frac{\kappa^2 n_{GJ}^2 e^2 \lambda^{3-1/3}}{2^{5/3}\pi^{14/3} d^2 R_*^{4/3}} \sqrt{\frac{mc^2\bar{h}}{2eE_0}} \times \int_0^{R_{PC}} \frac{dr r^{5/3}}{\sqrt{1 - r^2/R_{PC}^2}} \int_{\gamma_{min}(r,\omega)}^{\gamma_{max}} d\gamma \gamma^{7/2}. \tag{1}$$

γ_{max} is an effective value of the positron Lorentz-factor at which its radiation reflected from the surface of the star ceases to hit the telescope (figure 3). Here γ_{min} is of the order of 10^2. We assume that the initial angular width of the positron radiation diagram which is $\sim 1/\gamma_{min}$ exceeds the value of the minimal angle between the magnetic axis and the direction to the telescope. With the increase of the positron Lorentz-factor at lower altitudes the characteristic angle of its radiation diagram becomes less than θ_{min} (at $\gamma \sim \gamma_{max}$) and radiation ceases to be caught by the telescope.

The final estimation of the total reflected positron radiation flux for $\gamma_{min} \ll \gamma_{max}$ is the following:

$$J(\omega) \sim \frac{\kappa^2 n_{GJ}^2 e^2 \lambda^{3-1/3}}{40\pi^{14/3} d^2 R_*^{4/3}} \sqrt{\frac{mc^2\bar{h}}{2eE_0}} \gamma_{max}^{9/2} R_{PC}^{8/3}. \tag{2}$$

The electric field strength E_0 on the magnetic axis can be estimated as $E_0 \sim a\Omega R_*/cB \sim 10^7$ CGSE, where $\Omega = 2\pi/T$ is the angular frequency of the pulsar rotation, $B \sim 10^{11}$ Gs is the magnetic field strength and $a \sim 10^{-1} \div 10^{-2}$ is a small parameter which takes the effect of geometrical and gravitational factors into account (see [9]). Then, the estimation of the flux reduces to $J(\omega) \sim \kappa^2\lambda^{3-1/3}\gamma_{max}^{9/2} \times 10^{-40}$ W/(Hz×m^2) (at $R_{PC} = 10^4$ cm, $\bar{h} = 10^4$ cm, $d = 6 \times 10^{21}$ cm). As this expression shows, the parameter γ_{max} can be approximately estimated by the comparison of the calculated flux value with the observational radio emission of the Crab pulsar. Quite a good agreement is achieved if $\gamma_{max} \sim 10^3$ and $\kappa \sim 10^{-1}$ are chosen.

The expression also shows that in the interval of wavelengths considered here the radiation energy spectrum is defined by the term $\lambda^{3-1/3} \approx \lambda^{2.7}$, which is also in a quite good qualitative agreement with observational results (see refs in [6]).

Let us note that for $\lambda/L \ll 1$ the estimation $dN_{||}/dt \sim c/L$ might be more preferable. Here L is a spatial size of longitudinal inhomogeneities in the positron flow (for details see Appendix in [6]). In this case radiation spectrum is defined by the term $\lambda^{3.7}$ (instead of $\lambda^{2.7}$), which seems to be in a better accordance with the observations (see refs in [6]).

3. Estimation of the highest frequency at which the interpulse shift takes place

In the framework of our model the the positron Lorenz-factor value γ_{min} at the beginning of the effective path defines the minimal frequency ω of its radiation which can reflect from the surface and hit the telescope. The corresponding relation between γ_{min} and ω can be presented as $\omega \sim c\gamma_{min}^3/R$. The value of γ_{min} can be expressed through radiated frequency $\nu = \omega/(2\pi)$ and the distance r from the magnetic field axis as $\gamma_{min}(\nu, r) \sim \left(8\pi\nu R_*^2/3cr\right)^{1/3}$. Here we see that γ_{min} grows with the increase of ν and decrease of r. Therefore, at some value of $r = r_0(\nu)$ the magnitude of γ_{min} reaches the γ_{max} and the internal integral over γ in (1) tends to zero. In the region of polar cap $r < r_0$ in this case the considered radiation mechanism does not work. The values of r at which the moving positrons do not contribute to the observed radiation can be estimated as $r < r_0(\nu) \sim 8\pi\nu R_*^2/3c\gamma_{max}^3$.

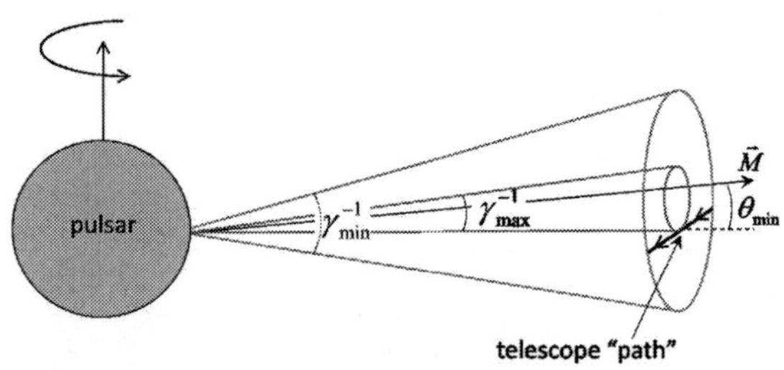

Figure 3. Schematic picture of angular regions (cones) of concentration of the reflected radiation by positrons at two extreme values of the considered Lorenz-factors. At $\gamma = \gamma_{max}$ radiation ceases to hit the telescope. Among other reasons is that the magnetic axis does not coincide with the ones of the cones due to its assumed inclination with respect to the surface normal

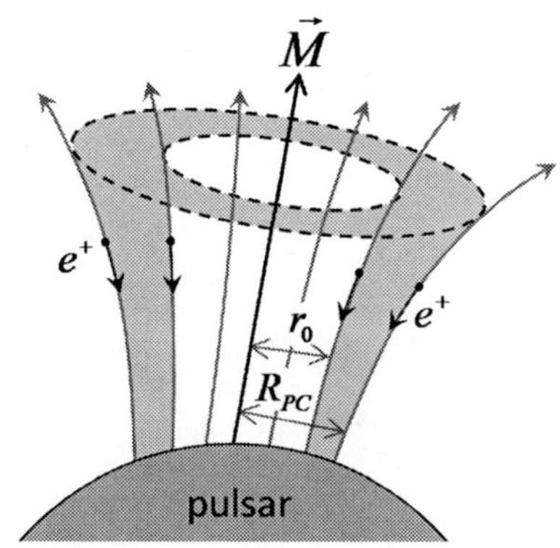

Figure 4. Region of space (orange hollow cone) contributing to the positron coherent radiation flux at $\nu < \nu_{max}$

Thus, the coherence leads to the formation of a hollow cone, restricted (in the vicinity of the star surface) from the inner and outer sides by the field lines situated on distances r_0 and R_{PC} respectively from the magnetic axis (figure 4). The frequency $\nu = \nu_{max}$, at which r_0 reaches the radius R_{PC} of the polar cap, is the one at which the considered radiation mechanism providing the interpulse shift disappears. It can be estimated as $\nu_{max} \sim 3cR_{PC}/(\gamma_{max}^3 8\pi R_*^2)$. Note that more accurate estimation of this frequency requires application of self-consistent models of the positron flow motion in the polar gap.

4. Stimulated Scattering and HF components

Appearance of the HF components in the same frequency range, as that in which the interpulse shift [1] takes place, permits us also to associate these components with reflection [4] from the neutron star surface. (See more detailed consideration of the problem with coherence and spectrum discussion in the Appendix of [6]). In the non-linear reflection model [5] (which is the model of induced Raman scattering by surface waves), the frequency shift of the HF

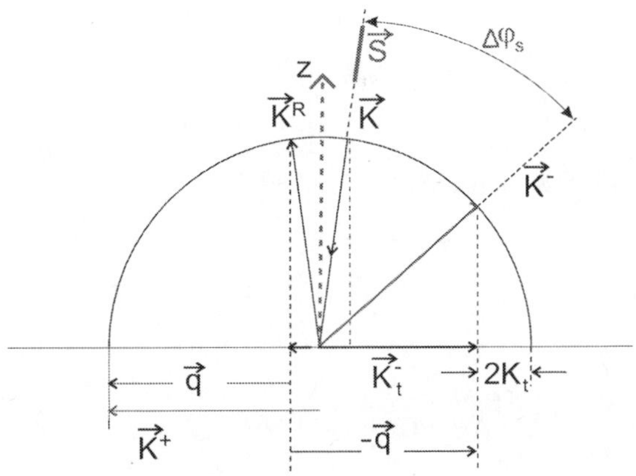

Figure 5. A scheme of the appearance of combinational Raman spectra of the first order at the IP (S pole) at stimulated scattering of radiation by returning positrons

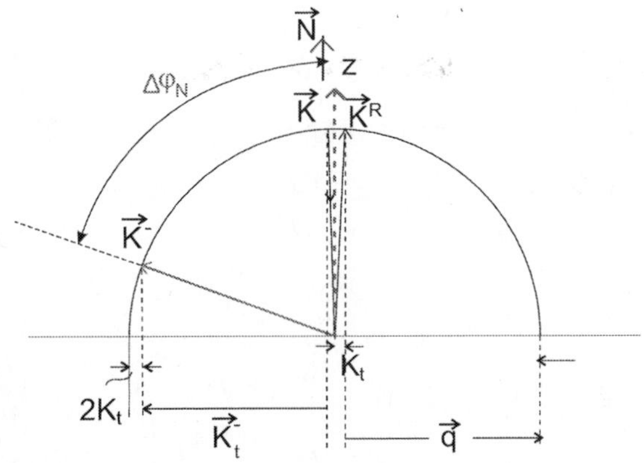

Figure 6. Appearance of the first-order Raman spectra at the MP (N pole). Here the direction of the phase shift is inverse compared to the S pole (schematically).

components appears in the natural way, regardless of the type of surface waves that induce the Raman scattering. (An independent treatment of the HFCs, not connected with the IP shift, see [7]).

5. Wood's anomaly in Mandel'stam-Raman electromagnetic fields

In our consideration it is only important that the reflection is associated with "Wood's anomaly" involving the diffracted waves grazing along the star surface (see [11, 12, 13] and refs in [5]). The grazing diffraction component has the largest amplitude with a large dielectric constant of the reflecting medium and makes the greatest contribution to the ray pressure swinging up the surface.

The corresponding estimates for the combinational reflected fields have the form:

$$E_{\pm} \approx (k\zeta)E_0, k_z \neq 0 \quad ; \quad E_{\pm} \approx \sqrt{\epsilon}(k\zeta)E_0, k_z = 0 \quad \epsilon \gg 1 \quad . \tag{3}$$

Here ζ is the amplitude of a surface Wood's wave and E_0 is the amplitude of the electromagnetic wave, generated by positrons, incident on the star surface.

In a conducting medium with complex dielectric permittivity, a "Wood's" resonance of a grazing diffracted component with the surface electromagnetic wave is possible. Since this corresponds to a small (of the surface impedance order) value of k_z/k, the kinematics presented here is practically the same. Dynamic conditions can differ significantly [14]. For the parameters of the neutron star in Crab with strong magnetic field, the surface may apparently consist of iron and have a sharp boundary (by courtesy of A.Potekhin, see refs in [15]).

In accordance with the scheme shown in figure 7, the change in frequency $\delta\omega$ leads to a change in the wave numbers, the tangential components of the wave vectors and angles in Raman scattering by the Wood's wave $q + \delta q$ (θ is the azimuth angle of the inclined magnetic moment):

$$\delta k = \delta\omega/V_g, \, \delta k_t = \delta k \sin\theta, \delta k_t^- = \delta k(1 - 2\sin\theta).$$

Here V_g is the group velocity of EM wave, angle θ^- in the equatorial plane (differing by a small angle from the "magnetic" one, where both the magnetic poles and the star center lie), indicates the position of the 1-st HF-component HFC1 corresponding to the direction of the diffraction

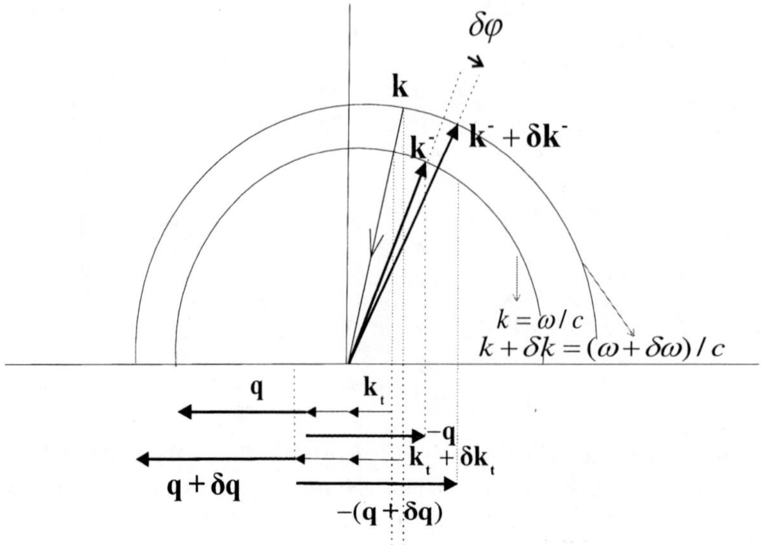

Figure 7. A scheme of drift of HFC1 position with the increase of frequency due to changing of the wave number of EM waves in the stimulated scattering model

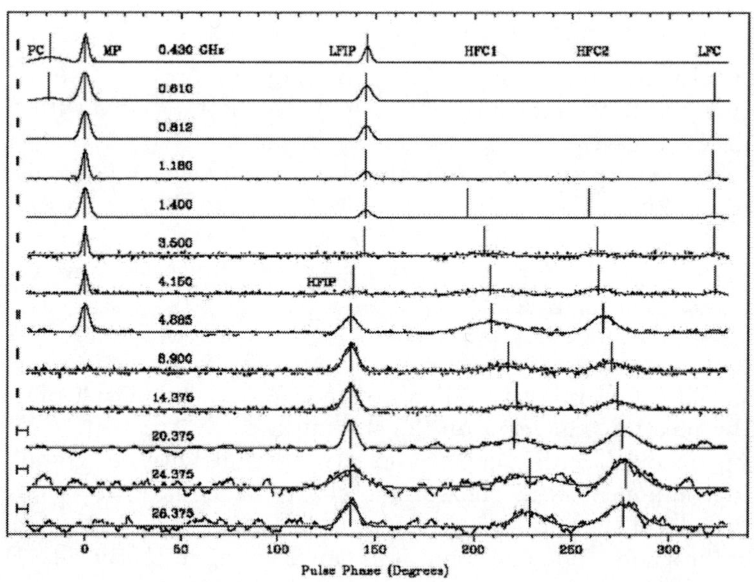

Figure 8. Frequency drift of the HF components. We also see the absence of the MP in this region and very wide HF pulses. By courtesy of the authors of [2]. ©AAS. Reproduced with permission.

maximum on a periodic structure, created by Wood's wave at frequency ω of EM wave emitted by returning positrons. Angle $\theta^{-} + \delta\varphi$ indicates the position of the same HF-component at

IOP Conf. Series: Journal of Physics: Conf. Series **932** (2017) 012020 doi:10.1088/1742-6596/932/1/012020

frequency $\omega + \delta\omega$. Other definitions are clear from the figure 7:

$$\sin(\theta^- + \delta\varphi) = (k_t^- - \delta k_t^-)/(k + \delta k).$$

For drift $\delta\varphi$ of the HFCI, an expression and an estimate in agreement with [2] follows:

$$\delta\varphi = \frac{\delta\omega}{kV_g} \frac{1 - 2\sin\theta - \sin\theta^-}{\cos\theta^-} \simeq \frac{\delta\omega}{\omega} \quad (4)$$

6. Drift of HF components

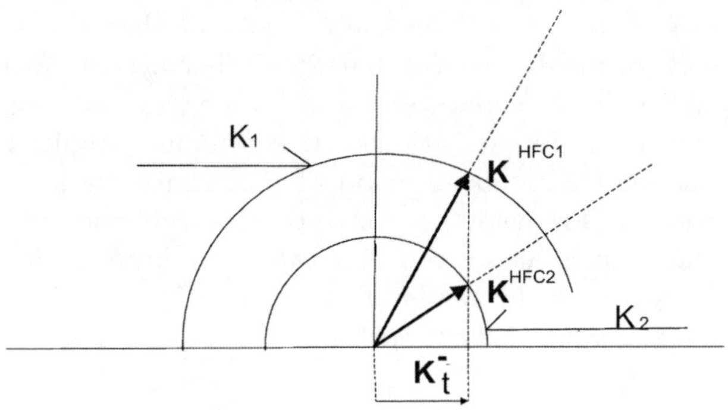

Figure 9. Two components can arise due to slow (k_1) and fast (k_2) waves, which are present in the magnetospheric plasma. Only the reflected Stokes waves are shown on the scheme. In reality, the waves, most likely, should arise not directly at reflection, but during the propagation process of the reflected Stokes wave in the anisotropic plasma.

In [5] it has been supposed that different HF components are produced by the opposite magnetic poles. An unidirectional drift, which is intrinsic to both components [2], stipulates one to associate them with the single magnetic pole of the star and, perhaps, with the birefringence in the magnetosphere. The observations to be explained, the second component should match over-horizon spread.

7. Summary

It is shown that the discussed new radiation mechanism can be applied for explanation of such unusual fact of the Crab pulsar radio emission as the shift of its interpulse at several GHz frequency. For this the assumption of the inclined magnetic field is used and returning positron flow is considered. The total flux of the coherent positron radiation is estimated and it (as well as the radiation spectrum) agrees with the results of observation quite well. In the framework of the proposed model the highest frequency at which the interpulse shift takes place is also roughly estimated.

The radiation of high-frequency components of the pulsar in the Crab Nebula can be considered also as a manifestation of an instability in the nonlinear reflection from the neutron star surface [5]. The discussed instability is a stimulated scattering by surface waves, predicted more than forty years ago and still observed nowhere and by no one. Drift of HFCs has a natural explanation in the framework of the stimulated scattering model.

The frequency drift of the components is very important in choosing the correct theoretical model. Particularly, the coincidence of its direction for both components is an argument in favor of the birefringence of the scattered wave in the anisotropic magnetized pulsar plasma. Returned motion of positrons, arising due to penetration of accelerating electric field of the gap in the pair plasma, has been considered in literature in connection with surface heating by the reverse current (see refs in [10]). The difference of a magnetic field from dipole one, which in particular leads to its slope, is also discussed with regard to its toroidal component as well (see refs ibidem). However, the low-frequency radiation of backflow positrons and reflected radiation from the surface of the pulsar have not been considered anywhere before except our papers.

Reflection leading to HF components formation may also be associated with the static periodic structures ("frozen" surface waves) in strong constant electric and magnetic fields [16, 17]. This possibility, as well as the drift of the components in these models, possibly associated with a change in the angle of incidence of radiation at the surface during the positron motion along the force line, and requires a separate study. The explanation of the large width of the HF components also requires an investigation of the dynamic conditions of the surface waves excitation. Possibly, a nonmonochromatic beam of positron radiation incident on the surface should be taken into account. However, the existence of a wide spectrum of the excited surface waves under Wood's conditions is also a consequence of the purely kinematic conditions when the considered waves do not lie in the plane of incidence. The interaction of these waves with each other, generating an analogue of wave turbulence, can also have a significant effect.

8. Acknowledgments

We are grateful to E Yu Bannikova, M Klochko, I Semenkina, N F Shul'ga, D P Barsukov, O M Ulianov, D M Vavriv and V V Zakharenko for help and useful discussion. The work of one of the authors (S V Trofymenko) was partially supported by the project 0117U004866 of the Ministry of Education and Science and by the project CO-1-8/2017 of the NAS of Ukraine.

References

[1] Moffett D and Hankins T 1996 *Astrophys. J.* **468** 779; arXiv:astro-ph/9604163
[2] Hankins T H, Jones G and Eilek J A 2015 *Astrophys. J.* **802** 130; arXiv:1502.00677v1
[3] Eilek J and Hankins T *Journal of Plasma Physics* **82**, article ID 635820302 (2016); arXiv:1604.02472
[4] Kontorovich V M and Trofymenko S V 2016 *AAPS* **7**; arXiv:1606.02966v2
[5] Kontorovich V M 2016 *Low Temperature Physics* **42** 672; arXiv:1701.02304v2
[6] Kontorovich V M and Trofymenko S V 2017 *JPSA* **7** 11; arXiv:1707.01584; contains more detailed bibliography
[7] Petrova S A 2010 *Radio Phys. and Radio Astron.* **1** 27
[8] Melrose D B 2017 *Rev. Mod. Plasma Phys.* DOI 10.1007/s41614-017-0007-0
[9] Beskin V S 2010 *Flows in Compact Astrophysical Objects* (Springer)
[10] Barsukov D P, Goglichidze O A and Tsygan A I 2016 *Astronomy Reports* **60** 586
[11] Gavrikov V K, Kats A V and Kontorovich V M 1969 *Sov. Phys. Doklady* **14** 564
[12] Gavrikov V K, Kats A V and Kontorovich V M 1970 *Sov. Phys. JETP* **31** 708
[13] Tymchenko M, Gavrikov V K, Spevak I S, Kuzmenko A A, and Kats A V 2015 *Appl. Phys. Lett.* **106** 261602
[14] Kats A V and Maslov V V 1972 *Sov. Phys. JETP* **35** 264
[15] Potekhin A Yu 2010 *Physics-Uspekhi* **180** 12 1279
[16] Shliomis M I 1974 *Sov. Phys. Uspekhi* **17** 153
[17] Kuznetsov E A and Spector M D 1976 *Sov. Phys. JETP* **44** 136

IOP Conf. Series: Journal of Physics: Conf. Series **932** (2017) 012021 doi:10.1088/1742-6596/932/1/012021

Hard X-ray quiescent emission in magnetars via resonant Compton upscattering

M G Baring[1], Z Wadiasingh[2], P L Gonthier[3] and A K Harding[4]

[1] Department of Physics and Astronomy, Rice University, Houston, TX 77251, U.S.A.

[2] Centre for Space Research, North-West University, Potchefstroom, South Africa

[3] Hope College, Department of Physics, 27 Graves Place, Holland, MI 49423, U.S.A

[4] Astrophysics Science Division, Code 663 NASA's Goddard Space Flight Center, Greenbelt, MD 20771, U.S.A.

E-mail: baring@rice.edu

Abstract. Non-thermal quiescent X-ray emission extending between 10 keV and around 150 keV has been seen in about 10 magnetars by RXTE, INTEGRAL, Suzaku, NuSTAR and Fermi-GBM. For inner magnetospheric models of such hard X-ray signals, inverse Compton scattering is anticipated to be the most efficient process for generating the continuum radiation, because the scattering cross section is resonant at the cyclotron frequency. We present hard X-ray upscattering spectra for uncooled monoenergetic relativistic electrons injected in inner regions of pulsar magnetospheres. These model spectra are integrated over bundles of closed field lines and obtained for different observing perspectives. The spectral turnover energies are critically dependent on the observer viewing angles and electron Lorentz factor. We find that electrons with energies less than around 15 MeV will emit most of their radiation below 250 keV, consistent with the turnovers inferred in magnetar hard X-ray tails. Electrons of higher energy still emit most of the radiation below around 1 MeV, except for quasi-equatorial emission locales for select pulse phases. Our spectral computations use a new state-of-the-art, spin-dependent formalism for the QED Compton scattering cross section in strong magnetic fields.

1. Introduction

A topical subset of the neutron star population is defined by magnetars. They are highly-magnetized stars that have historically been divided into two observational groups: Soft-Gamma Repeaters (SGRs) and Anomalous X-ray Pulsars (AXPs). Their extreme fields, which generally exceed the quantum critical value of $B_{\rm cr} = m_e^2c^3/(e^-\hbar) \approx 4.41 \times 10^{13}$ Gauss (where the electron cyclotron and rest mass energies are equal), are inferred from their X-ray timing properties, presuming that their rapid rotational spin down is due to magnetic dipole torques (e.g. [1]). Such a class of neutron stars with superstrong fields was postulated for SGRs by [2], and for AXPs by [3]. For recent reviews of magnetar science, see [4, 5, 6].

SGRs are transient sources that undergo repeated hard X-ray outbursts. Most of their ephemeral activity consists of short flares of subsecond duration in the 10^{38} erg/sec $< L < 10^{42}$ erg/sec range, often somewhat isolated in time, and sometimes occurring in storms. Yet three magnetars have exhibited giant supersecond flares of energies exceeding 10^{45} ergs: SGR 0525-66 in 1979, as mentioned above, SGR 1900+14 on August 27th, 1998 (e.g. see [7]), and SGR 1806-20 on December 27, 2004 (see [8]). SGRs also exhibit quiescent X-ray emission below 10 keV, of periods P in the range 2–12 sec. (e.g. [1, 9]). The AXPs are bright X-ray sources with periods

IOP Conf. Series: Journal of Physics: Conf. Series **932** (2017) 012021 doi:10.1088/1742-6596/932/1/012021

in the same range. Their quiescent signals below 10 keV are mostly thermal with steep power-law tails, and possess luminosities $L_X \sim 10^{33} - 4 \times 10^{35}\,\mathrm{erg\ s^{-1}}$ [10].[1] As with the SGRs, these L_X values far exceed their rotational power, perhaps being powered by their internal magnetic energy. Observations of outburst activity in AXP 1E 2259+586 [11], in AXP 1E1841-045 [12] and in others suggest that AXPs and SGRs are indeed very similar. This "unification paradigm" has garnered widespread support within the magnetar community over the last decade. There are also highly-magnetized pulsars that exhibit periods of magnetar-like bursting activity. The observational status quo of magnetars is summarized in the McGill Magnetar Catalog [13].[2]

An additional element of the magnetar phenomenon emerged following the discovery by INTEGRAL and RXTE of hard, non-thermal pulsed spectral tails in three AXPs [14, 15]. These luminous tails are extremely flat, extending up to 150 - 200 keV. Similar quiescent emission tails are seen in SGRs [16, 17]. In various magnetars, a turnover around 500 - 750 keV is implied by constraining *pre-2000* upper limits obtained by the COMPTEL instrument on the *Compton Gamma-Ray Observatory*. The need for such a spectral turnover is reinforced above 100 MeV by upper limits in *Fermi*-LAT data for around 20 magnetars [18, 19]. Magnetic inverse Compton scattering of thermal atmospheric soft X-ray seed photons by relativistic electrons is expected to be extremely efficient in magnetars, and thus is a prime candidate for generating the hard X-ray tails [20, 21]. This paper explores this model, highlighting some recent results from our ongoing program of modeling the resonant Compton hard X-ray emission in magnetars.

2. Hard X-rays in magnetars from magnetic inverse Compton scattering

The scenario for the generation hard X-ray tails considered in this paper is magnetic inverse Compton scattering of thermal atmospheric soft X-ray seed photons by relativistic electrons. This is extremely efficient in highly-magnetized pulsars because the scattering process is resonant at the electron cyclotron frequency and its harmonics, so that there the cross section in the electron rest frame exceeds the classical Thomson value of $\sigma_{\mathrm{T}} \approx 6.65 \times 10^{-25}\,\mathrm{cm}^2$ by $\sim 2-3$ orders of magnitude (e.g., [22, 23]). This efficiency is manifested in short cooling lengths, often less than 10^3 cm, for high speed electrons traversing a magnetar magnetosphere [24]. Provided there is a source of electrons with Lorentz factors $\gamma_e \gg 1$, single inverse Compton scattering events can readily produce the general character of hard X-ray tails [20, 21]. In particular, Baring & Harding [20] employed QED scattering cross sections in uniform fields, extending collision integral formalism for non-magnetic Compton upscattering that was developed by [25].

The spectra presented in [20] were characteristically flat, a consequence of the resonant cyclotron kinematics. These do not match observations, nor are they expected to since they integrate over all lines of sight in the uniform **B**. Non-uniform fields offer a different weighting of angular geometries, and when combined with cooling can steepen the spectrum considerably: see the magnetic Thomson investigation of [26]. In [20], we discerned that kinematic constraints correlating the directions and energies of upscattered photons yielded Doppler boosting and blueshifting along the local magnetic field direction. Therefore, the strong angular dependence of spectra computed for the uniform field case must extend to more complex magnetospheric field configurations. Consequently, emergent inverse Compton spectra in more complete models of hard X-ray tails will depend critically on an observer's perspective and the sampled locales of resonant scattering, both of which vary with the rotational phase of a magnetar. The construction of the resonant Compton upscattering model whose results are presented here is a geometrical extension of the work of [20] to dipolar magnetic field morphologies. Directed emission spectra have been generated for an array of observer perspectives and magnetic inclination angles α to the rotation axis; they serve as a basis for future calculations that

[1] See also the neutron star cooling site `http://neutronstarcooling.info/` for magnetars in a broader context.
[2] An on-line version can be found at `http://www.physics.mcgill.ca/~pulsar/magnetar/main.html`

will treat Compton cooling of electrons self-consistently. The scattering physics employs the state-of-the-art, spin-dependent magnetic Compton formalism developed by us in Gonthier et al. [27] that uses the preferred Sokolov and Ternov eigenstates of the Dirac equation. For details of the model, its kinematics and geometry and spectral characteristics, and their connection to observer perspectives, the reader is referred to the full exposition in Wadiasingh et al. [28].

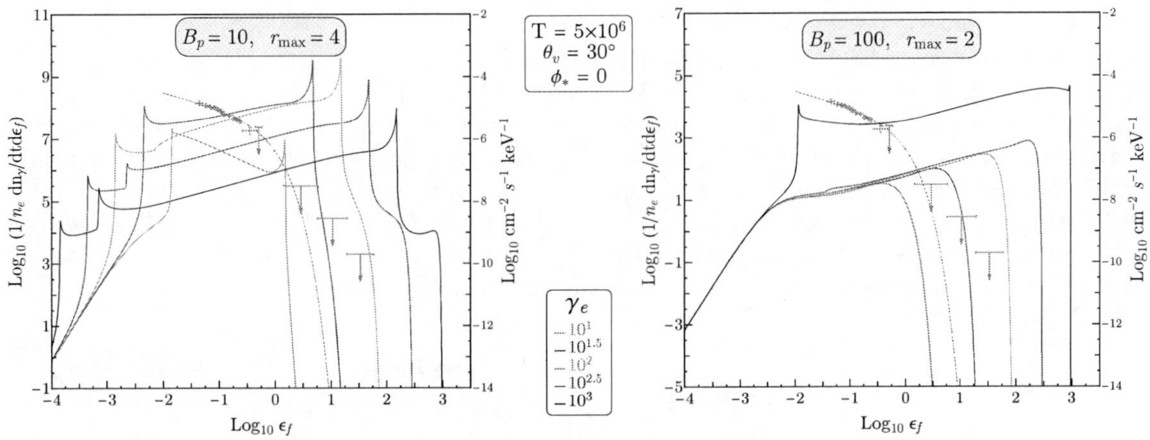

Figure 1. Spectra generated for meridional field loops, those where the line of sight to an observer is coplanar ($\phi_* = 0$) with the loops. The viewing angle of the observer is $\theta_v = 30°$ from the magnetic dipole axis. Results for various electron Lorentz factors are depicted. The left panel illustrates higher-altitude and lower-field directed spectra computed for $B_{\rm p} = 10$ and $r_{\rm max} = 4$ where the resonant interactions are readily sampled for Lorentz factors $\gamma_e > 10^2$. For the right panel with $r_{\rm max} = 2$ and $B_{\rm p} = 100$, the local field is much higher, precluding resonant interactions near equatorial regions unless Lorentz factors are much higher. Overlaid on the computed spectra are observational data points for AXP 4U 0142+61 (den Hartog et al. 2008b) along with a schematic $\varepsilon_f^{-1/2}$ power-law with a 250 keV exponential cutoff (black dotted curve).

Two elements of the extensive work in [28] are highlighted here. The first consists of selected but informative spectra, depicted in figure 1, computed for electrons of fixed Lorentz factors traversing individual field lines. They correspond to viewing angles coplanar with the field loops (meridional cases) that readily sample the Doppler boosting and beaming. The combination of γ_e and local field $B \sim B_{\rm p}/r_{\rm max}^3 \sim \gamma_e \varepsilon_f (1 + \cos\Theta_{{\rm B}n})$ essentially controls access to resonant interactions [20], and the value of the spectral index. Here ε_f is the upscattered photon energy in units of $m_e c^2$, and $\Theta_{{\rm B}n}$ is the observer's angle relative to $\mathbf{B}$ at the point of scattering. The coupling $B \sim \gamma_e \varepsilon_f (1 + \cos\Theta_{{\rm B}n})$ controls the directionality of emitted photons, with higher energies ε_f being beamed closer to the field lines. For much of the range in ε_f, the spectra that sample resonant interactions possess a characteristic scaling $dn/(dtd\varepsilon_f) \sim \varepsilon_f^{1/2}$, i.e. are extremely flat, even harder than the uniform field results in [20]. This approximate power-law dependence is a consequence of kinematics and magnetospheric geometry [28]. Bracketing these quasi-power-law bands are distinctive "horns" or cusps, distinguishing when resonant scatterings are and are not accessed. These appear both at low values ε_f in the soft X-rays/EUV, where they would be dominated by the surface emission signal (not shown), and also in the hard X-ray and gamma-ray domains. The narrow peaks of the horns are weighted images of the resonant differential cross section, enhanced by the beaming. Not all spectra possess frequency ranges where resonant interactions are accessible: for values of local B that are large, resonant interactions in the Wien peak are often not fully sampled, as is evident from computed spectra presented in the right panel of figure 1, and the $\gamma_e = 10$ example in the left panel.

Also illustrated in figure 1 are renormalized hard X-ray spectral data for one magnetar, to illustrate how the monoenergetic electron model from single field loops does not match observations. Modeling of hot thermal surface emission seeded by electron bombardment at loop footpoints is not undertaken. The computed spectra extend beyond the COMPTEL upper limits (green points) when $\gamma_e \gtrsim 30$. Therefore, lower Lorentz factors $\gamma_e \sim 10$ are desirable, and these a naturally generated by electron cooling [24]. High-energy attenuation mechanisms like photon splitting or magnetic pair creation may also be operating. In [28] we also exhibit spectra from integrations over field line azimuths, encompassing the non-meridional loops that dominate the contribution from a toroidal surface comprising dipolar field lines. These demonstrate steeper $dn/(dtd\varepsilon_f) \sim \varepsilon_f^0$ power-laws because the loops that do not provide tangents (i.e. **B** directions) at scattering locales that point toward an observer soften the spectrum. Even more interestingly, [28] illustrate that spectral arrays over such toroidal surfaces that span a range of altitudes yield an envelope that approximately matches the 4U0142+61 spectrum displayed in figure 1, provided that $\gamma_e \sim 10$. This suggests that models with more complete volumetric integrations and electron cooling incorporated will match the spectroscopy of hard X-ray tails.

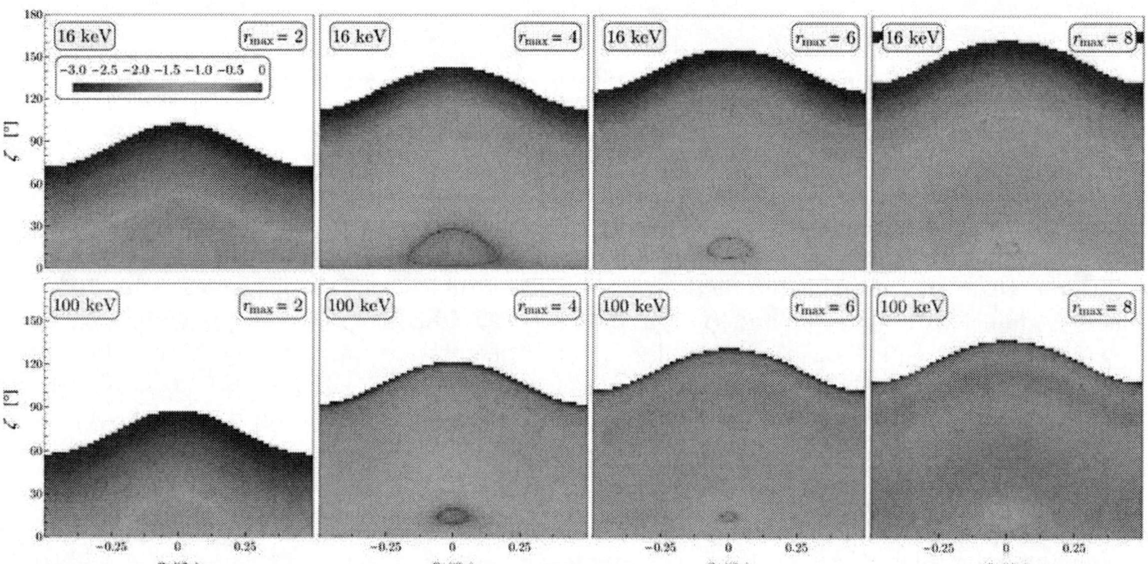

Figure 2. Normalized photon flux $\zeta - \Omega t/2\pi$ phase space maps for resonant Compton upscattering. These represent the logarithmically-scaled (base 10) intensity at energies 16 keV (top row) and 100 keV (bottom row), color coded according to the legend, as a function of spin phase $\Omega t/2\pi$ for each value of ζ on the ordinate. The intensity maps are for uncooled electrons with $\gamma_e = 10^2$, and a uniform surface temperature $T = 5 \times 10^6$ K. They are obtained for azimuthally-integrated bundles of field lines, i.e. a toroidal surface, with $B_p = 10$, and stellar magnetic inclination $\alpha = 15°$, and depict maps for maximum loop altitudes $r_{\max} = 2, 4, 6, 8$. Pulse profiles for a particular observer ζ are represented by horizontal cuts of the maps.

The spectral view so far has been for an instantaneous viewing perspective, i.e. fixed pulse phases. Another element delivered in [28] was the representation of how pulse profiles generated for spectra from toroidal surfaces vary with photon energy ε_f and the maximum surface altitude $r_{\max}$ (in units of $R_{\rm NS}$). These were expressed as sky maps in the $\zeta - \Omega t/2\pi$ plane, and an example is depicted in figure 2. The intensity scale is logarithmic, and is normalized so that the maximum in each row of panels is set to unity. Here, ζ is the angle between the viewer's direction and the magnetar spin axis, and for a particular choice of this angle, horizontal sections within each panel define an intensity trace with pulse phase. Generally, the pulsation profiles are of smooth, single-peaked character. Yet a symmetric double-peak structure of the profiles

IOP Conf. Series: Journal of Physics: Conf. Series **932** (2017) 012021 doi:10.1088/1742-6596/932/1/012021

in domains $\zeta \approx \alpha$ is evident, being manifested as sections of the red rings: these are realized when quasi-polar viewing is possible at select phases. The phase separation of this double-peak structure in domains $\alpha \approx \zeta$ shrinks at higher $r_{\rm max}$ and larger ε_f. This identifies a potentially potent observational diagnostic: comparing theoretical energy-dependent pulse profiles with observational ones can infer values for α and ζ in magnetars, an analogous protocol to that widely used in gamma-ray pulsar studies. To this end, [28] applied this to the observed phase separation of 0.4 between the two peaks in the pulse profile of 1E 1841-045 within the energy range of 20-35 keV. We found that for $\gamma_e \sim 10$ that would result from strong cooling, this suggests that $\alpha \lesssim 20^\circ$, an estimate that is quite similar to the value of $\alpha \sim 15^\circ$ inferred in the analysis of [29]. This determination would change if toroidal components to the equatorial field (scattering locales that dominate the spectral signal) yield twist angles $\Delta\phi \sim 1$, i.e., significantly larger than those found at colatitudes $\theta > 60^\circ$ in MHD simulations of field untwisting [30].

In summary, the sample results presented here provide an idea of the constraints imposed upon our model, and a taste of the promise of the resonant Compton upscattering picture in explaining the phenomenon of the hard X-ray tails in magnetar quiescent emission. They also suggest the usefulness of the model in probing the magnetic angle α, thus aiding in reducing uncertainties in the determination of magnetar field strengths using spin-down information.

Acknowledgments

MGB acknowledges support by the NASA Astrophysics Theory (ATP) and *Fermi* Guest Investigator Programs through grants NNX13AQ82 and NNX13AP08G. ZW is supported by the South African National Research Foundation. PLG thanks the Michigan Space Grant Consortium, the National Science Foundation (grant AST-1009731), and NASA ATP through grant NNX13AO12G. AKH. also acknowledges support through the NASA ATP program.

References

[1] Kouveliotou C *et al* 1998 *Nature* **393** 235
[2] Duncan R C Thompson C 1992 *ApJ* **392** L9
[3] Thompson C and Duncan R C 1996 *ApJ* **473** 332
[4] Turolla R, Zane S and Watts A L 2015 *Reports on Progress in Physics* **78** 116901
[5] Kaspi V M and Beloborodov A M 2017 *Ann. Rev. Astron. Astrophys.* **55** 261
[6] Coti Zelati F, Rea N, Pons J, Campana S and Esposito P 2017, *Preprint* 1710.04671
[7] Hurley K *et al* 1999 *Nature* **397** 41
[8] Palmer D M *et al* 2005 *Nature* **434** 1107
[9] Kulkarni S R *et al* 2003 *ApJ* **585** 948
[10] Viganò D *et al* 2013 *MNRAS* **434** 123
[11] Kaspi V M *et al* 2003 *ApJ* **588** L93
[12] Lin L *et al* 2011 *ApJ* **740** L16
[13] Olausen S A and Kaspi V M 2014 *ApJS* **212** 6
[14] Kuiper L, Hermsen W and Mendeź M 2004 *ApJ* **613** 1173
[15] den Hartog P R *et al* 2008a *A&A* **489** 245
[16] Götz D *et al* 2006 *A&A* **449** L31
[17] Enoto T *et al* 2010 *ApJ* **722** L162
[18] Abdo A A *et al* 2010 *ApJ* **714** 927
[19] Li J, Torres D F, de Oña Wilhelmi E, Rea N and Martin J 2017 *ApJ* **835** 54
[20] Baring M G and Harding A K 2007 *Astr. Space Sci.* **308** 109
[21] Fernández R and Thompson C 2007 *ApJ* **660** 615
[22] Daugherty J K and Harding A K 1986 *ApJ* **309** 362
[23] Gonthier P L *et al* 2000 *ApJ* **540** 907
[24] Baring M G, Wadiasingh Z and Gonthier P L 2011 *ApJ* **733** 61
[25] Ho C and Epstein R I 1989 *ApJ* **343** 227
[26] Beloborodov A M 2013 *ApJ* **762** 13
[27] Gonthier P L *et al* 2014 *Phys. Rev.* D **90** 043014
[28] Wadiasingh Z, Baring M G, Gonthier P L and Harding A K 2017 *ApJ* submitted
[29] An H *et al* 2015 *ApJ* **807** 93
[30] Chen A Y and Beloborodov A M 2017 *ApJ* **844** 133

IOP Conf. Series: Journal of Physics: Conf. Series **932** (2017) 012022 doi:10.1088/1742-6596/932/1/012022

Systematic study of magnetar outbursts

F Coti Zelati[1,2,3], N Rea[1,2], J A Pons[4], S Campana[3] and P Esposito[2]

[1] Institute of Space Sciences (ICE, CSIC–IEEC), Campus UAB, Carrer de Can Magrans, S/N, 08193, Barcelona, Spain
[2] Anton Pannekoek Institute for Astronomy, University of Amsterdam, Postbus 94249, NL-1090-GE Amsterdam, The Netherlands
[3] INAF – Osservatorio Astronomico di Brera, via Bianchi 46, I-23807 Merate (LC), Italy
[4] Departament de Física Aplicada, Universitat d'Alacant, Ap. Correus 99, E-03080 Alacant, Spain

E-mail: cotizelati@ice.csic.es

Abstract. We present the results of the systematic study of all magnetar outbursts observed to date through a reanalysis of data acquired in about 1100 X-ray observations. We track the temporal evolution of the luminosity for all these events, model empirically their decays, and estimate the characteristic decay time-scales and the energy involved. We study the link between different parameters (maximum luminosity increase, outburst peak luminosities, quiescent X-ray and bolometric luminosities, energetics, decay time-scales, magnetic field, spin-down luminosity and age), and reveal several correlations between different quantities. We discuss our results in the framework of the models proposed to explain the triggering mechanism and evolution of magnetar outbursts. The study is complemented by the Magnetar Outburst Online Catalog (http://www.magnetars.ice.csic.es), an interactive database where the user can plot any combination of the parameters derived in this work and download all reduced data.

1. Introduction

Magnetars are strongly magnetized (up to $B \sim 10^{14} - 10^{15}$ G) isolated X-ray pulsars with spin periods $P \sim 2 - 12\,$s and secular spin-down rates $\dot{P} \sim 10^{-15}$–10^{-11} s s^{-1}, whose emission is ultimately powered by the decay and the instability of their ultra-strong magnetic field [1–3]. Magnetars unpredictably undergo outbursts, in which the persistent X-ray luminosity increases by a factor of $\sim 10 - 1000$ up to $\sim 10^{35} - 10^{36}$ erg s^{-1}, and declines back to the quiescent level on a time-scale ranging from a few weeks up to several years.

Magnetar outbursts are likely triggered by local internal magnetic stresses that deform irreversibly part of the stellar crust and convert mechanically its magnetic energy into heat [4]. Part of the released heat is then conducted up to the surface and radiated, producing the observed thermal emission. The crustal displacements also implant a strong twist in the magnetosphere, likely confined to a bundle of current-carrying closed field lines anchored in the crust. The charged particles flowing along the field lines of the twisted bundle hit the star, providing an additional source of heat [5]. According to this scenario, the outburst evolution is regulated by the dissipation of the twist: the spatial extent of the bundle gets gradually more limited, the area on the star surface hit by the charges shrinks and the luminosity decreases.

Here we present the results of a systematic and homogeneous analysis of the spectral properties for 23 outbursts (from the very first active phases throughout their decays) from

17 magnetars using all the available data acquired by the *Swift*, *Chandra* and *XMM–Newton* X-ray observatories, as well as data collected in a handful of observations by the instruments aboard *Beppo*SAX, *ROSAT* and *RXTE*. This sums up to about 1100 observations, for a total dead-time corrected on source exposure time of more than 12 Ms.

2. Data analysis

We adopted standard procedures to extract source and background spectra and create or assign the response and auxiliary files starting from the raw *Swift*, *XMM–Newton* and *Chandra* data files publicly available. We performed the spectral analysis separately for data of different instruments (owing to known cross-calibration uncertainties) within XSPEC (v. 12.9.0; [6]).

To model the continuum emission, we fitted either a blackbody, a power-law, a blackbody plus a power-law, the superposition of two blackbodies or a resonant cyclotron scattering model (NTZ; [9,10]) to the spectral data sets. The NTZ model accounts for the repeated resonant cyclotron up-scatterings of thermal photons from the star surface onto the charges flowing in a twisted magnetosphere, and is based on three-dimensional Monte Carlo simulations. The topology of the magnetic field is assumed to be a globally twisted, force-free dipole in the model, and its parameters are the surface temperature (assumed to be the same over the whole surface), the bulk motion velocity of the charged particles in the magnetosphere (assumed constant through the magnetosphere), the twist angle and a normalization constant. In a few cases, the higher statistics quality available from *XMM–Newton* observations allowed us to probe more complicated models, such as the sum of three black-body components. The photo-electric absorption by the interstellar medium along the line of sight was described via the Tuebingen-Boulder model with cross-sections from [7] and chemical abundances from [8].

The absorbed and unabsorbed fluxes for the additive components and the total one (all in the 0.3–10 keV energy range) were computed for each fitted spectrum. Unabsorbed fluxes were converted to luminosities (as measured by an observer at infinity) assuming isotropic emission and the most reliable value for the distance of the source (e.g., [1]).

Figure 1 shows the temporal decays of the bolometric luminosities for all outbursts. We refer each curve to the epoch of the outburst onset, defined as the time of the first burst detection from the source, or of the giant flare in the case of SGR 1806−20. For XTE J1810−197 and 1E 1048.1−5937, for which no bursts were detected, we adopted the epoch of the observation where an increase in the X-ray flux was first measured as the reference epoch. Our bolometric luminosities do not take into account the emission of magnetars in the hard X-ray range, which in some sources was observed to give a relevant contribution to the total luminosity, especially during the initial phases (weeks) of the outburst (e.g., [11]). However we note that, if we consider all the hard X-ray observations of magnetar outbursts performed so far, our values for the bolometric fluxes are underestimated for a few sources (SGR 1806−20, 1E 2259+586, 1E 1547−5408, SGR 1745−2900, SGR 1935+2154, PSR J1119−6127 and 1E 161348−5055) mainly only at the outburst peak, yielding negligible differences in the estimated outburst energetics and decay timescale (see below).

2.1. Outburst decay time-scale and energetics

We modelled the decays of the X-ray luminosities of the single spectral components and of the bolometric luminosities using a constant (representing the quiescent level) plus one or more exponential functions. The number of required exponential functions was evaluated by means of the F-test, i.e., an additional exponential function was included only if it yielded an improvement in the fit of at least 3σ. L_q was fixed at the quiescent value or, in cases of non detections, constrained to be lower than the upper limit.

We estimated the outburst energy by integrating the best-fitting model for the bolometric light curves over the whole duration of the event, and extrapolating it to the quiescent value for

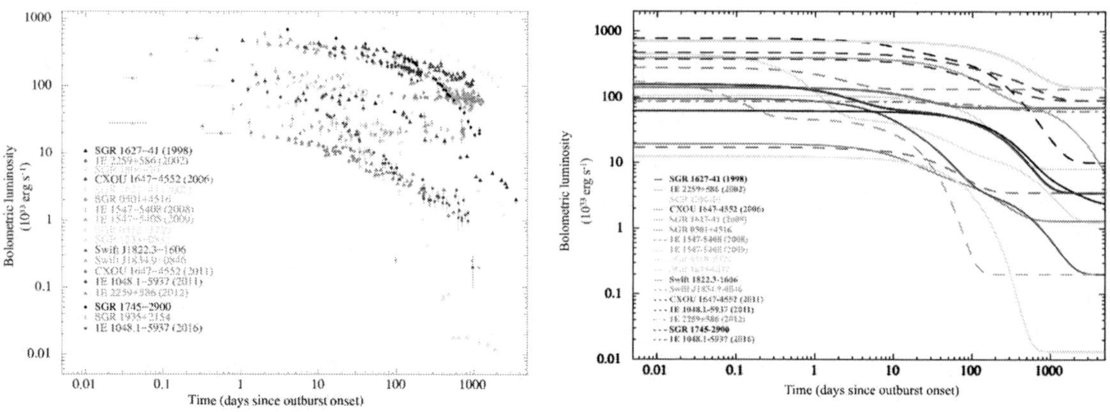

Figure 1. Left-hand panel: temporal evolution of the bolometric (0.01–100 keV) luminosities for the most densely monitored magnetar outbursts re-analysed in the present work. Right-hand panel: models reproducing the temporal evolution of the bolometric luminosities.

the cases where the observational campaign was not extended enough to follow completely the return to the pre-outburst state. For the sources that are still recovering from their outbursts (i.e., 1E 1547−5408, SGR 1745−2900, 1E 1048.1−5937, PSR J1119−6127 and 1E 161348−5055), we assumed no changes in the decay pattern down to quiescence for the estimate of the time-scales and energetics (in these cases the values derived should be considered only as upper limits).

3. Correlations

Our systematic analysis allows us to search for correlations between different parameters for all sources of our sample and their outbursts. Table 1 lists the significance for the correlations according to both the Spearman and Kendall τ rank correlation tests, and also reports on the shape of the correlation according to a power-law regression test. Figure 2 shows the results for some of the (anti-)correlations unveiled from our systematic study. To have a more complete sample, we also included SGR 1900+14, 4U 0142+61 and 1E 1841−045, for which re-brightenings or subtle variations around their persistent activity were reported throughout the last 15 years, as well as the other few magnetars (black stars), the central compact objects (grey crosses), the rotation-powered pulsars with evidence for a clear thermal component (red diamonds) and the X-ray dim isolated neutron stars (orange crosses) already reported by [12]. PSR J1119−6127 and the magnetars XTE J1810−197, 1E 1547−5408 and SGR 1745−2900, for which radio pulsed emission has been detected [13–15], are marked by black circles. Upper and lower limits are indicated by black arrowheads.

4. Results

We carried out the first systematic study of all sources experiencing magnetar-like outbursts up to the end of 2016, and for which extensive X-ray monitoring campaigns of their outbursts are available. We re-analysed in a coherent way about 1100 X-ray observations, adopting the same assumptions and spectral models throughout the whole sample. This work allows us to study possible correlations and anticorrelations between several different combinations of parameters, and put the results in the context of the models proposed to explain the triggering mechanism and evolution of magnetar outbursts.

Table 1. Results of the search for (anti-)correlations between different parameters. Letters in parentheses indicate the case of a correlation (c) or an anti-correlation (a). Values for the power-law index indicate the shape of the (anti-)correlation, and were estimated via a power-law regression test.

First parameter	Second parameter	Corr/Anticorr, Significance (σ) (c) or (a), Spearman / Kendall τ	PL index
Quiescent X-ray luminosity	Maximum luminosity increase	(a) , 5.7 / 4.9	-0.7
Dipolar magnetic field	Quiescent bolometric luminosity	(c) , 3.2 / 2.9	2.0
Dipolar magnetic field	Peak luminosity	(c) , 2.5 / 2.4	0.5
Dipolar magnetic field	Outburst energy	(c) , 3.7 / 3.3	1.0
Characteristic age	Outburst energy	(a) , 3.3 / 3.0	-0.4
Peak luminosity	Outburst energy	(c) , 4.0 / 3.7	1.4
Outburst energy	Decay time-scale	(c) , 3.9 / 3.6	0.5

4.1. On the relation between the outburst luminosity increase and the quiescent luminosity

It was already noted a few years ago that magnetars characterized by low quiescent luminosities ($L_q \sim 10^{31-33}$ erg s^{-1}) experience large luminosity increases when in outburst, whereas highly luminous sources in quiescence ($L_q \sim 10^{34-35}$ erg s^{-1}) undergo only subtle enhancements in luminosity during their outbursts ([16]; see also the top-left panel of Figure 2). This discovery also clarified that the distinction between 'transient' and 'persistent' sources within the magnetar population is deceptive, and only dependent on the initial quiescent luminosity of each source.

The anticorrelation between magnetars quiescent luminosities and their luminosity increases during outbursts is observed at a significance of 5.7σ (according to the Spearman test; see Table 2). This result suggests the existence of a limiting luminosity of $\sim 10^{36}$ erg s^{-1} for magnetar outbursts (regardless of the quiescent level of the source), and was interpreted in the framework of the internal crustal heating model as the observational manifestation of the self-regulating effect resulting from the strong temperature-dependence of the neutrino emissivity [16]: the surface photon luminosity for injected energies larger than $\sim 10^{43}$ erg reaches a limiting value of $\sim 10^{36}$ erg s^{-1} because the crust is so hot that most of the energy is released in the form of neutrinos before reaching the star surface. The observed anticorrelation is expected also in the scenario of the untwisting magnetospheric bundle, where the maximum theoretically expected luminosity might be of a few 10^{36} erg s^{-1} even for the generous case of a twist with $\psi \sim 1$ rad affecting a large part of the magnetospheric volume. The generally lower values observed for the peak luminosity are interpreted, in this model, as a consequence of the limited size of the current bundle and the twist [5].

The epoch of the outburst onset was defined throughout this study as the time of the first burst detection from the source (mostly with *Swift* BAT or *Fermi* GBM), or of the giant flare in the case of SGR 1806−20. This might represents a somewhat arbitrary choice, because the increase of the persistent flux during the time interval preceding the detection of magnetars bursting activity is usually missed by X-ray instruments. However, given the large sample, and the clear trend observed over several orders of magnitude, we do not expect to measure significantly different values for the outburst peak luminosity.

Different estimates on the time-scale of the luminosity increase were proposed in the past years. The internal crustal cooling models [16] show that the internal heat wave takes some time to propagate from the location in the crust where the energy is injected up to the surface layers. Therefore, the luminosity increase is not instantaneous but relatively fast, and might range from a few hours up to a few days depending on the depth of the region where heat is released. On the other hand, simplified one-dimensional models [17] show that the time-scale

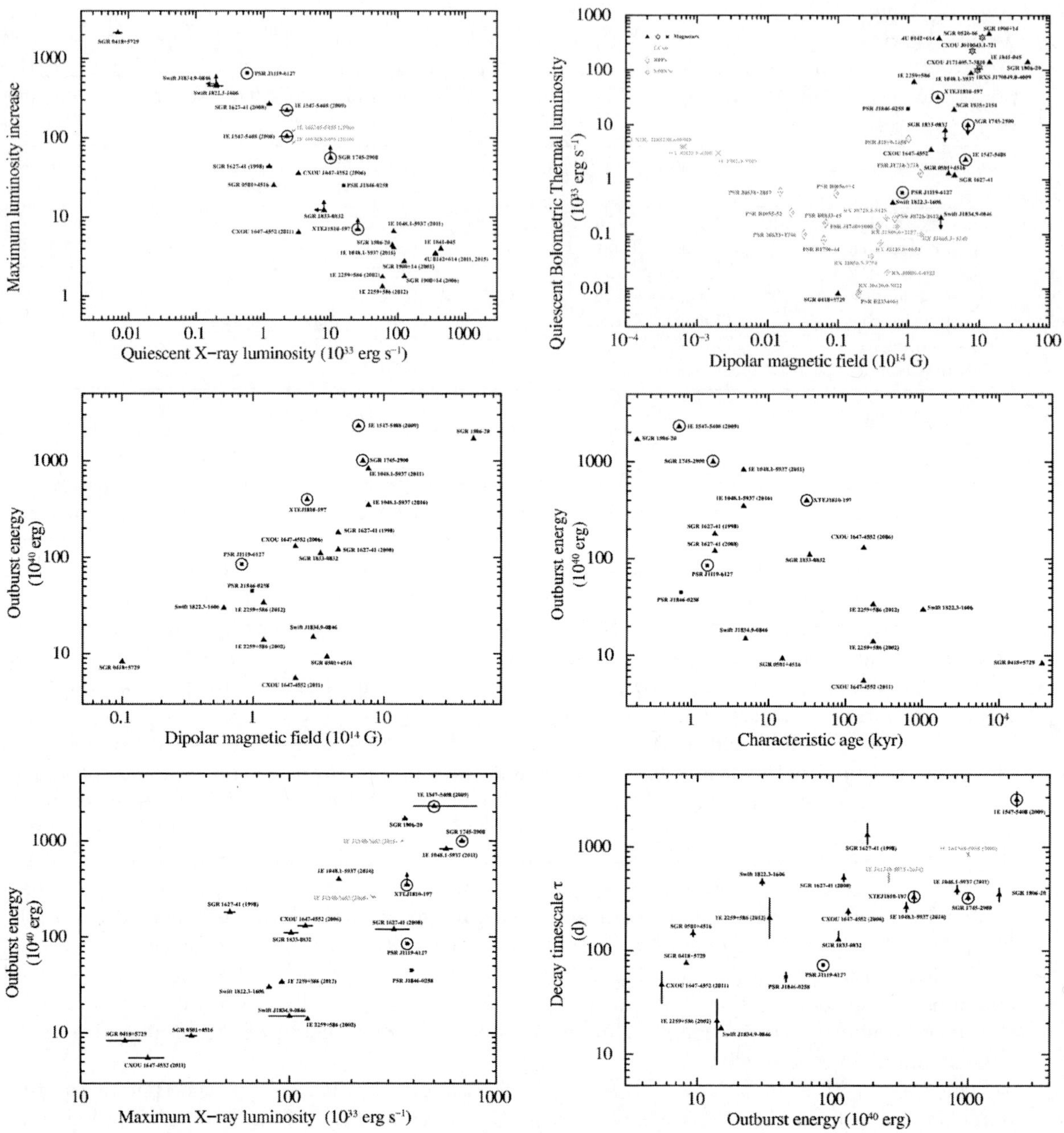

Figure 2. From left to right, top to bottom: quiescent X-ray luminosity versus maximum X-ray luminosity increase; quiescent bolometric luminosity relative to the thermal component versus the dipolar component of the magnetic field; total energy released during the outburst as a function of the dipolar component of the magnetic field; total energy released during the outburst as a function of the characteristic age; total energy released during the outburst versus maximum X-ray luminosity at the peak of the outburst; decay time-scale as a function of the total energy released during the outburst.

IOP Conf. Series: Journal of Physics: Conf. Series **932** (2017) 012022 doi:10.1088/1742-6596/932/1/012022

of magnetospheric twisting by a large thermoplastic wave (corresponding to the rise time of the outburst) can span from days to weeks. Within the large uncertainties, both models are compatible with a typical rise time of a few days.

4.2. On the quiescent luminosity versus the dipolar magnetic field

The top-right panel of Figure 2 shows the quiescent thermal bolometric luminosity as a function of the surface dipolar magnetic field. We observe a significant correlation (3.2σ according to the Spearman test) when including all sources belonging to the different classes considered in this study. The correlation is even more significant (3.9σ) after excluding the central compact objects, and is naturally explained in terms of magnetic field decay and Joule heating [18,12]. The central compact objects clearly depart from the general trend. The peculiar behavior of these objects might be explained in the framework of the 'hidden magnetic field' scenario: hypercritical accretion onto the neutron star surface during the initial stages of the star life can bury a magnetic field of a few 10^{13} G into the inner crust, yielding a strength for the external magnetic field that is significantly lower than the internal 'hidden' magnetic field. The large luminosity observed for these objects is most probably due to the toroidal and higher order multipolar components of the magnetic field trapped inside the crust [19–22]. The magnetic field might eventually re-emerge, after a few thousand years, settling on a value comparable to that at birth. If this picture is correct, we would expect a 'shift' of the central compact objects towards the right in the quiescent luminosity versus dipolar magnetic field diagram, as the CCOs get older. Some of the rotation powered pulsars also depart slightly from the observed trend (e.g., PSR J0538+2817, PSR B1055−52 and PSR J0633+1746). This might be possibly due to an additional contribution to the surface heating from slamming particles onto the stellar surface, as typically observed for pulsars with a high rotational energy loss rate.

The shape of the correlation can be approximated as $L_{q,bol} \propto B^2_{p,dip}$, in line with the dependence reported in a previous study using a reduced sample of sources [23].

4.3. On the dipolar magnetic field versus the outburst properties

We investigated possible correlations between the strength of the surface dipolar magnetic field and all the outburst parameters derived in the present study. There is no clear correlation between the magnetic field and either the outburst peak luminosity or the decay time-scale. Moreover, in a few cases the same source was observed to undergo two different outbursts with distinct properties. The correlation between the magnetic field and the outburst energetics is more evident (3.4σ according to the Spearman test; see the middle-left panel of Figure 2), and supports the idea that the energy reservoir of the outbursts is mainly provided by the dissipation of the magnetic field. We observe a sort of limiting energy as a function of age (see the middle-right panel of Figure 2). Young magnetars tend to experience more energetic outbursts than older magnetars, a characteristic that can be explained simply in terms of field decay. Previous estimates on the expected energetics distribution did not find a significant dependence of the energy of the events with age, but the fact that magnetic field decay limits the energy budget available for old magnetars, compared to young sources [24].

4.4. On the outburst energy versus other properties

The outburst energy correlates with the peak luminosity (at a significance of 4.0σ according to the Spearman test; see the bottom-left panel of Figure 2), but not with the quiescent X-ray luminosity ($< 2\sigma$). These results suggest that a larger luminosity at the peak of the outburst results in a larger energy released during the outburst event, regardless of the quiescent level of the source, and indicate similar patterns for the decay curves of magnetar outbursts. This is expected in both internal crustal cooling and untwisting bundle scenarios, since it only reflects the normalization of the decay curve.

The decay time-scale significantly correlates with the energetics (at a significance of 3.9σ according to the Spearman test; see the bottom-right panel of Figure 2): the longer the outburst, the more energetic. This suggests again that the decay pattern is similar from outburst to outburst. For example, we never observe a magnetar undergoing a rather weak outburst and then returning to quiescence over an extremely long time interval, or a magnetar showing an extremely powerful outburst and then rapidly decaying back to quiescence.

Acknowledgments
We are indebted to Santiago Serrano Elorduy and Martin Folger from the Institute of Space Sciences (CSIC–IEEC) for designing the Magnetar Outburst Online Catalogue. The scientific results reported in this study are based on observations obtained with *XMM–Newton*, an ESA science mission with instruments and contributions directly funded by ESA Member States and NASA, and *Swift*, a NASA/UK/ASI mission. This research has made extensive use of software provided by the *Chandra X-ray Center* in the application package CIAO. FCZ acknowledges funding in the framework of the Netherlands Organization for Scientific Research (NWO) Vidi award (PI: N. Rea) and the European COST Action MP1304 (NewCOMPSTAR), and is also supported by grants AYA2015-71042-P and SGR2014-1073.

References

[1] Olausen S A and Kaspi V M 2014 *Astrophys. J. Suppl. Ser.* **212** 6
[2] Turolla R, Zane S and Watts A 2015 *Rep. Prog. Phys.* **78** 116901
[3] Kaspi V M and Beloborodov A M 2017 *Annu. Rev. Astron. Astrophys.* in press
[4] Beloborodov A M and Levin Y 2014 *Astrophys. J.* **794** L24
[5] Beloborodov A M 2009 *Astrophys. J.* **703** 1044
[6] Arnaud K A 1996 *Astronomical Data Analysis Software and Systems V* (*ASP Conf. Ser.* vol 101) ed Jacoby G H and Barnes J (San Francisco: Astron. Soc. Pac.) p. 17
[7] Verner D A, Ferland G J, Korista K T and Yakovlev D G 1996 *Astrophys. J.* **465** 487
[8] Wilms J, Allen A and McCray R 2000 *Astrophys. J.* **542** 914
[9] Nobili L, Turolla R and Zane S 2008 *Mon. Not. R. Astron. Soc.* **386** 1527
[10] Nobili L, Turolla R and Zane S 2008 *Mon. Not. R. Astron. Soc.* **389** 989
[11] Kuiper L, Hermsen W, den Hartog P R and Urama J O 2012 *Astrophys. J.* **748** 133
[12] Viganò D, Rea N, Pons J A, Perna R, Aguilera D N and Miralles J A 2013 *Mon. Not. R. Astron. Soc.* **434** 123
[13] Camilo F, Kaspi V M, Lyne A G, Manchester R N, Bell J F, D'Amico N, McKay N P F and Crawford F 2000 *Astrophys. J.* **541** 367
[14] Camilo F, Ransom S M, Halpern J P and Reynolds J 2007 *Astrophys. J.* **666** 93
[15] Rea N *et al* 2013 *Astrophys. J.* **775** L34
[16] Pons J A and Rea N 2012 *Astrophys. J.* **750** L6
[17] Li X, Levin Y and Beloborodov A M 2016 *Astrophys. J.* **833** 189
[18] Pons J A and Geppert U 2007 *Astron. Astrophys.* **470** 303
[19] Geppert U, Page D and Zannias T 1999 *Astron. Astrophys.* **345** 847
[20] Viganò D and Pons J A 2012 *Mon. Not. R. Astron. Soc.* **425** 2487
[21] Shabaltas N and Lai D 2012 *Astrophys. J.* **748** 148
[22] Torres-Forné A, Cerdá-Durán P, Pons J A and Font J A *Mon. Not. R. Astron. Soc.* **456** 3813
[23] Pons J A, Link B, Miralles J A and Geppert U 2007 *Phys. Rev. Letters* **98** id. 071101
[24] Perna R and Pons J A 2011 *Astrophys. J.* **727** L51

On the spectrum and polarization of magnetar flare emission

R Taverna[1] **and R Turolla**[1,2]

[1]Department of Physics and Astronomy, University of Padova, via Marzolo 8, I-35131 Padova, Italy

[2]Mullard Space Science Laboratory, University College London, Holmbury St. Mary, Surrey RH5 6NT, UK

E-mail: taverna@pd.infn.it

Abstract. Bursts and flares are among the distinctive observational manifestations of magnetars, isolated neutron stars endowed with an ultra-strong magnetic field ($B \approx 10^{14}$–10^{15} G). It is believed that these events arise in a hot electron-positron plasma, injected in the magnetosphere, due to a magnetic field instability, which remains trapped within the closed magnetic field lines (the "trapped-fireball" model). We have developed a simple radiative transfer model to simulate magnetar flare emission in the case of a steady trapped fireball. We assume that magnetic Thomson scattering is the dominant source of opacity in the fireball medium, and neglect contributions from second-order radiative processes. The spectra we obtained in the 1–100 keV energy range are in broad agreement with those of available observations. The large degree of polarization ($\gtrsim 80\%$) predicted by our model should be easily measured by new-generation X-ray polarimeters, like *IXPE*, *XIPE* and *eXTP*, allowing one to confirm the model predictions.

1. Introduction

The most distinctive manifestation of magnetars, a particular kind of isolated neutron stars (NSs) observationally identified with Soft Gamma Repeaters (SGRs) and Anomalous X-ray pulsars (AXPs), is the emission of short ($\approx 10^{-2}$–1 s) energetic ($\approx 10^{36}$–10^{41} erg) X-ray bursts and longer ($\approx$ 1–50 s), even more energetic ($\approx 10^{41}$–10^{43}) intermediate flares. Furthermore, three SGRs have been observed to also emit giant flares, the most powerful events ever observed from compact objects [1, 2, for reviews]. According to the magnetar model, first developed by [3], the strong internal magnetic field of magnetars is believed to develop a large toroidal component, able to exert a strong magnetic stress on the conductive star crust. In contrast to what happens in 'normal' NSs, where this force can be balanced by the rigidity of the crust, in the case of magnetars the internal stresses are strong enough to displace single surface elements. As a result, the external magnetic field acquires in turn a non-zero toroidal component, becoming twisted, and this makes possible for charged particles to fill the magnetosphere, streaming along the closed field lines [4, 5]. Although the mechanisms that trigger magnetar bursting activity are still not completely clear, [6, 7] suggested a model according to which magnetar flares originate in the sudden re-arrangements of the external field lines that follow the crustal displacements triggered by the strong internal field. This injects in the magnetosphere an Alfvén pulse that

IOP Conf. Series: Journal of Physics: Conf. Series **932** (2017) 012023 doi:10.1088/1742-6596/932/1/012023

dissipates into an electron-positron pair plasma, which remains magnetically confined within the closed field lines (the "trapped fireball").

In this work we study the spectrum and polarization of the radiation emitted from a steady trapped fireball, providing simulations directly comparable with observations. Although our model relies on a series of simplifying assumptions, the simulated spectra we obtained in the 1–100 keV energy range, that can be suitably described in terms of the superposition of two blackbody components, turn out to be in broad agreement with the observations available so far [8, 9, 10]. Furthermore, the large degree of polarization expected for magnetar burst radiation can be easily measured by new-generation X-ray polarimeters, such as *IXPE*, *XIPE* and *eXTP*, allowing to confirm the model predictions. In section 2 we discuss the radiative processes which take place in the fireball medium, while in section 3 and 4 we deal with the problem of radiative transfer and describe the structure of our numerical codes. The results of our simulations are presented in section 5, while discussion and conclusions are reported in section 6.

2. Radiative processes in strong magnetic fields

We assume that magnetic scattering is the main source of opacity in the fireball medium. For the sake of simplicity, we consider scattering in the Thomson limit; the full case of Compton scattering will be addressed in a sequel work. In the presence of strong magnetic fields, photons emitted from the surface of a neutron star are expected to be linearly polarized in two normal modes, the ordinary (O) and the extraordinary (X), with the photon electric field oscillating either parallel or perpendicular to the $(\boldsymbol{k}, \boldsymbol{B})$ plane, with $\boldsymbol{k}$ the photon propagation direction and $\boldsymbol{B}$ the star magnetic field [11, 12, 13]. In the hypothesis (well verified for magnetar bursts) that the photon energy ε is much smaller than the electron cyclotron energy $\varepsilon_{\rm B} = m_{\rm e}c^2B/B_{\rm Q}$, with $m_{\rm e}$ the electron mass, scattering of photons onto electrons/positrons is non-resonant. In the electron rest frame, and neglecting the electron recoil, the scattering cross-sections depend on the polarization state of both the ingoing and the outgoing photons [14, 15, 16]:

$$\left[\frac{{\rm d}^2\sigma}{{\rm d}\varepsilon'{\rm d}\Omega'}\right]_{\rm OO} = \frac{3}{8\pi}\sigma_{\rm T}\left(1-\mu_{\rm Bk}^2\right)\left(1-\mu_{\rm Bk}'^2\right)\delta\,, \qquad \left[\frac{{\rm d}^2\sigma}{{\rm d}\varepsilon'{\rm d}\Omega'}\right]_{\rm OX} = \frac{3}{8\pi}\sigma_{\rm T}\left(\frac{\varepsilon}{\varepsilon_{\rm B}}\right)^2\mu_{\rm Bk}^2\nu_{\rm Bk}^2\delta\,,$$

$$\left[\frac{{\rm d}^2\sigma}{{\rm d}\varepsilon'{\rm d}\Omega'}\right]_{\rm XO} = \frac{3}{8\pi}\sigma_{\rm T}\left(\frac{\varepsilon}{\varepsilon_{\rm B}}\right)^2\mu_{\rm Bk}'^2\nu_{\rm Bk}^2\delta\,, \qquad \left[\frac{{\rm d}^2\sigma}{{\rm d}\varepsilon'{\rm d}\Omega'}\right]_{\rm XX} = \frac{3}{8\pi}\sigma_{\rm T}\left(\frac{\varepsilon}{\varepsilon_{\rm B}}\right)^2(1-\nu_{\rm Bk}^2)\delta\,, \tag{1}$$

where a prime labels the quantities after scattering, $\sigma_{\rm T}$ is the Thomson cross section, $\mu_{\rm Bk}$ is the cosine of the angle $\theta_{\rm Bk}$ between the photon direction and the local magnetic field, $\nu_{\rm Bk}$ is the cosine of the associated azimuth $\phi_{\rm Bk}$ and $\delta = \delta(\varepsilon - \varepsilon')$ is the Dirac δ-function. The previous expressions hold as far as the vacuum contributions in the dielectric tensor dominate over the plasma ones [17]. From equations (1) it appears clearly that all the cross-sections that involve X-mode photons are suppressed by a factor of $(\varepsilon/\varepsilon_{\rm B})^2 \propto (\varepsilon B_{\rm Q}/B)^2$ with respect to the O–O cross-section, which is essentially of the order of $\sigma_{\rm T}$. This implies that the medium becomes optically thin for extraordinary photons at much larger Thomson depths with respect to the ordinary ones.

Among the additional second-order processes that can take place in magnetized plasma there are thermal bremsstrahlung, photon splitting and double-Compton scattering. The bremsstrahlung between particles with the same charge becomes important only for high particle energies [18, above 300 keV, outside the range we considered our work], so that the only relevant bremsstrahlung process turns out to be the electron-positron branch, slightly enhanced with respect to the electron-ion one [19, 20]. Also in this case, however, the amplitude turns out to be negligible for magnetar burst radiation, allowing to neglect the contributions of thermal bremsstrahlung in our work. Moreover, although photon splitting can be important in a wider range of photon energies [21, 22], also in this case, the cross section of the process is much smaller than the Thomson cross section in the weak-field limit (and for photon energies $\lesssim$ 100 keV), while one should consider an additional suppression factor in the strong-field limit [23, 24]. For these reasons we can safely neglect also photon splitting effects in the model.

Therefore, double-Compton scattering turns out to be the unique process responsible for photon production in the fireball in our model. According to [25], for large scattering depths photons follow a Planck distribution at energies low enough to make double-Compton scattering dominant, while at higher energies scattering tends to establish a Bose-Einstein distribution. However, as shown [26], the photon chemical potential remains rather small, allowing to solve the photon transport assuming local thermal equilibrium (LTE) at large depth for both O- and X-mode photons [27].

3. Radiative transfer in the fireball atmosphere

We solved the radiative transfer equations for both the ordinary and the extraordinary photons in the geometrically thin surface layers of the fireball, which we term the atmosphere. The latter is divided into a number of patches, each labelled by the intensity of the magnetic field at the patch centre and by the angle θ_{B} between $\boldsymbol{B}$ and the local normal $\boldsymbol{z}$. Once the viewing geometry is set, i.e. the angles χ and ξ that the star rotation axis Ω makes with the observer line-of-sight (LOS) and the star magnetic axis, respectively), the contributions of all the patches in view are summed together to derive the spectral and polarization properties of the emitted radiation. We also assume that the patch dimensions are small enough with respect to the radial scale to neglect the curvature of the closed field lines that contain the fireball, so that we can treat the atmosphere in the plane-parallel approximation. Under these assumptions the radiative transfer equations assume a relatively simple form,

$$\mu_z \frac{\mathrm{d}n_i}{\mathrm{d}\tau} = C_i n_i - S_i \,, \qquad (2)$$

where n_i are the photon number intensities ($i = \mathrm{O}, \mathrm{X}$), $\mathrm{d}\tau$ is the infinitesimal Thomson depth, μ_{z} is the cosine of the angle between the photon direction and the local normal, C_i are terms that do not depend on τ and S_i represents the source terms. However, due to the suppression factor B^{-2} that affects the X-mode cross sections (see equations 1), integrating the equations for the Thomson depth could require long computational times. In particular, the X-mode photosphere will lie at different heights in the fireball atmosphere for different photon energies, while at the same time the medium remains optically thick for ordinary photons at all the photon energies. For these reasons, we adopted the approach used by [26], solving the radiative transfer equations in terms of the Rosseland mean optical depth τ_{R}, which is related to the Rosseland mean cross section $\sigma_{\mathrm{X}} = \sigma_{\mathrm{XX}} + \sigma_{\mathrm{XO}}$ of the X-mode photons.

4. Numerical implementation

We developed a specific FORTRAN code to solve the radiative transfer in the fireball atmosphere. We assumed a maximum optical depth $\tau_{\mathrm{R}}^{\max} = 1000$ at the base of the atmosphere, since this guarantees that the medium is optically thick for X-mode photons over the entire energy range 1–100 keV, up to $\tau_{\mathrm{R}}^{\min} = 0$ at the top. A good approximation for the temperature distribution in a scattering-dominated medium in the diffusion regime turns out to be $T = T_{\mathrm{b}}(1{+}0.75\tau_{\mathrm{R}})^{1/2}$ [26], where T_{b} is the bolometric temperature (we assumed $T_{\mathrm{b}} = 10$ keV). We checked a posteriori this provides a good approximation to the numerical solution one obtains solving the energy balance in the fireball also at small optical depths. Using this temperature distribution, one can simply relate the Thomson scattering depths τ_{O} and τ_{X} for O- and X-mode photons, respectively, to the Rosseland mean optical depth. In order to avoid too long runs, we decide to integrate the two radiative transfer equations for $\tau_{O,X} \leq 10$ only, taking $n_{O,X}$ as Planck distributions at the local temperature T otherwise. Since, for the chosen values of the parameters, the X-mode photosphere lies always at larger τ_R with respect to the O-mode one, we solve the transfer equation for n_X only in the case of $\tau_X \leq 10$ and $\tau_O > 10$, taking n_O as Planckian [26, 27]. For the sake of simplicity, we assume the external magnetic field as dipolar, with polar intensity $B_p = 10^{14}$ G; the fireball is then limited by the close field lines that reach a maximum distance R_{max} from the star, taken as twice the stellar radius $R_{NS} = 10$ km. For this reason, all the patches in which the fireball atmosphere is divided are characterized by an angle $\theta_B = 90°$.

In order to sum together the contributions coming from all the patches in view, we used an IDL ray-tracing code [27, 28, 29]. Once the viewing angles χ and ξ have been chosen, the code solves the terminator condition $\boldsymbol{z} \cdot \boldsymbol{\ell} = 0$, where $\boldsymbol{\ell}$ is the unit vector of the observer LOS, to determine the point of the fireball surface which enter into view at different rotational phases. Since in general the fireball terminator shape turns out to be quite complicate, we decide to neglect the general relativistic corrections in our code. Nevertheless, as we checked in previous works [28], general relativity marginally influences the behavior of the polarization observables at infinity, and the qualitative conclusions we discuss in the following hold in general. Finally, we considered two different emission geometries, the model a, in which photons come from the entire torus of the field lines that limit the trapped fireball, and the model b, in which the emitting region is only a portion of this torus. In this latter case, radiation emitted from the planar slices that limit the emitting part of the torus is also considered. QED effects on the polarization modes are treated as discussed in [28, 30]; in particular, since we assumed the vacuum contributions to the dielectric tensor as dominant with respect to the plasma ones, vacuum resonance effects turn out to be negligible in our model.

5. Results

The intensities calculated in the radiative transfer code and processed in the ray-tracer are then used to obtain the simulated spectra and the polarization observables (both phase-resolved and phase-averaged) as observed at infinity. Since observations of the 'burst forest' emitted by SGR 1900+14 in 2006 [8] suggest that the spectrum of the intermediate flares (and of normal bursts too) is thermal and well reproduced by the superposition of two blackbodies, we attempted to fit the phase-averaged spectra with two Planckian distributions (with temperatures T_1 and T_2 and normalizations A_1 and A_2), although comparing results from our simplified model with real observations could be premature. Figure 1 shows the phase-averaged total spectrum of the radiation emitted from the fireball in the case of a limited emitting region (model b) with angular opening $\Delta\phi = 90°$, for $\chi = 60°$ and $\xi = 30°$, together with the best fit and the single blackbody components [27, for further details]. Under our assumptions, the values of T_1 and T_2 as well as of the ratio A_2/A_1 turn out to be in broad agreement with the observations, although the temperature of the softer component (~ 2 keV) turns out to be somehow lower than the observed one (~ 5 keV). No significant difference is expected switching from model a to model b or changing the viewing geometry. Moreover, we found that our model is able to reproduce the typical pulsating profile of the observed magnetar flare decay tails [27].

The observed polarization signal, simulated taking into account both vacuum polarization and the geometrical effects due to the magnetic field topology [28], confirms that, according to our model, radiation collected from a magnetar flare is expected to be highly polarized in the extraordinary mode. In particular, the phase-resolved linear polarization fraction Π_L attains a value generally higher than 80 per cent for photon energies between 1–50 keV, while a decrease up to ~ 40 per cent is expected only at higher energies. This reflects the behavior of the scattering cross-sections (see equations 1), according to which X-mode photons dominate the spectrum at almost all the photon energies, where τ_X becomes comparable with τ_O, above ~ 50 keV. Such

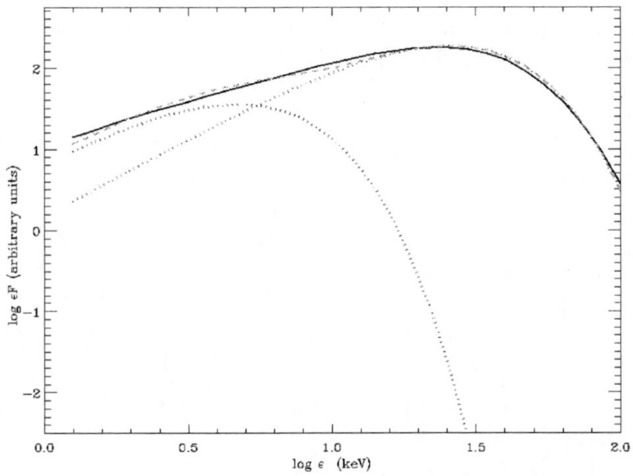

Figure 1. Phase-averaged photon spectrum (black solid line) of the radiation emitted in the case of model b (with $\Delta\phi = 90°$), fitted by the superposition of two blackbody components for $\chi = 60°$ and $\xi = 30°$. The best fit is marked by the light-blue dashed line, and the single blackbody components at themperatures T_1 (blue dotted line) and T_2 (red dotted line) are also shown.

a high polarization degree will be readily measurable by new-generation X-ray polarimeters like *IXPE*, *XIPE* and *eXTP*, as confimed by the simulations reported in Figure 2. Here, we refer to an event characterized by an X-ray flux $F_X = 4.68 \times 10^{-7}$ erg cm^{-2} s^{-1} in the 1–10 keV energy range and a duration $t_{\rm exp} = 1.737$ s, for different viewing geometries. It can be noted that both polarization fraction and angle measurements recover the values expected from the theoretical model with an acceptable degree of accuracy (within 1σ), although the errors on the polarization angle increase by decreasing the corresponding polarization degree.

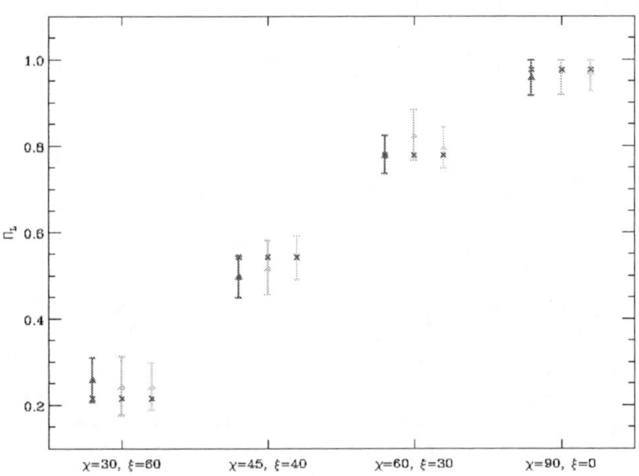

Figure 2. Simulation of the phase-averaged linear polarization fraction (triangles with error bars) of the *XIPE* (blue), *IXPE* (orange) and *eXTP* (green) polarimeters to the signal predicted by our theoretical model (in the case of model b) for a duration $t_{\rm exp} = 1.737$ s, X-ray flux $F_X = 4.68 \times 10^{-7}$ erg cm^{-2} s^{-1} in the 1–10 keV energy range and different viewing geometries (red crosses).

6. Conclusions

We have revisited the problem of modelling the spectral and polarization properties of the radiation emitted during magnetar flares in the context of the trapped-fireball model [6, 7]. Our code integrates the radiative transfer equations for both ordinary and extraordinary photons in the fireball atmospheric layer, divided into a number of different patches. This model generalizes the approach presented in [26], where only the case of $\boldsymbol{B}$ parallel to the patch normal was treated. The outputs of the radiative transfer code have then been reprocessed through a ray-tracing code, in order to obtain the spectra and the polarization observable distributions as measured by a distant observer. The results we obtained from the spectral analysis are compatible with observations, also in reproducing the typical pulse profile of magnetar flare decay tails. Finally, the model predicts a high degree of polarization, that promises to be readily measured by the recent polarimetric missions under development, allowing to verify the predictions of our theoretical model.

References

[1] Mereghetti S 2008 *A&A* Rev **15** 225
[2] Turolla R, Zane S and Watts A L 2015 *Rep. Prog. Phys.* **78** 11
[3] Duncan R C and Thompson C 1992 *ApJ* **392** L9
[4] Thompson C, Lyutikov M and Kulkarni S R 2002 *ApJ* **574** 332
[5] Nobili L, Turolla R and Zane S 2008 *MNRAS* **386** 1527
[6] Thompson C and Duncan R C 1995 *MNRAS* **275** 2
[7] Thompson C and Duncan R C 2001 *ApJ* **561** 980
[8] Israel G L *et al* 2008 *ApJ* **685** 1114
[9] Olive J-F *et al* 2004 *ApJ* **616** 1148
[10] Feroci M *et al* 2004 *ApJ* **612** 408
[11] Gnedin Yu N and Pavlov G G 1974 *Soviet Phys.-JETP Lett.* **38** 903
[12] Ho W C G and Lai D 2003 *MNRAS* **338** 233
[13] Lai D, Ho W C G, van Adelsberg M, Wang C and Heyl J S 2010 *X-ray Polarimetry: A New Window in Astrophysics* (Cambridge: Cambridge University Press)
[14] Herold H 1979 *Phys. Rev.* D **19** 2868
[15] Ventura J 1979 *Phys. Rev.* D **19** 1684
[16] Mészáros P 1992 *High-Energy Radiation from Magnetized Neutron Stars* (Chicago: University Chicago Press)
[17] Harding A K and Lai D 2006 *Rep. Prog. Phys.* **69** 2631
[18] Haug E 1975 *Zeitschrift Naturforschung Teil* A **30** 1099
[19] Svensson R 1982 *ApJ* **258** 335
[20] Haug E 1985 *A&A* **148** 386
[21] Adler S L, Bahcall J N, Callan C G and Rosenbluth M N 1970 *Phys. Rev. Lett.* **25** 1061
[22] Adler S L 1971 *Ann. Phys.* NY **67** 599
[23] Stoneham R J 1979 *J. Phys.* A **12** 2187
[24] Bulik T 1998 *Acta Astronomica* **48** 695
[25] Lightman A 1981 *ApJ* **244** 392
[26] Lyubarsky Y E 2002 *MNRAS* **332** 199
[27] Taverna R and Turolla R 2017 *MNRAS* **469** 3610
[28] Taverna R, Turolla R, González Caniulef D, Zane S, Muleri F and Soffitta P 2015 *MNRAS* **454** 3254
[29] Zane S and Turolla R 2006 *MNRAS* **366** 727
[30] Taverna R, Muleri F, Turolla R, Soffitta P, Fabiani S and Nobili L 2014 *MNRAS* **438** 1686

IOP Conf. Series: Journal of Physics: Conf. Series **932** (2017) 012024 doi:10.1088/1742-6596/932/1/012024

Vacuum birefringence and X-ray polarimetry in transient magnetars

D González-Caniulef[1], S Zane[1], R Taverna[2], R Turolla[2], X Song[1] and K Wu[1]

[1] Mullard Space Science Laboratory, University College London, Holmbury St. Mary, Dorking, Surrey, RH5 6NT, UK

[2] Department of Physics, University of Padova, via Marzolo 8, 35131 Padova, Italy

E-mail: `denis.caniulef.14@ucl.ac.uk`

Abstract. Recent optical polarimetry observations of an X-ray dim isolated neutron star, RX J1856.5−3754, showed a first evidence for QED vacuum birefringence induced by a strong magnetic field. This important result can be confirmed by performing systematically polarimetry observations in the X-ray band for other strongly magnetized neutron stars, such as transient or persistent magnetars. We computed the phase averaged polarization fraction (PF) and polarization angle (PA) expected in the thermal emission from transient magnetars in the soft X-ray energy band. We found that the detection of a PF higher than 60% is a strong evidence for vacuum birefringence. We also found that a steady change in the PA measured from transient magnetars during their outburst decay (up to 23 degrees for a magnetospheric untwisting of 0.5 rad) is a strong signature of vacuum birefringence. This latter detection would also provide an independent check of the magnetospheric untwisting model for these sources. Simulations show that these measurements are achievable by future polarimetric missions such as XIPE and IXPE with 20−380 ks of observational time, and with eXTP with 3−60 ks.

1. Introduction

According to quantum electrodynamics (QED), a strong magnetic field polarizes the vacuum, altering the dielectric and magnetic permeability tensors. This produces what is known as vacuum birefringence, predicted more than 80 years ago by Heisenberg and Euler (1936), that has been searched for in terrestrial laboratories but never detected. Neutron stars (NSs), with strong magnetic fields, are unique sources where we can study vacuum birefringence [1, 2, 3]. Recently, an optical polarimetry observation of an X-ray dim isolated neutron star (XDINS), RX J1856.5−3754 (hereafter RX J1856, with magnetic field $B \sim 10^{13}$ G), performed with the VLT, showed a phase-averaged polarization fraction, PF $\sim$ 16%, consistent with a potential signal of vacuum birefringence [4]. Future missions, such as the Imaging X-ray Polarimetry Explorer (IXPE [5]), the X-ray Imaging Polarimeter Explorer (XIPE [6]), and the enhanced X-ray Timing and Polarimetry mission (eXTP [7]), will allow us to perform X-ray polarimetry in transient magnetars (TMs, with magnetic fields $B \sim 10^{14}$ G). In these sources, the X-ray luminosity may increase up to $\sim$ 3 orders of magnitude [8], providing the conditions for a further potential detection of vacuum birefringence. In this work, we model and study the polarization properties of TMs, extending our previous work on XDINSs [3].

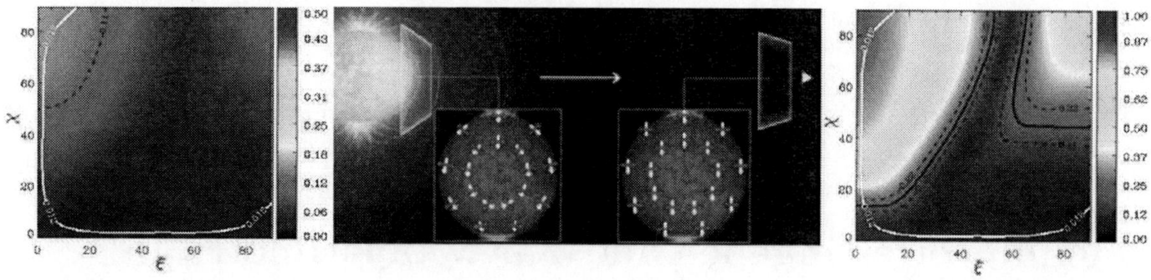

Figure 1. Left Panel: Phase-averaged PF for the radiation emitted from the NS surface (without vacuum birefringence) computed for different values of the angles ξ (magnetic axis and spin axis) and χ (line of sight and spin axis). The radiation is computed for a magnetized NS atmosphere, with $B \sim 10^{13}$ G as in the case of RX J1856, in the B-band for the VLT. The white line corresponds to the constraints on the viewing angles from the 1% X–ray pulsed fraction observed for RX J1856. Notice the [0.0–0.5] scale for the colorbar. **Central Panel**: Change of the electric field of the radiation as photons propagate through the NS magnetosphere. The subpanels show the direction of the photon E-field (red arrows) and the star B-field (white arrows) close to the surface (left subpanel) and after photons propagate through the magnetosphere with vacuum birefringence (right subpanel). Image credit: ESO. **Right Panel**: Same as in the left panel but accounting for vacuum birefringence. The black solid line corresponds to the PF observed with VLT and the black dashed lines mark the associated errors [4]. Notice the [0.0–1.0] scale for the colorbar.

2. XDINSs & Polarization Boost

Photons in a strong magnetic field are linearly polarized in two normal modes: the ordinary (O) and the extraordinary (X) modes, with the photon E-field oscillating either parallel or perpendicular to the plane of the local B-field and the photon propagation direction k. In a simplified picture, under vacuum birefringence, photons propagating in the NS magnetosphere rotate their electric field to re-adapt the normal modes according to the local direction of the magnetic field. This effect operates up to the so called adiabatic radius, which for XDINSs is located at a distance of a few NS radii [2]. As photons leave the star surface and travel along the line of sight (Figure 1, central panel), the magnetic field of the star becomes more uniform, which in turn forces a more uniform distribution of the electric field of the radiation, increasing the observed PF with respect to that produced according the B-field topology at the star surface [2, 3]. Figure 1 shows the PF expectations for RX J1856 in the optical band for **a)** the radiation emitted from the star surface, where the non-uniform B-field and the rotation of the Stokes parameters produce a relatively low PF (left panel), and **b)** the one obtained at the adiabatic radius (when vacuum birefringence is operating), where the B-field is more uniform and the small rotation of the Stokes parameters can produce a relatively high PF (right panel). If we consider vacuum birefringence (plus the constraints on the viewing angles given by the $\sim$ 1% X-ray pulsed fraction and those discussed in [9]), the results of our simulations for different emission models (atmosphere, condensed surface and 100% polarized blackbody radiation) turn out to be in agreement with the observed PF$\sim$ 16%. If vacuum birefringence is neglected, the measured PF and the constraints on the viewing angles discussed in [9] cannot be reproduced. Therefore, this suggests a first detection of the signal of vacuum birefringence [4].

3. Transient Magnetars

TMs are relatively hot NSs with outburst activity in which their X-ray flux can increase up to 3 orders of magnitude within hours and then decay on a time-scale of months [8]. The outbursts

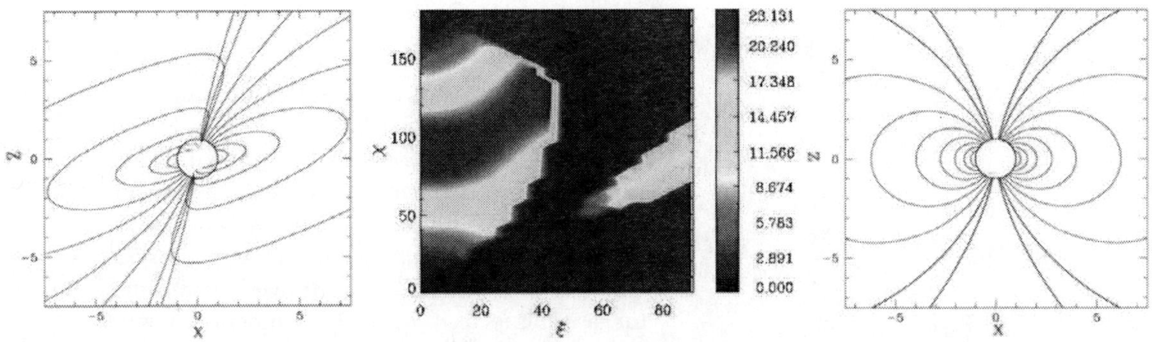

Figure 2. Left Panel: Example of a globally twisted magnetosphere with $\Delta\Phi = 2$ rad. **Central Panel**: Expected variation of the phase-averaged PA for TMs during a magnetospheric untwisting of 0.5 rad (as expected during the outburst decay, for details see Sec. 3), in the $2-6$ keV energy range. **Right Panel**: A fully untwisted magnetosphere, i.e. a perfect dipole. © The left and right panels are reproduced by permission of the AAS from the original ref. [10].

and subsequent flux decay observed in these sources are explained in the magnetar model in terms of catastrophic instabilities leading to a reconfiguration of the magnetosphere [10]. In particular, the internal toroidal field may transfer some helicity to the external magnetic field via plastic deformations or starquakes of the solid NS crust, producing a twisted magnetosphere (Figure 2, left panel). This twisted field evolves and gradually untwists [11, 12], dissipating magnetic energy until, eventually, a nearly dipolar, stable configuration is reached (Figure 2, right panel).

We model the polarization properties of TMs and study the phase-averaged PF and PA during the TM flux decay. For the TM mass and radius we assume canonical values $M = 1.5M_{\odot}$ and $R = 12$ km, respectively. Based on observational studies of representative sources such as XTE J1810−197 and CXOU J164710.2−455216 [13], we consider a TM with one hot polar cap, located in one of the magnetic poles, covering 15% of the star surface. Following [13], and for simplicity, we assume that the TM has a globally twisted magnetosphere. We assume that, at the flux peak, the magnetosphere has a twist angle $\Delta\Phi = 1.0$ rad and the hot polar cap has a uniform temperature of $T = 1.0$ keV. When quiescence is reached, the magnetosphere has untwisted by $\Delta\Phi = 0.5$ rad and the polar cap cooled down to $T = 0.5$ keV. These values are reminiscent of those inferred by [13], while analyzing the outburst decay of XTE J1810−197 and CXOU J164710.2−455216. We also assume that the X-ray thermal radiation is produced in a magnetized atmosphere composed of fully-ionized hydrogen. The intensities for the O-mode and X-mode are obtained by solving the radiative transfer equation, and assuming a polar magnetic field $B_p = 10^{14}$ G, perpendicular to the surface. These intensities are, then, used to model the thermal emission from the hot polar cap, neglecting the variation of the magnetic field over the relatively small extension of the emitting region. For computing the polarization quantities, we follow the procedure previously discussed for XDINSs [3], in which the Stokes parameters are computed at the adiabatic radius [2].

The results of our simulations can be summarized as follows: **a)** Similar to previous results obtained for XDINSs [3], the super-strong magnetic field surrounding TM can boost the observed PF of the thermal radiation, via the effect of vacuum birefringence, up to $\sim 99\%$ (considering the most favourable viewing geometry $\chi \sim 90°$ and $\xi \sim 0°$). **b)** When vacuum birefringence is operating and the TM has an untwisting magnetosphere, the value of PA can change up to 23 degrees (from the onset until quiescence, Figure 2, central panel), offering an interesting test for the magnetar model. **c)** If vacuum birefringence is not present and the polar cap shrinks during

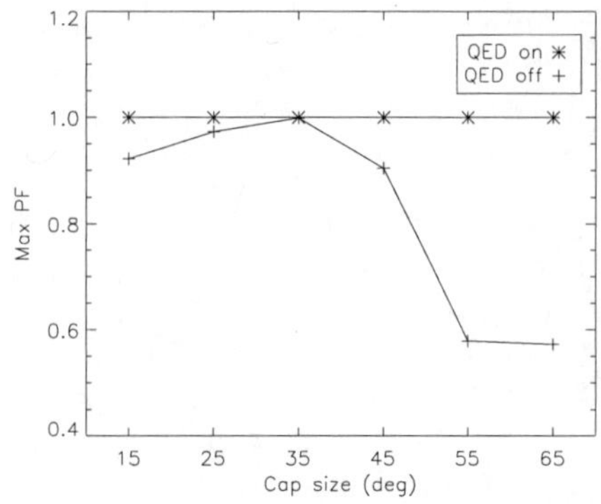

Figure 3. Maximum phase-averaged PF for different sizes of the polar cap (semi-angle of the polar cap). The line with asterisks symbols (*) corresponds to the case in which vacuum birefringence is operating. The line with crosses symbols (+) shows the case in which vacuum birefringence is not present. The radiation at the star surface is computed for a magnetized, pure-H atmosphere with $B_p = 10^{14}$ G and temperature $T = 0.5$ keV.

the TM flux decay, we found that the PF may increase from $\sim 65\%$ up to $\sim 99\%$ (Figure 3). Instead, if vacuum birefringence is operating, the PF stays almost constant independently on the cap size. This is an interesting result as it represents a further test on the magnetar scenario for TMs and the detection of a nearly constant PF as the decay proceeds will provide a further evidence for the presence of vacuum birefringence.

4. Future X-ray polarimetry missions

A number of missions devoted to X-ray polarimetry are currently under development, all with an estimated launch in the next decade: IXPE (NASA), XIPE (ESA), and eXTP (CAS/CNAS). All of them will be able to observe in 2–6 keV energy range, relevant for the study of TMs. Simulations (based on parameters for XTE J1810−197 and CXOU J164710.2−455216 [14, 15]) show that in order to observe a PF$\sim$ 70% and variation of the PA$\sim$ 10 deg, at the onset of the outburst (with a maximum flux $\sim 10^{-11}$ erg/cm^2/s), the observation time required by eXTP and XIPE are $\sim$ 3 ks and $\sim$ 20 ks, respectively. At the end of the X-ray flux decay, when the TM is in quiescence (minimum flux $\sim 5 \times 10^{-13}$ erg/cm^2/s), the observation time required to detect the same PF and PA by eXTP and XIPE are $\sim$ 60 ks and $\sim$ 380 ks, respectively. IXPE has a similar sensitivity of XIPE, therefore the requested observing time is of the same order. IXPE will likely allow to detect 1-2 TM during the planned mission lifetime.

Acknowledgments

DGC is supported by a CONICYT Becas-Chile fellowship 72150555.

References

[1] Heyl J and Shaviv N 2002 *Phys. Rev.* D **66** 023002
[2] Taverna R, Turolla R, González Caniulef D, Zane S, Muleri F and Soffitta P 2015 *MNRAS* **454** 3254
[3] González Caniulef D, Zane S, Taverna R, Turolla R and Wu K 2016 *MNRAS* **459** 3585
[4] Mignani R *et al* 2017 *MNRAS* **465** 492
[5] Weisskopf M *et al* 2013 *Proceedings of the SPIE* **8859** 885908
[6] Soffitta P *et al* 2016 *Proc. SPIE* **9905** 990515
[7] Jahoda K, Kouveliotou C and Praxys Team 2015, *Am. Astron. Soc. Meeting Abstr* **225** 338.40
[8] Rea N and Esposito P 2011 *ASSP* **21** 247
[9] Ho W 2007 *MNRAS* **380** 71
[10] Thompson C, Lyutikov M and Kulkarni S 2002 *ApJ* **574** 332
[11] Beloborodov A 2009 *ApJ* **703** 1044
[12] Kaspi V and Beloborodov A 2017 *ARA&A* **55** 261
[13] Albano A *et al* 2010 *ApJ* **722** 788
[14] Bernardini F *et al* 2009 *A&A* **498** 195
[15] An H, Kaspi V, Archibald R and Cumming A 2013 *ApJ* **763** 82

IOP Conf. Series: Journal of Physics: Conf. Series **932** (2017) 012025 doi:10.1088/1742-6596/932/1/012025

Fast radio bursts and their possible neutron star origins

J W T Hessels[1,2]

[1] ASTRON, the Netherlands Institute for Radio Astronomy, Postbus 2, 7990 AA, Dwingeloo, The Netherlands
[2] Anton Pannekoek Institute for Astronomy, University of Amsterdam, Science Park 904, 1098 XH Amsterdam, The Netherlands.

E-mail: `j.w.t.hessels@uva.nl`

Abstract. The discovery of the 'Lorimer Burst', a little over a decade ago, ignited renewed interest in searching for short-duration radio transients (Lorimer *et al* 2007 *Science* **318** 777). This event is now considered to be the first established Fast Radio Burst (FRB), which is a class of millisecond-duration radio transients (Thornton *et al* 2013 *Science* **341** 53). The large dispersive delays observed in FRBs distinguish them from the individual bright pulses from Galactic pulsars, and suggests that they originate deep in extragalactic space. Amazingly, FRBs are not rare: the implied event rate ranges up to many thousands of events per sky, per day (Champion *et al* 2016 *MNRAS* **460** L30). The fact that only two dozen FRBs have been discovered to date is a consequence of the limited sensitivity and field of view of current radio telescopes (Petroff *et al* 2016 *PASA* **33** e045). The precise localization of FRB 121102, the first and currently only FRB observed to repeat (Spitler *et al* 2014 *ApJ* **790** 101; Spitler *et al* 2016 *Nature* **531** 202; Scholz *et al* 2016 *ApJ* **833** 177), has led to the unambiguous identification of its host galaxy and thus proven its extragalactic origin and large energy scale (Chatterjee *et al* 2017 *Nature* **541** 58; Tendulkar *et al* 2017 *ApJL* **834** L7; Marcote *et al* 2017 *ApJL* **834** L8). It remains unclear, however, whether all FRBs are capable of repeating [many appear far less active (Petroff *et al* 2015 *MNRAS* **454** 457)] or whether FRB 121102 implies that there are multiple sub-classes. Regardless, the repetitive nature of FRB 121102 and its localization to within a star-forming region in the host galaxy (Bassa *et al* 2017 *ApJL* **843** L8) imply that the bursts might originate from an exceptionally powerful neutron star – one necessarily quite unlike any we have observed in the Milky Way. In these proceedings, I give a very brief introduction to the FRB phenomenon and focus primarily on the insights that FRB 121102 has provided thus far.

1. Introduction

Radio pulsar surveys have made great strides in the last decade. Thanks to increased computational power, higher time and frequency resolution can be achieved. This is enabling, e.g., the discovery of many more millisecond pulsars. Furthermore, long dwell times and larger fields-of-view are being employed to search for sporadically emitting sources like the rotating radio transients [1]. Larger fields of view and more on-sky time also open the prospect of discovering other types of millisecond-duration radio flashes. For example, given the scientific importance of gamma-ray bursts [2], it is natural to ponder and hope: are there also similar sirens of extreme (astro)physics to be found at radio wavelengths?

Unfortunately, there is no sufficiently sensitive all-sky radio monitor yet available. Compared to the all-sky monitoring routinely done in X-rays and γ-rays, a typical 'wide-field' radio telescope capable of offering (adequately sensitive) high-time-resolution burst searches has a field of view of only $\sim 1-10$ sq deg. In fact, the 64-m Parkes telescope, which has found the vast majority of the roughly two dozen FRBs published to date, requires ~ 10 days of observations to discover a single FRB, despite their inferred large event rate. Again thanks to increasing computational power, however, truly wide-field radio telescopes like CHIME [3], UTMOST [4], ASKAP [5] and APERTIF [6] are beginning to collect high-time-resolution data, and promise to detect up to dozens of FRBs per day. Such searches may also identify even rarer types of short-duration radio transients.

Adding to the computational challenges, short-duration radio waves are strongly affected by propagation effects induced by the intervening magneto-ionized medium between source and observer [7]. In particular, dispersion causes astrophysical radio pulses to arrive later at lower frequencies. This requires correction ('de-dispersion') in wide-band radio observations. Such effects are not purely a hindrance, however; they can also encode valuable information about the source's properties, its local environment, and its distance [8]. For example, the measurement of a relatively high Faraday rotation measure in one FRB implied that it is associated with a dense, magnetized plasma [9]. Furthermore, the dispersion measure (DM= $\int_0^d n_e(l)dl$) is the integrated column density of ionized material along the line of sight, and can be used as a proxy for distance [10]. This is how the extragalactic distances of FRBs are inferred in the (majority of) cases where no direct distance measurement is available. Dispersion is also critical for separating genuine astrophysical FRBs from a strong background of man-made radio frequency interference (RFI), particularly because some artificial signals can appear (at first glance) similar to FRBs [11].

A decade after the discovery of the 'Lorimer Burst' [12], the FRB phenomenon remains enigmatic. The field has gained great momentum, however, in the last few years, thanks in large part to many new discoveries and observational insights (see [13] for a broader review than provided here). The new wide-field radio telescope 'FRB factories', as well as prospects for directly localizing bursts in real time [14], promise major advances in the *coming* years as well.

Many theories for the physical nature of the FRBs have been proposed [15, 16, 13] — including both cataclysmic events involving neutron star collision (or collapse), along with non-cataclysmic scenarios involving a young and/or highly magnetized neutron star. While some FRBs appear to be one-off events — arguably supporting a cataclysmic origin — the sporadically repeating FRB 121102 has also recently been discovered [17]. It is thus currently unclear whether there is a single type of FRB, or whether we are seeing multiple source populations [18]. In any case, it is very plausible that exotic manifestations of neutron stars are at least part of the puzzle. After all, several types of millisecond-duration radio emission are already known to be created in neutron star magnetospheres: i. canonical, magnetic polar cap pulsar radio emission; ii. giant pulses, like those seen in the Crab; and iii. radio pulses from magnetars, which arguably originate from yet another physical process [19]. In these proceedings I tell the story of the discovery and follow-up study of 'The Repeating FRB', FRB 121102. The repetitive nature of this particular FRB has allowed deep, multi-wavelength campaigns that have delivered many key insights.

2. The Repeater FRB 121102

Until 2014, all reported FRBs had been discovered using the 64-m Parkes radio telescope and its multi-beam receiver. The discovery of FRB 121102 (also known as the 'Spitler Burst'; figure 1) in the PALFA pulsar survey [20], which uses the 305-m Arecibo telescope [23, 24], was the first non-Parkes detection. At the time, the discovery greatly bolstered the astrophysical interpretation of the FRBs because it helped dispel concerns that the phenomenon was some pernicious source of local interference, like the so-called Perytons [25, 11].

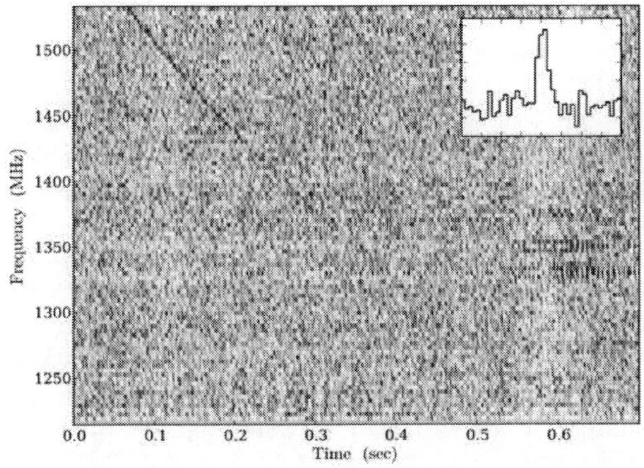

Figure 1. The first observed burst from FRB 121102 [20]. The main panel shows the characteristic dispersive delay and burst intensity as a function of frequency and time. The inset is the burst light curve after dedispersion and integrating across the observed band. (Credit: L. Spitler)

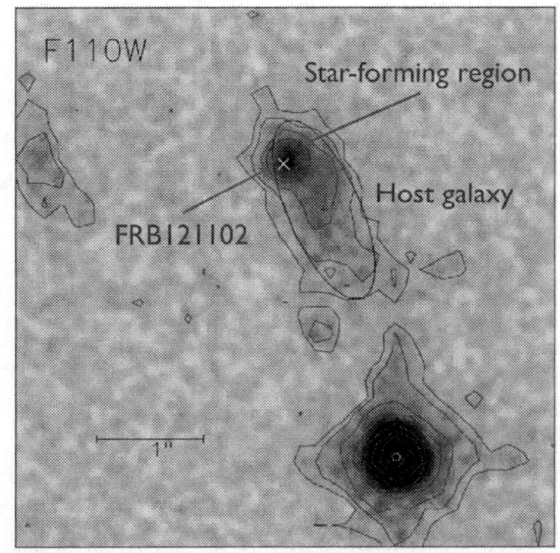

Figure 2. Localization of FRB 121102 within a star-forming region in its host galaxy [21]. The position of the EVN burst and persistent source localization is shown by the white 'x' [22]. (Adapted from [21]

Surprisingly, the burst appeared to have an inverted spectral index, becoming much brighter towards the top of the observing band. This had not been seen in previous FRBs, and was explained as possibly coming from the detection of the burst in a side-lobe of the receiver sensitivity pattern (though this required fine-tuning [26]). The low Galactic latitude of the burst ($b = -0.2°$) left open the possibility that FRB 121102 could be a intermittent Galactic pulsar like the rotating radio transients (RRATs) [1]. However, since the PALFA survey is restricted to Galactic latitudes $|b| < 5°$, by definition it can only find such sources along the Galactic plane. Furthermore, FRB 121102's DM$= 557\,\mathrm{pc\,cm^{-3}}$ is roughly $3\times$ larger than the maximum predicted Galactic contribution along this line of sight [10], meaning that an unmodelled HII region would need to be invoked to explain the apparent discrepancy. The lack of observed pulsars with anomalously high DMs in this part of the sky provided an additional argument that FRB 121102 was something different.

A follow-up observing campaign was planned in order to search for repeat bursts at the same sky location and DM. Though no other FRB had been seen to repeat, despite dedicated follow-up [27], the possibility of a RRAT origin for FRB 121102 gave extra motivation for continued monitoring. Surprisingly, 10 more bursts were detected in just a few hours of additional observations [17] (for comparison, the discovery observation lasted only $\sim 200\,\mathrm{s}$). In a grid of pointing positions that covered the initial discovery position and side-lobes of the receiver, these additional bursts all had consistent sky position and DM, thereby demonstrating that they were coming from the same source, and that it was following the sidereal rate. The detection of multiple bursts immediately ruled out a cataclysmic origin for FRB 121102; whatever source is producing the bursts needs to survive the energetic events that produce them. Interestingly, the new sample of bursts showed a broad range of burst widths, morphologies, and spectra [17]. Though the refined localization indicated that the first burst had indeed been detected in a receiver side-lobe, the spectral volatility could not be ascribed to an instrumental effect. Rather, it would have to be intrinsic to the emission process itself or the result of propagation in the intervening magneto-ionized material.

Multi-wavelength observations of the field found no evidence for Galactic structure capable of explaining FRB 121102's anomalously high DM [18]. In tens of hours of further monitoring, however, only 6 more bursts were detected, using the Green Bank Telescope (GBT) and Arecibo. This suggested that the initial follow-up campaign had been rather fortunate in finding 10 bursts, and that the activity level of FRB 121102 is highly variable (in fact 6 of the 10 bursts originally presented were detected within a single 10-min time window [17]).

The $\sim 2'$ localization precision of the Arecibo redetections was completely insufficient to identify a host galaxy. Rather, precise and direct localization of FRB 121102 was achieved after tens of hours of monitoring using the Very Large Array (VLA) in a fast-dump (5 ms visibilities) mode. During a period of pronounced source activity in September 2016, these observations captured 9 bursts, which were directly imaged and localized to sub-arcsecond precision [28]. Unlike indirect methods that attempt to associate an FRB with a multi-wavelength counterpart or afterglow [29], this method leaves no ambiguity about the association.

With a precision localization in hand, it soon became apparent that FRB 121102 has both persistent radio and optical counterparts. The VLA continuum maps identified a $\sim 200\,\mu$Jy persistent radio source at the exact location of the bursts. Very-long-baseline radio interferometric (VLBI) observations using the European VLBI Network (EVN) and the Very Long Baseline Array (VLBA) also detected this source, and showed that it is compact on milli-arcsecond angular scales. Archival Keck, and newly acquired Gemini data identified a weak (~ 25 mag) optical counterpart, possibly just barely resolved.

Additional EVN observations, including Arecibo, detected one bright and three weaker bursts. These observations stored the raw voltage signals from each radio dish, which allowed offline correlation to be done at high time resolution during the specific times of the bursts. With a longer-baseline interferometric array, it was possible to improve the precision of the burst localization by another order of magnitude, achieving a ~ 12 mas burst position uncertainty (dominated by limited uv coverage)[22]. These observations also detected the persistent radio counterpart and showed that it must be < 0.7 pc in extent and < 40 pc from the source of the bursts. This established that there is either a physical link between the source of the bursts and the source of the persistent radio emission or, perhaps, that the persistent radio source is *directly* responsible for the bursts themselves.

Using the 8-m Gemini North, optical spectroscopy of the persistent optical counterpart detected strong emission lines and was able to measure a redshift: $z = 0.193$ (luminosity distance ~ 1 Gpc). This confirmed FRB 121102's extragalactic origin beyond any doubt, and established the first robust association of an FRB with a host galaxy[30]. Importantly, this also establishes the energy scales of the bursts and persistent source: $L_{\rm Burst} \sim 10^{40-42} \times (\Delta\Omega_b/4\pi)\,{\rm erg\,s^{-1}}$ and $L_{\rm Persistent} \sim 7\times10^{38}\times(\Delta\Omega_p/4\pi)\,{\rm erg\,s^{-1}}$. Here $\Delta\Omega_b$ is the burst emission solid angle (which could be small) and $\Delta\Omega_p$ is the persistent emission solid angle, almost certainly different than $\Delta\Omega_b$, and potentially isotropic. Surprisingly, the host galaxy is a low-metallicity dwarf with a total stellar mass $\sim 10^8\,{\rm M_\odot}$. Interestingly, superluminous supernovae (SLSNe) and long gamma-ray bursts (LGRBs) are also preferentially found in dwarf galaxies similar to FRB 121102's host [22, 30, 31]. This presents the tantalizing possibility that FRB 121102 is a young and extreme neutron star created during such an explosion. This hypothesis is now being tested via targeted searches for millisecond-durations radio bursts towards the sites of known SLSNe and LGRBs. More recent *Hubble* observations show that FRB 121102 is close to the center (figure 2) of an active star-forming region (or the blend of several unresolved star-forming regions) [21]. This accounts for the host galaxy's optical emission lines, and further strengthens the association of FRB 121102 to (massive) star formation.

While the repetition of FRB 121102 has thus had huge practical advantages for its localization, this also affords deep, multi-wavelength observing campaigns that can more

IOP Conf. Series: Journal of Physics: Conf. Series **932** (2017) 012025 doi:10.1088/1742-6596/932/1/012025

precisely characterize the properties of the radio bursts themselves and search for prompt emission at higher energies. Using coherent dedispersion, it is possible to recover a much higher effective time resolution, which does not suffer from dispersion smearing within individual frequency channels. This has demonstrated that FRB 121102 bursts are often multi-peaked and can have structure at $< 1\,\mathrm{ms}$ timescales ([17, 18]; also Hessels et al, in prep.). While the bursts are often narrow-band ($\sim 100 - 200\,\mathrm{MHz}$ at central observing frequencies of $\sim 1.4\,\mathrm{GHz}$), multi-telescope observations show that they are also occasionally broadband (spanning several GHz) [32]. The degree to which this is intrinsic to the burst emission mechanism or due to extrinsic propagation effects remains unclear. While there are certainly precedents for narrow-band features in the bursts of the Crab pulsar [33] it has also been suggested that the spectral features can be due to plasma lensing, which has the added advantage that it can boost the observed brightness of the bursts and lessen the energy required to explain the bursts [8]. Multi-wavelength observations, simultaneous with radio burst detections, have thus far failed to detect optical or X-ray counterparts to the bursts or persistent radio source [34, 35]. Unfortunately, given the large distance to FRB 121102, this is only mildly constraining on possible magnetar models [35].

In summary, there are several lines of evidence that suggest that FRB 121102 originates from a young and highly magnetized neutron star. Several models invoking 'super-giant' radio pulses, magnetar flares, etc. have been proposed [36, 37, 38, 39]. While the precedents of millisecond-duration radio pulses from neutron stars make that a logical (and conservative) conclusion, the extreme energetics and lack of direct Galactic analogue leave various open questions: e.g. are the FRB 121102 bursts rotationally or magnetically powered? Models have also been proposed in which a neutron star can produce sufficiently luminous bursts only because of an external influence (e.g. [40]). Ultimately, one should remain open-minded because a definitive proof of the neutron star scenario is still lacking: e.g. the detection of a well-defined periodicity that could only plausibly be ascribed to a rotating neutron star.

An equally compelling puzzle is whether FRB 121102 has the same physical origin as the remaining FRB population [17, 41]. Deep searches for repeats from other FRBs have so far detected no additional bursts [27], and thus FRB 121102 has to (at least) be atypically active compared to other sources; this requires explanation either via intrinsic source properties or external effects. That said, FRB 121102 shares many of the other characteristics of the general population, and the spectral and temporal properties of other FRBs suggest some commonality [42, 43]. There are high chances that more repeating FRBs will be discovered in the coming years by the aforementioned 'FRB factories', and that FRBs will begin to be precisely localized in real time. Thus, it appears that clarity is within sight.

Acknowledgments
I heartily thank my various collaborators for their great FRB research accomplishments in the last few years. In particular, I would like to thank the following colleagues for their help in preparing the talk on which this proceedings contribution is based; in alphabetical order: Cees Bassa, Shami Chatterjee, Jim Cordes, Kelly Gourdji, Casey Law, Benito Marcote, Daniele Michilli, Zsolt Paragi, Emily Petroff, Paul Scholz, and Laura Spitler. On a personal note, I also thank the dedicated staff of the Arecibo Observatory for their continued professionalism and hard work – especially in the aftermath of a major natural disaster in Puerto Rico.

References

[1] McLaughlin M A *et al* 2006 *Nature* **439** 817–20
[2] Gehrels N and Mészáros P 2012 *Science* **337** 932–6
[3] Amiri M *et al* 2017 *ApJ* **844** 161
[4] Caleb M *et al* 2016 *MNRAS* **458** 718–25
[5] Bannister K W *et al* 2017 *ApJL* **841** L12

[6] Maan Y and van Leeuwen J 2017 *Preprint* 1709.06104
[7] Rickett B J 1990 *ARA&A* **28** 561–605
[8] Cordes J M, Wasserman I, Hessels J W T, Lazio T J W, Chatterjee S and Wharton R S 2017 *ApJ* **842** 35
[9] Masui K *et al* 2015 *Nature* **528** 523–5
[10] Cordes J M and Lazio T J W 2002 *Preprint*
[11] Petroff E *et al* 2015 *MNRAS* **451** 3933–40
[12] Lorimer D R, Bailes M, McLaughlin M A, Narkevic D J and Crawford F 2007 *Science* **318** 777–80
[13] Petroff E 2017 *Preprint* 1709.02189
[14] Law C J, Bower G C, Burke-Spolaor S, Butler B, Lawrence E, Lazio T J W, Mattmann C A, Rupen M, Siemion A and VanderWiel S 2015 *ApJ* **807** 16
[15] Macquart J P *et al* 2015 *Advancing Astrophysics with the Square Kilometre Array (AASKA14)* 55
[16] Katz J I 2016 *Modern Physics Letters A* **31** 1630013
[17] Spitler L G *et al* 2016 *Nature* **531** 202–5
[18] Scholz P *et al* 2016 *ApJ* **833** 177
[19] Lorimer D R and Kramer M 2004 *Handbook of Pulsar Astronomy* (Cambridge: Cambridge University Press)
[20] Spitler L G *et al* 2014 *ApJ* **790** 101
[21] Bassa C G *et al* 2017 *ApJL* **843** L8
[22] Marcote B *et al* 2017 *ApJL* **834** L8
[23] Cordes J M *et al* 2006 *ApJ* **637** 446–55
[24] Lazarus P *et al* 2015 *ApJ* **812** 81
[25] Burke-Spolaor S, Bailes M, Ekers R, Macquart J P and Crawford III F 2011 *ApJ* **727** 18
[26] Kulkarni S R, Ofek E O and Neill J D 2015 *Preprint* 1511.09137
[27] Petroff E *et al* 2015 *MNRAS* **454** 457–62
[28] Chatterjee S *et al* 2017 *Nature* **541** 58–61
[29] Keane E F *et al* 2016 *Nature* **530** 453–6
[30] Tendulkar S P *et al* 2017 *ApJL* **834** L7
[31] Metzger B D, Berger E and Margalit B 2017 *ApJ* **841** 14
[32] Law C J *et al* 2017 *Preprint* 1705.07553
[33] Hankins T H and Eilek J A 2007 *ApJ* **670** 693–701
[34] Hardy L K *et al* 2017 *MNRAS* **472** 2800–7
[35] Scholz P *et al* 2017 *ApJ* **846** 80
[36] Pen U L and Connor L 2015 *ApJ* **807** 179
[37] Connor L, Sievers J and Pen U L 2016 *MNRAS* **458** L19–23
[38] Cordes J M and Wasserman I 2016 *MNRAS* **457** 232–57
[39] Lyubarsky Y 2014 *MNRAS* **442** L9–13
[40] Zhang B 2017 *ApJL* **836** L32
[41] Petroff E, Barr E D, Jameson A, Keane E F, Bailes M, Kramer M, Morello V, Tabbara D and van Straten W 2016 *PASA* **33** e045
[42] Champion D J *et al* 2016 *MNRAS* **460** L30–4
[43] Ravi V *et al* 2016 *Science* **354** 1249–52

IOP Conf. Series: Journal of Physics: Conf. Series **932** (2017) 012026 doi:10.1088/1742-6596/932/1/012026

Properties of Konus-Wind SGR bursts

A V Kozlova on behalf of the Konus-Wind team
Ioffe Institute, 26 Polytekhnicheskaya, St. Petersburg, 194021, Russia
E-mail: ann_kozlova@mail.ioffe.ru

Abstract. We present preliminary results of systematic temporal and spectral analysis of the Konus-*Wind* observational data on short and intermediate SGR bursts. We conclude that the burst energy spectra are equally well described, in the $20 - 200$ keV range, by both power-law with an exponential cutoff, and double black-body functions. We also discuss energetics and durations of the bursts, distributions of the spectral parameters and correlations between them.

1. Introduction

Soft gamma repeaters (SGRs) are a rare, enigmatic class of astrophysical objects. There are 25 sources discovered during more than 30 years of observations [1]. The SGRs are believed to be isolated neutron stars with rotation periods of $\sim 5 - 12$ s and inferred surface magnetic field strength $B \sim 10^{14} - 10^{15}$ G. They are characterized by sporadic and unpredictable periods of bursting activity in soft gamma-rays (< 200 keV). These periods of activity may last from few days to few years and are separated by long periods of quiescence, where a much weaker persistent X-ray source observed. The bursting activity is quite diverse and consists of three types of events: the most intense, giant flares, characterized by short, hard initial pulse having a huge energy release of $\sim 10^{44}$–10^{46} erg followed by a long-duration decaying tail; the most frequent short duration bursts with peak luminosity $L_{max} < 10^{42}$ erg/s; and intermediate bursts with energies and luminosities between the short bursts and giant flares.

From its launch in 1994, Konus-*Wind* (KW) has detected ~ 250 bursts from 6 SGRs. In next section we present the results of systematic temporal and spectral analyses of 201 short and intermediate bursts from 5 SGRs. Giant flares and burst series were excluded from our analysis since they require a special treatment. Also, we do not include SGR 0501+4516 because of a very low burst statistics. The results of KW study of this SGR have been already published in [2].

2. Results and Discussion

Our analyses include 106 bursts from SGR 1806-20, 47 bursts from SGR 1900+14, 14 bursts from SGR 1627-41, 22 bursts from SGR 1550-5418, and 12 bursts from SGR 1935+2154. For each studied SGR we investigate the distribution of T_{90} parameter, defined as the time during which 90 % of the total burst counts have been accumulated. SGR 1806-20, with the greatest number of intermediate bursts, have a longer average burst duration of ~ 0.7 s compared to ~ 0.2 s for the rest. All durations fall well within the same order of magnitude, indicating a similar origin for the bursts across the SGR population.

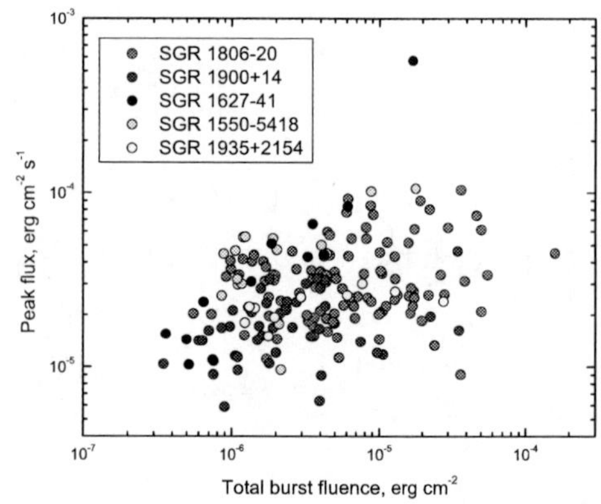

Figure 1. Peak flux vs total burst fluence (both in 20–200 keV range).

Motivated by our previous work [3] and results obtained in other studies (e.g.[4, 5]), two model functions were used to fit time-integrated spectra: the power-law with an exponential cutoff (CPL) model, parametrized as E_p: $f(E) \propto E^{\alpha} \exp(-(2+\alpha)E/E_p)$, where α is the power-law (PL) photon index and E_p is the peak energy in the νF_ν spectrum and a sum of two black-body (BB) functions with the normalization proportional to the surface area (2BB). Our spectral analysis shows that the KW-measured SGR spectra are equally well described, in the 20–200 keV range, by both CPL and 2BB models. For each SGR, a mean $\Delta\chi^2$ between those model fits is less than 4. A detailed study of the bright burst from SGR 1935+2154 [3] showed that such a difference in χ^2 does not allow to favour any of the models even in the case of a high count statistics.

The obtained burst energetics is shown in figure 1 with the peak flux vs the total fluence distribution for each SGRs. These parameters are calculated using the best-fit model flux in the 20 to 200 keV range. In the spectral fits with the CPL model, we find E_p value distributions peak between 20 and 40 keV similar for all studied SGRs. The observed variety in α distributions between different SGR sources ($\alpha \sim -1.2$ for SGR 1806-20, $\alpha \sim -0.5$ for SGR 1900+14 and SGR 1550-5418, $\alpha \sim -0.2$ for SGR 1627-41, $\alpha \sim 0.2$ for SGR 1935+2154) confirms the results obtained in previous studies [5, 6, 7] and could arise from differences in value of magnetic field, geometry, or plasma temperature. The temperatures and emission areas we find for the 2BB model are similar among SGRs: low-kT $\sim 3-7$ keV, $R_{low-kT} \sim 13-100$ km and high-kT $\sim 11-16$ keV, $R_{high-kT} \sim 3-10$ km. These values are consistent with those found in others studies [6, 8, 9]. The 2BB temperatures are distributed in rather narrow ranges for all of the bursts, despite orders of magnitude difference in energy release. This suggests that these parameters depend weakly on either the burst magnitude or morphology.

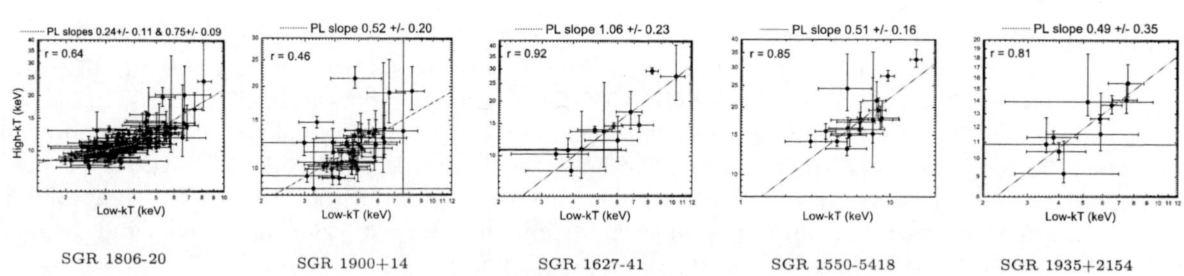

Figure 2. Correlations between temperatures of the low-kT BB vs high-kT BB.

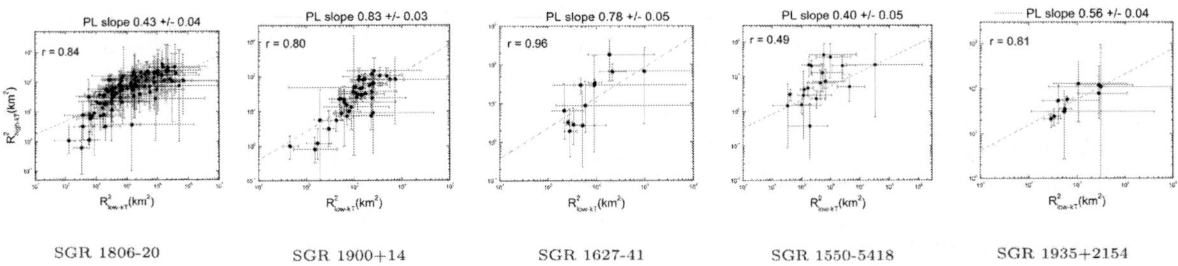

Figure 3. Correlations between the emitting areas corresponding to the low-kT and high-kT BBs.

In figures 2, 3 we show that 2BB temperatures are well correlated, and the same is true for their emission areas. For SGR1806-20 we find that the temperatures of the low-kT BB vs high-kT BB are best fit with a broken power law (BPL) with the break located at ~ 4 and 11 keV. This result is consistent with the time-resolved spectroscopy of GBM bursts from SGR 1550-5418 [10]. Other SGRs distributions do not require BPL and show only a positive correlation between the low- and high-kTs. The areas of the emitting regions of the low- and high-kT BBs are positively correlated for all SGRs. It is consistent with the reports on time-integrated analysis in the literature (e.g. [9]). The slope of PL fit for low-kT vs high-kT distribution obtained for SGR 1550-5418 (see figure 3) is much shallower than the 1.86 ± 0.09, previously reported in [5]. However, we note that this difference could arise from the $8 - 200$ keV GBM energy range, which is softer than energy range covered in our work. We also investigated (see figure 4) the relation between the emission area and temperature for both BB components simultaneusly. All SGRs showed a similar trend: the area of low-kT BB decreases with kT at a

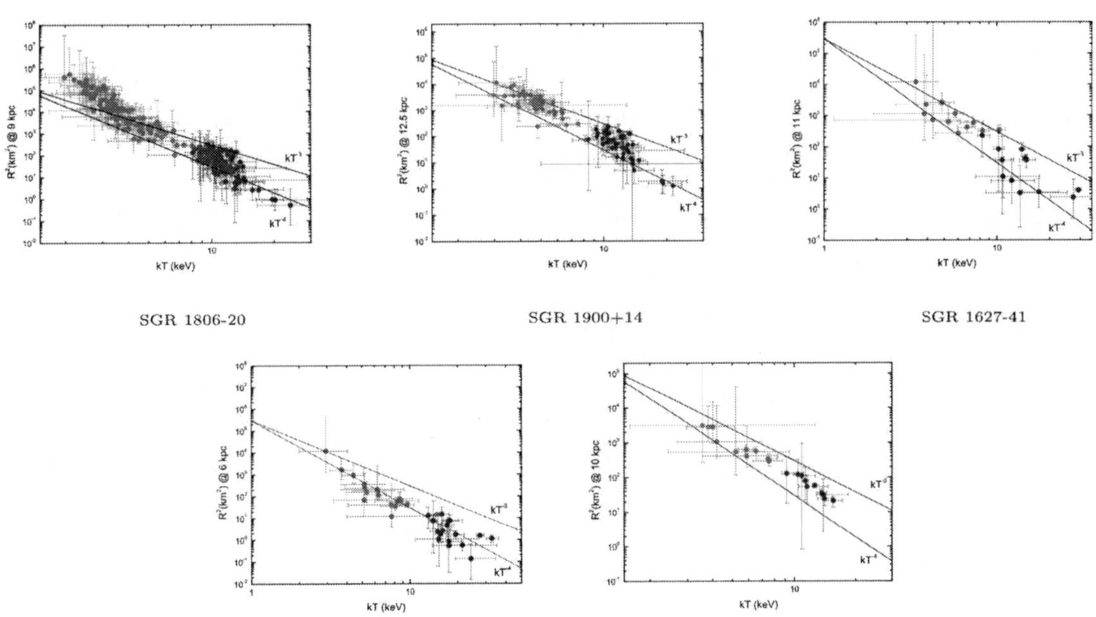

Figure 4. Correlations between the radii of the emission areas vs temperature for both BBs simultaneusly.

slower pace than the area of the high-kT BB. A further study of such similarities, distinctions, and individual features should lead to a deeper understanding of the emission mechanism taking place during SGR bursts.

Acknowledgments
This work is supported by RFBR grant 15-02-00532.

References

[1] Olausen S A and Kaspi V M 2014 *ApJS* **212** 6
[2] Aptekar R L, Cline T L, Frederiks D D, Golenetskii S V, Mazets E P and Pal'shin V D 2009 *ApJL* **698** L82–L85
[3] Kozlova A V *et al.* 2016 *MNRAS* **460** 2008–2014
[4] Lin L, Kouveliotou C and van der Horst A J 2011 *ApJ* **739** 87
[5] van der Horst A J *et al.* 2012 *ApJ* **749** 112
[6] Feroci M, Caliandro G A, Massaro E, Mereghetti S and Woods P M 2004 *ApJ* **612** 408–413
[7] Lin L *et al.* 2012 *ApJ* **756** 54
[8] Olive J F, Hurley K, Sakamoto T, Atteia J L, Crew G, Ricker G, Pizzichini G, Barraud C and Kawai N 2004 *ApJ* **616** 1148–58
[9] Nakagawa Y *et al.* 2007 *PASJ* **59** 653–78
[10] Younes G *et al.* 2014 *ApJ* **785** 52

IOP Conf. Series: Journal of Physics: Conf. Series **932** (2017) 012027 doi:10.1088/1742-6596/932/1/012027

High time resolution multi-band photo-polarimetric observations of the binary millisecond redback pulsar J1023+0038 with the BTA

Yu A Shibanov[1,2], **G M Beskin**[3,5], **S V Karpov**[3,5], **V L Plokhotnichenko**[3], **D A Zyuzin**[1], **A F Kholtygin**[4], **V V Sokolov**[3] **and Yu V Baryshev**[4]

[1] Ioffe Institute, Politekhnicheskaya 26, St. Petersburg, 194021, Russia

[2] Peter the Great Saint Petersburg Polytechnic University, Politekhnicheskaya 29, St. Petersburg, 195251, Russia

[3] Special Astrophysical Observatory, Nizhnij Arkhyz, Karachaevo-Cherkessia, 369167, Russia

[4] Saint Petersburg State University, Universitetskij pr. 2 8, St. Petersburg, 198504, Russia

[5] Kazan Federal University, 16a Kremlyovskaya St., Kazan 420008, Russia

E-mail: `shib@astro.ioffe.ru`

Abstract. We briefly report first results of high time resolution optical multi-band panoramic photo-polarimetric observations of the eclipsing binary millisecond redback pulsar J1023+0038 obtained in February 2017 with the 6 m BTA telescope. The time resolution was varied from 10 to 120 ms depending on observational mode. Our data show that the pulsar still remained in the low-mass X-ray binary stage, characterised by rapid flaring at time scales of 10–100 s with amplitudes of 0.2–0.5 mag. We resolved a fine structure of the flares at time scales of 0.1–10 s. The polarimetry at the time scale of 0.1 s shows no polarization with an upper limit of 2%–4% for the linear polarisation degree in flaring and quiet stages, while at a 10 minute scale averaging it is about 1.5% at 3σ significance. We shortly outline implications of the results.

1. Introduction

Rotation powered millisecond pulsars (RMSPs) were discovered in the radio about 35 yr ago, the first one was PSR B1937+21 [1]. Two independent groups immediately suggested that RMSPs were spun-up ("recycled") by accretion in close binary systems [2, 3] – the mechanism proposed 6 yr earlier [4]. For a long time this idea was only supported by the fact that most RMSPs are in binary systems with ordinary stellar companions. The discovery of 401 Hz X-ray coherent pulsations from the accretion powered neutron star (NS) in the low mass X-ray binary (LMXB) SAX J1808.4-3658 in 1998 [5] was the first direct evidence of the accretion spin-up.

Compact RMSPs binaries with binary periods P_b <1 day show two distinct sub-classes: "black widows" (BWs; [6]) with sub-stellar companion masses $M<0.1M_\odot$ and "redbacks" with $M>0.1M_\odot$ bloated companions close to filling their Roche lobes [7]. The LMXB–RMSP connection was finally firmly established by the recent discovery of three redbacks, PSR J1023+0038 [8], XSS J12270−4859 [9, 10], and PSR J1824−2452I [11], directly showing transformations between the accretion and the RMSP stages. Among these three, PSR J1023+0038 (hereafter J1023) is the most studied and intriguing object. It showed the LMXB–RMSP transition in 2003 [8] and

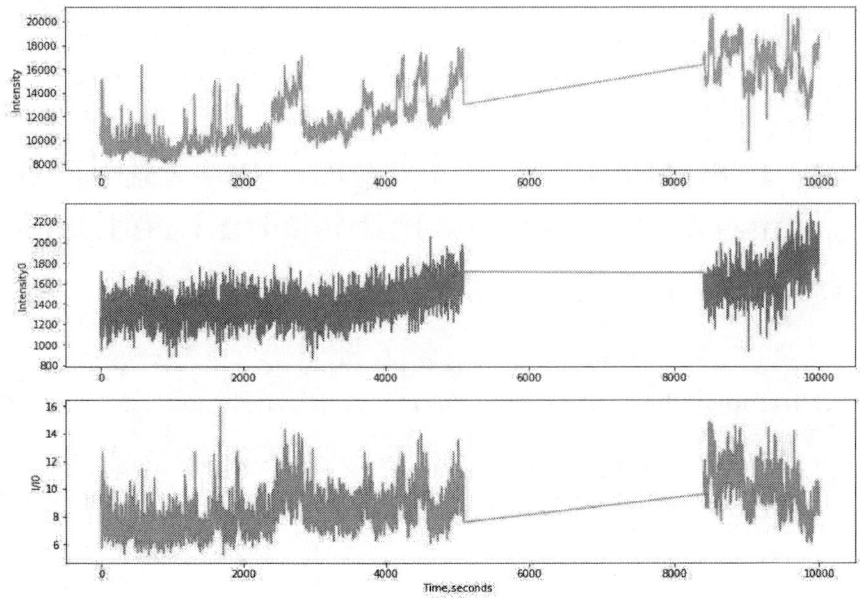

Figure 1. Total light curves in counts per time bin of J1023 (*top*) and the comparison star (*middle*) in the B band, and their ratio (*bottom*). The time bin is 1.2 s.

then suddenly returned back to a low-luminosity accretion stage in 2013 [12], demonstrating that the transition itself is a complicated multi-stage process. J1023 is an eclipsing binary 1.64 ms pulsar with the $P_b = 4.754$ h and an $\approx 0.2M_\odot$ non-degenerate G-class companion. The parallax-based distance is 1.37 kpc [13]. The object is being monitored in different spectral domains. Its redback nature follows from the modulation of its optical flux and colour with the period P_b due to heating of the companion's face by the pulsar [14]. The modulation of the X-ray flux with P_b is interpreted as the presence of an intra-binary shock, which is supported by γ-ray observations [15]. In the accretion stage, J1023 shows sporadic switches between high and low X-ray luminosity modes with durations from minutes to hours, accompanied by occasional bright flares [16]. Coherent X-ray pulsations with the pulsar period, possibly caused by channelled accretion onto magnetic poles of the NS, are detected in the high luminosity mode [17]. However, the implied accretion rate appears to be too low to provide the matter penetration through the centrifugal barrier of the pulsar. It is also intriguing, that the spin-down rates in the LMBX and RMSP stages are almost the same [18], suggesting that the pulsar wind continues to operate. This is supported by the detection of radio brightening episodes during the low X-ray mode intervals [19].

In the optical, J1023 is a relatively bright object, $V \approx 16.7$, with a low extinction, $A_V \approx 0.22$. First optical high time resolution observations in 2015 also showed occasional flares and sharp dips at 0.1–1 magnitude levels and 20 s–10 min time scales [14]. Implied spatial scales can be as small as $0.3R_\odot$, which is comparable with the J1023 accretion disk size. However, many flares and dip fronts were not resolved due to a limited time resolution > 0.3 s. Higher time resolution observations are necessary to study the variability nature.

2. Observations

J1023 was observed on February 17, 2017 during 3.5 h with the 6 m BTA telescope using the Multicolor Panoramic Photo-Polarimeter (MPPP) instrument [20]. Conditions were clean with

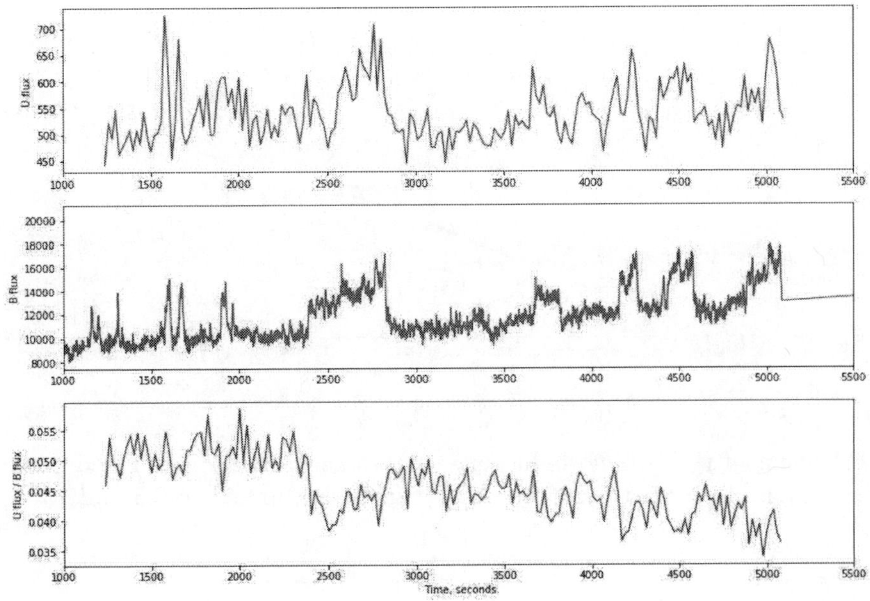

Figure 2. Comparison of the U (*top*) and B (*middle*) band light curve fragments at time bin sizes of 20 and 1.2 s, respectively. The U/B flux ratio is shown in the *bottom* panel.

seeing of $\approx 1''.7$. The MPPP allows one to register photons from the target in four photometric bands and/or to obtain three Stokes parameters simultaneously. Photons can be registered by two detectors: the microchannel-plate-based position-sensitive photon counter (PSD) providing the time resolution up to 100 ns, and the low readout noise PhotonMax 512B EMCCD from Princeton Instruments with the resolution up to 1 ms. In our observations we employed, a "wide-field" regime using the EMCCD, where the area of $40''$ with the target and a comparison star was observed in the B band with 0.12 s exposures, and simultaneous PSD observations with 1 μs resolution in the U band. We also used the polarimetric EMCCD mode [21, 22] in B and V bands, in which only the target was observed in four polarization orientations simultaneously.

3. Results

In figure 1, the total background-subtracted light curve of J1023 in the B band binned to 1.2 s resolution is shown. The gap in the data corresponds to the polarimetric observations. The data for a stable comparison star and the intensity ratio of J1023 to the star are presented as well to demonstrate the stability of the instrument performance. Several bright flares of different durations are seen in the light curve of J1023 when the emission intensity sharply increases by a factor of 1.5. Some of them are unresolved at the selected time binning. They are not instrument artifacts as in that case they would also be seen in the star light curve at $(4-6)\sigma$ levels. Unfortunately, in the U band, where we had much higher time resolution, J1023 is much fainter and the light curve is much noisier. However, the main features seen in B are also resolved in U, particularly when we apply coarser binning (figure 2). The U/B ratio shows that J1023 becomes redder during bright intervals, which is consistent with the results obtained in [14] in 2015.

We resolved, for the first time, very sharp bright flares with durations and rising times of about 0.3 and 0.1 s, respectively. An example in figure 3 demonstrates that such flares could hardly be detected at a factor of 2–3 lower resolution (red and blue lines). During some short

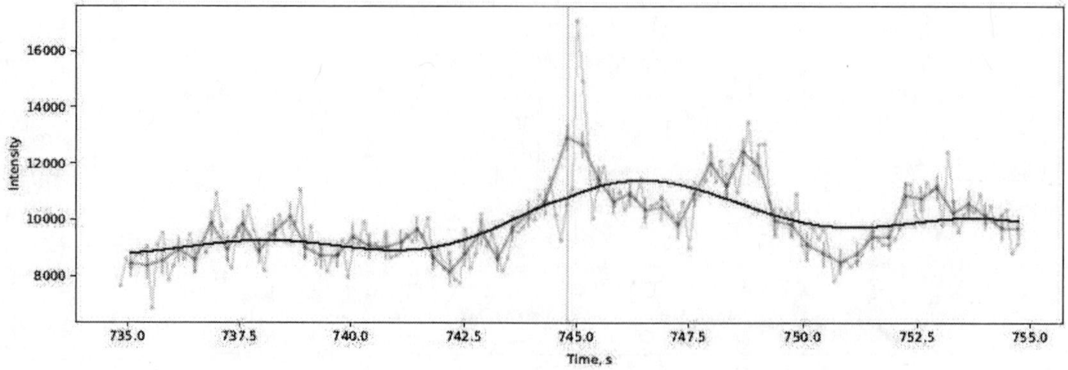

Figure 3. Fragment of the B band light curve with a sharp flare. The green curve is shown at the native resolution of 0.12 s, the red and blue curves are binned to 0.3 and 10 s, respectively.

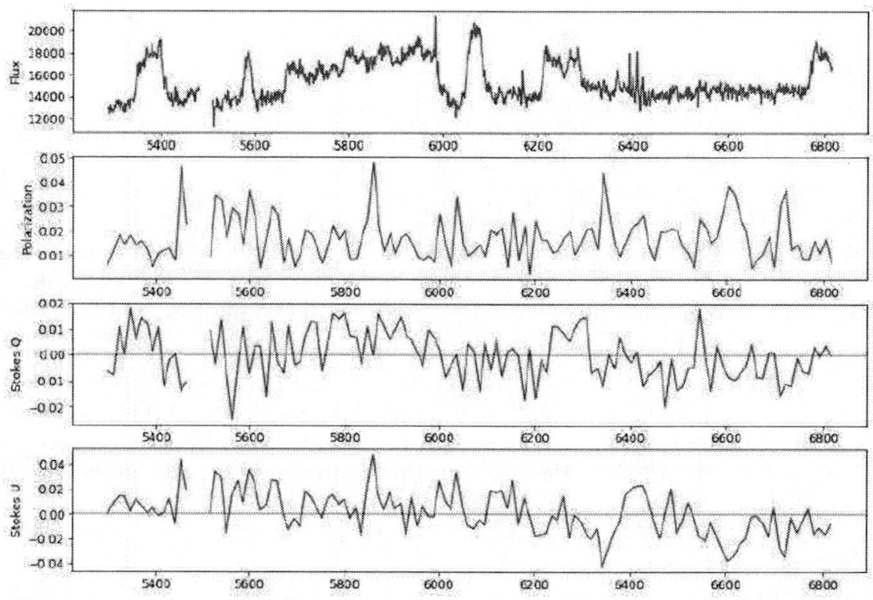

Figure 4. From top to bottom: the B band flux, the polarization degree, and normalised Stokes parameters Q and U of J1023 *vs* time. The polarisation curves are binned to 12 s.

intervals of $\sim$ 100 s, J1023 showed signatures of a periodic flux modulation. Using the Lomb-Scargle periodogram method, we found a quasi-periodicity with the period of $\approx$ 7.6 s.

Our polarisation data (figure 4) indicate a linear polarisation degree $p = (1.5 \pm 0.5)\%$ using 10 min averaging. However, this should be considered with caution. No obvious polarisation-intensity correlation is found, and only upper limits $p \leq 2\%$ and 4% can be derived at 10 and 0.1 s resolutions, which are consistent with the radio upper limit of $\approx$ 4% found in [23].

4. Discussion and Conclusions

Increasing only by a factor of three the time resolution as compared to the previous observations [14] allowed us to resolve, for the first time, bright flares in the optical emission of J1023 with durations of $\approx$ 100 ms. This implies the presence of bright regions with sizes smaller than the accretion disk and agrees well with the predictions of accretion models for various propeller regimes [24]. For J1023, they suggest a variability scale of 50–100 ms related to an interplay between accretion/outflow episodes. Thus, J1023 is likely to be in an unstable propeller regime. Our polarisation upper limits reject the presence of a bright jet. Optimisation of the MPPP observational setup for J1023 will allow us to better constrain the lowest variability scale and to confirm independently the pulsar optical pulsations recently reported in [25]. Coordinated optical–X-ray–radio high time resolution observations would be very useful in this respect.

Acknowledgments

The work is performed in the framework of the Russian Government Program of Competitive Growth of the Kazan Federal University. Development of hardware for high temporal resolution observations on the BTA and data analysis are supported by the Russian Science Foundation grant No 14-50-00043 and by the RFBR projects No 17-52-45048 and 16-02-00604A.

References

[1] Backer D C, Kulkarni S R, Heiles C, Davis M M and Goss W M 1982 *Nature* **300** 615–8
[2] Alpar M A, Cheng A F, Ruderman M A and Shaham J 1982 *Nature* **300** 728–30
[3] Radhakrishnan V and Srinivasan G 1982 *Current Science* **51** 1096–9
[4] Bisnovatyi-Kogan G S and Komberg B V 1974 *SvA* **18** 217–21
[5] Wijnands R and van der Klis M 1998 *Nature* **394** 344–6
[6] Fruchter A S, Gunn J E, Lauer T R and Dressler A 1988 *Nature* **334** 686–9
[7] Roberts M S E 2013 *Neutron Stars and Pulsars: Challenges and Opportunities after 80 years* (*IAU Symposium* vol 291) ed van Leeuwen J pp 127–32
[8] Archibald A M *et al* 2009 *Science* **324** 1411–4
[9] Bassa C G *et al* 2014 *MNRAS* **441** 1825–30
[10] Roy J *et al* 2015 *ApJ* **800** L12
[11] Papitto A *et al* 2013 *Nature* **501** 517–20
[12] Halpern J P, Gaidos E, Sheffield A, Price-Whelan A M and Bogdanov S 2013 *The Astronomer's Telegram* **5514**
[13] Deller A T *et al* 2012 *ApJ* **756** L25
[14] Shahbaz T, Linares M, Nevado S P, Rodríguez-Gil P, Casares J, Dhillon V S, Marsh T R, Littlefair S, Leckngam A and Poshyachinda S 2015 *MNRAS* **453** 3461–73
[15] Takata J *et al* 2014 *ApJ* **785** 131
[16] Patruno A, Archibald A M, Hessels J W T, Bogdanov S, Stappers B W, Bassa C G, Janssen G H, Kaspi V M, Tendulkar S and Lyne A G 2014 *ApJ* **781** L3
[17] Archibald A M *et al* 2015 *ApJ* **807** 62
[18] Jaodand A, Archibald A M, Hessels J W T, Bogdanov S, D'Angelo C R, Patruno A, Bassa C and Deller A T 2016 *ApJ* **830** 122
[19] Bogdanov S, Deller A T, Miller-Jones J C A, Archibald A M, Hessels J W T, Jaodand A, Patruno A, Bassa C and D'Angelo C 2017 *Preprint* 1709.08574
[20] Plokhotnichenko V L, Beskin G M, de Bur V G, Karpov S V, Bad'in D A, Lyubetskaya Z V, Lyubetskij A P and Pavlova V V 2009 *Astrophysical Bulletin* **64** 308–16
[21] Beskin G, Karpov S, Plokhotnichenko V, Stepanov A and Tsap Y 2017 *Stars: From Collapse to Collapse* (*Astronomical Society of the Pacific Conference Series* vol 510) ed Balega Y Y *et al* p 303
[22] Beskin G, Karpov S, Plokhotnichenko V, Stepanov A and Tsap Y 2017 *PASA* **34** e010
[23] Deller A T, Moldon J, Miller-Jones J C A, Patruno A, Hessels J W T, Archibald A M, Paragi Z, Heald G and Vilchez N 2015 *ApJ* **809** 13
[24] Romanova M M, Blinova A A, Ustyugova G V, Koldoba A V and Lovelace R V E 2017 *Preprint* 1704.08336
[25] Ambrosino F *et al* *Nature Astronomy* **1** 266

IOP Conf. Series: Journal of Physics: Conf. Series **932** (2017) 012028 doi:10.1088/1742-6596/932/1/012028

MHD simulations of resistive viscous accretion disk around millisecond pulsar

M Čemeljić, V Parthasarathy and W Kluźniak

Nicolaus Copernicus Astronomical Center, ul. Bartycka 18, PL-00-716 Warsaw, Poland

E-mail: miki@camk.edu.pl

Abstract. We perform MHD simulations of a thin resistive and viscous accretion disk around a neutron star with the surface dipolar magnetic field of 10^8 Gauss. The system evolution is followed during the interval of 500 millisecond pulsar rotations. Matter is accreted through a stable accretion column from the disk onto the star. We also show propagation of the stellar wind through the corona. Analysis of the mass accretion flux and torques on the star shows that the disk reaches the quasi-stationary state.

1. Introduction

In the interaction of a neutron star (NS) with its close companion star, an accretion disk is formed around the NS. Properties of a binary system depend on the type of companion star, the neutron star mass and the magnetic field strength and geometry. Kluźniak & Kita suggested a hydro-dynamical model for the accretion disk [1], with viscosity parameterized by Shakura & Sunyaev α-prescription [2]. We extend this model to the non-ideal MHD, and include the magnetosphere in the innermost part of a star-disk system. One example of such object is a millisecond pulsar: $M = 1.4M_\odot$, $R \sim 10$ km, $B \sim 10^8$ Gauss, $P = 0.05$ sec (5 msec), $\rho_0 = 4.62 \times 10^{-6}$ g/cm^3, $\dot{M}_0 = 10^{-9} M_\odot$/yr.

2. Numerical setup

We use the PLUTO v.4.1 code [3,4] in spherical coordinates with logarithmically stretched grid in radial direction to perform axisymmetric 2D star-disk simulations in $\Theta = [0, \pi/2]$ half-plane. The resolution is $R \times \Theta = [109 \times 50]$ grid cells, with the maximal radius of 30 stellar radii. Following [5] we set up the disk as in [1] assuming that the corona is in a hydrostatic equilibrium. The viscosity and the resistivity are parameterized as $\alpha c^2/\Omega$ (Shakura & Sunyaev), where c is the sound speed, Ω is the Keplerian speed and α is a free parameter, which values are between 0 and 1. We use a split-field method in which only changes from the initial stellar magnetic field evolve in time while the initial stellar field is held constant. Our initial setup is shown in figure 1.

The equations we solve using the PLUTO code are:

$$\frac{\partial \rho}{\partial t} + \nabla \cdot (\rho \boldsymbol{v}) = 0, \quad \frac{\partial \rho \boldsymbol{v}}{\partial t} + \nabla \cdot \left[\rho \boldsymbol{v}\boldsymbol{v} + \left(P + \frac{\boldsymbol{B}\cdot\boldsymbol{B}}{8\pi} \boldsymbol{I} \right) - \frac{\boldsymbol{B}\boldsymbol{B}}{4\pi} - \boldsymbol{\tau} \right] = -\rho \nabla \Psi_G$$

$$\frac{\partial E}{\partial t} + \nabla \cdot \left[\left(E + P + \frac{\boldsymbol{B}\cdot\boldsymbol{B}}{8\pi} \right) \cdot \boldsymbol{v} - \frac{(\boldsymbol{v}\cdot\boldsymbol{B})\boldsymbol{B}}{4\pi} \right] = -\rho \nabla \Psi_G \cdot \boldsymbol{v}$$

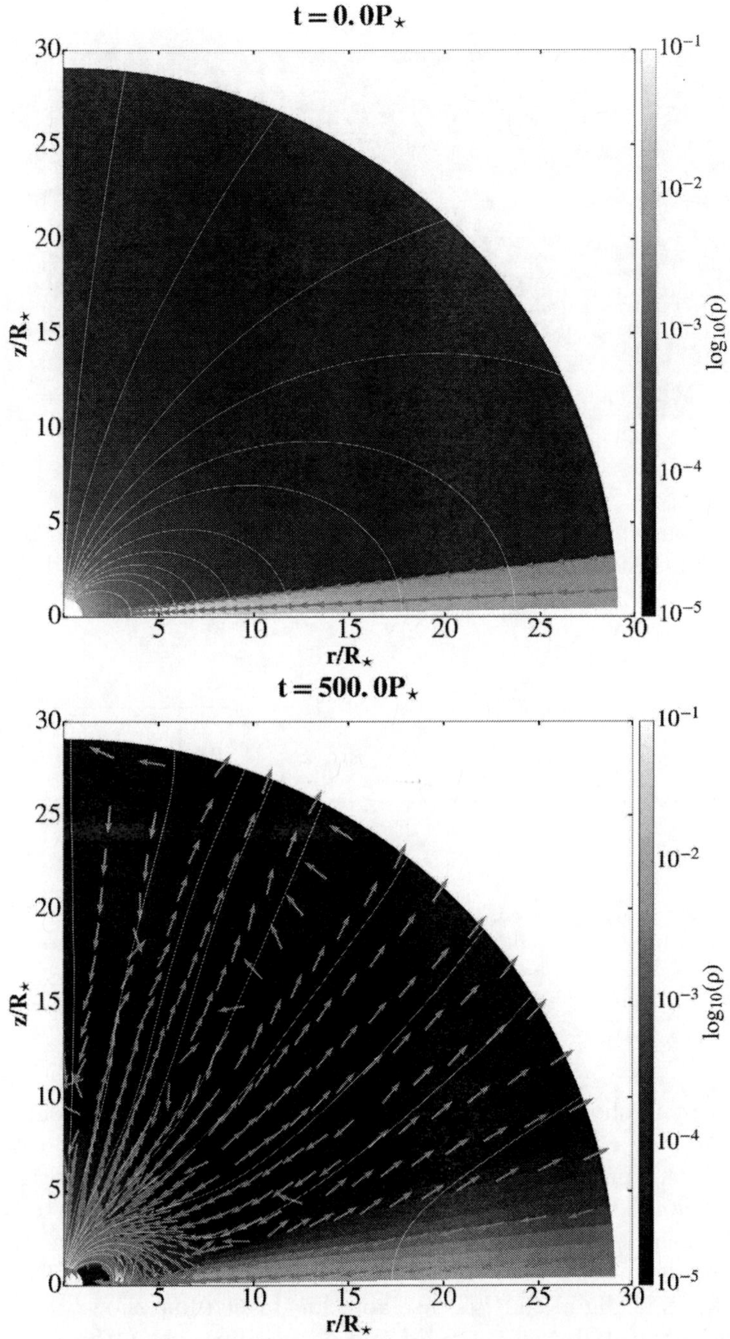

Figure 1. Simulated matter density distribution is shown with the logarithmic color grading. The initial setup is in the top panel and the system after 500 pulsar rotations is shown in the bottom panel. Magnetic field lines are plotted by solid lines and the poloidal velocity distribution is displayed by blue vectors (that are not normalized).

$$\textit{with} \quad E = \frac{P}{\gamma - 1} + \rho \frac{\boldsymbol{v}^2}{2} + \frac{\boldsymbol{B}^2}{8\pi} \quad \textit{and} \quad \frac{\partial B}{\partial t} + \nabla \times (\boldsymbol{B} \times \boldsymbol{v} - \eta_m \boldsymbol{J}) = 0 \ ,$$

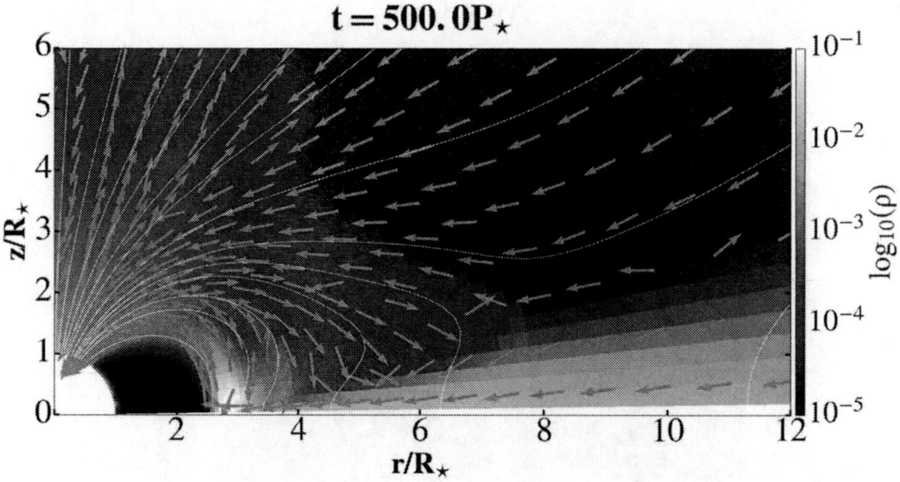

Figure 2. Central part of the system after 500 pulsar rotations is presented to visualize the accretion column and the magnetic field lines connected to the disk beyond the corotation radius $R_{cor} = 4.65\ R_*$.

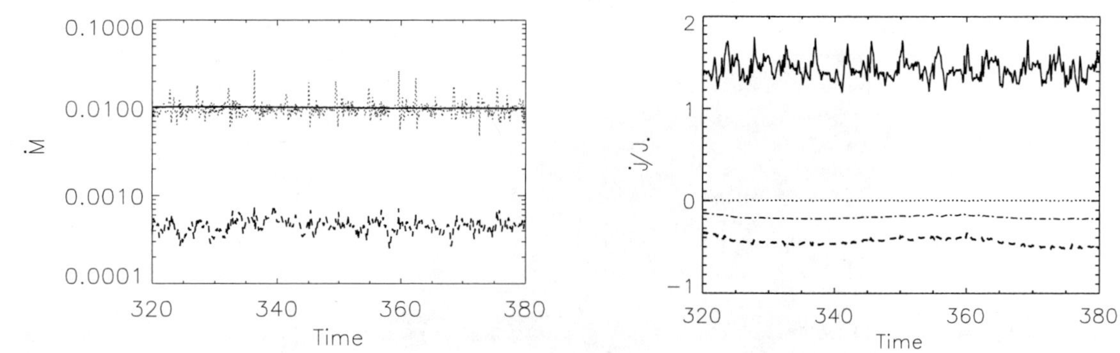

Figure 3. The left panel shows the time dependence of the mass flux. The mass flux is measured in the code units $M_0 = \rho_0 R_*^2 v_{K,*}$, and the time is measured in the pulsar rotation periods. The time interval during the quasi-stationary state is shown. Solid line shows the mass flux through the disk at $R = 12R_*$. This mass flux is distributed onto the star (dotted line) and into the stellar wind (thin dashed line). Right panel shows the torque acting on the stellar surface in the units of the stellar angular momentum for the same time interval. Dotted line shows the kinetic torque which is negligible in this case and dot-dashed line shows the magnetic torque on the star produced by the stellar wind. Solid and dashed lines show the magnetic torque acting on the star surface from the field lines below and beyond the corotation radius. Positive torque spins-up and negative spins-down the star.

where η_m and τ are the resistivity and the viscous stress tensor respectively. Following [5] we remove the Ohmic and the viscous heating terms in the PLUTO energy equation to prevent the thermal thickening of the accretion disk. The resistive and the viscous terms are still present in the momentum and the induction equations. In the boundary conditions we assume the stellar surface to be a rotating perfect conductor so the electric field is zero in the stellar reference frame and the flow velocity is parallel to the magnetic field. We additionally prescribe the rotation of the matter atop the star to match the magnetic field evolution.

3. Results

We obtain long-lasting solutions for the millisecond pulsar as shown in figure 1. In figure 2 we show a zoom into this solution to show the accretion column and magnetospheric region in more detail.

Mass fluxes and torques obtained in our star-disk interaction simulation are shown in figure 3. Most of the mass from the disk is accreted onto the star and only about 1/100 of it goes into the stellar wind. The torque of mass infalling onto the star depends on the origin of the mass: when it comes from the region beyond the corotation radius it slows down the star and when it comes from the region below the corotation radius it spins-up the star.

4. Conclusions

We present the preliminary results of the long-lasting numerical simulations of a star-disk system with magnetospheric interaction in the case of a millisecond pulsar. The disk in our viscous and resistive MHD simulations reaches the quasi-stationary state. In further work we will investigate properties and stability of such a disk. To completely address the stability of the disk and the accretion column the 3D simulations should also be performed.

Acknowledgments

We thank A Mignone and his team of contributors for the possibility to use the PLUTO code, especially C Zanni for help with modifications. MČ thanks to ASIAA/TIARA, Taiwan for the possibility to use their HPC Linux clusters.

References

[1] Kluźniak W and Kita D 2000 *Preprint* astro-ph/0006266
[2] Shakura N I and Sunyaev R A 1973 *A&A* **24** 337
[3] Mignone A, Bodo G, Massaglia S, Matsakos T, Tesileanu O, Zanni C and Ferrari A 2007 *ApJS* **170** 228
[4] Mignone A, Zanni C, Tzeferacos P, van Straalen B, Colella P and Bodo G 2012 *ApJS* **198** 7
[5] Zanni C and Ferreira J 2009 *A&A* **508** 1117

IOP Conf. Series: Journal of Physics: Conf. Series **932** (2017) 012029 doi:10.1088/1742-6596/932/1/012029

On the origin of pulsing X-ray emission of AE Aqr

V I Globina[1] **and N R Ikhsanov**[1,2,3]

[1] Central Astronomical Observatory of RAS at Pulkovo, Pulkovskoe sh. 65-1, St.Petersburg, Russia

[2] Saint Petersburg State University, Universitetskaya emb. 7-9, St.Petersburg, Russia

[3] Special Astrophysical Observatory, Nizhnij Arkhyz, Russia

E-mail: gvi1109@gmail.com, ikhsanov@gaoran.ru

Abstract. The cataclysmic variable AE Aquarii is a low-mass close binary system containing a red dwarf and a 33 s rotation period magnetic white dwarf which operates as a rotation-powered pulsar. The 33 s pulsations are detected in the optical, UV and X-rays. The 16.5 s harmonic is also present in the optical and UV. This pulsing emission comes from two hot spots ($T_p \sim 26000$ K) located in the regions of magnetic poles on the white dwarf surface. The nature of the X-ray pulsations of AE Aqr is still under discussion. No 16.5 s harmonic in X-ray is observed and the luminosity of the pulsing component in X-rays is significantly smaller than the luminosity of the pulsing component in the optical and UV. We suggest that the source of pulsing X-ray emission is also located in the magnetic pole region at the surface of the white dwarf and can be associated with a hot spot ($T \sim 10^6 - 10^7$ K). This spot is heated by the backflowing charged particles. However, the source of particles responsible for the heating the X-ray spot differs from the source of particles responsible for the heating the area emitting in the optical an UV.

1. Introduction

The cataclysmic variable AE Aquarii is located at a distance of 91.4 ± 3.3 pc [1] and contains a red and a white dwarf. The orbital period of the system is about 9.88 hr. The discovery of the 33 s coherent oscillations in the optical and X-ray emission of the system made it possible to determine the spin period of the white dwarf $P_s \simeq 33.08$ s. The 16.5 s harmonic in the optical and UV also was detected. According to observations with the Hubble Space Telescope, UV and optical oscillations are identified with two hot spots ($T_p \sim 26000$ K) located in the regions of magnetic poles on the white dwarf surface with the area of about $10^{16}\,\mathrm{cm}^2$. However, X-ray oscillations with a period of 16.5 s have not been detected. This peculiarity results in the problem of the origin of pulsing X-ray emission from AE Aqr.

The X-ray emission of cataclysmic variables is usually interpreted as a result of accretion on the white dwarf surface. But this interpretation cannot be applied to AE Aquarii. It is recognized that the white dwarf in this system operates as the rotation-powered pulsar [2]. This star is steadily spinning down at the rate $\dot{P}_0 = (5.64 \pm 0.02) \times 10^{-14}\,\mathrm{s\,s}^{-1}$ which implies that the spin-down power exceeds the bolometric luminosity of the system by a factor of 5. Such feature is typical for rotation-powered pulsars but is very unusual for white dwarfs in cataclysmic variables. The only object with similar properties is currently known is AR Scorpii [3].

It also appears that the white dwarf in AE Aqr operates as a magnetic propeller. MHD simulation suggests that the material streaming into the Roche lobe of the white dwarf through the L1 point is leaving the system after its interaction with the magnetic field of the white dwarf [4]. It should be noted, however, that the mass injection into the magnetosphere of the white dwarf occurs deep inside its light cylinder and the propeller action contributes into the energy budget of the system by only a few per cent [5].

All available indirect methods of determination of the magnetic field of the white dwarf lead to the estimate of its dipole magnetic moment $\mu \sim 10^{34}\,\mathrm{G\,cm^3}$ [6]. It indicates that the strength of the dipole component of the surface magnetic filed ranges from 50 MG (at the magnetic equator) to 100 MG (in the region of magnetic poles). In this case, the Alfvén radius of the white dwarf, $r_\mathrm{A} = \left(\mu^2/\dot{M}(2GM_\mathrm{wd})^{1/2}\right)^{2/7} \simeq 4.2 \times 10^{10}\,\mathrm{cm}$, significantly exceeds its corotation radius, $r_\mathrm{cor} = \left(GM_\mathrm{wd}/\omega_\mathrm{s}^2\right)^{1/3} \simeq 1.4 \times 10^9\,\mathrm{cm}$. The light cylinder radius is $1.6 \times 10^{11}\,\mathrm{cm}$. Here M_wd and ω_s are the mass and the angular velocity of the white dwarf, $\dot{M}$ is the mass-transfer rate. The centrifugal force at the magnetospheric boundary under these conditions prevents matter from accreting onto the surface of the white dwarf. Observational confirmations of this conclusion are the following. Studies of the Doppler H_α tomogram suggest that in spite of the very intensive mass transfer between the red and white dwarf there is no developed accretion disk in the system (see [5] and references therein). Instead, the matter is expelled from the system due to its interaction with the magnetic field of the rapidly spinning white dwarf. All data collected in X-ray observations also argues against an accretion nature of X-ray emission from AE Aqr (see next section).

All of the above indicate that pulsing X-ray emission cannot be explained by means of accretion process. There is an another mechanism based on the interaction of a stream of matter lost by the red dwarf with the white dwarf magnetospere. We explore a situation in which magnetic poles are heated by backflowing charged particles accelerated in the white dwarf magnetosphere. We suggest that the initial acceleration occurs in the current sheet at the magnetospheric boundary in the region of interaction between the stream and the magnetic field of the white dwarf. The particle injected into the magnetosphere are then accelerated due to electric potential inside the rotating magnetic field of the white dwarf.

2. Properties of X-ray emission of AE Aqr

The X-ray light curve of AE Aqr is characterized by flares that are in good correlation with optical and UV flares [7]. Flaring occurs almost permanently and is separated by relatively short quiescent phases. On the timescale from a few minutes to an hour the luminosity of the object in UV can change by an order of magnitude.

The observed X-ray emission of AE Aqr consists of 33 s pulsing component and non-pulsing component. Contribution of the pulsing component is approximately 16% during quiescence and decreases down to 7% in flares [8]. This indicates that the contribution of the pulsing component into the flare emission is relatively small. According to [9], both pulsing and non-pulsing flux increase as the flare advances, but the net increase of pulsing flux to the quiescence is very small and corresponds to about 3% – 4% of the increase in non-pulsing flux. The total X-ray luminosity of the system in $0.4-10\,\mathrm{keV}$ is $\sim 7\times10^{30}\,\mathrm{erg\,s^{-1}}$ during quiescence and reaches $\sim 2\times10^{31}\,\mathrm{erg\,s^{-1}}$ during flares [7]. It is a factor of a few smaller than the luminosity of the system in the UV. This indicates that the pulsing emission in UV and X-rays can unlikely be explained in terms of the same mechanism. In particular, traditional interpretation of pulsing UV emission as a result of reemission of the X-ray photons in this particular case is not applicable.

The spectrum of X-ray emission of AE Aqr is much softer than typical spectra of accretion-powered white dwarfs. The X-rays are dominated by photons in the range $0.1 - 1$ keV [7]. According to [8], the linear size of the source of non-pulsing X-ray emission is larger than the size the white dwarf by almost two orders of magnitude. Moreover, the plasma density in this source is about $10^{11}\,\mathrm{cm^{-3}}$ that is a few orders less than the typical density in the accretion column of cataclysmic variables. This data argues against an accretion nature of X-ray emission from AE Aqr.

3. Particle acceleration in the white dwarf magnetosphere

We consider a situation in which the hot spot responsible for pulsing X-ray emission of AE Aqr is a result of heating the surface of the white dwarf by backflowing relativistic particles accelerated in its magnetosphere. The X-ray spot is located at the magnetic pole of the white dwarf but its size is smaller than the size of the spots responsible for the optical and UV pulsations. This may indicate that particles responsible for the origin of the X-ray spot are more energetic and accelerated in a local region inside the area of the open field lines. This situation can be expected if the source of these particles is located inside the light cylinder of the white dwarf. In particular, this can be the region of interaction between the magnetic field of the white dwarf and the stream of material which is flowing into the Roche lobe of the white dwarf through the L1 point. According to results of numerical simulations [4], the interaction occurs at a distance of $\sim 50R_{\rm wd}$ from the surface of the white dwarf. The acceleration scenario in this case can be divided into two steps. The initial (first step) acceleration may occur in the current sheet at the region of interaction between the magnetic field of the white dwarf and the stream of matter. The next (second) step is acceleration of the injected particles into the magnetosphere by the electric field generated according to the conventional scenario of rotation-powered pulsars.

3.1. Acceleration in the current sheet

The linear velocity of matter streaming through the Roche lobe of the white dwarf under the conditions of interest is significantly smaller than the linear velocity at the magnetospheric boundary. It therefore appears that the interaction between the stream and the magnetosphere leads to stretching the field lines in the magnetopause in the azimuthal direction and, hence, to a generation of the toroidal components of the magnetic field at the magnetospheric boundary. As a result of this interaction, a local current sheet (or a number of current sheets) in this region occurs. Rapid flaring of AE Aqr indicates that the field configuration at the magnetospheric boundary is unstable and, hence, an electric field in the region of flaring turbulization of current sheets is generated. The energy of particles accelerated in the electric field can be evaluated as $\varepsilon_0 \simeq eE_0a_0$, where E_0 and a_0 are the electric field and the thickness of the current sheet. We assume that the thickness a_0 is a critical thickness that would be required for turbulization of current sheet plasma [11]. It can be estimated as

$$a_0 \approx \frac{B_{\rm A}c}{4\pi neu_{\rm cr}}. \tag{1}$$

Here $B_{\rm A}$ is the dipole magnetic field strength at the magnetospheric boundary, n is plasma density and $u_{\rm cr}$ is critical velocity at which plasma becomes turbulized. Using results of numerical simulations of plasma flow in AE Aqr [4] one find the following number of the parameters: $B_{\rm A} \approx 200\,{\rm G}$, $n \approx 10^{11}\,{\rm cm}^{-3}$ and $u_{\rm cr} = v_{\rm Te} = \sqrt{kT/m_{\rm e}} \simeq 3.7 \times 10^7\,{\rm cm/s}$ for $T \simeq 10^4\,{\rm K}$. Also we assume that the electric field generated in the current sheet is of order of the magnetic field at the magnetospheric boundary. As a result, the thickness of the current sheet and the particle energy are $a_0 \approx 300\,{\rm cm}$, $\varepsilon_0 \approx 2.6 \times 10^{-5}\,{\rm ergs} \simeq 16\,{\rm MeV}$.

3.2. Acceleration inside the magnetosphere

As the second step, we consider acceleration of the injected particles inside the magnetosphere. Efficiency of this acceleration phase can be achieved due to initial pre-acceleration of particles in the current sheet. A detailed description of the acceleration mechanism is beyond the scope of this paper. Here we just explore if the energy budget associated with the voltage in the magnetosphere of the fast rotating white dwarf is sufficient to power the luminosity of the pulsing X-ray component. To answer this question we use expression for the electric potential in the magnetosphere of the white dwarf from [12]. Particle energy acquired in this electric field is $\varepsilon_{\rm p} \approx 50\,{\rm ergs} \simeq 3 \times 10^7\,{\rm MeV}$. In this case Lorentz factors of electrons and protons are $\gamma_e \sim 10^7$ and $\gamma_p \sim 10^4$, respectively. Maximum particle energy is constrained by the radiative losses of relativistic electrons and protons caused by the curvature radiation. The radiative losses of protons are significantly less than of electrons. It therefore appears that relativistic protons might be more effective in heating the pole regions. The required density of heating particles is about $10^9\,{\rm cm}^{-3}$, which represents about 1% of the stream density in the region of its interaction with the magnetosphere of the white dwarf.

Acknowledgments
We would like to thank anonymous referee for useful comments. GVI would like to thank the organizing committee of the conference "Physics of Neutron Stars – 2017" for kind hospitality and financial support. NRI acknowledges support of the Russian Scientific Foundation under the grant no. 14-50-00043.

References
[1] Ramsay G, Schreiber M R, Gänsicke B T and Wheatley P J 2017 *Astron. and Astrophys.* **604A** p 107
[2] de Jager O C, Meintjes P J, O'Donoghue D and Robinson E L 1994 *Monthly. Not. Roy. Astron. Soc.* **267** p 577
[3] Marsh T R *et al* 2016 *Nature* **537** p 374
[4] Isakova P B *et al* 2016 *Astron. Rep.* **60** p 498
[5] Ikhsanov N R and Beskrovnaya N G 2012 *Astron. Rep.* **56** p 595
[6] Ikhsanov N R 1998 *Astron. and Astrophys.* **338** p 521
[7] Choi C-S, Dotani T and Agrawal P C 1999 *Astrophys. J.* **525** p 399
[8] Itoh K, Okada S, Ishida M and Kunieda H 2006 *Astrophys. J.* **639** p 397
[9] Choi C-S and Dotani T 2006 *Astrophys. J.* **646** p 1149
[10] Eracleous M, Horne K, Osborne J P and Clayton K L 1995 *Proc. Magnetic Cataclysmic Variables* (ASP Conference Series) vol 85, ed D Buckley and B Warner p 392
[11] Pustil'nik L A 1975 *Sov. Astron.* **19** p 195
[12] Ikhsanov N R and Biermann P L 2006 *Astron. and Astrophys.* **445** p 305

IOP Conf. Series: Journal of Physics: Conf. Series **932** (2017) 012030 doi:10.1088/1742-6596/932/1/012030

Interaction of the jet from the neutron star with the interstellar medium

S O Kiikov

South Ural State University (National Research University), Prospect Lenina 76, Chelyabinsk, 454080, Russia

E-mail: kiikov@susu.ru

Abstract. The interaction between the hypersonic plasma jet from the accreting neutron star and the ambient interstellar medium is studied. It is assumed that the jet is launched from the accretion disk via the open magnetic field anchored in the disk. The analytical investigation for the structure of the working surface of the jet is carried out. The estimates of the volume stream functions in the region of the interaction between the jet and the interstellar medium are derived. The obtained results allow to examine the distribution of the plasma velocity fields in the interaction region.

1. Introduction

The compact astrophysical objects are known to produce jets that transport enormous amounts of mechanical energy and momentum into the external medium [1, 2]. As a result of the jet interaction with the interstellar medium the inhomogeneous structures are created. The investigations of plasma parametres in the interaction region plays an important role in understanding of the structure properties. Despite the intensive researches the interaction process is not enough studied. In this work the analytical investigation of the structure of the interaction region between the jet from the accreting neutron star and the surrounding interstellar plasma is performed.

2. Basic Equations

Let us assume that the jet from the accreting neutron star is hypersonic and the interstellar medium is rested before the interaction. The jet penetration into the interstellar gas leads to the formation of the structure including the termination and bow shock waves, and the contact discontinuity between these shocks. The region between the termination and bow shock waves is usually named the working surface of the jet. The plasma flow between the shock waves is supposed to be quasi-steady and axisymmetrical. The shock waves and contact discontinuity are considered as the spheres with the common center. The density of the flow near the contact discontinuity is accepted to be constant. It is also believed that the shock waves are very strong.

Let us introduce the volume stream function ψ that satisfies relations given in [3]:

$$v_r = \frac{1}{r^2 \sin\theta}\frac{\partial\psi}{\partial\theta}, \qquad v_\theta = \frac{1}{r\sin\theta}\frac{\partial\psi}{\partial r}, \tag{1}$$

where r and θ are the spherical coordinates, v_r and v_θ are the radial and tangential components of plasma velocity, respectively. Then the equations of the flow are, from [4]:

$$\frac{\partial}{\partial r}\left(\frac{1}{\sin\theta}\frac{\partial\psi_i}{\partial r}\right)+\frac{\partial}{\partial\theta}\left(\frac{1}{r^2\sin\theta}\frac{\partial\psi_i}{\partial\theta}\right)=A_ir^2\sin\theta, i=1,2; \tag{2}$$

where ψ_1 is the volume stream function in the region between the contact discontinuity and bow shock wave, and ψ_2 is the volume stream function in the region between the termination shock wave and the contact discontinuity, A_1 and A_2 are the constants. From the equation (2) we have the solutions:

$$\psi_i=\left(\frac{1}{r}B_i+C_ir^2+\frac{1}{10}A_ir^4\right)\sin^2\theta, i=1,2; \tag{3}$$

where A_1, A_2, B_1, B_2, C_1 and C_2 are the unknown constants. These constants we can determine from the following conditions:

$$\psi_1|_{r=r_{bs}}=\frac{1-\gamma}{2(\gamma+1)}v_{bs}r_{bs}^2\sin^2\theta, \qquad \left.\frac{\partial\psi_1}{\partial r}\right|_{r=r_{bs}}=v_{bs}r_{bs}\sin^2\theta, \tag{4}$$

$$\psi_1|_{r=r_c}=0, \tag{5}$$

$$\psi_2|_{r=r_d}=-\frac{1-\gamma}{2(\gamma+1)}(v_{jet}-v_{bs})r_d^2\sin^2\theta, \qquad \left.\frac{\partial\psi_2}{\partial r}\right|_{r=r_d}=-(v_{jet}-v_{bs})r_d\sin^2\theta, \tag{6}$$

$$\psi_2|_{r=r_c}=0, \tag{7}$$

$$\Omega|_{r=r_{bs}}=\frac{4}{\gamma^2-1}\frac{v_{bs}}{r_{bs}}\sin\theta, \qquad \Omega|_{r=r_d}=-\frac{4}{\gamma^2-1}\frac{(v_{jet}-v_{bs})}{r_d}\sin\theta, \tag{8}$$

where γ is the adiabatic index, r_d is the distance between the termination shock wave and the common center of curvature of the contact discontinuity and the bow shock wave, r_c and r_{bs} are the radii of curvature of the contact discontinuity and the bow shock wave respectively; v_{jet} is the central jet velocity and v_{bs} is the bow shock velocity; Ω is the vorticity that is given by

$$\Omega=\frac{1}{r}\left(\frac{\partial(rv_\theta)}{\partial r}-\frac{\partial v_r}{\partial\theta}\right). \tag{9}$$

3. Estimates

For $\gamma=\frac{5}{3}$ from the equations (1)–(9) we obtain the approximate expressions for the volume stream functions:

$$\psi_1\approx\left[\frac{r_c}{r}-\left(\frac{r}{r_c}\right)^4\right]v_{bs}r_c^2\sin^2\theta, \tag{10}$$

$$\psi_2\approx-\left[\frac{r_c}{r}-\left(\frac{r}{r_c}\right)^4\right](v_{jet}-v_{bs})r_c^2\sin^2\theta. \tag{11}$$

The estimates (10) and (11) describe the structure of the streamlines in the vicinity of the contact discontinuity and allow to examine the distribution of the plasma velocity fields in the interaction region depending on the collimation angle and distance from the neutron star. These calculations are valid under the assumption of the constant density approximation near the stagnation point.

IOP Conf. Series: Journal of Physics: Conf. Series **932** (2017) 012030 doi:10.1088/1742-6596/932/1/012030

4. Conclusions

We performed a simplified analytical investigation for the plasma flow in the regions between the termination shock wave, contact discontinuity and bow shock wave that are formed due to the interaction of the jet from the accreting neutron star with surrounding interstellar medium. We obtained the estimates for the volume stream functions that allow one to study the distribution of the plasma velocity fields in the interaction region between the jet and the ambient medium. The obtained results allow one to investigate the distribution of the plasma velocity fields in the interaction region depending on the distance from the accreting neutron star and the collimation angle of the jet.

References

[1] Beskin V S 2009 *MHD Flows in Compact Astrophysical Objects* (Berlin: Springer)
[2] Beskin V S 2010 *Us. Fiz. Nauk* **180** 1241
[3] Landau L D, Lifshitz E M 1987 *Course of Theoretical Physics, Vol. 6, Fluid Mechanics* (New York: Pergamon)
[4] Milne-Thomson L M 1962 *Theoretical Hydrodynamics* (London: Macmillan and Co LTD.)

IOP Conf. Series: Journal of Physics: Conf. Series **932** (2017) 012031 doi:10.1088/1742-6596/932/1/012031

How young the accretion-powered pulsars could be?

M V Kostina[1] and N R Ikhsanov[1,2,3]

[1] Saint Petersburg State University, St. Petersburg, Russia
[2] Pulkovo Observatory, St. Petersburg, Russia
[3] Special Astrophysical Observatory, Nizhnij Arkhyz, Russia

E-mail: maria@astro.spbu.ru, ikhsanov@gaoran.ru

Abstract. A question about the age of accretion-powered X-ray pulsars has recently been reopened by a discovery of the X-ray pulsar SXP 1062 in the SMC. This High Mass X-ray Binary (HMXB) contains a neutron star rotating with the period of 1062 s and is associated with a supernova remnant of the age $\sim 10^4$ yr. An attempt to explain the origin of this young long-period X-ray pulsar within the traditional scenario of three basic states (ejector, propeller and accretor) encounters difficulties. Even if this pulsar were born as a magnetar the spin-down time during the propeller stage would exceed 10^4 yr. Here we explore a more circuitous way of the pulsar spin evolution in HMXBs, in which the propeller stage in the evolutionary track is avoided. We find this way to be possible if the stellar wind of the massive companion to the neutron star is magnetized. The geometry of plasma flow captured by the neutron star in this case differs from spherically symmetrical and the magnetospheric radius of the neutron star is smaller than that evaluated in the convention accretion scenarios. We show that the age of an accretion-powered pulsar in this case can be as small as $\sim 10^4$ years without the need of invoking initial magnetic field in excess of 10^{13} G.

1. Introduction

The age of neutron stars (NSs) identified with accretion-powered pulsars in High Mass X-ray Binaries (HMXBs) is expected to exceed 10^5 yr [13]. A newly formed NS needs this time to spin-down from its initially short period up to a critical period, $P_{\rm cr}$, at which the accretion of material onto its surface starts. Observations, however, suggest that the accretion-powered NS in HMXBs can be significantly younger. In particular, the 1062 s X-ray pulsar SXP 1062 has recently been identified with a Supernova Remnant of the age $\sim 10^4$ yr [2, 3, 12]. This discrepancy cannot be resolved within the framework of traditional scenario without invoking additional assumptions about properties of the NS itself [1, 11], or its environment [7]. In this paper, we show that under certain conditions evolution of a NS can follow a simplified scenario, skipping the propeller stage. The time required for a NS to come into accretor stage this way is significantly smaller than that expected in traditional scenarios.

2. Spin-down phase

A newly formed NS is presumed to rotate rapidly and be strongly magnetized. Initially it appears to be a powerful ejector of relativistic wind which prevents the surrounding material from penetrating to within the Bondi radius of the NS and, correspondingly, from reaching the stellar surface. The NS, therefore, starts its evolution with a non-accreting spin-down phase at which its spin period monotonically increases. As the spin period of the NS reaches

IOP Conf. Series: Journal of Physics: Conf. Series **932** (2017) 012031 doi:10.1088/1742-6596/932/1/012031

a critical value, $P_{\rm cr}$, the accretion of material onto its surface starts and the star switches on as an accretion-powered pulsar. The critical period, $P_{\rm cr}$, is usually defined by equating the magnetospheric radius, $r_{\rm m}$, with the corotation radius, $r_{\rm cor} = (G\,M_{\rm NS}/\omega_{\rm s}^2)^{1/3}$, where $M_{\rm NS}$ is the mass and $\omega_{\rm s}$ is the angular velocity of the NS. The initial spin-down phase in the general case can be divided into ejector and propeller states (see e.g. [6] and references therein).

2.1. Ejector state

The rotational rate of NS in the ejector state decreases by the conventional spin-powered pulsar energy-loss mechanism expressed by equation for the magneto-dipole radiation. The spin-down power in this case is converted predominantly into the energy of relativistic wind ejected from the magnetosphere of the NS. The star remains in the ejector state as long as the pressure of the relativistic wind at its the Bondi radius exceeds the ram pressure of the surrounding matter. The spin-down time of NS in a course of the ejector state can be evaluated as (see e.g. [7] and references therein)

$$\tau_{\rm ej} \simeq 8 \cdot 10^5 \ {\rm yr} \times f_{\rm m}^{-1/2} I_{45}\, \dot{M}_{16}^{-1/2}\, v_7^{-1/2}\, \mu_{30}^{-1}, \tag{1}$$

where $f_{\rm m}$ is a dimensionless parameter of order unity which is according for plasma contribution into the processes of energy release in the magnetosphere of NS, I_{45} is the moment of inertia of the NS in units 10^{45} g cm^2, $\dot{M}_{16}$ is the mass transfer rate between the system components in units 10^{16} g s^{-1}, v_7 is the velocity of NS in the frame of surrounding material in units 10^7 cm s^{-1} and μ_{30} is the dipole magnetic moment of NS in units 10^{30} G cm^3. This indicates that the spin-down time of a strongly magnetized ($\mu \gg 10^{30}$ G cm^3) NS, which moves through a relatively fast stellar wind of its massive companion can be significantly shorter than a million years.

2.2. Propeller state

The ejector spin-down ceases as the pressure of relativistic wind at the Bondi radius decreases up to the ram pressure of surrounding material. The surrounding gas under these conditions penetrates to within the Bondi radius and moves towards the NS forming the accretion flow. The interaction between the flow and the magnetic field of the NS leads to formation of the magnetosphere. The radius of the magnetosphere, $r_{\rm m}$, depends on the parameters of the NS, the mass accretion rate and on the geometry and parameters of the accretion flow itself. If $r_{\rm m} < r_{\rm cor}$ the star switches on as an accretion-powered pulsar. Otherwise, it remains to spin-down in the propeller state up to a moment when the corotation radius reaches the radius of the magnetosphere.

The NS in the propeller state is spinning-down due to the interaction between its magnetosphere and surrounding matter. The mechanism of this interaction is rather complicated. However, the spin-down power of NS in this state can be limited to $L_{\rm ns} \leq L_{\rm pr}^{\rm max}$, where

$$L_{\rm pr}^{\rm max} = \frac{\dot{M} G M_{\rm ns}}{r_{\rm m}} = \dot{M}\omega_{\rm s}^2 r_{\rm m}^2 \left(\frac{\omega_{\rm k}(r_{\rm m})}{\omega_{\rm s}}\right)^2 = \omega_{\rm s} K_{\rm sd} \left(\frac{\omega_{\rm k}(r_{\rm m})}{\omega_{\rm s}}\right)^2, \tag{2}$$

$\omega_{\rm k}(r_{\rm m}) = \left(G M_{\rm ns}/r_{\rm m}^3\right)^{1/2}$ is the Keplerian angular velocity at the magnetospheric radius of NS and $K_{\rm sd} = I\dot{\omega}_{\rm s}$ is the spin-down torque exerted to NS from matter surrounding its magnetospheric boundary (note, that the spin-down power of a star with the moment of inertia I which rotates with the angular velocity $\omega_{\rm s}$ and brakes at the rate $\dot{\omega}_{\rm s}$ is $L_{\rm sd} = I\omega_{\rm s}\dot{\omega}_{\rm s}$).

The spin-down timescale of the NS in the propeller state can be evaluated as $\tau_{\rm pr} \geq E_{\rm rot}(\omega_{\rm ej})/L_{\rm pr}^{\rm max}$, where $E_{\rm rot} = I\,\omega_{\rm ej}^2/2$ is the rotational energy of the NS and $\omega_{\rm ej}$ is the angular velocity of the NS at the moment when it switches into the propeller state. Hence,

$$\tau_{\rm pr} \geq 10^6 \ {\rm yr} \times I_{45}\, \dot{M}_{16}^{-11/14}\, \mu_{30}^{-3/7}\, v_7^{1/2}\, m^{-8/7}, \tag{3}$$

where m is the mass of the NS in units $1.4\,M_{\odot}$. Thus, the spin-down time of the NS under the conditions of interest significantly exceeds recently evaluated age of SXP 1062.

3. The direct ejector to accretor state transition

For a NS to switch its state directly from ejector to accretor the spin period at which the matter starts to penetrate to within the Bondi radius should be equal the critical period $P_{\rm cr}$. In the conventional quasi-spherical or Keplerian disk accretion scenarios in which the magnetospheric radius of the NS is close to the Alfvén radius $r_{\rm A} = \left(\mu^4/2GM_{\rm ns}\dot{M}^2\right)^{1/7}$, this situation can unlikely be realized (see fig. 1). It could occur only if the mass accretion rate onto the NS were in excess of $10^{24}\,{\rm g\,s^{-1}}$ (see eq. (10) in [6]), which significantly exceeds the typical mass-loss rate of O/B-type stars.

The direct transition of a NS from the ejector to accretor state could be also expected if accretion onto a NS were realized according to a scenario in which the magnetospheric radius of the NS is smaller than the Alfvén radius. Such a scenario has recently been proposed for wind-fed HMXBs in which the stellar wind of massive component is magnetized. As recently shown in [9], the structure of the accretion flow onto a NS would significantly deviate from the traditional quasi-spherical and Keplerian disk if the magnetic field in the stellar wind of optical companion at the Bondi radius is $B \geq B_{\rm min}$, where

$$B_{\rm min} \simeq 10^{-3}\,{\rm G} \times \xi_{0.2}^{5/8}\,\mu_{30}^{1/4} m^{-7/8}\,P_{100}^{-5/8}\,\dot{M}_{16}^{3/8} v_7^{3/2}. \qquad (4)$$

Here $\xi_{0.2} = \xi/0.2$ is a dimensionless parameter accounting for density and velocity gradient in the material which the NS captures at its Bondi radius and P_{100} is the orbital period of the binary system in units 100 days. If this condition is satisfied the NS would accrete matter from a non-keplerian magnetic disk which is referred to as Magnetic self-Levitating Disk (see e.g. [7, 8, 9] and references therein). Numerical study of such accretion flow has been made in [5] (see also references therein). The magnetospheric radius of the NS in this case is [9]

$$r_{\rm ma} = \left(\frac{c\,m_{\rm p}^2}{16\sqrt{2}\,e\,k_{\rm B}}\right)^{2/13} \frac{\alpha_{\rm B}^{2/13}\,\mu^{6/13}(GM_{\rm ns})^{1/13}}{T_0^{2/13}\,\dot{M}^{4/13}}, \qquad (5)$$

where $m_{\rm p}$ is the proton mass, $k_{\rm B}$ is the Boltzmann constant and T_0 is the temperature of the matter at the inner radius of the ML-disk. The parameter $\alpha_{\rm B} = D_{\rm eff}/D_{\rm B}$ is the ratio of the effective diffusion coefficient at the magnetospheric boundary, $D_{\rm eff}$, to the Bohm diffusion coefficient. Under the conditions of interest the interchange instabilities of the magnetospheric boundary are suppressed by the magnetic field shear [10] and the value of $\alpha_{\rm B}$ does not exceed unity.

Evaluating the ratio

$$\frac{r_{\rm ma}}{r_{\rm A}} = 0.16 \cdot \alpha_{\rm B}^{2/13}\,m^{20/91}\,\mu_{30}^{-10/91}\,T_6^{-2/13}\,\dot{M}_{16}^{-2/91} \qquad (6)$$

one finds that the magnetospheric radius of a NS accreting from the ML-disk under the conditions of interest is significantly small than the Alfvén radius. Here $T_6 = T_0/10^6$ K [4].

Thereby, within the framework of the MLA scenario, the direct ejector to accretor transition of a NS can be realized at substantially lower values of $\dot{M}$ (see fig. 2)

$$\dot{M} \gtrsim 9 \cdot 10^{16}\,{\rm g\,s^{-1}} \times f_{\rm m}^{-13/11} \alpha_{\rm B}^{12/11}\,\mu_{30}^{10/11}\,v_7^{13/11}\,T_6^{-12/13}\,m^{-20/11} \qquad (7)$$

For $\alpha_{\rm B} \sim 0.01$ (which is a typical case of the Earth magnetosphere, see [8] and references therein) the mass accretion rate at which the direct ejector to accretor state transition is expected lies in the range $\dot{M} \sim 10^{15} \div 10^{16}\,{\rm g\,s^{-1}}$, which is typical for HMXBs.

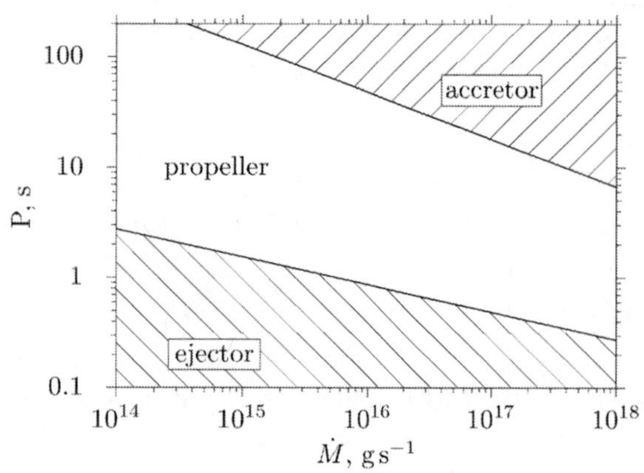

Figure 1. Three possible states of a NS in a HMXB with a non-magnetic optical companion (i.e. $B < B_{\rm min}$) for the quasi-spherical or Keplerian disk accretion scenarios. The dipole magnetic moment of the NS is $\mu = 10^{31}$ G cm^3.

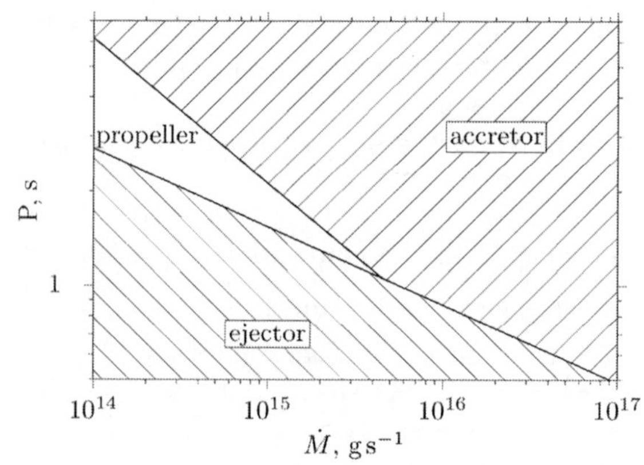

Figure 2. Three possible states of a NS in a HMXB with a magnetic optical companion (i.e. $B \geq B_{\rm min}$) for the Magnetic Levitation Accretion (MLA) scenario. The dipole magnetic moment of the NS is $\mu = 10^{31}$ G cm^3.

4. Conclusions

Our basic conclusion is that the direct transition of a NS from the ejector to accretor state can be realized if the NS accretes matter from the ML-disk. In this case the spin-down time of the NS to a moment when its switches as an accretion-powered pulsar is comparable to the spin-down time in the ejector state (for a discussion see also [7, 8, 9]). It, therefore, appears that the age of accretion-powered pulsars in HMXBs can be as young as $\tau_{\rm ej}$ provided the magnetic field in their plasma environment exceeds $B_{\rm min}$ expressed by Eq. (4).

Acknowledgments

We are grateful to an anonymous referee for useful comments and suggested improvements. MVK thanks the Organizers for kind hospitality and financial support. NRI acknowledges support of the Russian Scientific Foundation under the grant no. 14-50-00043.

References

[1] Fu L and Li X-D 2012 *ApJ* **757** 171
[2] Haberl F, Sturm R, Filipović M D, Pietsch W and Crawford E J 2012 *A&A* **537** L1
[3] Hénault-Brunet V, Oskinova L M, Guerrero M A, Sun W, Chu Y-H, Evans C J, Gallagher III J S, Gruendl R A and Reyes-Iturbide J 2012 *MNRAS* **420** L13
[4] Hickox R C, Narayan R, Kallman T R 2004 *ApJ* **614** 881
[5] Igumenshchev I V 2008 *ApJ* **677** 317
[6] Ikhsanov N R 2001 *A&A* **375** 944
[7] Ikhsanov N R 2012 *MNRAS* **424** L39
[8] Ikhsanov N R and Finger M N 2012 *ApJ* **753** 1
[9] Ikhsanov N R and Mereghetti S 2015 *MNRAS* **454** 3760
[10] Ikhsanov N R and Pustil'nik L A 1996 *A&A* **312** 338
[11] Popov S B and Turolla R 2012 *MNRAS* **421** L127
[12] Sturm R, Haberl F, Oskinova L M, Schurch M P E, Hénault-Brunet V, Gallagher III J S and Udalski A 2013 *A&A* **556** A139
[13] Urpin V, Konenkov D and Geppert U 1998 *MNRAS* **299** 73

IOP Conf. Series: Journal of Physics: Conf. Series **932** (2017) 012032 doi:10.1088/1742-6596/932/1/012032

Flaring activity of the SFXT IGR J16418−4532

D Poliakov[1,2†], **V Aitov**[2] **and N Ikhsanov**[1,2,3*]
[1]Pulkovo Observatory, St. Petersburg, Russia
[2]Saint Petersburg State University, St. Petersburg, Russia
[3]Special Astrophysical Observatory, Nizhnij Arkhyz, Russia

E-mail: †polyakovdmi93@gmail.com, *ikhsanov@gaoran.ru

Abstract. Supergiant fast X-ray transients (SFXTs) are a sub-class of wind-fed High Mass X-ray Binaries (HMXB) in which the normal companion is a supergiant. These systems were collected in a sub-class because of short flares (a few hours duration) in which the X-ray luminosity increases by a few orders of magnitude. One of the members of SFXTs is the X-ray 1212 s pulsar IGR J16418−4532, which is characterized by a high quiescent X-ray luminosity and flaring on a short timescale. We show that the degenerate component of the system is either a magnetar which accretes matter from a Keplerian disk of quasi-spherical flow, or a regularly magnetized neutron star which rotates near spin equilibrium and accretes matter from a non-Keplerian magnetic disk.

1. Introduction

IGR J16418−4532 is a High Mass X-ray Binary system (HMXB) with a short eclipsing orbit of $P_{\rm orb} \sim 3.7$ days [1] which belongs to the Supergiant fast X-ray transients SFXTs subclass of wind-fed HMXBs [2, 3]. The optical component of this system is an O8.5 I supergiant of the mass $M_{\rm OB} = 31.5 M_{\odot}$, radius $R_{\rm OB} = 21.4 R_{\odot}$, luminosity $L_{\rm OB} = 4.47 \times 10^5 L_{\odot}$, and the effective temperature $T_{\rm eff} = 32274$ K [4], which underfills its Roche lobe [3]. A distance to the system is ~ 13 kpc [5]. The neutron star (NS) in this system manifests itself as an X-ray pulsar with the period 1212 ± 6 s [6]. The X-ray luminosity of the system in the quiescence, $L_{\rm qt} \sim 10^{36}\,{\rm erg\,s^{-1}}$, is unusually high for SFXTs.

2. Flaring activity

The X-ray luminosity of the system in $18 - 100$ keV during outbursts lies in the range $(2 - 40) \times 10^{36}\,{\rm erg\,s^{-1}}$ [6]. The orbital phase distribution of the outbursts is uniform, apart the eclipse region where no outbursts were detected. The duration of the outbursts varies from a few minutes up to a few hours [3, 7]. The total duration of these outbursts places the active duty cycle of IGR J16418−4532 at 6.14% [3]. The raising/decline time of the outbursts lies in the range $100 - 1000$ s [3].

3. Equilibrium rotation

As recently shown by Ikhsanov & Mereghetti [8], the equilibrium period of a NS in a wind-fed HMXB lies within the range $P_{\rm min}^{\rm eq} \leq P_{\rm eq} \leq P_{\rm max}^{\rm eq}$, where

$$P_{\rm min}^{\rm eq} \simeq 14\,{\rm s} \times \mu_{30}^{6/7}\, \dot{\mathfrak{M}}_{15}^{-3/7}\, m^{-5/7}, \qquad (1)$$

is the minimum possible equilibrium period, which corresponds to the case of accretion from a Keplerian disk, and

$$P_{\rm max}^{\rm eq} \simeq 20\,{\rm s} \times P_{\rm orb(d)}^{15/7}\, \xi_{0.2}^{-15/7}\, \beta_0^{-12/7}\, c_6^{24/7}\, m^{-8/7} \tag{2}$$

is the maximum possible equilibrium period corresponding a situation in which the angular velocity of the accreting matter is smaller than the Keplerian angular velocity at the mangertospheric radius $\omega_{\rm s}(r) < \omega_{\rm k}(r_{\rm m}) = \left(GM_{\rm ns}/r_{\rm m}^3\right)^{1/2}$. Here μ_{30} is the dipole magnetic moment $\left(\mu = (1/2)B_{\rm ns}R_{\rm ns}^3\right)$ of a NS with the surface magnetic field $B_{\rm ns}$ and the radius $R_{\rm ns}$ in units $10^{30}\,{\rm G\,cm^3}$ and m is its mass, $M_{\rm ns}$, in units $1.4\,M_{\odot}$. The parameter $\dot{\mathfrak{M}}_{15}$ represents the mass accretion rate onto the surface of the NS in units $10^{15}\,{\rm g\,s^{-1}}$, which can be evaluated from the observed X-ray luminosity, $L_{\rm x}$, of the pulsar as $\dot{\mathfrak{M}} = L_{\rm x}R_{\rm ns}/GM_{\rm ns}$. $P_{\rm orb(d)}$ is the orbital period of the binary system measured in days. $\xi = \xi/0.2$ is the dimensionless parameter accounting for dissipation of angular momentum due to density and velocity gradients in a gas with no magnetic field which is falling free towards the NS in a quasi-spherical fashion [9]. c_6 is the speed of sound, $c_{\rm s,0}$, in the gas captured by neutron star from the stellar wind of its massive companion at the Bondi radius, $r_{\rm G} = 2GM_{\rm ns}/v_{\rm rel}^2$, in units $10^6\,{\rm cm\,s^{-1}}$ (here $v_{\rm rel}$ is the velocity of NS in the frame of surrounding material). Finally, $\beta_0 = \beta(r_{\rm G}) = 8\pi\rho_0 c_{\rm s,0}^2/B_{\rm f0}^2$ is the ratio of thermal to magnetic pressure in the material of the density ρ_0 which the NS captures at its Bondi radius, and $B_{\rm f0} = B_{\rm f}(r_{\rm G})$ is the magnetic field of the accreting matter itself.

The lines $P_{\rm min}^{\rm eq} = P_{\rm min}^{\rm eq}(P_{\rm orb})$ and $P_{\rm max}^{\rm eq} = P_{\rm max}^{\rm eq}(P_{\rm orb})$ are shown in Figure 1, which is the $P_{\rm s}$ vs. $P_{\rm orb}$ diagram. The position of IGR J16418−4532 on the diagram is marked by asterisk. It is located well above the line of the minimum possible period and is rather close to the maximum period of a NS which accretes from a magnetized non-Keplerian disk (a so-called the Magnetic Levitating disk, or ML-disk) in which the material is confined by its own magnetic field. Formation of the ML-disk can proceed if $R_{\rm sh} > \max\{r_{\rm A},\, r_{\rm circ}\}$, where $r_{\rm circ}$ is the circularization radius, $r_{\rm A}$ is the Alfvén radius. This is valid if the NS velocity relative to wind of its optical component satisfies the inequality $v_{\rm kd} < v_{\rm rel} < v_{\rm ma}$, where

$$v_{\rm ma} \simeq 550\,{\rm km\,s^{-1}} \times \beta_0^{-1/5}\, \mu_{30}^{-6/35}\, m^{12/35}\, \dot{\mathfrak{M}}_{16}^{3/35}\, c_6^{2/5}. \tag{3}$$

and

$$v_{\rm kd} \simeq 430\,{\rm km\,s^{-1}} \times \xi_{0.2}^{3/7}\, \beta_0^{1/7}\, m^{3/7}\, P_{\rm orb(d)}^{-3/7}\, c_6^{-2/7}. \tag{4}$$

This indicates that the accretion process in IGR J16418−4532 can unlikely be explained in terms of the Keplerian disk accretion scenario. For this scenario to realize the surface magnetic field of the neutron star should satisfy the condition $B_{\rm ns} \geq B_{\rm kd}$, where

$$B_{\rm kd} \sim 3 \times 10^{14}\,{\rm G} \,\times\, R_6^{-3}\, m^{5/6}\, \dot{\mathfrak{M}}_{16}^{1/2} \left(\frac{P_{\rm s}}{1000\,{\rm s}}\right)^{7/6}, \tag{5}$$

is the solution of equation $P_{\rm s} = P_{\rm min}^{\rm eq}$ for the parameters of IGR J16418−4532. Also, one can obtain similar condition $B_{\rm ns} \leq B_{\rm ml}$ for the Magnetic Levitation accretion scenario, where

$$B_{\rm ml} \sim 3.4 \times 10^{12}\,{\rm G} \,\times\, R_6^{-3}\, m^2\, \dot{\mathfrak{M}}_{16}^{1/2}\, c_6^{7/3} \left(\frac{v_{\rm rel}}{500\,{\rm km\,s^{-1}}}\right)^{-35/6} \beta_0^{-7/6}, \tag{6}$$

is the solution of equation $R_{\rm sh} = r_{\rm A}$ for the parameters of IGR J16418−4532. Here $R_{\rm sh}$ is the Shvartsman radius [10]. These conditions indicate that accretion process in this pulsar should be constructed with the quasi-spherical or Magnetic Levitation accretion scenarios. A possibility for the NS in IGR J16418−4532 is far from the equilibrium rotation can be rather rejected since the spin-up time of the NS in this case will be less than 30 years.

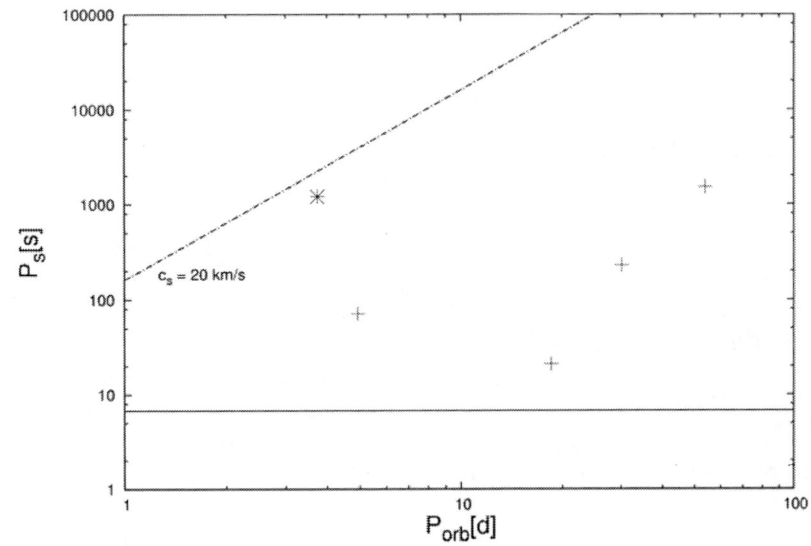

Figure 1. P_s vs. P_{orb} diagram. The solid line corresponds the minimum equilibrium period, $P_{min}^{eq} = P_{min}^{eq}(P_{orb})$, for $\mu = 10^{30}\,\mathrm{G\,cm^3}$ and $\dot{\mathfrak{M}} = 5\times10^{15}\,\mathrm{g\,s^{-1}}$. The dash-dotted lines indicate the maximum equilibrium period, $P_{max}^{eq} = P_{max}^{eq}(P_{orb})$, for $\beta_0 = 1$ and $c_{s,0} = 20\,\mathrm{km\,s^{-1}}$. The position of IGR J16418−4532 is marked by the asterisk, the crosses denote several other of presently known pulsing SFXTs

4. Flaring timescale

The observed flaring in SFXTs is associated with the sporadic variations of the mass accretion rate onto the surface of NS. This can be a reason of variations of the mass capture rate by the NS from the wind of its companion or/and instabilities of the accretion flow inside the Bondi radius or/and variations of the rate at which the accreting material enters the magnetic field of the NS at its magnetospheric boundary. All of these possibilities implies variations of the radius of the magnetosphere of the NS. The magnetospheric radius decreases during flares and increases to its initial value as the mass-transfer rate towards the magnetospheric boundary decreases and the system switches back to the quiescence.

The characteristic time on which the system can change its state from quiescence to flaring within this approach is limited to $t_{rf} \geq \tau_a(r_m)$, where $\tau_a(r_m) = r_m/v_a(r_m)$ is the Alfvén time, $v_a(r_m) = 2\mu/r_m^3\sqrt{4\pi\rho(r_m)}$ is the Alfvén velocity and $\rho(r_m)$ is the density of matter in the magnetopause at the magnetospheric boundary. If a NS accretes matter from a quasi-spherical flow or/and from a hot turbulent envelope [11], then the Alfvén time in the magnetopause is comparable to the dynamical time $t_{ff}(r) = r^{3/2}/(2GM_{ns})^{1/2}$, which under the conditions of interest is only a fraction of a second. The raise time of flares in this case reflects the characteristic time of variations of the mass-transfer rate in the accretion flow beyond the magnetospheric boundary.

If a NS accretes matter form a disk the Alfvén velocity in the magnetopause is comparable with the speed of sound in the plasma at the magnetospheric boundary, which is significantly smaller than the dynamical (free-fall) velocity. This indicates that the characteristic time on which the system can switch into flaring is limited to $t_{rf} \geq r_m/c_s(r_m)$. The magnetospheric radius of a NS accreting from a disk can be limited as $r_{ma} < r_m < r_A$, where (see [8] and references therein)

$$r_{\rm ma} = 2 \times 10^8\,{\rm cm}\ \times\ \alpha_{\rm B}^{2/13} \mu_{30}^{6/13} L_{36}^{-4/13} m^{5/13} R_6^{-4/13} T_6^{-2/13} \tag{7}$$

is the radius of the magnetosphere of a NS accreting from the ML-disk, and

$$r_{\rm A} \simeq 6 \times 10^8\,{\rm cm}\ \times\ \mu_{30}^{4/7}\ m^{1/7}\ L_{36}^{-2/7}\ R_6^{-2/7} \tag{8}$$

is the Alfvén radius which is defined by equating the magnetic pressure of the dipole magnetic field of the NS with the ram pressure of the free-falling spherical flow. Here $\alpha_{\rm B}$ is the efficiency parameter, L_{36} is the X-ray luminosity of the pulsar in units of $10^{36}\,{\rm erg\,s^{-1}}$, and T_6 is the temperature of material in the magnetopause at the magnetospheric boundary in units 10^6 K, which is normalized following Hickox et al [12]. Consequently, if the NS in IGR J16418−4532 accrtes matter from a disk, then the raising time of flares can hardly be less than a minute and about 100 s if the temperature of material in the disk at its inner radius is about 10^4 K.

5. Conclusion

Assuming that the NS in IGR J16418−4532 rotates close to the equilibrium period we showed that accretion in this system can hardly be explain by the accretion from Keplerian disk since this scenario requires supercritical NS magnetic field. Obtained restrictions of the characteristic time on which the system can switch into flaring are consistent with observed flaring properties. This allows to investigate the mechanism of flaring formation within Magnetic Levitation accretion scenario in IGR J16418−4532.

Acknowledgments

VA thanks organizers for kind hospitality and financial support. NI acknowledges support of the Russian Scientific Foundation under the grant no. 14-50-00043.

References

[1] Levine A M, Bradt H V, Chakrabarty D, Corbet R H D and Harris R J 2011 *The Astrophysical Journal Supplement* **196** 6
[2] Romano P 2015 *Journal of High Energy Astrophysics* **7** 126–36
[3] Drave S P, Bird A J, Sidoli L, Sguera V, McBride V A, Hill A B, Bazzano A and Goossens M E 2013 *MNRAS* **433** 528–42
[4] Martins F, Schaerer D and Hillier D J 2005 *Astron. Astrophys.* **436** 1049–65
[5] Rahoui F, Chaty S, Lagage P O and Pantin E 2008 *Astron. Astrophys.* **484** 801–13
[6] Sidoli L, Mereghetti S, Sguera V and Pizzolato F 2012 *MNRAS* **420** 554–61
[7] Ducci L, Sidoli L and Paizis A 2010 *MNRAS* **408** 1540–50
[8] Ikhsanov N R and Mereghetti S 2015 *MNRAS* **454** 3760–5
[9] Ruffert M 1999 *Astron. Astrophys.* **346** 861–77
[10] Shvartsman V F 1971 *SvA* **15** 377
[11] Ikhsanov N R 2001 *Astron. Astrophys.* **375** 944–9
[12] Hickox R C, Narayan R and Kallman T R 2004 *ApJ* **614** 881–96

IOP Conf. Series: Journal of Physics: Conf. Series **932** (2017) 012033 doi:10.1088/1742-6596/932/1/012033

Period distribution of pulsars in the Magellanic Clouds: Propeller line versus Equilibrium period

A S Tanashkin[1] **and N R Ikhsanov**[1,2,3]

[1] Saint Petersburg State University, Saint Petersburg, Russia
[2] Pulkovo Observatory, Saint Petersburg, Russia
[3] Special Astrophysical Observatory, Nizhnij Arkhyz, Russia

E-mail: `artyom.tanashkin@gmail.com`, `ikhsanov@gaoran.ru`

Abstract. A majority of accretion-powered X-ray pulsars in wind-fed High Mass X-ray Binaries (HMXBs) located in the Magellanic Clouds are observed to be transient X-ray sources. They are characterized by short luminous outbursts, while spending most of the time in quiescence. The quiescent states of the pulsars in the diagram "Pulsar Period vs. X-ray Luminosity" fall on a line with the slope -0.43. The same slope is expected for the propeller line which separates stars in the accretor state from stars in the propeller state. We show, however, that a line with the same slope would also be expected if rotation of the pulsars is close to equilibrium.

1. Introduction

A majority of HMXBs in the Magellanic Clouds (LMC and SMC) are transient X-ray sources which occasionally exhibit short powerful outbursts while spending most of the time in quiescence. As recently recognized [1] the quiescent states of some transients in the logarithmic diagram "X-ray Luminosity vs. Pulsation Period" fall on the straight line with the slope $-2.32 \pm 0.09\,(1\sigma)$ (this corresponds $P_{\rm s} \propto L_{\rm x}^{3/7}$, where $P_{\rm s}$ is the period and $L_{\rm x}$ is the X-ray luminosity of the pulsar, see figure 1). This finding was initially discussed in terms of the propeller line which separates accretion-powered sources from those in which the neutron star is in the propeller state. This line is expected to be of the same slope [2]. We show, however, that positions of pulsars in the diagram would be located on the same line if the neutron stars rotate with the equilibrium period.

2. Equilibrium line

The equation governing spin evolution of an accretion-powered neutron star reads

$$2\pi I\dot{\nu} = K_{\rm su} + K_{\rm sd}, \tag{1}$$

where I is the moment of inertia and $\nu = 1/P_{\rm s}$ is the rotation frequency of the neutron star, which rotates with the period $P_{\rm s}$. According to this equation the period of an accretion-powered pulsar evolves to the equilibrium period $P_{\rm eq}$ (equilibrium rotation), which is defined by equating the spin-up, $K_{\rm su}$, with the spin-down, $K_{\rm sd}$, torque exerted to the neutron star from the surrounding material.

The time required for the neutron star to reach equilibrium rotation depends on the parameters of the star itself and parameters of the accretion flow. The star switches on as an accretion-powered pulsar at the moment of its transition from the propeller to accretor state. The transition occurs as the spin

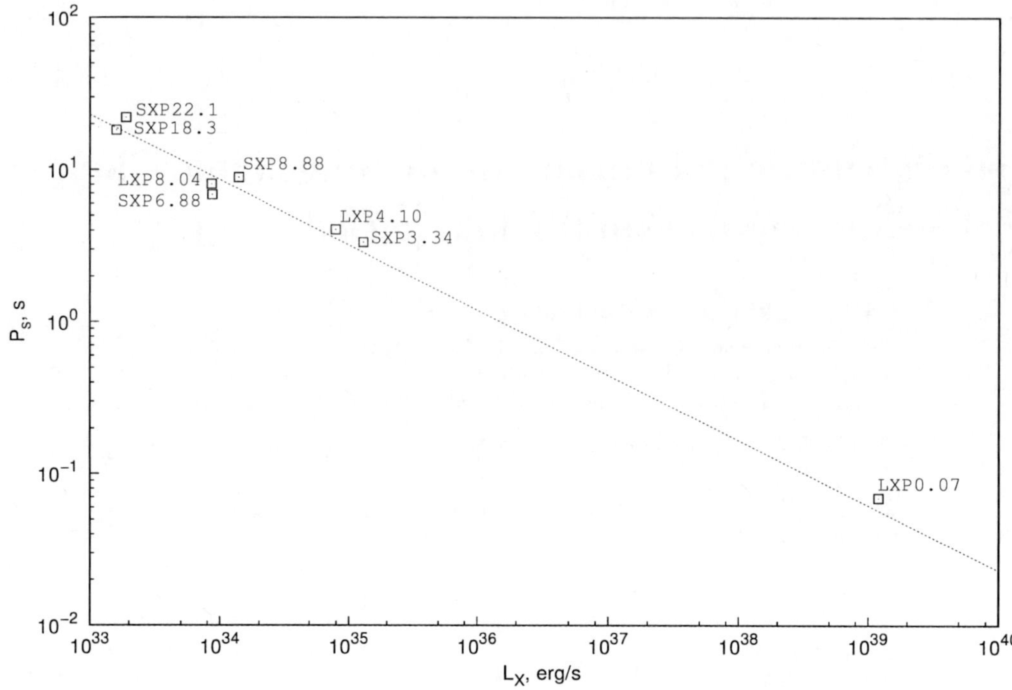

Figure 1. The $P_{\rm s}$ vs. $L_{\rm x}$ diagram. Squares represent individual sources. The dashed line demonstrates the slope of the $P_{\rm min}^{\rm eq} = P_{\rm min}^{\rm eq}(L_{\rm x})$ relation. The position of the line corresponds to the value of the surface magnetic field $B \sim 3 \times 10^{11}$G.

period increases up to the value $P_{\rm pr}$, at which the corotation radius, $r_{\rm cor} = \left(GM_{\rm ns}/\omega_{\rm s}^2\right)^{1/3}$ of the star reaches its magnetospheric radius, $r_{\rm m}$. Here $M_{\rm ns}$ is the mass and $\omega_{\rm s} = 2\pi\nu$ is the angular velocity of the neutron star. If the star accretes from a Keplerian disk its equilibrium period is close to $P_{\rm pr}$. Otherwise, the equilibrium period exceeds $P_{\rm pr}$ and the star remains to spin-down after the state transition up to the moment when the condition $K_{\rm sd} \sim K_{\rm su}$ is satisfied. This would occur on the timescale

$$t_{\rm sd} \sim \frac{\nu}{2|\dot{\nu}|} \sim \frac{\pi I}{2P_{\rm pr}|K_{\rm sd}(P_{\rm pr})|}. \tag{2}$$

The spin-down torque exerted to an accreting neutron star is (see [3] and references therein)

$$K_{\rm sd} = \frac{\mu^2}{(r_{\rm m} r_{\rm cor})^{3/2}} \left(\frac{\Omega(r_{\rm m})}{\omega_{\rm s}} - 1\right), \tag{3}$$

where $\Omega(r_{\rm m})$ is the angular velocity of material interacting with the magnetic field of the neutron star at its magnetospheric boundary and μ is the dipole magnetic moment of the neutron star. Assuming $\Omega(r_{\rm m})/\omega_{\rm s} \ll 1$ one can approximate the absolute value of the spin-down torque exerted to the neutron star at the moment of its transition into the accretor state (i.e. $r_{\rm m} = r_{\rm cor}$) as $\sim \mu^2/r_{\rm cor}^3$. Putting this to equation (2) yields

$$t_{\rm sd} \sim 500\,{\rm yr} \times I_{45}\, m\, \mu_{30}^{-2}\, P_{\rm s}, \tag{4}$$

where I_{45} is the moment of inertia, m is the mass and μ_{30} is the dipole magnetic moment of the neutron star in units $10^{45}\,{\rm g\,cm^2}$, $1.4\,M_{\odot}$, and $10^{30}\,{\rm G\,cm^3}$, respectively. A similar timescale of evolution of the pulsar period has been derived from studies of short-term spin-up/spin-down episodes observed in several accretion-powered pulsars in HMXBs (see [4] and references therein). Our estimate, therefore, favors a hypothesis that most of accretion-powered pulsars observed in current epoch are close to spin equilibrium and rotate with the equilibrium period $P_{\rm eq} \geq P_{\rm min}^{\rm eq}$, where (see [3] and references therein)

$$P_{\rm min}^{\rm eq} \sim 15\,{\rm s} \times \mu_{30}^{6/7} m^{-2/7} R_6^{-3/7} L_{35}^{-3/7}. \tag{5}$$

Here R_6 is the radius of the neutron star in units 10^6 cm and L_{35} is the X-ray luminosity of the pulsar in units $10^{35}\,{\rm erg\,s^{-1}}$.

Thus, the slope of the line defined by equation $P_{\rm eq} = P_{\rm eq}(L_{\rm X})$ is the same as the slope of the propeller line (i.e. $P \propto L_{\rm X}^{-3/7}$). In this light, the result reported in [1] seems not to be very unusual. The points of quiescent states of transient pulsars rotating with equilibrium period are expected to fall on the straight line with the slope close to $-2.32 \pm 0.09\,(1\sigma)$ in the "X-ray Luminosity vs. Pulsation Period" ($L_{\rm X}$ vs. $P_{\rm s}$) diagram.

3. Discussion

The dashed line in figure 1 corresponds the equilibrium period line for a neutron star with the surface magnetic field $B \sim 3 \times 10^{11}$ G. The corotation radius of the neutron stars under these conditions is comparable to the Alfvén radius, $r_{\rm A} = \mu^{4/7} (0.5\,GM_{\rm ns})^{1/7} (L_{\rm X} R_{\rm ns})^{-2/7}$, which is defined by equating the magnetic pressure due to the dipole magnetic field of the star with the ram pressure of free falling material. This may give an incorrect impression that both of the interpretations, i.e. propeller line and equilibrium line, are identical. To avoid a confusion we should note that the radius of the magnetosphere of an accreting neutron star can under certain conditions be significantly smaller than Alfvén radius. This situation, in particular, is expected if the magnetic field in the material which the neutron star captures at its Bondi radius exceeds the critical value expressed by equation (15) in [3]. In this case the neutron star accretes matter from a so-called Magnetic Levitating disk (ML-disk) which is approaching the neutron star up to the distance [4, 3]

$$r_{\rm ma} = \left(\frac{c\, m_{\rm p}^2}{16\sqrt{2}\, e\, k_{\rm B}} \right)^{2/13} \frac{\alpha_{\rm B}^{2/13} \mu^{6/13} (GM_{\rm ns})^{5/13}}{T_0^{2/13} L_{\rm X}^{4/13} R_{\rm ns}^{4/13}}. \tag{6}$$

Here c is the speed of light, $m_{\rm p}$ is the mass of a proton, e is the electron charge, $k_{\rm B}$ is the Boltzmann constant, T_0 is the temperature of the matter at the inner radius of the ML-disc and $\alpha_{\rm B}$ is the dimensionless efficiency parameter. The ratio

$$\frac{r_{\rm ma}}{r_{\rm A}} = 0.14\, \alpha_{\rm B}^{2/13} m^{22/91} \mu_{30}^{-10/91} T_6^{-2/13} L_{33}^{-2/91} R_6^{-2/91} \tag{7}$$

under the conditions of interest is significantly smaller than unity. Here T_6 is the temperature of the matter at the inner radius of the ML-disc in units 10^6 K and L_{33} is the X-ray luminosity of the pulsar in units 10^{33} erg s^{-1}. It, therefore, appears that neutron stars in the systems presented in figure 1 within the Magnetic Levitation Accretion scenario are in the accretor state and their rotation is close to equilibrium.

Acknowledgments
We are thankful to the anonymous referee for his valuable comments. AST thanks organizers of the conference "Physics of Neutron Stars 2017" for kind hospitality and financial support. NRI acknowledges support of the Russian Scientific Foundation under the grant no. 14-50-00043.

References

[1] Christodoulou D M, Laycock S, Yang J and Fingerman S 2016 *ApJ* **829** 30–40
[2] Stella L, White N E and Rosner R 1986 *ApJ* **308** 669–79
[3] Ikhsanov N R and Mereghetti S 2015 *MNRAS* **454** 3760–65
[4] Ikhsanov N R, Likh Y S and Beskrovnaya N G 2014 *Astron. Rep.* **58** 376–85

IOP Conf. Series: Journal of Physics: Conf. Series **932** (2017) 012034 doi:10.1088/1742-6596/932/1/012034

Equation of state for neutron stars. Some recent developments

P Haensel and M Fortin

N. Copernicus Astronomical Center, Bartycka 18, 00-716 Warszawa, Poland

E-mail: haensel@camk.edu.pl

Abstract. Calculations using the chiral effective field theory (ChEFT) indicate that the four-body force contribution to the equation of state (EOS) of pure neutron matter (PNM) at the nuclear density n_0 is negligibly small. However, the overall uncertainty in the EOS of PNM at n_0 remains $\sim 20\%$. Relativistic mean field (RMF) calculations with in-medium scaling, and including hyperons and Δ resonances, can be made consistent with recent nuclear and astrophysical constraints. Dirac-Brueckner-Hartree-Fock calculations with some medium dependence of the nuclear interaction yield neutron star (NS) models with hyperonic cores consistent with 2 $M_\odot$ stars and agreeing with the saturation parameters of nuclear matter. Many unified EOS for the NS crust and core were calculated, and are reviewed here. The effect of the finite size of baryons on the EOS, its treatment via the excluded-volume approximation, and its relevance for the hypothetical hybrid-star twins at ~ 2 $M_\odot$ are dicussed.

1. Introduction

The equation of state is necessary for constructing NS models. In particular, the EOS determines the maximum allowable mass for NS, $M_{\rm max}$[EOS]. The discovery of two radio pulsars with $M \simeq 2$ $M_\odot$ [1, 2] yields a very important constraint: $M_{\rm max}[{\rm EOS}] > 2$ $M_\odot$. This means that the nuclear forces acting between nucleons (or more generally - hadrons) push up $M_{\rm max}$ by a factor of at least ~ 3 compared to the maximum allowable mass of 0.7 $M_\odot$ for the EOS of a free Fermi gas of neutrons [3].

A convenient unit for the NS matter baryon (number) density $n_{\rm b}$ is the standard nuclear density $n_0 = 0.16$ fm^{-3}, corresponding to the mass-energy density unit $\varepsilon_0/c^2 = 2.8 \times 10^{14}$ g cm^{-3}. Matter inside NS is strongly degenerate, and therefore the pressure depends only on the density, $P = P(\varepsilon)$ or equivalently $P = P(n_{\rm b})$.

The theoretical calculation of the $P(\varepsilon)$ relation for NS is a challenge, because one is dealing with a strongly interacting, dense many-particle system. For $\varepsilon \lesssim \frac{1}{2}\varepsilon_0$, corresponding to the NS crust, nuclear theory methods are still valid, but some uncertainties persist. For NS cores one expects $\frac{1}{2}\varepsilon_0 < \varepsilon < 10\varepsilon_0$, with theoretical uncertainties growing rapidly with ε. Does the deconfinement of quarks occur inside the cores of massive NS? The answer cannot be given by the theory alone, because even at $10\varepsilon_0$ we are dealing with a non-perturbative QCD regime. Fortunately, NS exist and are available for observations! For $\varepsilon \gtrsim 100\varepsilon_0$ reliable perturbative QCD calculations are possible [4], but this is very far from the $\sim 10\varepsilon_0$ expected at the center of the most massive NS.

In the present paper we review some calculations of the EOS of NS done in the last five years. In section 2 we review calculations based on ChEFT. Section 3 is devoted to RMF models, with

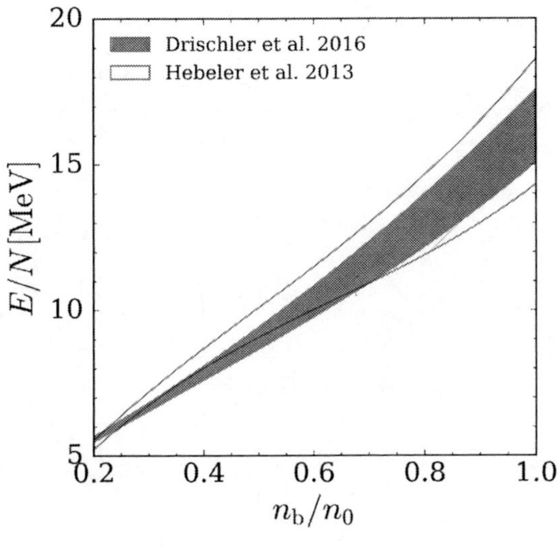

Figure 1. Energy per nucleon calculated from the nucleon 2N+3N force obtained using ChEFT in the $\nu = 4$ approximation. Two methods for the solution of the A-body problem were applied [7]. The blue band results from uncertainties in the experimental constants in the contact forces, the variation of the Λ-cutoff, and errors in the numerical solution of the many-body problem. Based on results of [7, 8].

a scaling of masses and/or coupling constants in the dense NS medium. EOS with hyperons obtained using the relativistic theory of baryon matter with a one-boson-exchange interaction, is described in section 4. In section 5 new unified EOS for the NS crust and core are reviewed. In section 6 we describe a model dealing with the effect of the finite size of baryons on the EOS and its relevance to hypothetical quark-core twins with $M \simeq 2\ \mathrm{M}_{\odot}$. Conclusions are summarized in section 7.

2. ChEFT and the EOS

ChEFT is used since the 1990s to construct two-body and many-body nuclear forces, within a diagrammatic scheme, guided by chiral symmetry, ordered by a power counting, and with a regularization of the momentum integrations (review in [5]). It considers pion-exchange-mediated long and intermediate range components of the force, while short-range components are represented by contact (zero-range) interactions. The strengths of the contact forces are determined experimentally (fitting scattering data and few-nucleon bound states).

The nuclear interaction operator $\hat{V}$ acting within an A-nucleon system is the sum of contributions of diagrams ordered by powers of the parameter Q/Λ, where Q is the nucleon (N) momentum scale within the system and Λ is the cut-off in the momentum space needed to regularize the integrations over momenta. For a few-N systems one may put $Q \simeq m_\pi c/\hbar = 0.7\ \mathrm{fm}^{-1}$ and $\Lambda \simeq 500\ \mathrm{MeV}/(\hbar c) = 2.5\ \mathrm{fm}^{-1}$. Then $Q/\Lambda \sim 1/3$ and ordering by $(Q/\Lambda)^\nu$ looks promising. Let the nuclear force calculated at an assumed Λ be $\hat{V}(\Lambda)$ and Q/Λ be small. Then the sum of the $\nu = 0$ and $\nu = 2$ terms is a reasonable approximation; it reminds the effective range theory approximation for the 2N scattering (the $\nu = 1$ contribution vanishes due to parity conservation). Notice that $\hat{V}$ is not an observable, but it is used to calculate the observables. The $\nu = 0, 2, 3$ diagrams give a complete $2N$-force, but they yield an incomplete $3N$-force, while adding $(Q/\Lambda)^4$ diagrams completes the $3N$ force and the $4N$ force. In this way one obtains $\hat{V} \simeq \hat{V}_{2N}(\Lambda) + \hat{V}_{3N}(\Lambda) + \hat{V}_{4N}(\Lambda)$. One has still to calculate the ground state energy of the A-body system under consideration. The simplest case relevant for NS is pure neutron matter (PNM). At $n_\mathrm{b} = n_0$ the $4N$ forces contribute only -0.2 MeV to the energy per nucleon E (e.g., [6]), and therefore the $2N + 3N$ approximation seems excellent (this was already known since long time for $A = 4$). However, uncertainties in $\hat{V}(\Lambda)$ and in the A-body calculations of E grow quite rapidly with increasing n_b (figure 1). The best calculations for PNM give $E(n_0)$ and $P(n_0)$ with a 20% and 30% uncertainty, respectively. Viewing that Q grows with density, and this is

IOP Conf. Series: Journal of Physics: Conf. Series **932** (2017) 012034 doi:10.1088/1742-6596/932/1/012034

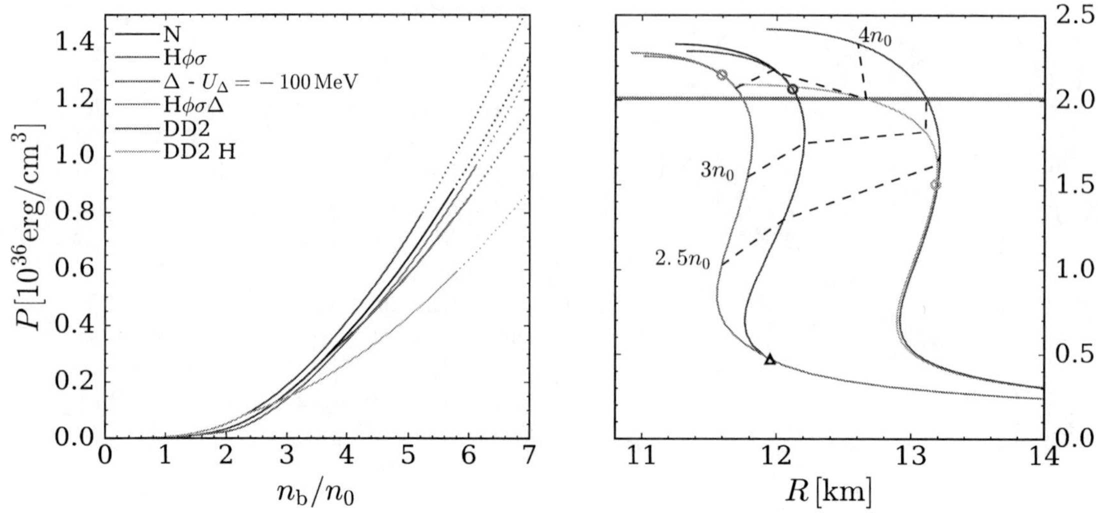

Figure 2. EOS and $M(R)$ for the RMF models with in-medium scaling of the hadron coupling constants and/or masses. Dotted segments in the EOS (left panel) are above central density of stable NS. Dashed segments in the right panel connect NS models with same n_c, whose value is indicated. See text for details.

accompanied with a decreasing precision of the A-body calculation of E, the ChEFT does not seem promising for the precise calculation of the EOS of the NS core.

3. RMF models with scaled masses and couplings

Within the RMF model baryon (B) fields are coupled to meson (M) fields, and their coupling constants and masses g_{BM}, m_B, and m_M, respectively, enter the effective Lagrangian density $\mathcal{L}$. The EOS is then calculated using static solutions of the field equations, obtained in the mean-field approximation.

Two different approaches can be chosen to reach consistency of an RMF model with nuclear and astrophysical constraints. The first approach assumes a minimal structure of $\mathcal{L}$, with only quadratic terms in the vector M-fields, while g_{BM} and m_M are scaled with the scalar meson field σ [9, 10] ($\mathcal{L}$ should remain a scalar!). The second way relies on enriching the structure of $\mathcal{L}$ by adding terms which are quartic in the vector fields: new RMF models of this kind will be briefly reviewed in section 5.

In the first approach a dimensionless scalar field f is introduced instead of σ. Specifically for $B = N = n, p$, one has $f = (g^\star_{\sigma N}/m_N)\sigma$, where $g^\star_{\sigma N}$ is a renormalized (scaled) coupling constant. The scaling with f was motivated, *inter alia*, by an expected partial restoration of the chiral symmetry at high n_b and the Brown-Rho hadron mass scaling [9]. Because $\mathcal{L}$ was quadratic in the vector mesons fields ω_μ, ϕ_μ, ρ^i_μ (the index i reflects the isovector nature of ρ), the equations of motion for the vector fields are linear in the corresponding source-densities (baryon density, isospin density, strangeness density). Therefore, what one really had to solve was a highly nonlinear equation of motion for f only. In the static case under consideration it is $\partial E[n_n, n_p, f]/\partial f = 0$. The EOS and the resulting $M - R$ relation for a purely nucleonic core is plotted in figure 2 (label N). The extension to include the hyperons H (label H$\phi\sigma$ in figure 2) assumed also a scaling with f, using experimental potential wells U_H for hyperons in symmetric nuclear matter at n_0 to get $g_{\sigma H}$ at n_0, and applying the SU(6) symmetry for the

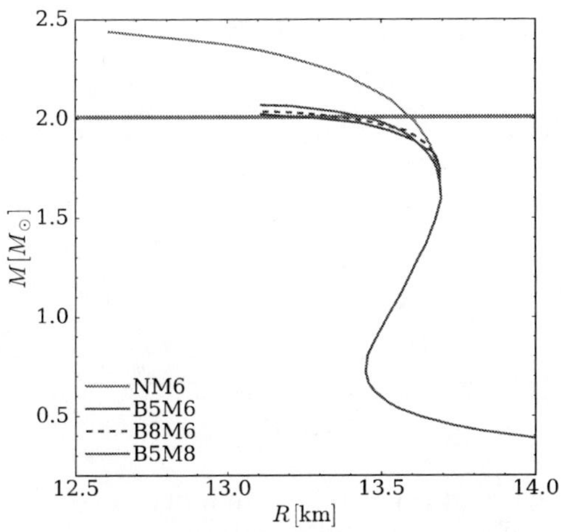

Figure 3. DBHF calculations. Figure adapted from figure 2 of [13]. Notation: Mk - number k of species of included mesons; Bk - number k of species of included baryons; N - nucleons only.

vector-mesons coupling to H. The vector meson field ϕ^μ, coupled to the H only, was added to get an additional repulsion between H. The radius of $1.0 - 2.0\ \mathrm{M}_\odot$ NS is ~ 12 km. Noticeable is also the very small decrease of $M_{\rm max}$ due to the presence of a hyperon core at highest masses. All these features are due to the fact that hyperons are hardly present, even for $M > 2\ \mathrm{M}_\odot$, in NS cores.

It has been suggested that the presence of Δ particles (resonances) in NS cores could strongly soften the EOS, leading to the $M_{\rm max} < 2\ \mathrm{M}_\odot$ (Δ paradox [11]). The extension of the medium-scaled RMF model to the NHΔ plasma has shown that for a sufficiently deep potential well for Δ in symmetric nuclear matter ($-100 < U_\Delta < -50$ MeV), the decrease of $M_{\rm max}$ caused by the Δs is very small, albeit with Δ^- appearing already in $1.0\ \mathrm{M}_\odot$ NS (in figure 2 curves Δ and H$\phi\sigma\Delta$ when hyperons are included, obtained for $U_\Delta = -100$ MeV). On the other hand, the radius R of $1.0 - 2.0\ \mathrm{M}_\odot$ NS shrinks due to the Δ softening of the EOS by some 0.5 km. Summarizing, the EOS presented in [10] and references therein yield $M_{\rm max} \simeq 2.3\ \mathrm{M}_\odot$ with H and Δ in NS cores, and produce $R \simeq 12$ km for $1.0 - 2.0\ \mathrm{M}_\odot$ NS.

It is worth to mention new extensions of older RMF models with density-dependent g_{NM} [12] to NH cores (DD2 H version of the purely nucleonic DD2 model in figure 2). It is argued there that the Lorentz invariance of $\mathcal{L}$ was respected, because $n_{\rm b} = \sqrt{j^\mu_{\rm b} j_{{\rm b}\mu}}$, where $j^\mu_{\rm b}$ is the total baryon four-current $j^\mu_{\rm b} = \sum_B j^\mu_B$. However, this is true only if all baryon species have the same macroscopic (bulk) motion.

4. Hyperon cores in the DBHF theory

The relativistic extension of the Brueckner-Hartree-Fock (BHF) theory of nuclear matter (Dirac-BHF = DBHF) is applied in [13] to calculate the EOS, starting from one-boson-exchange (OBE) interactions known as Bonn OBE models. The relativistic Bonn OBE NN interactions fit a few thousand of NN data (i.e., phase shifts, ^{2}H) equally well as the (non-relativistic) Argonne or Urbana NN potentials. The numerical calculations reported in [13] were a real *tour de force*: the complexity of the DBHF theory is much higher than that of the BHF one. To get agreement with saturation parameters of nuclear matter the Bonn OBE coupling constant $g_{N\sigma}$ was weakly scaled with the scalar component of the mass operator $g_{N\sigma} \longrightarrow g^\star_{N\sigma}(\Sigma^{\rm s}_N)$ (this was an analogue of the $g_{N\sigma}$ scaling with σ in [9, 10]). The required enhancement of $g_{N\sigma}$ is of 2% at n_0. This minute scaling yields a Bonn A* interaction with the correct saturation parameters of nuclear matter. The $3N$ forces needed not to be added to the $2N$ ones because their repulsive effect was reproduced by the density dependence of the vector component of the mass operator, $\Sigma^{\rm v}_N$).

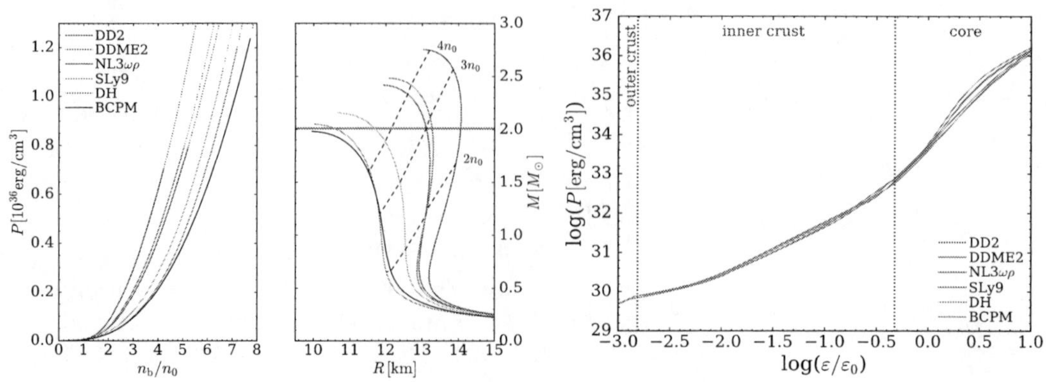

Figure 4. Examples of unified EOS (left and right panels) and their $M(R)$ (middle). The right panel (RMF models in red, others in grey) illustrates the persisting uncertainties in the EOS for the inner crust and outer core. See text for details.

This yielded also a universal repulsion between the baryons. The generalization of the Bonn A* to the NH matter was done using the SU(6) symmetry for the coupling of the vector mesons to hyperons while the fitting to U_H was used to get $g_{H\sigma}$. The $M(R)$ curves for different versions of the Bonn A* interaction are shown in figure 3. Results are only weakly dependent on the number of hyperon species and meson species included. The central density at $M_{\max}$ is only about $4n_0$, while $M_{\max} \simeq 2.03 - 2.08\ \mathrm{M}_\odot$ (see figure 3). The radius for $1.0 - 1.9\ \mathrm{M}_\odot$ is $R > 13.5$ km.

5. New unified EOS for crust and core

The new unified EOS are based on the same nuclear interaction model for both core and the inner crust. Therefore, these EOS give a physically meaningful transition between the crust and the liquid core. This is important for the correct determination of R and the crustal mass and moment of inertia. The many-body problem for the inner crust is solved using the energy density functional (EDF) method (overview, e.g., in [14]). For the uniform n-p-e plasma of the outer core the energy per nucleon becomes a function of the nucleon densities $E(n_n, n_p) = E^{\mathrm{N}}_{\mathrm{bulk}}(n_n, n_p) + E_e(n_e)$. For the inner crust the EDF method allows for the calculation of the non-uniform nucleon structure that minimizes E. In this way, additional "finite size" terms resulting from the surface tension ($E^{\mathrm{N}}_{\mathrm{surf}}$), the spin-orbit coupling ($E^{\mathrm{N}}_{\mathrm{s-o}}$), and the Coulomb interaction (E_{Coul}), appear in the expression for E, apart from the bulk (size-independent) and the electron contributions: $E = E^{\mathrm{N}}_{\mathrm{bulk}} + E^{\mathrm{N}}_{\mathrm{surf}} + E^{\mathrm{N}}_{\mathrm{s-o}} + E_{\mathrm{Coul}} + E_e(n_e)$.

The recent BCPM (Barcelona-Catania-Paris-Madrid) EDF [15] is based on a realistic potential v_{12} A_{18} model supplemented with an effective density-dependent potential $\widetilde{v}_{12}$ resulting from the averaging of the $3N$ UIX potential over the third nucleon coordinate. Similarly as the UIX $3N$, $\widetilde{v}_{12}$ contains two phenomenological parameters, c_1 and c_2, whose values are adjusted to get correct saturation parameters of symmetric nuclear matter: $E_{\mathrm{s}} = -16.0$ MeV and $n_{\mathrm{s}} = 0.16\ \mathrm{fm}^{-3}$. Then three additional phenomenological parameters (c_3, c_4, c_5) appear in $E^{\mathrm{N}}_{\mathrm{surf}}$, and the sixth one, c_6, measures the strength of $E^{\mathrm{N}}_{\mathrm{s-o}}$. The fitting of 467 known spherical nuclei yielded an optimal set of c_3, c_4, c_5, c_6, with a r.m.s. deviation with respect to experimental binding energies $\sigma_{\mathrm{Bind}} = 1.3$ MeV. The BCPM EOS is very similar to the DH one [16], yields a very similar $M(R)$ relation, but the Z of the proton clusters in the inner crust is ~ 30 instead of ~ 40 for the DH model. Moreover, so-called pasta layers composed of nonspherical nuclear clusters (see e.g., [17]) are present in the bottom sector of the inner BCPM crust, while there

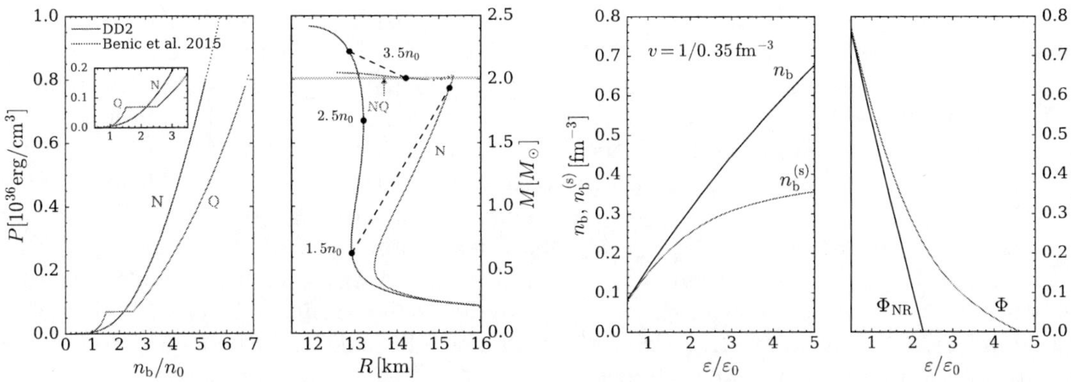

Figure 5. Left panel: Nucleon (N) and quark (Q) EOS from [20]. $\Phi_{\rm NR}$ makes DD2-EV extremely stiff. Right panel: Based on unpublished results of J.L. Zdunik (2015). See text for details.

were no pastas in the DH crust. These differences are most probably due to different $E^{\rm N}_{\rm surf}$, which were obtained using different methods in DH and BCPM models.

Nine new unified EOS based on the RMF models were derived in [18] and four of them shown to be consistent with a set of modern constraints both from nuclear experiments and theoretical calculations. The presence of pasta layers in the inner crust was conditioned by the model-dependent surface energy, $E^{\rm N}_{\rm surf}$, which should not be too high. Otherwise, the crust with spherical proton clusters undergoes direct phase transition to the uniform n-p-e plasma of the core (like for DH).

The $P(\varepsilon)$ and $M(R)$ relations for new unified EOS are shown in figure 4, where we limited ourselves to NS with nucleon cores. Actually, some new RMF EOS could be extended by including the hyperons, and keeping $M_{\rm max} > 2\ {\rm M}_\odot$ [18]. We illustrate the still persisting uncertainties in the EOS of the inner crust and outer core ($n_{\rm b} < 2\ n_0$) in figure 4. The apparent convergence of the RMF EOS and non-relativistic EOS at $10\varepsilon_0$ results actually from the crossings of the RMF and non-relativistic EOS lines. The crossings are due to $({\rm d}P/{\rm d}\varepsilon)_{\rm RMF} \longrightarrow 1$ (from below) for $\varepsilon \gg \varepsilon_0$, while no such asymptotic behaviour of $P(\varepsilon)$ is exhibited by the non-relativistic models.

6. Finite size of baryons and quark-core twins

In contrast to quarks and leptons, baryons have a finite size. In spite of this fact, they are usually treated in the many-body theories as point-like objects. Their finite size can be accounted for in the excluded volume (ExVol) approach, where they are treated as non-penetrable spheres of radius $r_N (N = n, p)$. The ExVol construction can be consistently implemented within the RMF model of baryons and mesons (for review, see [19]). Let us denote the volume fraction available for baryons by Φ. Then the Fermi momenta of nucleons moving within the volume available to them are calculated from $k_{{\rm F}N} = \left(3\pi^2 n_N/\Phi\right)^{1/3}$. Treating nucleons non-relativistically, we get $\Phi_{\rm NR} = 1 - n_{\rm b}v$, where v is the excluded volume per nucleon. Problems appear when we require the ExVol construction to keep $\mathcal{L}$ a Lorentz scalar. We define v, as well as r_N, in the rest-frame of nuclear matter. It is easy to show that the excluded volume per nucleon (measured in the matter rest-frame) is $v = \frac{1}{2}\frac{4\pi}{3}(2r_N)^3 = 4V_N$. As nucleons are moving, the nucleon volume measured in the (macroscopic) matter rest-frame depends on its speed. This is why the correct Lorentz invariant available volume fraction is $\Phi = 1 - n_{\rm b}^{\rm (s)}v$, where $n_{\rm b}^{\rm (s)}$ is the scalar density of nucleons, the source for the scalar field.

Benić et al. [20] started with the RMF DD2 model with density-dependent coupling constants [12]. Then they implemented a non-relativistic ExVol scheme to get DD2-EV, still fitting the same nuclear matter saturation parameters. Their non-relativistic $\Phi_{\rm NR}$ with large $v = 2.86$ fm^3 corresponds to $r_N = 0.55$ fm and this means a hard-core radius of 1.1 fm. Consequently, $\Phi_{\rm NR} = 0$ at $n_{\rm crit} = 0.35$ fm^{-3} (see right panel of figure 5), $P(n_{\rm crit}) \longrightarrow +\infty$. Therefore in this model the quark deconfinement N$\longrightarrow$Q has to occur before the $n_{\rm crit}$ singularity. This (not-so-fine) tuning resulted in the 1st order N$\longrightarrow$Q phase transition at $P_{\rm NQ}$, with a large jump in ε, $\varepsilon_{\rm Q}/\varepsilon_{\rm N} > \frac{3}{2}(1 + P_{\rm NQ}/\varepsilon_{\rm N})$. Consequently, an unstable segment of $M(R)$ separated N-stars from more compact NQ-stars, twins with $M \simeq 2$ M$_\odot$, see left panel of figure 5. It is the use of $\Phi_{\rm NR}$ that makes DD2-EV extremely stiff. Using physical Φ would make the EOS softer and then the twin-NQ stars would most probably go away.

7. Conclusions

Progress in calculating the NS EOS is very slow. The precise EOS for PNM for $n_{\rm b} \lessapprox n_0$ has still not be obtained, except for $n_{\rm b} \ll n_0$. On the positive side, ChEFT calculations show that the 4BF contribution at n_0 is negligibly small. There seems to be no hope to calculate precise EOS using ChEFT for $n_{\rm b} > n_0$ because of the very slow convergence at supra-nuclear densities. Many new unified EOS are now available; some EOS-dispersion in the inner crust and in the outer NS core is present. The scaling of the in-medium effective nuclear Lagrangian with scalar quantities makes EOS more flexible, and allows for the H-puzzle and the Δ-puzzle to be solved, keeping agreement with nuclear constraints. Complete DBHF calculations of the EOS including hyperons were performed, and $M_{\rm max} > 2.0$ M$_\odot$ was obtained. However, to reproduce the saturation parameters of nuclear matter, an in-medium scaling of the σ meson coupling to nucleons had to be introduced. The relativistic version of the excluded-volume approximation of the baryon size effect on the EOS is still to be performed. This is in particular still needed in the context of the hypothetical hybrid-twins at $M \sim 2$ M$_\odot$.

Acknowledgments

This work was supported by the Polish National Science Centre (NCN) grant UMO-2014/13/B/ST9/02621.

References

[1] Fonseca E *et al* 2016 *Astrophys. J.* **832** 167
[2] Antoniadis J *et al* 2013 *Science* **340** 6131
[3] Oppenheimer J R and Volkoff G M 1939 *Phys. Rev.* **55** 540
[4] Kurkela A, Fraga E S, Schaffner-Bielich J and Vuorinen A 2014 *Astrophys. J.* **789** 127
[5] Machleidt R and Entem D R 2011 *Phys. Rep.* **503** 1
[6] Hebeler K, Krebs H, Epelbaum E, Golak J and Skibinski R 2015 *Phys. Rev.* C **91** 044001
[7] Drischler C, Carbone A, Hebeler K and Schwenk A 2016 *Phys. Rev.* C **94** 054307
[8] Hebeler K, Lattimer J M, Pethick C J and Schwenk A. 2013 *Astrophys. J.* **773** 11
[9] Maslov K A, Kolomeitsev E E and Voskresensky D N 2016 *Nucl. Phys.* A **950** 64
[10] Kolomeitsev E E, Maslov K A and Voskresensky D N 2017 *Nucl. Phys.* A **961** 106
[11] Drago A, Lavagno A, Pagliara G and Pigato D 2014 *Phys. Rev.* C **90** 065809
[12] Typel S and Wolter H H 1999 *Nucl. Phys.* A **656** 331
[13] Katayama T and Saito K 2015 *Phys. Lett.* B **747** 43
[14] Walecka J D 2004 *Theoretical Nuclear and Subnuclear Physics* (Imperial College Press), Chapter 25
[15] Sharma B K, Centelles M, Vinas X, Baldo M and Burgio B F 2015 *Astron. Astrophys.* **584** A103
[16] Douchin F and Haensel P 2001 *Astron. Astrophys.* 380 151
[17] Chamel N and Haensel P 2008 *Living Reviews in Relativity* **11** 10
[18] Fortin M *et al* 2016 *Phys. Rev.* C **94** 035804
[19] Typel S 2016 *Eur. Phys. J.* A **52** 16
[20] Benić S, Blaschke D, Alvarez-Castillo D E, Fischer T and Typel S 2015 *Astron. Astrophys.* **577** A40

Neutron star equation of state and uncertainty on the radius determination

M Fortin

Nicolaus Copernicus Astronomical Center of the Polish Academy of Sciences, Bartycka 18, Warsaw 00-716, Poland

E-mail: fortin@camk.edu.pl

Abstract. The new generation of X-ray telescopes will provide simultaneous measurements of the radius and mass of a number of neutron stars (NSs), potentially constraining their equation of state (EoS). However the use of non-unified EoSs, that are based on different nuclear models for the crust and core, can introduce an uncertainty on the NS radius of the order of the precision expected from these instruments. Two solutions to this problem are presented. First a large number of unified EoSs and second an approximate approach to the NS crust that allows to obtain with a very large precision the relation between the NS mass and radius with no need to use an EoS for the crust. Correlations between the NS radius and nuclear parameters, which are properties of the EoS that can be indirectly measured in laboratory, are discussed. These open the possibility of constraining the NS radius with experiments on Earth. Finally, the appearance of hyperons in the core of NSs is examined. They are included in the EoS models consistently with available measurements on the hyperon properties. Because of the scarcity of these experimental data one can not exclude that hyperons appear at high densities in NSs in spite of the observations of massive NSs.

1. Introduction

Although NSs have been observed for the first time 50 years ago, relatively little is known about the properties of matter inside them. The so-called EoS which relates the pressure to the density in their interior is one of the many mysteries of these objects. Indeed, the neutron-rich cold dense matter that composes NSs is not accessible in terrestrial laboratories and the modeling of a many-body system of strongly interacting particles is a complex and unsolved problem (see eg. [1, 2]). However the determination of some astrophysical NS properties like the gravitational mass M and circumferential radius R may constrain the EoS and unveil the properties of NS matter. For the time being, the only constraint comes from the observation of the two pulsars with masses close to $2\,M_\odot$ [3, 4]. More constraints in the form of simultaneous determinations of the mass and radius of a number of NSs are expected in the near future thanks to the newly-operating NICER [5], Athena [6] and LOFT-like [7] missions.

2. Non-unified equations of state and radius uncertainty

The NS interior consists of two main parts: the core and the crust. The latter is solid and composed of a lattice of nuclei, permeated by a gas of electrons and of free neutrons at high density. It extends from the surface to the core-crust transition at which the nuclei disappear, at a density $n_{\rm cc} \simeq 0.5 n_0$, where $n_0 = 0.16$ fm^{-3} is the nuclear saturation density. The central part

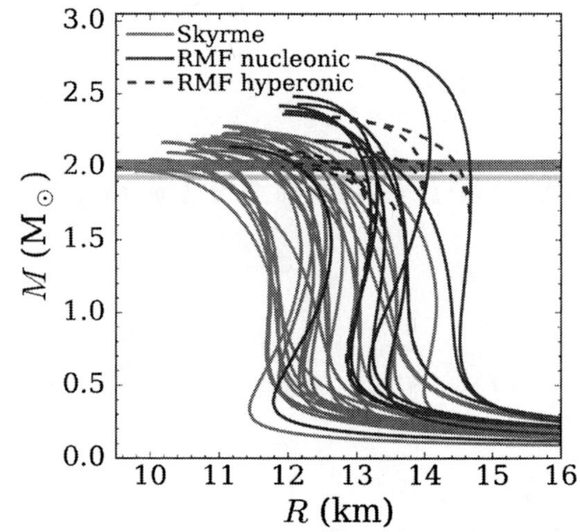

Figure 1. $M - R$ relations for the unified EoSs presented in [8]. 33 EoSs with purely nucleonic cores (solid lines), and 15 ones with hyperonic cores (dashed lines), all consistent with the existence of $2\,M_\odot$ NSs are obtained using the Skyrme (red) and the RMF (blue) parametrizations.

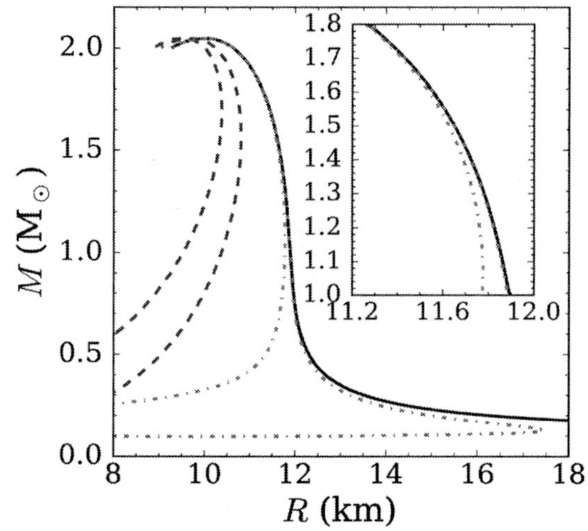

Figure 2. $M - R$ relations obtained with the approximate approach (in red). The $M(R_{\rm core})$ relation (in blue) together with eq (1) is employed, using only an EoS for the core. The black line corresponds to the $M-R$ relation for a unified EoS. See text for details.

of the NS, the core, is liquid and formed of free nucleons, in addition to electrons and muons, and possibly more exotic form of matter like hyperons at high densities.

Calculating an EoS for the NS crust is much less straightforward than for the core, as the crust is non-uniform and composed of atomic nuclei. Consequently there are much less crust EoSs available compared to core ones. In particular very few so-called unified EoSs, in the sense that the properties of both the core and the crust have been calculated based on the same nuclear model, have been developed. Thus a widely-used approach to obtain an EoS for the entire NS interior consists in using two EoSs, one for the crust and another one for the core based on different nuclear models, and connecting (or 'gluing') them together. However, as shown in [8] the various core-crust gluing procedures reported in the literature result in differences in the radius for a $1.0\,M_\odot$ NS as large as ~ 1 km! This is of the same order as the expected precision on radii measurements from the new generation of X-ray telescopes (NICER, Athena, ...). This problem of the core-crust gluing has therefore to be solved or at least handle so that constraints on the EoS can be correctly obtained from NS joint mass-radius observational determinations.

3. Unified equations of state and approximate approach to the crust

Two strategies can be used to tackle the problem posed by the gluing of a crust EoS to one for the core.

The first one consists in using unified EoSs. In [8] a set of 33 nuclear models based on two nuclear approaches: Skyrme and Relativistic Mean Field (RMF) parametrizations, are used to obtain 33 EoSs with purely nucleonic cores, and 15 with hyperons appearing at high density. They are all consistent with the existence of $2\,M_\odot$ NSs and describe the core and the crust in a unified fashion (for the RMF models, the outer crust is not consistently calculated but the gluing between the outer and inner crust does not significantly affect the star radius for masses of astrophysical interest: $M > 1\,M_\odot$). They cover a large part of the $M - R$ plane as shown in figure 1. In addition the 33 nuclear models are confronted to a set of modern nuclear constraints both experimental (on the incompressibility K, the symmetry energy, and its slope L, all evaluated at saturation) and theoretical (calculations for pure neutron matter from [9, 10]). A set of 4 nuclear models, 4 nucleonic and 4 hyperonic EoSs are then selected. For reference, they give $R_{1.4} = 13.10 \pm 0.65$ km for the radius of a $1.4\,M_{\odot}$ NS. EoS tables are made available online to the NS community as supplemental material to Ref. [8].

In case the crust EoS is not available for the nuclear model used for the core EoS, the second strategy presented in [11] relies on an approximate approach to the crust. Since the mass of the crust is small compared to the total mass of the star, one can rewrite the TOV equations for hydrostatic equilibrium in General Relativity and obtain the following relation between the radius of the core $R_{\rm core}$ and the total radius of the NS:

$$\frac{\sqrt{1-2GM/Rc^2}}{\sqrt{1-2GM/R_{\rm core}c^2}} = \frac{\mu_{\rm cc}}{\mu_0} \qquad (1)$$

With the help of eq (1) one can then obtain the relation between the mass and radius of a NS using only an EoS for the core. In other words, with this approach, no crust EoS is required. First for a given core EoS, for densities larger than $n_{\rm cc}$, (see below), one solves the TOV equations to obtain the relation between M and $R_{\rm core}$. Then using eq (1) one can obtain the relation between the total radius of the star R and M. Two quantities enter eq (1). The baryon chemical potential at zero pressure μ_0 is given for a non-accreted crust by ^{56}Fe as the minimum of energy is obtained for this element: $\mu_0 = 930.4\,$MeV [1]. The baryon chemical potential at the core-crust transition $\mu_{\rm cc}$, is usually not known for a given core EoS. However for a wide range of values of L and a large set of EoSs it has been determined in [12] that $0.38 \leq n_{\rm cc}/n_0 \leq 0.63$. Consequently, taking $n_{\rm cc} = 0.5 n_0$ appears reasonable. Then the chemical potential at the transition is $\mu_{\rm cc} = (P(n_{\rm cc}) + \rho(n_{\rm cc}))/n_{\rm cc}$, where $P(n_{\rm cc})$ and $\rho(n_{\rm cc})$ are the pressure and mass-energy density obtained using the core EoS. For $M > 1\,M_{\odot}$ the approximate approach gives values of R with a very good precision: $< 0.2\%$. In figure 2 the unified EoS from [13], with $n_{\rm cc} = 0.48 n_0$, is plotted in black. The blue dotted lines correspond to the $M(R_{\rm core})$ relations obtained when solving the TOV equations for two values of $n_{\rm cc} = n_0/2$ and n_0 from right to left. The red dot-dashed lines are obtained using eq (1). Similar and simple formulas applicable for a NS with an accreted crust and for the crust thickness are also obtained in Ref. [11].

4. Nuclear parameters and radii

In addition to astrophysical observations, measurements thanks to laboratory experiments of nuclear parameters, that describe the properties of the NS EoS, can constrain the NS radius. Indeed correlations between various nuclear parameters and R have been investigated extensively. For example, in line with numerous previous works, the linear correlation between L and the NS radius has been confirmed in [8] using 33 unified EoSs but it was pointed out that it becomes weaker with the increase of the NS mass. The existence of a strong correlation between R and a linear combination of L and the slope M of the incompressibility K obtained for 44 EoSs (24 of the Skyrme-type, 18 RMF ones, and 2 microscopic calculations, all consistent with $2\,M_{\odot}$ NSs) is reported in [14]. This correlation is found to be almost independent of the NS mass, for $0.6 < M < 1.8\,M_{\odot}$. Using experimental constraints from giant monopole resonance and other nuclear properties on L and M, one then obtain $R_{1.4} = 11.09 - 12.86$ km.

5. Experimentally calibrated hyperonic EoSs

Hyperons were discovered in laboratory in the early 1950s and are studied experimentally since then although relatively little is known about them [15]. The binding energy of ~ 40 Λ-hypernuclei (ie. nuclei with a Λ hyperon) has been measured in laboratory and the observation of

Ξ-hypernuclei indicates an attractive Ξ-nucleon interaction. However, there is no indication from nuclear experiments of the existence of Σ-hypernuclei up to now, suggesting that the Σ-nucleon interaction is most probably repulsive. Several double Λ-hypernuclei, that is an hypernuclei with two Λ hyperons, have been observed but so far only a single one has been unambiguously identified: ${}^{6}_{\Lambda\Lambda}$He, and its properties measured.

In the late 1950s it was suggested that hyperons could be present in NS cores. However their appearance in the NS interior usually leads to the softening of the EoS (smaller pressure at high density than for a nucleonic EoS) and thus to a lower value of the NS maximum mass. Therefore the existence of massive NSs can appear difficult to reconcile with the presence of hyperons in the NS core. This is the so-called hyperon puzzle.

In [16], for 6 RMF parametrizations, single and double Λ-hypernuclei are modeled and the couplings for the hyperons that enter the EoS model are calibrated by fitting the measured binding energy of such nuclei instead of employing widely-used values for the hyperon potentials. It is pointed out that the usual values taken in the literature for the Λ-potentials are not consistent with the measured hypernuclei properties. The other hyperons are then included in the models by taking into account the uncertainties existing on the interaction between the nucleons and the Σ and Ξ hyperons. Owing to the lack of experimental data on the Ξ and Σ hyperons it is then possible to obtain $2\,M_\odot$ NSs with hyperons. Thus there is so far no answer to the hyperon puzzle.

6. Conclusion

A careful treatment of the NS EoS at the core-crust transition is necessary to properly constrain the properties of NS matter thanks to incoming joint measurements of the mass and radius of a number of NSs from the new generation of X-ray satellites. Unified EoS using the same nuclear model for the core and the crust such as the ones presented in [8] or the approximate and yet very accurate approach to the crust developed in [11] can be used. Measurements of nuclear parameters from laboratory experiments can also provide constraints on the NS radius [8, 14]. Finally, as shown in [16], more measurements of the hyperon properties are necessary in order to confirm or exclude the presence of hyperons inside the core of massive NSs.

Acknowledgments

This work was supported by the Polish National Science Centre (NCN) grant UMO-2014/13/B/ST9/02621 and the New-Compstar COST Action MP1304.

References

[1] Haensel P, Potekhin A Y and Yakovlev D G 2007 *Neutron Stars I. Equation of State and Structure* (New York: Springer)
[2] Oertel M, Hempel M, Klähn T and Typel S 2017 *Reviews of Modern Physics* **89** 015007
[3] Antoniadis J *et al* 2013 *Science* **340** 448
[4] Fonseca E *et al* 2016 *Astrophys. J.* **832** 167
[5] Arzoumanian Z *et al* 2014 *Proc. SPIE* **9144** 914420
[6] Motch C *et al* 2013 *Preprint* 1306.2334
[7] Feroci M *et al* 2012 *Proc. SPIE* **8443** 84432D
[8] Fortin M *et al* 2016 *Phys. Rev.* C **94** 035804
[9] Gandolfi S, Carlson J and Reddy S 2012 *Phys. Rev.* C **85** 032801
[10] Hebeler K, Lattimer J M, Pethick C J and Schwenk A 2013 *Astrophys. J.* **773** 11
[11] Zdunik J L, Fortin M and Haensel P 2017 *Astron. Astrophys.* **599** A119
[12] Ducoin C, Margueron J, Providência C and Vidaña I 2011 *Phys. Rev.* C **83** 045810
[13] Douchin F and Haensel P 2001 *Astron. Astrophys.* **380** 151
[14] Alam N *et al* 2016 *Phys. Rev.* C **94** 052801
[15] Gal A, Hungerford E V and Millener D J 2016 *Reviews of Modern Physics* **88** 035004
[16] Fortin M, Avancini S S, Providência C and Vidaña I 2017 *Phys. Rev.* C **95** 065803

IOP Conf. Series: Journal of Physics: Conf. Series **932** (2017) 012036 doi:10.1088/1742-6596/932/1/012036

Neutron star dynamics under time dependent external torques

M A Alpar and E Gügercinoğlu

Faculty of Engineering and Natural Sciences, Sabancı University, Orhanlı -Tuzla, İstanbul 34956, Turkey

E-mail: alpar@sabanciuniv.edu

Abstract. The two component model of neutron star dynamics describing the behaviour of the observed crust coupled to the superfluid interior has so far been applied to radio pulsars for which the external torques are constant on dynamical timescales. We recently solved this problem under arbitrary time dependent external torques. Our solutions pertain to internal torques that are linear in the rotation rates, as well as to the extremely non-linear internal torques of the vortex creep model. Two-component models with linear or nonlinear internal torques can now be applied to magnetars and to neutron stars in binary systems, with strong variability and timing noise. Time dependent external torques can be obtained from the observed spin-down (or spin-up) time series, $\dot{\Omega}(t)$.

1. The two component model with constant external torques

The two component model for neutron star dynamics was proposed immediately after the observation of the first pulsar glitches [1]. This model and its extension to nonlinear vortex creep coupling between the crust and the superfluid [2] have been applied with remarkable success to the postglitch relaxation data from radio pulsars, for which the external torques are constant on dynamical timescales. This talk presents, in summary, our recent solution of the two component models for linear and nonlinear coupling for the general case of arbitrary time dependent external torques [3]. These general solutions are applicable to the timing behaviour of neutron stars in X-ray binaries and magnetars, as well as timing noise from radio pulsars.

Postglitch exponential relaxation on timescales of days to weeks supported the theoretical expectations of superfluidity in the neutron star interior. As exponential relaxation is produced in linear systems, the original two component model of Baym et al [1] posits a linear coupling between the crust and the superfluid interior. The equations of the linear two component model are

$$I_{\rm c}\dot{\Omega}_{\rm c} + I_{\rm s}\dot{\Omega}_{\rm s} = N_{\rm ext}, \qquad (1)$$

$$\dot{\Omega}_{\rm s} = -\frac{\Omega_{\rm s} - \Omega_{\rm c}}{\tau_0} = -\frac{\omega}{\tau_0}. \qquad (2)$$

$I_{\rm c}$ and $I_{\rm s}$ denote the moments of inertia and $\Omega_{\rm c}$ and $\Omega_{\rm s}$ are the rotation rates of the crust and the superfluid, respectively. The rotational velocity lag between the two components is denoted by $\omega \equiv \Omega_{\rm s} - \Omega_{\rm c}$. The internal torque that the superfluid exerts on the crust is

$$N_{\rm int} \equiv -I_{\rm s}\dot{\Omega}_{\rm s}, \qquad (3)$$

while $N_{\rm ext}$ denotes the external torque on the neutron star. The crust (normal matter) rotation rate $\Omega_{\rm c}$ obeys the Navier Stokes Equation, but is already in rigid body rotation on superfluid timescales. The superfluid rotation rate $\Omega_{\rm s}$ does not obey the Navier Stokes Equation. Physically, the superfluid spins down by a flow of quantized vortices away from the rotation axis. Under many kinds of physical interactions between the vortices and normal matter, the vortex current, and the superfluid rate are linear in the angular velocity lag ω.

Under a constant external torque $N_{\rm ext}$, the system has a steady state,

$$\begin{aligned} \dot{\omega} &= 0, \\ \dot{\Omega}_{\rm s} &= \dot{\Omega}_{\rm c} = \frac{N_{\rm ext}}{I}. \end{aligned} \tag{4}$$

The steady state value ω_∞ of the lag is determined by

$$\frac{\omega_\infty}{\tau_0} = -\frac{N_{\rm ext}}{I}. \tag{5}$$

The linear response to any glitch induced offset $\delta\omega(0)$ from steady state is exponential relaxation

$$\dot{\Omega}_{\rm c} = \dot{\Omega}_{\rm c}(0) - \frac{I_{\rm s}}{I}\frac{\delta\omega(0)}{\tau}e^{-t/\tau}. \tag{6}$$

Several components i of the exponential relaxation are indeed observed following pulsar glitches. An analysis of a postglitch relaxation yields $I_{\rm s,i}/I \sim 10^{-3} - 10^{-2}$ for each component of the exponential relaxation. This points at the crust superfluid. With the effective masses (entrainment) [4] taken into account, $I_{\rm s,i}/I \lesssim 10^{-1}$, pointing to the crust + outer core [5]. The core superfluid is already coupled tightly to the crust; the core is effectively a part of the crust, dynamically, on glitch and postglitch relaxation timescales [6]. An analysis in terms of the *two* component model is enough: Since $I_{\rm s,i}/I \ll 1$, the different superfluid components i with moments of inertia $I_{\rm s,i}$ can be handled with the crust in separate two component models and then the response of the crust to each can be superposed.

2. Vortex creep and the nonlinear two component model

A superfluid component with vortex pinning can spin down by the thermally activated flow (creep) of vortices against pinning potentials. The vortex creep model gives the spindown rate

$$\dot{\Omega}_{\rm s} = -\frac{4\Omega_{\rm s}v_0}{r}\exp\left(-\frac{E_{\rm p}}{kT}\right)\sinh\left(\frac{\omega}{\varpi}\right) \equiv -f(\omega), \tag{7}$$

where

$$\varpi \equiv \frac{kT}{E_{\rm p}}\omega_{\rm cr}, \tag{8}$$

to replace (2). Here, $E_{\rm p}$ is the pinning energy and T is the temperature. The distance r of the vortex lines from the rotation axis is approximately equal to the neutron star radius R in the crust superfluid where creep takes place against the pinning sites in the crust lattice. The microscopic vortex velocity v_0 around pinning centres is $\approx 10^7$ cm/s [7]. The two component model is now nonlinear in the lag ω.

The values of the parameters, in particular of $E_{\rm p}/kT$, can be such that (7) requires $\sinh(\omega/\varpi) \gg 1$, so that $\sinh(\omega/\varpi) \simeq (1/2)\exp(\omega/\varpi)$. Equations (1) and (7) then yield

$$\dot{\Omega}_{\rm c} = \frac{N_{\rm ext}}{I_{\rm c}} + \frac{I_{\rm s}}{I_{\rm c}}\frac{\varpi}{2\tau_1}\exp(\omega/\varpi), \tag{9}$$

where

$$\tau_{\mathrm{l}} \equiv \frac{kT}{E_{\mathrm{p}}} \frac{R\omega_{\mathrm{cr}}}{4\Omega_{\mathrm{s}} v_0} \exp\left(\frac{E_{\mathrm{p}}}{kT}\right). \tag{10}$$

For nonlinear coupling, the steady state value of the lag, ω_∞, is given by (4) and (7).

The solution for the crust spindown rate under a constant external torque $N_{\mathrm{ext}} = I\dot{\Omega}_\infty$ is

$$\dot{\Omega}_{\mathrm{c}}(t) = \frac{I}{I_{\mathrm{c}}}\dot{\Omega}_\infty - \frac{I_{\mathrm{s}}}{I_{\mathrm{c}}}\dot{\Omega}_\infty \left[1 + \left[\exp\left(\frac{t_0}{\tau_{\mathrm{nl}}}\right) - 1\right] \exp\left(-\frac{t}{\tau_{\mathrm{nl}}}\frac{I}{I_{\mathrm{c}}}\right)\right]^{-1}, \tag{11}$$

with a nonlinear creep relaxation time

$$\tau_{\mathrm{nl}} \equiv \frac{kT}{E_{\mathrm{p}}} \frac{I\omega_{\mathrm{cr}}}{N_{\mathrm{ext}}} = \frac{kT}{E_{\mathrm{p}}} \frac{\omega_{\mathrm{cr}}}{|\dot{\Omega}|_\infty}, \tag{12}$$

and recoupling (waiting) timescale

$$t_0 \equiv \frac{I\delta\omega}{N_{\mathrm{ext}}} = \frac{\delta\omega}{|\dot{\Omega}|_\infty}. \tag{13}$$

Nonlinear creep regions are responsible for glitches through vortex unpinning avalanches leading to a decrease $\delta\Omega_{\mathrm{s}}(r)$ in those superfluid regions through which the unpinned vortices move rapidly during the glitch. Angular momentum conservation leads to $\delta\Omega_{\mathrm{s}} = (I_{\mathrm{c}}/I_{\mathrm{s}})\Delta\Omega_{\mathrm{c}} \gg \Delta\Omega_{\mathrm{c}}$, and $\delta\omega \cong \delta\Omega_{\mathrm{s}}$. Because of the very sensitive dependence of the creep rate on $\delta\omega$ this glitch induced offset from the steady state lag ω_∞ will stop creep altogether in the superfluid regions effected. As the same external torque is now acting on less moment of inertia, the observed spindown rate of the crust will suffer a glitch associated jump

$$\frac{\Delta\dot{\Omega}_{\mathrm{c}}}{\dot{\Omega}_{\mathrm{c}}} \cong \frac{I_{\mathrm{s}}}{I_{\mathrm{c}}}. \tag{14}$$

Creep restarts after a waiting time $t_0 \cong \delta\Omega_{\mathrm{s}}/|\dot{\Omega}|_\infty$. This "Fermi function" response was predicted as a standard signature of the vortex creep response to a glitch involving vortex unpinning (see figure 4 in [2]) and later observed in the Vela pulsar [8]. The common, almost ubiquitous form of nonlinear response is the constant second derivative $\ddot{\Omega}_{\mathrm{c}}$ interglitch timing behaviour observed in the Vela and other pulsars [9,10]. This is the response to a uniform density of vortices unpinned at the glitch leading to a range of waiting times throughout the superfluid regions. Such power law behaviour is characteristic of nonlinear dynamics.

Two component models with linear or nonlinear internal torques have so far been applied to post-glitch or inter-glitch timing behaviour of radio pulsars [9,11–14], where the external torque, with a secular (characteristic) timescale $\tau_{\mathrm{c}} \equiv \Omega_{\mathrm{c}}/2|\dot{\Omega}_{\mathrm{c}}| \sim 10^3 - 10^6$ yr is constant for timescales of observed postglitch relaxation.

While constant secular external torques are characteristic for radio pulsars, neutron stars in X-ray binaries, magnetars and transients exhibit strong variations in the observed spin-down or spin-up rates. This indicates variable external torques, including strong torque noise.

3. Linear two-component model with a time dependent external torque

The two component model postglitch response with a linear internal torque and a time varying external torque is described by

$$\dot{\omega} = -\frac{\omega}{\tau} - \frac{N_{\mathrm{ext}}(t)}{I_{\mathrm{c}}}. \tag{15}$$

The relaxation time τ is determined by the physics of the internal torque. The solution is

$$\begin{aligned} \omega(t) &= e^{-t/\tau}\left[\omega(0) - \frac{1}{I_{\rm c}}\int_0^t e^{t'/\tau} N_{\rm ext}(t')\,{\rm d}t'\right], \\ \dot{\Omega}_{\rm c}(t) &= \frac{N_{\rm ext}(t)}{I_{\rm c}} + \frac{I_{\rm s}}{I}\frac{e^{(-t/\tau)}}{\tau}\left[\omega(0) - \frac{1}{I_{\rm c}}\int_0^t e^{(t'/\tau)} N_{\rm ext}(t')\,{\rm d}t'\right]. \end{aligned} \tag{16}$$

4. Vortex Creep Model with a Time Dependent External Torque

For vortex creep under a time dependent external torque (1) and (7) lead to

$$\dot{\omega} = -\frac{I\varpi}{2I_{\rm c}\tau_{\rm l}}e^{\omega/\varpi} - \frac{N_{\rm ext}(t)}{I_{\rm c}}. \tag{17}$$

In terms of $y \equiv \exp(-\omega/\varpi)$, equation (17) becomes

$$\frac{{\rm d}y}{{\rm d}t} - \frac{N_{\rm ext}(t)}{I_{\rm c}\varpi}y - \frac{I}{2I_{\rm c}\tau_{\rm l}} = 0. \tag{18}$$

This has an integration factor $\exp\left(-\frac{X(t)}{I_{\rm c}\varpi}\right)$ where

$$X(t) = \int_0^t N_{\rm ext}(t')\,{\rm d}t'. \tag{19}$$

The solution for the angular velocity lag ω exhibits an exponential dependence on the postglitch lag $\omega(0)$ and on $X(t)$, the cumulative angular momentum transfer by the external torque:

$$\begin{aligned} e^{-(\omega/\varpi)} &= e^{-\omega(0)/\varpi}\exp\left(\frac{X(t)}{I_{\rm c}\varpi}\right) \\ &+ \exp\left(\frac{X(t)}{I_{\rm c}\varpi}\right)\int_0^t \frac{I}{2I_{\rm c}\tau_{\rm l}}\exp\left(-\frac{X(t')}{I_{\rm c}\varpi}\right){\rm d}t'. \end{aligned} \tag{20}$$

From equations (1), (17), and (20) the response of nonlinear creep to a glitch in the presence of a time dependent external torque is obtained,

$$\begin{aligned} \dot{\Omega}_{\rm c}(t) &= \frac{N_{\rm ext}(t)}{I_{\rm c}} \\ &+ \frac{I_{\rm s}}{I_{\rm c}}\frac{\varpi}{2\tau_{\rm l}}\left[\exp\left(\frac{\frac{X(t)}{I_{\rm c}} - \omega(0)}{\varpi}\right) + \exp\left(\frac{X(t)}{I_{\rm c}\varpi}\right)\int_0^t \frac{I}{2I_{\rm c}\tau_{\rm l}}\exp\left(-\frac{X(t')}{I_{\rm c}\varpi}\right){\rm d}t'\right]^{-1}. \end{aligned} \tag{21}$$

The nonlinear creep timescale and the waiting time are now time dependent, involving the running time average $\langle N_{\rm ext}(t)\rangle \equiv X(t)/t$ of the external torque:

$$\tau_{\rm nl}(t) \equiv \frac{kT}{E_{\rm p}}\frac{I\omega_{\rm cr}}{\langle N_{\rm ext}(t)\rangle}, \tag{22}$$

$$t_0(t) \equiv \frac{I\delta\omega(0)}{\langle N_{\rm ext}(t)\rangle}. \tag{23}$$

5. Applications

Equations (16) and (21) show that the observed $\dot{\Omega}_c(t)$ gives $N_{\rm ext}(t)/I_c$ to lowest order in $I_s/I_c \lesssim 0.1$. The residuals after a first fit to the $\dot{\Omega}_c(t)$ time series will be I_s/I_c times a certain convolution of the external torque with the internal torque, as defined in (16) or (21), allowing for a consistency check of the model.

Let us consider three kinds of time dependent external torques; (i) an exponentially decaying external torque, (ii) power law time dependence and (iii) timing noise.

(i) Taking an exponentially decaying torque with time scale τ_d added to the preglitch external torque N_0, we have

$$N_{\rm ext}(t) = N_0 + \delta N e^{-t/\tau_d} = I\dot{\Omega}_\infty + \delta N e^{-t/\tau_d}. \tag{24}$$

For linear internal torques, (16) gives

$$\begin{aligned}\dot{\Omega}_c(t) &= \dot{\Omega}_\infty + \frac{\delta N}{I_c} e^{-t/\tau_d}\left[1 - \frac{I_s}{I}\frac{\tau_d}{\tau_d - \tau}\right] \\ &+ \frac{I_s}{I} e^{-t/\tau}\left(\frac{\omega(0)}{\tau} + \frac{I}{I_c}\dot{\Omega}_\infty + \frac{\delta N}{I_c}\frac{\tau_d}{\tau_d - \tau}\right).\end{aligned} \tag{25}$$

For vortex creep, the solution is obtained by substituting in (21) the integrated angular momentum transfer

$$X(t) = N_0 t + \delta N \tau_d[1 - e^{-t/\tau_d}]. \tag{26}$$

(ii) For a power law torque with the index α added to the preglitch torque we have

$$N_{\rm ext}(t) = N_0 + \frac{\delta N t_0^\alpha}{(t + t_0)^\alpha} = I\dot{\Omega}_\infty + \frac{\delta N t_0^\alpha}{(t + t_0)^\alpha}. \tag{27}$$

Then (16) gives

$$\begin{aligned}\dot{\Omega}_c(t) = \dot{\Omega}_\infty &+ \frac{\delta N}{I_c}\frac{t_0^\alpha}{(t + t_0)^\alpha}\left[1 - \frac{I_s}{I}\frac{(t + t_0)}{(1 - \alpha)\tau}\right] \\ + \frac{I_s}{I} e^{-t/\tau}&\left(\frac{\omega(0)}{\tau} + \frac{I}{I_c}\dot{\Omega}_\infty + \frac{\delta N}{I_c}\frac{t_0}{(1 - \alpha)\tau} + \frac{I_s}{I}\frac{\delta N}{I_c}\frac{t_0^\alpha}{(1 - \alpha)\tau^2}\int_0^t \frac{e^{t'/\tau}\,{\rm d}t'}{(t + t_0)^{\alpha - 1}}\right),\end{aligned} \tag{28}$$

for linear internal torques.

The integrated angular momentum transfer $X(t)$ is

$$X(t) = N_0 t + \frac{\delta N t_0}{\alpha - 1}\left[1 - \frac{t_0^{\alpha - 1}}{(t + t_0)^{\alpha - 1}}\right], \tag{29}$$

defining the response with nonlinear internal torques through Equation (21).

(iii) Take white torque noise:

$$N_{\rm ext} = \sum_i \alpha_i \delta(t - t_i), \tag{30}$$

where α_i are amplitudes of torque variations. For linear coupling equations (16) and (30) lead to the power spectrum

$$\begin{aligned}P(f) &= \frac{1}{\sqrt{2\pi}}\left[\frac{2\langle\alpha^2\rangle}{I_c^2} + \left(\frac{I_s/I}{1 + (2\pi\tau f)^2}\right)\left(\frac{\langle\alpha\rangle\langle\omega\rangle}{I_c} - \frac{\langle\alpha^2\rangle}{I_c^2}\right)\right. \\ &+ \left.\left(\frac{(I_s/I)^2}{1 + (2\pi\tau f)^2}\right)\left(\frac{2\langle\alpha^2\rangle}{I_c^2} + 2\langle\omega^2\rangle - \frac{\langle\alpha\rangle\langle\omega\rangle}{I_c}\right)\right],\end{aligned} \tag{31}$$

where $\langle\alpha\rangle$ is the mean external torque variation amplitude and $\langle\omega\rangle$ denotes the mean value of the angular velocity lag. This result was obtained earlier for the linear two component model [15–17]. To lowest order we find the model power spectrum, a constant $P(f)$ for white noise. To order I_s/I we find the power spectrum of the integrated process, which is flat at low f and a random walk spectrum, $P(f) \propto f^{-2}$, at high frequencies $f \gg \tau^{-1}$. The first fit will give the strength of the noise process, the term proportional to $\langle\alpha^2\rangle$. The residuals to order I_s/I will constrain $\langle\alpha\rangle\langle\omega\rangle$ and $\langle\omega^2\rangle$. The term proportional to $(I_s/I)^2$ can be neglected to a good approximation. Noise processes other than white noise can be handled similarly for the linear two component model, by inserting the noise model in (16) and calculating the power spectrum. For the nonlinear two component model, equation (21) does not lead to explicit expressions for any time dependent external torque, but a well defined algorithm can be constructed to obtain the power spectrum, as for the times series in cases (i) and (ii).

Acknowledgments
This work was supported by the Scientific and Technological Research Council of Turkey (TÜBİTAK) under the grant 113F354. M.A.A. is a member of the Science Academy (Bilim Akademisi), Turkey.

References
[1] Baym G, Pethick C J, Pines D and Ruderman M 1969 *Nature* **224** 872
[2] Alpar M A, Anderson P W, Pines D and Shaham J 1984 *ApJ* **276** 325
[3] Gügercinoğlu E and Alpar, M A 2017 *MNRAS* **471** 4827
[4] Chamel N 2012 *Phys. Rev. C* **85** 035801
[5] Gügercinoğlu E and Alpar, M A 2014 *ApJ* **788** L11
[6] Alpar M A, Langer S A and Sauls J A 1984b *ApJ* **282** 533
[7] Gügercinoğlu E and Alpar M A 2016 *MNRAS* **462** 1453
[8] Buchner S and Flanagan C 2008 *American Institute of Physics Conference Series* vol 983, ed C Bassa, Z Wang, A Cumming and V Kaspi p 145
[9] Akbal O, Alpar MA, Buchner S and Pines D 2017 *MNRAS* **469** 4183
[10] Yu M *et al* 2013, *MNRAS* **429** 688
[11] Alpar M A, Chau H F, Cheng K S and Pines D 1993 *ApJ* **409** 345
[12] Alpar M A, Chau H F, Cheng K S and Pines D 1996 *ApJ* **459** 706
[13] Akbal O, Gügercinoğlu E, Şaşmaz Muş S and Alpar M A 2015 *MNRAS* **449** 933
[14] Gügercinoğlu E 2017, *MNRAS* **469** 2313
[15] Lamb F K, Pines D and Shaham J 1978a *ApJ* **224** 969
[16] Lamb F K, Pines D and Shaham J 1978b *ApJ* **225** 582
[17] Baykal A, Alpar M A and Kızıloğlu Ü 1991, *A&A* **252** 664

IOP Conf. Series: Journal of Physics: Conf. Series **932** (2017) 012037 doi:10.1088/1742-6596/932/1/012037

Glitches as probes of neutron star internal structure and dynamics: Effects of the superfluid-superconducting core

Erbil Gügercinoğlu

Faculty of Engineering and Natural Sciences, Sabanci University, Orhanli, Tuzla, 34956 Istanbul, Turkey

E-mail: egugercinoglu@gmail.com

Abstract. Glitches, sudden spin-up of pulsars with subsequent recovery, provide us with a unique opportunity to investigate various physical processes, including the crust-core coupling, distribution of reservoir angular momentum within different internal layers, spin-up in neutral and charged superfluids and constraining the equation of state of the neutron star (NS) matter. In this work, depending on the dynamic interaction between the vortex lines and the nuclei in the inner crust, and between the vortex lines and the magnetic flux tubes in the outer core, various types of relaxation behavior are obtained and confronted with the observations. It is shown that the glitches have strong potential to deduce information about the cooling behavior and interior magnetic field configuration of NSs. Some implications of the relative importance of the external spin-down torques and the superfluid internal torques for recently observed unusual glitches are also discussed.

1. Introduction

Pulsars are rapidly rotating NSs. Being exceptionally stable rotators, they rival the best atomic clocks. This fact alone allows to obtain a precise timing solution after observing frequency ν and spin-down rate $\dot{\nu}$. It enables one to predict NS rotational evolution with high accuracy. However, pulsars, especially the young ones, show stochastic wandering, i.e. noise in their phase, rotation period and spin-down rate [1], rendering some level of fluctuations in measurements. Moreover, regular long-term spin-down of pulsars is observed to be occasionally interrupted by sudden spin-ups in their rotation rates $\Omega = 2\pi\nu$, i.e. glitches. Glitches appear in timing data as fractional changes in the rotation rate with $\Delta\Omega/\Omega = 10^{-11} - 10^{-5}$. They are usually accompanied by larger jumps in the spin-down rate, $\Delta\dot{\Omega}/\dot{\Omega} = 10^{-4} - 10^{-1}$ [2–4]. Both changes tend to recover fully or partially on timescales of the order of several days to a few years.

Glitch monitoring demonstrated for the first time that NSs contain superfluid at the highest observable densities and temperatures; since sudden spin-up with long relaxation give indirect evidence for bulk superfluidity in NS matter [5]. Recent observation of decline in surface flux of the 330 yr old Cas A NS also confirmed a transition into superfluid/superconducting state in the core with temperature $T \leq 10^{8-9}$ K [6]. Anderson and Itoh [7] in their seminal paper explained glitches in terms of catastrophic unpinning of vortex lines, i.e. topological defects within a superfluid carrying angular momentum, from ambient normal matter crystal in the NS crust. The standard model, i.e. the vortex creep model by Alpar et al. [8] invokes the interaction of vortex lines of the NS inner crust with the crystal nuclei to account for the glitch occurrence and post-glitch relaxation. In the vortex creep model, the coupling between the superfluid and crust is maintained by the thermally activated migration of vortex lines at a presumed lag $\omega \equiv \Omega_s - \Omega_c$ between the superfluid and crustal rotational rates. At the time of a glitch, vortex lines rip off from pinning centers and impart their angular momentum to the crust. As a consequence, the crust spins-up and the lag decreases. Hence the coupling weakens. This results in decoupling of that superfluid region from the externally decelerating magnetospheric dipole and/or wind torques. Since the external torque is now acting on a smaller moment of inertia, the spin-down rate increases following a glitch. Post-glitch recovery appears as recoupling of glitch affected superfluid region to the crust.

Various phases of glitch observations have strong potential of constraining and extracting information about NS internal structure and dynamics. In this paper, it is shown how the occurrence, magnitudes and relaxations of glitches observed from canonical radio pulsars and magnetars can be used to probe different physical processes inside NSs.

2. Glitch occurrence as neutron star probe

Surprisingly, the origin of pulsar glitches still remains controversial although it has been intensively studied for about half of an century. Nevertheless, the occurrence of glitches may shed light on the internal structure of NSs. It appears that there are at least two different behaviors and spin-up timescales related to the glitches. The well-known quasi-periodic giant glitcher, the Vela pulsar, undergoes large spin-ups with a timescale less than 40 s [9]. Its glitches initially show abrupt fall off from a large value. Conversely, the glitches typical for the Crab pulsar are somewhat weaker and display a day long comparatively slow, extended spin-up [10]. In the literature, different mechanisms based on superfluid dynamics were proposed to trigger glitches. For instance, crustquake induced vortex unpinning and avalanches from high density vortex traps [11] may account for the Crab and the Vela glitching behaviors, respectively. While studying crustquakes, one can investigate two important parameters, the strain-stress relation and the yield strength of the NS crust [12]. On the other hand, by investigating vortex traps, one can study the impact of crust breaking and impurity on the distribution of vortex lines within the crust [13]. Also, as a crustal-driven mechanism, coupling of vortex oscillations, i.e. kelvons, to lattice phonons was proposed to account for rapid angular momentum exchange between crustal superfluid and charged normal matter [14]. Through this mechanism one can clarify the dependence of phonon spectrum and angular momentum reservoir efficiency on the temperature evolution of the crust. Furthermore, interaction of crustal magnetic field with flux tubes in the outer core may give rise to extra pinning barrier for vortex lines [15]. This mechanism may help to understand the effects of transition into superconducting state and expulsion of magnetic field in the form of Meissner currents [16].

3. Glitch magnitudes as neutron star probe

In the vortex creep model, as well as in other models, glitch magnitudes in the rotation rate $\Delta\Omega/\Omega$ and in the spin-down rate $\Delta\dot{\Omega}/\dot{\Omega}$ are directly related to the superfluid region(s) involved in glitches. Thus, the glitch magnitudes are a measure of the glitch affected crustal superfluid moment of inertia and the equation of state [8,17,18]. This picture has been challenged theoretically [19]. Recent band theory calculations by Chamel [20] revealed that Bragg scattering of free neutrons on crystal somewhat reduces the mobility of crustal superfluid neutrons and, accordingly, the angular momentum reservoir [21,22]. This crustal mass entrainment effect suggests involvement of the core superfluid in the glitches. Gügercinoğlu and Alpar [23,24] have conducted the first principles calculations of glitches in the superfluid-superconducting core.

Vortex creep across flux tubes model [25] describes post-glitch relaxation of radio pulsars and magnetars in terms of dynamical interaction of vortex lines and toroidally arranged flux tubes in the outer core. In NSs, vortex lines' orientation is parallel to the rotation axis while flux tube array acquires very complicated poloidal plus toroidal structure inherited from the progenitor star after superconducting phase transition. Vortex lines inevitably intersect with toroidal flux tubes while the interaction with poloidal flux tubes is highly dependent on inclination angle. Due to non-zero core temperature, vortex lines slowly migrate out of pinning potentials sustained by flux tubes (i.e. creep) in accordance with the stellar spin-down. Model predictions can be summarized as follows:

- Only the toroidal field region (a rather small fraction of the core) participates in glitches via decoupling from external torque. This fact explains why glitch magnitudes are tiny.
- NS core response to each glitch in the spin down rate $\dot{\Omega}$ is exponential recovery,

$$\Delta\dot{\Omega}_{\rm c}(t) = -\frac{I_{\rm tor}}{I}\frac{\Delta\Omega}{\tau_{\rm tor}}\exp\left(-\frac{t}{\tau_{\rm tor}}\right), \tag{1}$$

with the toroidal field relaxation timescale [23]

$$\begin{aligned}\tau_{\rm tor} &\simeq 60\left(\frac{10^{-10}\,{\rm rad\,s^{-2}}}{|\dot{\Omega}|}\right)\left(\frac{T}{10^8\ {\rm K}}\right)\left(\frac{10\ {\rm km}}{R}\right)x_{\rm p}^{1/2}\\ &\times\left(\frac{m_{\rm p}}{m_{\rm p}^*}\frac{10^{14}\ {\rm g\ cm^{-3}}}{\rho}\frac{B_\phi}{10^{14}\ {\rm G}}\right)^{1/2}\ {\rm days},\end{aligned} \tag{2}$$

where $I_{\rm tor}/I$ is fractional moment of inertia of the toroidal field region, T is the temperature, B_ϕ is the strength of the toroidal field component, ρ is the mass density, $m_{\rm p}^*(m_{\rm p})$ is the effective (bare) mass of protons, $x_{\rm p}$ is the proton fraction, and R is the radius of the location of the toroidal field region.

- As a pulsar ages relaxation timescale (2) becomes longer and the glitches resemble step-like changes which are supported by observations [2,3].

With entrainment in the crustal superfluid taken into account, the angular momentum balance gives [23]

$$\frac{\Delta\Omega_{\rm c}}{\delta\Omega_{\rm s}} \leq \frac{m_{\rm n}}{m_{\rm n}^*}\frac{I_{\rm cs}}{I - I_{\rm cs} - I_{\rm tor}}, \tag{3}$$

where $\delta\Omega_{\rm s}$ and $I_{\rm cs}$ denote change in the superfluid rotation rate and the crustal superfluid moment of inertia, respectively. $m_{\rm n}^*/m_{\rm n} > 1$ characterizes the mass entrainment factor [20]. In the NS core, achieving critical threshold for vortex unpinning from the potential sustained by flux tubes is more likely than the case for NS crust if relevant superfluid instabilities occur [24]. Then condition (3) becomes

$$\frac{\Delta\Omega_{\rm c}}{\delta\Omega_{\rm s}} \leq \frac{m_{\rm n}}{m_{\rm n}^*}\frac{I_{\rm cs} + \alpha I_{\rm tor}}{I - I_{\rm cs} - I_{\rm tor}}, \tag{4}$$

where α is the fraction of the toroidal field region in which vortex lines unpinned from the flux tubes participated in the glitch. Thus, glitch magnitude observations yield constraints for the non-dissipative crustal entrainment effect.

4. Glitch relaxation as neutron star probe

Canonical radio pulsars. Temporal evolution of the change in the spin-down rate at the time of a glitch is well described by the fit function [2,3]

$$\Delta\dot{\Omega}(t) = \Delta\dot{\Omega}_{\rm d}{\rm e}^{-t/\tau_{\rm d}} + \Delta\dot{\Omega}_{\rm p} = -Q\,\Delta\Omega\,\tau_{\rm d}^{-1}{\rm e}^{-t/\tau_{\rm d}} + \Delta\dot{\Omega}_{\rm p}. \tag{5}$$

Here $\Delta\dot{\Omega}_{\rm d} = -\Delta\Omega_{\rm d}/\tau_{\rm d}$ and $Q = \Delta\Omega_{\rm d}/\Delta\Omega$ is the ratio of the exponentially decaying fraction to the total glitch magnitude $\Delta\Omega = \Delta\Omega_{\rm d} + \Delta\Omega_{\rm p}$. Upon comparing the fit function (5) with vortex creep across flux tubes model prediction, equations (1) and (2), one arrives at three possibilities [25]:

(i) If $\tau_{\rm tor} \approx \tau_{\rm d}$, then $I_{\rm tor}/I \sim Q$.

(ii) If $\tau_{\rm tor} \gg \tau_{\rm d}$ but $Q \ll 1$ then the relaxation of the toroidal field region is not completed yet and one has only an upper bound $I_{\rm tor}/I \leq 1 - Q$.

(iii) If $\tau_{\rm tor} \ll \tau_{\rm d}$, then the prompt response of the toroidal field region is missed from the observations.

Then, post-glitch observations of pulsars enable us to infer macroscopic properties like core temperature and internal magnetic field configuration, and microscopic traits like equation of state parameters and strength of the superfluid-superconductor coupling.

In the inner crust, a linear regime of creep of vortices against nuclei also results in an exponential post-glitch response with a decay timescale

$$\tau_1 = \frac{kT}{E_{\rm p}} \frac{R\omega_{\rm cr}}{4\Omega_{\rm s} v_0} \exp\left(\frac{E_{\rm p}}{kT}\right), \tag{6}$$

where $v_0 \approx 10^7$ cm/s is microscopic vortex velocity around the nuclei [8,24]. From crustal response one can deduce pinning strength and lattice properties, for which there exist early analytical estimates [26] and recent numerical computations [27].

Magnetars. Magnetar glitches differ from radio pulsar glitches in many respects. Magnetars, through displaying unstable spin-down and burst like activities following their glitches [4], provide strong evidence that both magnetospheric processes and internal superfluid play an important role in their glitch events as reflected by anomalous Q values [28]. Magnetars' different behavior from radio pulsars in general may be related to their unusual properties. Due to magnetic field decay, magnetar spin-down rates are somewhat lower than those of canonical radio pulsars [29]. As a result of the magnetic field decay, magnetars also have higher surface temperatures [30]. In strong and twisted magnetic field configurations, the toroidal field component transfers some of its magnetic energy to the poloidal field so that in magnetars $B_\phi \leq 0.01 B_{\rm p}$ [31]. In order to fit the relaxation timescale (2) to post-glitch observations it is argued that magnetars should have a cooler core [25]. If the density of the core exceeds a critical threshold, the powerful direct Urca process operates. Magnetar surface thermal emission behavior can be explained by a cooler core and a heater in the crust [32]. Therefore, magnetars may have a massive core allowing enhanced cooling processes. This, in turn, implies that magnetars may be originated from massive progenitors in which dynamo processes endowed them with stronger magnetic fields than those of canonical pulsars.

Glitches with external torque variation. High magnetic field pulsars exhibit glitches which have much in common with magnetars. PSR J1119–6127 is observed to have additional pulse components emergent with glitch [33]. PSR J1846–0258 showed violent emission changes and burst activity right after the glitch [34]. These peculiar glitches have some implications for NS structure [35]. Gügercinoğlu and Alpar [28] developed the vortex creep model [8] by taking time varying external torques into account in order to answer these issues. Variable external torque and superfluid coupling results in extra terms in the Q values [28]. An exponentially decaying torque on timescale $\tau_{\rm d}$ added to the pre-glitch level $N_0 = I\dot{\Omega}_\infty$, with $\dot{\Omega}_\infty$ being long-term spin-down rate, yields $N_{\rm ext}(t) = N_0 + \delta N \exp(-t/\tau)$. For this particular choice of time varying external torque post-glitch spin-down rate the evolution is [28]

$$\Delta\dot{\Omega}_{\rm c}(t) = \frac{\delta N}{I_{\rm c}}\,{\rm e}^{-t/\tau}\left(1 - \frac{I_{\rm s}}{I}\frac{\tau}{\tau - \tau_{\rm d}}\right) + {\rm e}^{-t/\tau_{\rm d}}\,\frac{I_{\rm s}}{I}\left(\frac{\omega(0)}{\tau_{\rm d}} + \frac{I}{I_{\rm c}}\dot{\Omega}_{\infty} + \frac{\delta N}{I_{\rm c}}\frac{\tau}{\tau - \tau_{\rm d}}\right). \quad (7)$$

The same formalism can be applied to sources showing precession like oscillations [36] and anomalous braking indices [37] as well.

5. Conclusions

Pulsar glitch observations enable us to infer NS internal structure in many ways:

- Pulsar glitch observations can be used to place stringent constraints on the equation of state [17,22].
- Post-glitch exponential decay provide indirect measure for magnetic field configuration [25].
- Magnetar glitch observations are best explained by a core in which the direct Urca cooling operates [25].
- Glitches with external torque variation implies a strong coupling between the internal superfluid and spinning down or up magnetospheric or accretion torques [28].

References

[1] Hobbs G, Lyne A G and Kramer M 2010 *Mon. Not. Roy. Astron. Soc.* **402** 1027
[2] Espinoza C M, Lyne A G, Stappers B W and Kramer M 2011 *Mon. Not. Roy. Astron. Soc.* **414** 1679
[3] Yu M et al. 2013 *Mon. Not. Roy. Astron. Soc.* **429** 688
[4] Dib R and Kaspi V M 2014 *Astrophys. J.* **784** 37
[5] Baym G, Pethick C J, Pines D and Ruderman M 1969 *Nature* **224** 872
[6] Shternin P S et al. 2011 *Mon. Not. Roy. Astron. Soc.* **412** L108
[7] Anderson P W and Itoh N 1975 *Nature* **256** 25
[8] Alpar M A, Anderson P W, Pines D, and Shaham J 1984 *Astrophys. J.* **276** 325
[9] Dodson R G, McCulloch P M and Lewis D R 2002 *Astrophys. J. Lett.* **564** L85
[10] Lyne A G et al. 2015 *Mon. Not. Roy. Astron. Soc.* **446** 857
[11] Cheng K S, Alpar M A, Pines D, and Shaham J 1988 *Astrophys. J.* **330** 835
[12] Chugunov A I and Horowitz C J 2010 *Mon. Not. Roy. Astron. Soc.* **407** L54
[13] Warszawski L, Melatos A and Berloff N G 2012 *Phys. Rev. B.* **835** 104503
[14] Epstein R I and Baym G 1992 *Astrophys. J.* **387** 276
[15] Sedrakian A and Cordes J M 1999 *Mon. Not. Roy. Astron. Soc.* **307** 365
[16] Jones P B 2006 *Mon. Not. Roy. Astron. Soc.* **365** 339
[17] Link B, Epstein R I and Lattimer J M 1999 *Phys. Rev. Lett.* **83** 3362
[18] Antonelli M and Pizzochero P M 2017 *Mon. Not. Roy. Astron. Soc.* **464** 721
[19] Chamel N and Carter B 2006 *Mon. Not. Roy. Astron. Soc.* **368** 796
[20] Chamel N 2013 *Phys. Rev. Lett.* **110** 011101
[21] Andersson N, Glampedakis K, Ho W C G and Espinoza C M 2012 *Phys. Rev. Lett.* **109** 241103
[22] Delsate T, Chamel N, Gürlebeck N, Fantina A F, Pearson J M and Ducoin C 2016 *Phys. Rev. D.* **94** 023008
[23] Gügercinoğlu E and Alpar M A 2014 *Astrophys. J. Lett.* **788** L11
[24] Gügercinoğlu E and Alpar M A 2016 *Mon. Not. Roy. Astron. Soc.* **462** 1453
[25] Gügercinoğlu E 2017 *Mon. Not. Roy. Astron. Soc.* **469** 2313
[26] Alpar M A 1977 *Astrophys. J.* **213** 527
[27] Seveso S, Pizzochero P M, Grill F and Haskell B 2016 *Mon. Not. Roy. Astron. Soc.* **455** 3952
[28] Gügercinoğlu E and Alpar M A 2017 *Mon. Not. Roy. Astron. Soc.* **471** 4827
[29] Dall'Osso S, Granot J and Piran T 2012 *Mon. Not. Roy. Astron. Soc.* **422** 2878
[30] Beloborodov A M and Li X 2016 *Astrophys. J.* **833** 261
[31] Fujisawa K and Kisaka S 2014 *Mon. Not. Roy. Astron. Soc.* **445** 2777
[32] Kaminker A D, Potekhin A Y, Yakovlev D G and Chabrier G 2009 *Mon. Not. Roy. Astron. Soc.* **395** 2257
[33] Weltevrede P, Johnston S and Espinoza C M 2011 *Mon. Not. Roy. Astron. Soc.* **411** 1917
[34] Livingstone M A, Kaspi V M and Gavriil F P 2010 *Astrophys. J.* **710** 1710
[35] Akbal O, Gügercinoğlu E, Şaşmaz Muş S and Alpar M A 2015 *Mon. Not. Roy. Astron. Soc.* **449** 933
[36] Ashton G, Jones D I and Prix R 2017 *Mon. Not. Roy. Astron. Soc.* **467** 164
[37] Ou Z W, Tong H, Kou F F and Ding G Q 2016 *Mon. Not. Roy. Astron. Soc.* **457** 3922

Precession and glitches in the framework of three-component model of neutron star

D P Barsukov[1,2], **O A Goglichidze**[2], **K Y Kraav**[1] **and M V Vorontsov**[1]
[1] Peter the Great St. Petersburg Polytechnic University, 195251, Saint Petersburg, Russian Federation
[2] Ioffe Institute, 194021, Saint Petersburg, Russian Federation
E-mail: goglichidze@gmail.com

Abstract. We consider the pulsar rotation assuming that the neutron star consists of crust component (which rotation is observed) and two core components. One of the core components contains pinned superfluid which can, for some reasons, suddenly inject small fraction of stored angular momentum in it. In the framework of this simple model the star can demonstrate glitch-like events together with long period precession (with period $1 - 10^4$ years).

1. Introduction

Some radio pulsars demonstrate periodic variations in pulse timing and beam shape. The most favorable explanation to these phenomena is the precession of the neutron star with periods $T_p \sim 10^2$ days [1, 2]. There are pulsar characteristics and behavior features which can be interpreted as a manifestation of precession with much larger periods $T_p \sim 10^2 - 10^4$ years [3, 4]. Another type of timing irregularities is pulsar glitches which are the sudden increase of pulsar angular velocity Ω with subsequent smooth recovery. Standard model relates these events with avalanche-like unpinning of superfluid neutron vortices pinned to the nuclei in the inner crust or proton flux tubes in the neutron star core. However, as it was first pointed out by Shaham [5] and later confirmed by more detailed researches [6, 7], the pinning of neutron superfluid should dramatically decrease the period of precession making it very far from observed values. We propose a simple phenomenological model which allows the same pulsar to precess with a long period and demonstrate glitch-like timing features.

2. Three-components model

Let us assume that a neutron star consists of three dynamically distinguished components. For the sake of simplicity all components are supposed to rotate as rigid bodies.

The c-component is the outer component. It rotates with angular velocity $\vec{\Omega}_c = \vec{\Omega}$:

$$d_t \vec{M}_c = \vec{K}_{\rm ext} + \vec{N}_{rc} + \vec{N}_{gc}, \tag{1}$$

where $\vec{M}_c = I_c \vec{\Omega} + I_c \hat{\epsilon} \vec{\Omega}$ is the c-component angular momentum, I_c is one of its principal moments of inertia. Tensor $\hat{\epsilon}$ is the effective oblateness tensor. It describes the total departure of c-component mass distribution and electromagnetic field energy distribution from spherical symmetry. It is assumed that $\epsilon_{\alpha\beta} \ll 1$. Vector $\vec{K}_{\rm ext}$ is the external electromagnetic torque.

Note that so-called anomalous torque is taken into account by tensor $\hat{\epsilon}$ [8]. Therefore, torque $\vec{K}_{\rm ext}$ contains only the terms $\propto \Omega^3$.

The g-component is an inner component. It consists of normal matter characterized by moment of inertia I_g and rotating with angular velocity $\vec{\Omega}_g$. It also contains superfluid matter with angular momentum $\vec{L}_g$ pinned to the normal matter:

$$d_t\vec{M}_g = \vec{N}_{cg} + \vec{N}_{rg}, \quad d_t\vec{L}_g = \vec{\Omega}_g \times \vec{L}_g, \tag{2}$$

where $\vec{M}_g = I_g\vec{\Omega}_g + \vec{L}_g$ is the total angular momentum of the g-component.

The r-component is the other inner component. It rotates with angular velocity $\vec{\Omega}_r$:

$$d_t\vec{M}_r = \vec{N}_{cr} + \vec{N}_{gr} \tag{3}$$

where $\vec{M}_r = I_r\vec{\Omega}_r$ is the angular momentum of the r-component, I_r is its moment of inertia. The components interact with each other by torques $\vec{N}_{ij}$, where $i,j = c,g,r$ and, obviously, $\vec{N}_{ij} = -\vec{N}_{ji}$.

If $\vec{K}_{\rm ext} = 0$ and $\hat{\epsilon} = 0$, system of equations (1)-(3) has a simple equilibrium solution: $\vec{\Omega}_g = \vec{\Omega}_r = \vec{\Omega} = {\rm const}$, $\vec{L}_g = L_g\vec{e}_\Omega$, where $\vec{e}_\Omega = \vec{\Omega}/\Omega$. Further we will consider only small corrections to this equilibrium state. Introducing notations $V^{||} = (\vec{e}_\Omega \cdot \vec{V})$ and $\vec{V}^{\perp} = \vec{V} - \vec{e}_\Omega\,(\vec{e}_\Omega \cdot \vec{V})$ applied to arbitrary vector $\vec{V}$ and holding only the linear in $\vec{\mu}_{ij} = \vec{\Omega}_j - \vec{\Omega}_i$ and $\vec{L}_g^{\perp}$ terms one can rewrite system of equations (1)-(3) in the following form:

$$\dot{\Omega} = R^{||}_{gc} + R^{||}_{rc} + S^{||}, \tag{4}$$

$$\dot{\mu}^{||}_{cg} = R^{||}_{cg} + R^{||}_{rg} - R^{||}_{gc} - R^{||}_{rc} - S^{||}, \tag{5}$$

$$\dot{\mu}^{||}_{cr} = R^{||}_{cr} + R^{||}_{gr} - R^{||}_{gc} - R^{||}_{rc} - S^{||}, \tag{6}$$

$$\dot{\omega}^{||}_{g} = 0, \tag{7}$$

$$\Omega\dot{\vec{e}}_\Omega = \vec{R}^{\perp}_{gc} + \vec{R}^{\perp}_{rc} + \vec{S}^{\perp}, \tag{8}$$

$$\dot{\vec{\mu}}^{\perp}_{cg} = (\omega^{||}_g - \Omega)\,[\vec{e}_\Omega \times \vec{\mu}^{\perp}_{cg}] - [\vec{\Omega} \times \vec{\omega}^{\perp}_g] + \vec{R}^{\perp}_{cg} + \vec{R}^{\perp}_{rg} - \vec{R}^{\perp}_{gc} - \vec{R}^{\perp}_{rc} - \vec{S}^{\perp}, \tag{9}$$

$$\dot{\vec{\mu}}^{\perp}_{cr} = -[\vec{\Omega} \times \vec{\mu}^{\perp}_{cr}] + \vec{R}^{\perp}_{cr} + \vec{R}^{\perp}_{gr} - \vec{R}^{\perp}_{rc} - \vec{R}^{\perp}_{gc} - \vec{S}^{\perp}, \tag{10}$$

$$\dot{\vec{\omega}}^{\perp}_g = -\frac{\omega^{||}_g}{\Omega}\left(\vec{R}^{\perp}_{gc} + \vec{R}^{\perp}_{rc} + \vec{S}^{\perp} + [\vec{\Omega} \times \vec{\mu}^{\perp}_{cg}]\right), \tag{11}$$

where $\vec{R}_{ij} = \vec{N}_{ij}/I_j$, $\vec{S} = \vec{K}_{\rm ext}/I_c - \vec{\Omega} \times \hat{\epsilon}\vec{\Omega}$, $\vec{\omega}_g = \vec{L}_g/I_g$, and the point denotes the time derivative in the c-component frame of reference. We assume that vectors $\vec{R}_{ij}$ linearized in μ_{ij} can be represented as

$$\vec{R}_{ij} = -\left(\alpha_{ij}\mu^{||}_{ij}\vec{e}_\Omega + \beta_{ij}\vec{\mu}^{\perp}_{ij} + \gamma_{ij}[\vec{e}_\Omega \times \vec{\mu}^{\perp}_{ij}]\right), \tag{12}$$

where α_{ij}, β_{ij} and γ_{ij} are the phenomenologically introduced interaction constants.

3. Quasi-stationary rotation

Let us first assume that glitches do not occur. In this case, if the largest internal relaxation time-scale $\tau_{rel} = \max\left(1/\alpha_{ij}, 1/\beta_{ij}, 1/\gamma_{ij}\right)$ is much smaller than the period of precession T_p, and assumption (12) is valid, the neutron star will rotate in quasi-stationary regime. It means that, the departures of internal components rotation from c-component rotation are determined by instant values of $\vec{\Omega}$ and $\dot{\vec{\Omega}}$ and do not depend on prehistory (see detail in [9]). In other words, one can neglect time derivative terms in equations (5), (6), (9)-(11) because of their quadratic smallness. Solving equations (5), (6), (10) and (11) for $\mu^{||}_{cr}$, $\mu^{||}_{cg}$ $\vec{\mu}^{\perp}_{cr}$ and $\vec{\mu}^{\perp}_{cg}$, and substituting the obtained expressions into equations (4) and (8), one can obtain

$$\tilde{I}_{\rm tot}\dot{\Omega} = K^{||}_{\rm ext}, \tag{13}$$

$$I_c\Omega\dot{\vec{e}}_\Omega = \frac{(1+\Gamma)(\vec{K}_{\rm ext} - I_c\vec{\Omega}\times\hat{\epsilon}\vec{\Omega}) + B\vec{e}_\Omega \times (\vec{K}_{\rm ext} - I_c\vec{\Omega}\times\hat{\epsilon}\vec{\Omega})}{(1+\Gamma)^2 + B^2}, \qquad (14)$$

where $\tilde{I}_{\rm tot} = I_c + I_g + I_r$ and in the weak interaction limit ($\alpha_{ij}, \beta_{ij}, \gamma_{ij} \ll \Omega$) equals to

$$B \approx \frac{\beta_{gc} + \beta_{rc}}{\Omega} \ll 1, \qquad \Gamma \approx \frac{\gamma_{gc} + \gamma_{rc}}{\Omega} \ll 1. \qquad (15)$$

One can note that according to equation (13) the neutron star is braked as if it is a solid body with moment of inertia $\tilde{I}_{\rm tot}$ (which does not include the moment of inertia of pinned superfluid). It is a general feature of quasi-stationary approximation [10, section 2.3].

Equation (14) describes the change in the orientation of angular velocity vector $\vec{\Omega}$ relative to the c-component. The first and the third terms on the right-hand side arise due to action of external torque $\vec{K}_{\rm ext}$, the second term makes vector $\vec{\Omega}$ precess about a star principal axis and the fourth term describes the damping of the precession due to internal energy dissipation. Using equation (14), one can estimate precession period and precession damping time-scale as

$$T_p \sim \frac{2\pi}{\epsilon\Omega}, \qquad \tau_d \sim \frac{2\pi}{\epsilon\Omega B} \gg T_p, \qquad (16)$$

where ϵ is characteristic oblateness. The external torque $\vec{K}_{\rm ext}$ tends to minimize the energy loss caused by electromagnetic radiation. It leads to increase or decrease of precession amplitude depending on the orientation of star dipole moment relative to the principal axes [11]. One can see that expression for precession period T_p does not contain L_g. This seems to contradict to Shaham who has obtained that T_p equals rather to $L_g/\tilde{I}_{\rm tot}\Omega$ [5]. The reason for this discrepancy is that we allow the g-component (which containing the pinned superfluid) to move relative to the c-component which rotation is observed. It is required that $\beta_{cg} \ll \Omega$ to ensure the validity of expressions (16). The Shaham's result can be reproduced by passing to limit $\beta_{cg} \to \infty$ in equations (9)-(11) so that $\vec{\mu}^{\perp}_{cg} \to 0$.

4. Glitch-like event

Since superfluid is pinned, the lag between rotations of superfluid and normal fractions of g-component increases during the retardation of star rotation. When critical lag is reached the superfluid unpins and then repins such that some small fraction of its angular momentum is transferred from superfluid into the normal fraction. In present work we do not specify the physical mechanism of glitch triggering. However, we assume that the glitch occurs and then relaxes at time-scales much less the the time-scales of quasi-stationary evolution. It allows us to neglect the effects of external torque and star oblateness considering these processes.

The solution of equations (4)-(6) for Ω with the following initial conditions: $\Omega = \Omega_0$, $\mu^{||}_{cr} = 0$, $\mu^{||}_{cg} = \Delta L_g/I_g$ and zeroth $S^{||}$ has the form

$$\Omega(t) = \Omega_0 + \Delta\Omega\left(1 - e^{-p_+ t} - Q(1 - e^{-p_- t})\right), \qquad (17)$$

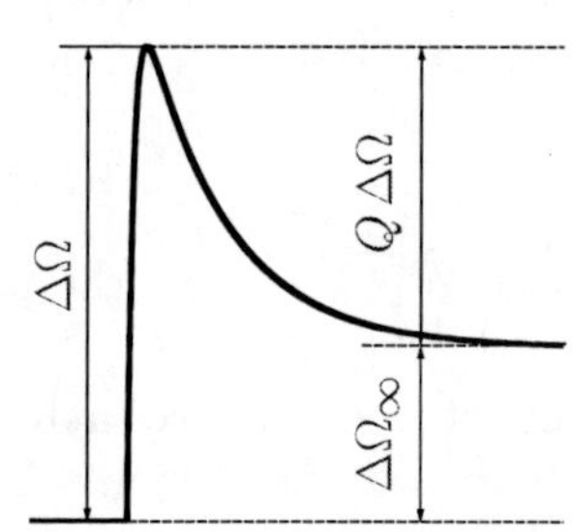

Figure 1. The sketch of the glitch-like behavior of Ω.

where $\Delta\Omega = \Delta\Omega_\infty/(1-Q)$, $\Delta\Omega_\infty = \Delta L_g/\tilde{I}_{\rm tot}$, $Q = (\tilde{I}_{\rm tot}\alpha_{cg} - I_c p_+)/(\tilde{I}_{\rm tot}\alpha_{cg} - I_c p_-)$ and the coefficients p_+ and p_- ($p_+ > p_-$) are the roots of equation

$$\begin{aligned} p^2 \; &- \; (\alpha_{cg} + \alpha_{rg} + \alpha_{cr} + \alpha_{gr} + \alpha_{gc} + \alpha_{rc})\, p \, + \\ &+ \; (\alpha_{gc} + \alpha_{rg} + \alpha_{cg}) \cdot (\alpha_{cr} + \alpha_{gr} + \alpha_{rc}) + (\alpha_{rc} - \alpha_{rg}) \cdot (\alpha_{gr} - \alpha_{gc}) = 0. \end{aligned} \qquad (18)$$

The meaning of parameters $\Delta\Omega$, Ω_∞ and Q becomes clear if one looks at the sketch of the solution in figure 1. We suppose that the interaction between the c- and g-components is the strongest one ($\alpha_{cg} \gg \alpha_{rc}, \alpha_{rg}$). In this case,

$$p_+ \approx \left(1 + \frac{I_g}{I_c}\right)\alpha_{cg}, \quad p_- \approx \frac{\tilde{I}_{\rm tot}}{I_c + I_g}\left(\alpha_{cr} + \alpha_{gr}\right) \quad \text{and} \quad Q \approx \frac{I_r}{\tilde{I}_{\rm tot}}. \tag{19}$$

If one wants to relate solution (17) with observed pulsar glitches, then $1/p_+$ and $1/p_-$ should be interpreted as glitch growth (< 30 s [12]) and relaxation ($1 - 10^2$ days [13]) times respectively. One can see that the glitch growth rate is proportional to α_{cg} while the subsequent relaxation is governed by the r-component. Thus, on the one hand, the interaction between the c- and g-components should be strong enough to ensure rapid spin-up. But simultaneously it should be not too much strong, namely $\beta_{cg} \ll \Omega$, to long period precession could exist It is natural to assume that $\alpha_{cg} \sim \beta_{cg}$.

5. Discussion

Let us speculate a little about possible nature of formally introduced components. The c-component can be associated with neutron star crust and a part of core charged particles which is strongly coupled with the crust ($I_c \sim 10^{-2}-10^{-1} I_{\rm ns}$). The role of g-component can be performed by tangles of closed flux tubes which could be formed after protons became superconductive from chaotic small-scale magnetic field. Alternatively it can be a torus composed of closed flux tubes. [14]. If the characteristic cross-section $S_{\rm tor}$ of the region occupied by the toroidal field is of the order of 1 km^2, then $I_g \sim \rho_p S_{\rm tor} r_{\rm ns}^3 \sim 10^{-3} I_{ns}$, where ρ_p is the proton mass density. Some part of superfluid neutron vortices presented in the core should be pinned to the flux tubes. On the one hand, this interaction prevents the tangles or torus from collapse. On the other hand, when the critical rotational lag is reached, the vortices unpin from the flux tubes triggering the glitch. Since the flux tubes are closed in the core, the g-component, being magnetically decoupled, can rotate with angular velocity different from $\vec{\Omega}$ [15]. The r-component can be represented by the part of core neutron superfluid which is not pinned to the g-component and a part of normal matter coexisting with it and weakly coupled with c-component ($I_r \sim I_{\rm ns}$). The interaction coefficients values which are needed to make the model able to explain the observed glitches time-scales can be estimated with expressions (19). The weakest side of the presented simple model is the value of recovery fraction Q. According to (19) it is of the order of unity. It is not so bad for young pulsars but mature ones demonstrate rather recovery fractions several orders less than unity [13]. However, we believe that it is a consequence of the oversimplifying of the model and further researches will allow us to avoid this discrepancy.

References

[1] Ashton G, Jones D I and Prix R 2016 *MNRAS* **458** 881–99
[2] Kerr M, Hobbs G, Johnston S and Shannon R M 2016 *MNRAS* **455** 1845–54
[3] Arzamasskiy L, Philippov A and Tchekhovskoy A 2015 *MNRAS* **453** 3540–53
[4] Biryukov A, Beskin G and Karpov S 2012 *MNRAS* **420** 103–17
[5] Shaham J 1977 *ApJ* **214** 251–60
[6] Sedrakian A, Wasserman I and Cordes J M 1999 *ApJ* **524** 341–60
[7] Link B 2006 *A&A* **458** 881–884 (*Preprint* `arXiv:astro-ph/0608319`)
[8] Goglichidze O A, Barsukov D P and Tsygan A I 2015 *MNRAS* **451** 2564–74
[9] Barsukov D P, Goglichidze O A and Tsygan A I 2013 *Astronomy Reports* **57** 26–39
[10] Barsukov D P, Goglichidze O A and Tsygan A I 2014 *MNRAS* **444** 1318–33
[11] Goldreich P 1970 *ApJ* **160** L11
[12] Dodson R, Lewis D and McCulloch P 2007 *Astrophys. Space. Sci.* **308** 585–9
[13] Lyne A G, Shemar S L and Smith F G 2000 *MNRAS* **315** 534–42
[14] Gügercinoğlu E and Alpar M A 2014 *ApJ* **788** L11
[15] Glampedakis K and Lasky P D 2015 *MNRAS* **450** 1638–50

Binary and ternary ionic compounds in the outer crust of accreted neutron stars

N Chamel

Institut d'Astronomie et d'Astrophysique, CP-226, Université Libre de Bruxelles, 1050 Brussels, Belgium

E-mail: nicolas.chamel@ulb.ac.be

Abstract. The outer crust of an accreted neutron star is thought to contain a large distribution of different nuclear species resulting from the burying of ashes of X-ray bursts and superbursts. By analysing the stability of multicomponent Coulomb crystals against phase separation, it is found that various binary and ternary ionic compounds could be formed.

1. Introduction

Many properties of the crust of a neutron star depend to a large extent on its structure, which in turn have implications for various astrophysical phenomena, such as Crab like pulsar sudden spin-ups, thermal relaxation in transiently accreting stars, giant flares from soft gamma-ray repeaters, and gravitational-wave emission (see, e.g., Ref. [1]).

The outermost region of a neutron star is expected to be made of iron ^{56}Fe, the end-product of stellar nucleosynthesis. The properties of compressed iron can be probed in terrestrial laboratories up to pressures of order 10^{14} dyn cm^{-2} (densities of a few tens of g cm^{-3}) with nuclear explosions and laser-driven shock-wave experiments. Under these conditions, iron has an hexagonal close-packed structure [2]. Ab initio calculations predict various structural phase transitions at higher pressures. Although such pressures are tremendous according to terrestrial standards, they still remain negligibly small compared with those prevailing in a neutron star. Deeper in the star, at a density $\rho_{\rm eip} \approx 2 \times 10^4$ g cm^{-3}, the interatomic spacing becomes comparable with the atomic radius. At densities $\rho \gg \rho_{\rm eip}$, atoms are crushed into a dense Coulomb plasma of atomic nuclei in a charge compensating background of highly degenerate electrons. The denser regions of the crust are expected to be stratified into a succession of layers with different compositions. The onset of neutron emission at densities $\rho_{\rm drip} \sim 10^{11}$ g cm^{-3} delimits the boundary between the outer and inner regions of the crust (see, e.g., Ref. [3]).

It has long been assumed that each layer of the outer crust consists of a body-centered cubic (bcc) crystal made of a single nuclear species [4]. The stratification of the crust (see, e.g. Refs. [5, 6, 7, 8]) is mainly a consequence of the cold catalysed matter hypothesis [9, 10]. Because neutron stars are formed from gravitational core-collapse supernova explosions with temperatures exceeding 10^{10} K, their interior is thought to be in full thermodynamic equilibrium. As the star cools down by emitting neutrinos, the distribution of nuclides shrinks, and eventually reduces to one or two species at absolute zero temperature (see, e.g. Ref. [11]). However, nuclear processes in the outer envelope of the star are likely to be quenched after the crystallisation (see, e.g., Ref. [1] and references therein). Moreover, the constitution of neutron-star crusts may differ from that of cold catalysed matter due to the fallback of material from the envelope ejected during the supernova explosion and due to the accretion of matter from a stellar companion.

In the presence of different nuclear species, various solid structures may form. We have recently studied the formation of multinary compounds in dense stellar matter, and analysed their stability [8]. We have found that substitutional binary compounds with cesium chloride structure are generally present at the interface between adjacent layers in the outer crust of a cold nonaccreted neutron star [8], as first shown in Ref. [12]. In this paper, we pursue our investigations in accreted neutron-star crusts.

2. Formation of multinary ionic compounds in accreted neutron-star crusts

The accretion of matter from a stellar companion onto a neutron star forms an hydrogen-rich envelope. Stable hydrogen burning produces helium, and depending on the accretion rate, unstable helium burning converts within seconds all the envelope into nuclides in the nickel-cadmium range, leading to X-ray bursts. Less frequent but more energetic are superbursts presumably triggered by unstable carbon burning. Ashes of these thermonuclear explosions may be further processed as they sink deeper inside the star (see, e.g., Ref. [1]).

The equilibrium structure of such multicomponent plasmas remains highly uncertain. According to classical molecular dynamics cooling simulations of hot plasmas, nuclei tend to crystallize into a bcc lattice, as in nonaccreted crusts [13]. However, the system may have not fully relaxed to its true equilibrium state due to the very slow dynamics of crystal growth. Moreover, these simulations were performed at fixed ion number densities, and may not represent the conditions prevailing in neutron stars since the density does not necessarily vary continuously throughout the stellar interior contrary to the pressure. Recently, a different approach has been followed to find the ground-state configuration (at zero temperature and fixed pressure) using genetic algorithms [14]. But such unconstrained global searches of the equilibrium structure remains computationally costly, and for this reason have been restricted to the ternary systems O-Fe-Se, and Fe-As-Se considering polycrystalline mixtures of stoechiometric compounds at two different values of the pressure.

A thermodynamically favorable condition for a multinary compound made of nuclei with charges $\{Z_i\}$ to exist in the crust of a neutron star is the stability against the separation into pure (bcc) phases [8] (unstable compounds might still exist depending on how fast ions diffuse). Indeed, coexisting phases separate and settle to different depths due to gravity. As shown in Ref. [8], the stability condition of any compound at zero temperature is independent of the stellar environment (neglecting electron exchange and correlation, electron charge polarisation, and quantum-zero point motion of ions about their equilibrium positions), and is embedded in the following inequality:

$$\mathcal{R}(\{Z_i\}) \equiv \frac{C}{C_{\rm bcc}} f(\{Z_i\}) \frac{\bar{Z}}{\overline{Z^{5/3}}} > 1\,, \tag{1}$$

where $\mathcal{R}$ is the ratio of the electrostatic energy of the compound to that of coexisting bcc phases, $f(\{Z_i\})$ is the dimensionless structure function of the compound, C the corresponding structure constant[1], $C_{\rm bcc} = -1.444231$ the structure constant of a bcc lattice [15], and overlined symbols indicate mean values (e.g. $\bar{Z} \equiv \sum_i \xi_i Z_i$, where ξ_i denotes the relative abundance of nuclei i). For compounds made of isotopes with the same charge Z, the structure function reduces to $f(Z) = Z^{2/3}$. For multicomponent systems with different charges $\{Z_i\}$, the quantity $\mathcal{R}$ only depends on the charge ratios $\{Z_i/Z_j\}$ with $i \neq j$. In particular, the function $\mathcal{R}(Z_1, Z_2, Z_3)$ of a ternary compound can be expressed as [8]

$$\mathcal{R}(q,p) = \frac{C}{C_{\rm bcc}} \frac{\eta_1 + \eta_2 q^2 + \eta_3 p^2 + \eta_{12} q + \eta_{13} p + \eta_{23} qp}{(\xi_1 + \xi_2 q + \xi_3 p)^{1/3}(\xi_1 + \xi_2 q^{5/3} + \xi_3 p^{5/3})}\,, \tag{2}$$

[1] The structure constant C is related to the Madelung constant C_M by $C = C_M(4\pi/3)^{1/3}$.

IOP Conf. Series: Journal of Physics: Conf. Series **932** (2017) 012039 doi:10.1088/1742-6596/932/1/012039

where $q \equiv Z_2/Z_1$ and $p \equiv Z_3/Z_1$. The (dimensionless) coefficients η_1, η_2, η_3, η_{12}, η_{13}, and η_{23} are determined by the spatial arrangement of ions. The function $\mathcal{R}(Z_1, Z_2)$ of a binary compound can be easily inferred from Eq. (2). A compound may be ordered or disordered depending on its composition (see, e.g. Ref. [16]). However, the existence of amorphous solids in accreted neutron-star crusts was found to be incompatible with the observed thermal emission [17]. Therefore, we focus on crystalline compounds.

As shown in Ref. [8], stellar compounds having the same structure as auricupride ($AuCu_3$), fluorite (CaF_2), and tungsten carbide (WC) are not thermodynamically favored ($\mathcal{R} \leq 1$). Structures similar to that of rocksalt (NaCl), cesium chloride (CsCl), and cubic perovskite ($BaTiO_3$), shown in Fig. 1, can lead to thermodynamically favored compounds depending on the composition (including binary compounds based on the cubic perovskite structure with $Z_1 = Z_3$ or $Z_1 = Z_2$). We have also considered the binary compound (3) proposed by Dyson [18]. The values for the structure constants are indicated in Table 1. The structure function of Dyson's compound was calculated as follows [19]: $\eta_1 = C_{\rm bcc}/(4^{4/3}C_{\rm sc})$, $\eta_2 = \eta_1+1/2$, $\eta_{12} = 1-\eta_1-\eta_2$, and $C_{\rm sc} = -1.418649$ is the structure constant of a simple cubic lattice taken from Ref. [12]. Dyson's compound is found to be thermodynamically favored for small charge ratios $q < 0.0181571$, with $\mathcal{R}$ reaching 1.0003 for $q \simeq 0.0076$. The ratio $\mathcal{R}$ obtained from Eq. (2) for the CsCl structure coincides with the inverse of Eq.(5) given in Ref. [20] with $q = \alpha$, $C = (4\pi/3)^{1/3}\zeta_2$, $\eta_1 = \eta_2 = \zeta_1/(2^{4/3}\zeta_2)$, and $\eta_{12} = 1-2\eta_1$. We have analysed the stability against phase separation of binary and ternary compounds that could potentially exist in the crust of an accreted neutron star using the nuclear abundancies given in Table I of Ref. [13]. Thermodynamicallly favored compounds are listed below.

- Rocksalt structure: **AgNe**.
- Cesium chloride structure: **AgCa**, AgTi, AgCr, AgFe, **AgCo**, AgNi, AgZn, AgGe, AgAs, AgSe, **AgKr**, **KrCa**, KrTi, KrCr, KrFe, **KrCo**, KrNi, KrZn, KrGe, KrAs, KrSe, SeCa, SeTi, SeCr, SeFe, SeCo, SeNi, SeZn, SeGe, SeAs, AsCa, AsTi, AsCr, AsFe, AsCo, AsNi, AsZn, AsGe, GeCa, GeTi, GeCr, GeFe, GeCo, GeNi, GeZn, ZnCa, ZnTi, ZnCr, ZnFe, ZnCo, ZnNi, NiCa, NiTi, NiCr, NiFe, NiCo, **CoCa**, CoTi, CoCr, CoFe, FeCa, FeTi, FeCr, CrCa, **CrTi**, TiCa.
- Perovskite structure: $AgNeO_3$.

Binary compounds indicated in boldface are consistent with nuclear abundancies: they could accommodate all available nuclei with the corresponding charges Z_1 and Z_2. Compounds with CsCl structure appear to be dynamically stable (against lattice vibrations) since $q \leq 3.6$ [20, 21]. With $q = 4.7$, AgNe might be dynamically unstable. This analysis might suggest that the crust of an accreted neutron star would generally consist of a mixture of various compounds and pure bcc crystals. However, the coexistence of different phases appears unlikely due to gravitational settling, as argued in Ref. [8] (see also Ref. [14]). This warrants further studies.

Table 1. Structure constants of the stellar compounds shown in Fig. 1.

	C	η_1	η_2	η_3	η_{12}	η_{13}	η_{23}
NaCl	-1.418649	0.403981	0.403981	0	0.192038	0	0
CsCl	-1.444231	0.389821	0.389821	0	0.220358	0	0
Dyson	-1.418649	0.160330	0.660330	0	0.179340	0	0
$BaTiO_3$	-1.36588	0.121479	0.121479	0.514083	0.0686701	0.149645	0.0246441

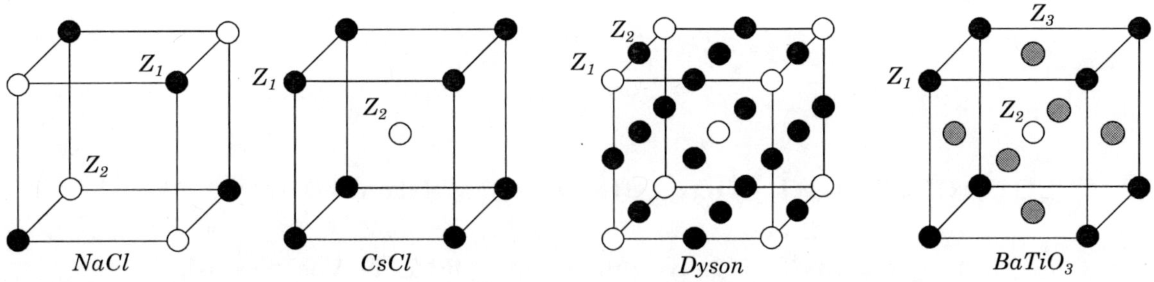

Figure 1. Stellar compounds considered in this work with examples of terrestrial compounds. The structure labelled "Dyson" was proposed by Freeman Dyson in 1971 [18].

3. Conclusions

We have found that contrary to catalysed neutron-star crusts, various compounds could potentially exist in the crust of an accreted neutron star. If confirmed, their presence may have important implications for the thermal and mechanical properties of the crust [20, 21, 22]. However, our analysis has been restricted to a few binary and ternary compounds at zero temperature, and we have ignored the influence of electron exchange, electron polarisation, and quantum zero-point motion of ions. Moreover, the structure of the crust might depend on the cooling history of the star, and on the crystallisation process. This deserves further studies.

Acknowledgments

This work was mainly financially supported by Fonds de la Recherche - FNRS (Belgium). Partial support comes also from the COST Action MP1304 "NewCompStar". The author thanks D G Yakovlev and A A Kozhberov for valuable discussions.

References

[1] Chamel N and Haensel P 2008 *Living Rev. Relativ.* **11** 10
[2] Stixrude L 2012 *Phys. Rev. Lett.* **108** 055505
[3] Chamel N, Fantina A F, Zdunik J L and Haensel P 2015 *Phys. Rev. C* **91** 055803
[4] Ruderman M 1968 *Nature* **218** 1128
[5] Pearson J M, Goriely S and Chamel N 2011 *Phys. Rev. C* **83** 065810
[6] Kreim S, Hempel M, Lunney D and Schaffner-Bielich J 2013 *Int. J. Mass Spec.* **349-350** 63
[7] Utama R, Piekarewicz J and Prosper H B 2016 *Phys. Rev. C* **93** 014311
[8] Chamel N and Fantina A F 2016 *Phys. Rev. C* **94** 065802
[9] Harrison B K, Wakano M and Wheeler J A 1958 *Onzième Conseil de Physique Solvay* (Stoops, Brussels, Belgium) pp 124-146
[10] Harrison B K, Thorne K S, Wakano M and Wheeler J A 1965 *Gravitation Theory and Gravitational Collapse* (The University of Chicago Press)
[11] Gulminelli F and Raduta A R 2015 *Phys. Rev. C* **92** 055803
[12] Jog C J and Smith R A 1982 *Astrophys. J.***253** 839
[13] Horowitz C J and Berry D K 2009 *Phys. Rev. C* **79** 065803
[14] Engstrom T A, Yoder N C and Crespi V H 2016 *Astrophys. J.* **818** 183
[15] Baiko D A, Potekhin A Y and Yakovlev D G 2001 *Phys. Rev. E* **64** 057402
[16] Igarashi T, Nakao N and Iyetomi H 2001 *Contrib. Plasma Phys.* **41** 319
[17] Shternin P S, Yakovlev D G, Haensel P and Potekhin A Y 2007 *Mon. Not. R. Astron. Soc.* **382** L43
[18] Dyson F J 1971 *Ann. Phys.* **63** 1
[19] The coefficients introduced by Dyson [18] are given by $P_3 = -C_{\rm bcc}2^{1/3}$ and $\gamma = 2\left(2^{2/3}C_{\rm sc}/C_{\rm bcc} - 1\right)$.
[20] Kozhberov A A and Baiko D A 2015 *Phys. Plasmas* **22** 092903
[21] Kozhberov A A and Baiko D A 2012 *Contrib. Plasma Phys.* **52** 153
[22] Igarashi T and Iyetomi H 2003 *J. Phys. A* **36** 6197

Δ resonances and charged ρ mesons in neutron stars

K A Maslov[1,2], **E E Kolomeitsev**[2,3] **and D N Voskresensky**[1,2]
[1]National Research Nuclear University (MEPhI), Kashirskoe sh. 31, 115409 Moscow, Russia
[2]Joint Institute for Nuclear Research, Joliot-Curie 6, 141980 Dubna, Moscow region, Russia
[3]Matej Bel University, Tajovského 40, 97401 Banská Bystrica, Slovakia

E-mail: maslov@theor.mephi.ru

Abstract. We study the equation of state of cold and dense baryon matter within the relativistic mean-field framework with hadron masses and coupling constants dependent on the mean scalar field. We include $\Delta(1232)$ isobars into previously developed models with hyperons and consider the possibility of charged ρ-meson condensation. The Δ-isobars, being included with realistic values of the attractive in-medium potential, do not lead to a strong decrease of the maximum predicted neutron star (NS) mass, and our models thus resolve the "Δ-resonance puzzle". The charged ρ^--meson condensation leads to a substantial maximum NS mass decrease. However, the observational constraint on the minimal value of the maximum NS mass remains fulfilled in all our models under consideration. The decrease of NS mass can be lowered, if one assumes a limited decrease of the ρ-meson effective mass at high densities.

1. Introduction

The description of NS properties requires the knowledge of the equation of state (EoS) of cold hadronic matter, which can be conveniently constructed within the relativistic mean-field (RMF) framework. A realistic phenomenological EoS has to fulfill a large set of experimental constraints [1]. The most challenging problem is to pass the constraint on the pressure in the isospin-symmetrical matter (ISM) (so-called "flow constraint" [2]), which favors a rather soft EoS, and the constraint that the maximum NS mass should be above the measured mass $M = 2.01 \pm 0.04\, M_\odot$ [3] ($M_\odot$ is the solar mass) of the pulsar PSR J0348+0432, that favors a stiff EoS in beta equilibrium matter (BEM). In standard RMF models hyperons and Δ-isobars may appear in NS cores, which results in a decrease of the maximum NS mass below the observed limit, cf. [4]. As was shown in [5], the softening of an EoS of NS matter can also be the result of a ρ^--meson condensation. In [6, 7] we proposed a RMF model with scaled hadron masses and coupling constants (labeled there as MKVOR), which allows to solve the hyperon puzzle and fulfills the maximum mass constraint, the flow constraint and various other constraints, but the Δs and meson condensates were not included. In [8] we included Δ into consideration. Here we show the results of the MKVOR-based model with hyperons together with the inclusion of Δs and the condensation of ρ^- mesons.

2. The RMF model with scaled hadron masses and couplings

The framework we use [9] is a generalization of the non-linear Walecka model, in which the effective coupling constants $g^*_{mb} = g_{mb}\chi_{mb}(\sigma)$ and hadron masses $m^*_i = m_i\Phi_i(\sigma)$ depend on the scalar field σ through the scaling functions $\chi_{mb}(\sigma)$ and $\Phi_i(\sigma)$. Here $m = \{\sigma, \omega, \rho, \phi\}$ denotes mesons, $b = (N, H, \Delta)$ lists baryon species, where N denotes nucleons $\{p, n\}$, H stands for hyperons $\{\Lambda, \Sigma, \Xi\}$ and Δ denotes Δ-isobars, and index i runs through all hadrons (m, b). The expression for the energy density is obtained in the standard mean-field approximation, and disregarding the possibility of a charged ρ-meson condensation it renders as:

$$E[\{n_b\},\{n_l\},f] = \sum_b E_{\rm kin}\big(p_{{\rm F},b}, m_b\,\Phi_b(f), s_b\big) + \sum_{l=e,\mu} E_{\rm kin}(p_{{\rm F},l}, m_l, s_l) + \frac{m_N^4 f^2}{2C_\sigma^2}\eta_\sigma(f)$$

$$+\frac{1}{2m_N^2}\Big[\frac{C_\omega^2 n_V^2}{\eta_\omega(f)} + \frac{C_\rho^2 n_I^2}{\eta_\rho(f)} + \frac{C_\phi^2 n_S^2}{\eta_\phi(f)}\Big], \quad E_{\rm kin}(p_{\rm F}, m, s) = (2s+1)\int_0^{p_{\rm F}} \frac{p^2 dp}{2\pi^2}\sqrt{p^2+m^2}, \qquad (1)$$

$$n_V = \sum_b x_{\omega b} n_b\,, \quad n_I = \sum_b x_{\rho b} t_{3b} n_b\,, \quad n_S = \sum_H x_{\phi H} n_H\,,$$

where $f = g_{\sigma N}\chi_{\sigma N}(\sigma)\sigma/m_N$ is the dimensionless scalar field variable, and we define the coupling constant ratios $x_{mb} = g_{mb}/g_{mN}$. The isospin projection of baryon b is t_{3b}, and $p_{{\rm F},j} = (6\pi^2 n_j/(2s_j+1))^{1/3}$ is the Fermi momentum of a fermion j, where s_j is the fermion spin and n_j denotes the density of a species j, $j = (b, l)$, l labels leptons. If the finite size effects are negligible (infinite matter), the meson coupling constants, masses and scaling functions enter the energy density only in combinations $C_M = g_{MN} m_N/m_M$, $M = \sigma, \omega, \rho$, $C_\phi = g_{\omega N}\, m_N/m_\phi$,

$$\eta_\omega(f) = \frac{\Phi_\omega^2(f)}{\chi_{\omega N}^2(f)}\,, \eta_\rho(f) = \frac{\Phi_\rho^2(f)}{\chi_{\rho N}^2(f)}\,, \eta_\phi(f) = \frac{\Phi_\phi^2(f)}{\chi_{\phi H}^2(f)}\,, \eta_\sigma(f) = \frac{\Phi_\sigma^2[\sigma(f)]}{\chi_{\sigma N}^2[\sigma(f)]} + \frac{2\,C_\sigma^2}{m_N^4 f^2} U[\sigma(f)]\,.$$

Here the scaling function $\eta_\sigma(f)$ is expressed via the self-interaction potential $U(\sigma)$ entering the Lagrangian of the model. Explicit expressions for the scaling functions $\eta_m(f)$ and values of the parameters are given in [6, 7, 8]. The version of the MKVOR model suitable for the inclusion Δs was labeled in [8] as MKVOR*. We use $\chi_{Mb} = \chi_{MN}$, $\chi_{\phi H} = 1$.

The vector-meson coupling ratios with baryons $x_{\omega B}$ and $x_{\rho B}$ are chosen following the quark SU(6) symmetry. The baryon coupling ratios with the scalar field $x_{\sigma B}$ are deduced from the potentials $U_B(n_0) = C_\omega^2 m_N^{-2} x_{\omega B} n_0/\eta_\omega(f(n_0)) - x_{\sigma B}(m_N + m_N^*(n_0))$ in ISM at the saturation baryon density $n = n_0$. There are large uncertainties in fixing of the Δ potential $U_\Delta(n_0) \equiv U_\Delta$ in the literature. We use $U_\Delta = -50\,{\rm MeV}$, which results from calculations [10]. The details on parameters of all included baryon species can be found in [8].

3. Inclusion of ρ^- condensation

The ρ-meson contribution to the Lagrangian is [5, 9]

$$\mathcal{L}_\rho = -\frac{1}{4}\vec{R}_{\mu\nu}\vec{R}^{\mu\nu} + \frac{1}{2}m_\rho^2\Phi_\rho^2\vec{\rho}_\mu\vec{\rho}^{\,\mu} - \sum_b g_{\rho b}\chi_{\rho b}\vec{\rho}_\mu \vec{j}^{\,\mu}_{I,b}\,, \quad \vec{j}^{\,\mu}_{I,b} = \bar{\psi}_b\gamma^\mu \vec{t}_b \psi_b\,, \qquad (2)$$

$$\vec{R}_{\mu\nu} = \partial_\mu\vec{\rho}_\nu - \partial_\nu\vec{\rho}_\mu + g'_\rho\chi'_\rho(f)[\vec{\rho}_\mu\times\vec{\rho}_\nu] + \mu_{{\rm ch},\rho}\delta_{\nu 0}[\vec{n}_3\times\vec{\rho}_\mu] - \mu_{{\rm ch},\rho}\delta_{\mu 0}[\vec{n}_3\times\vec{\rho}_\nu]\,,$$

where $(\vec{n}_3)^a = \delta^{a3}$ is the unit vector in the isospin space, $\mu_{{\rm ch},\rho}$ is the chemical potential for charged mesons, and g'_ρ and $\chi'_\rho(f)$ are the non-Abelian coupling constant and its scaling function, respectively. From the hidden local symmetry arguments it follows that $g'_\rho = g_{\rho N}$, which we adopt here, and for simplicity we consider $\chi'_\rho(f) = 1$. The new ansatz for the ρ meson field includes non-zero $\rho_0^{(3)}$ and $\rho_i^\pm = (\rho_i^{(1)} + \rho_i^{(2)})/\sqrt{2}$, $i = 1, 2, 3$ components. It can be shown that the condition $\rho_i^{(+)}\rho_j^{(-)} - \rho_i^{(-)}\rho_j^{(+)} = 0$ is required for minimization of the energy. This implies that the ratio $\rho_i^{(+)}/\rho_i^{(-)}$ is constant and independent of the spatial index i. Thus we assume $\rho_i^{(-)} = a_i\,\rho_c$ and $\rho_i^{(+)} = a_i\,\rho_c^\dagger$, where $\vec{a} = \{a_i\}$ is the spatial unit vector, and

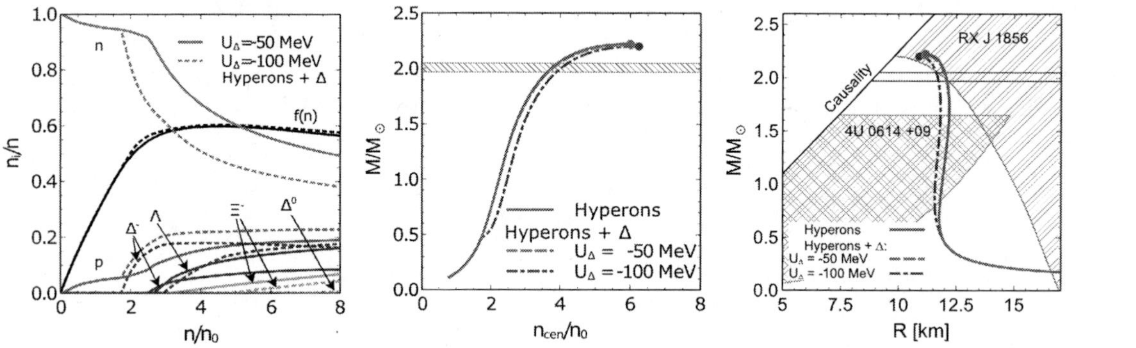

Figure 1. Particle fractions as functions of baryon density in BEM (left panel), NS mass as a function of its central density (middle panel) and the mass-radius curves (right panel) for the MKVOR* model with hyperons and Δs included.

ρ_c is a complex amplitude of the charged ρ meson field. With this ansatz the equations for minimization of the thermodynamic potential for the $\rho_0^{(3)}$ and ρ_c fields have two solutions. One of them is the traditional one with only $\rho_0^{(3)}$ being non-zero. The second solution is $\rho_0^{(3)} = \frac{\mu_{\mathrm{ch},\rho} - m_\rho \Phi_\rho}{g_\rho \chi'_\rho}$, $|\rho_c|^2 = \frac{(-n_I - n_\rho)\theta(-n_I - n_\rho)}{2\, m_\rho\, \eta_\rho^{1/2}\, \chi'_\rho}$, where $n_\rho = a\,(m_\rho\, \Phi_\rho - \mu_{\mathrm{ch},\rho})$, $a = \frac{m_N^2 \eta_\rho^{1/2}\, \Phi_\rho}{C_\rho^2 \chi'_\rho}$. The ρ charged density is $n_{\mathrm{ch},\rho} = -2 m_\rho \Phi_\rho(f) |\rho_c|^2 < 0$. The contribution of the charged ρ^- meson condensate to the energy density is then given by

$$\Delta E_{\mathrm{ch},\rho}[\{n_b\}; f] = -\frac{C_\rho^2}{2\, m_N^2\, \eta_\rho} (n_I + n_\rho)^2 \theta(-n_I - n_\rho) - \mu_{\mathrm{ch},\rho} n_{\mathrm{ch},\rho}\,, \tag{3}$$

where $\theta(-n_I - n_\rho) = 1$ for $n_I + n_\rho < 0$ and zero otherwise. The charge neutrality condition is $\sum_b Q_b n_b - n_e - n_\mu + n_{\mathrm{ch},\rho} = 0$ and the relations between chemical potentials in BEM are $\mu_e = \mu_\mu = \mu_{\mathrm{ch},\rho}$, $\mu_b = \mu_n - Q_b \mu_l$. All equations are solved self-consistently with the equation of motion for the scalar field $\partial E/\partial f = 0$. The pressure is given by $P = \sum_{j=b,l,\{\mathrm{ch},\rho\}} \mu_j n_j - E$.

4. Numerical results

In Fig. 1 we show the results of the inclusion of Δs into our model (MKVOR*) with hyperons. In the BEM the appearance of Δ^- becomes favorable at $n \simeq 2.5\,n_0, 1.73\,n_0$ for $U_\Delta = -50\,\mathrm{MeV}, -100\,\mathrm{MeV}$, correspondingly. We find that despite the considerable amount of Δs in NS matter at large densities, the reduction of the NS mass does not exceed $0.2\,M_\odot$, and the decrease of the maximum NS mass is smaller than $0.05\,M_\odot$. The mass-radius relation is also affected moderately. The radius of a $1.5\,M_\odot$ star decreases by 0.5 km if Δs are included.

We study the ρ^- condensation for various ρ-meson effective mass scaling functions $\Phi_\rho(f)$, shown in the left panel of Fig. 2. They are labeled by the minimum values $\Phi_{\rho,\min}$. Case $\Phi_{\rho,\min} = 0$ corresponds to universal scaling, $\Phi_m(f) = \Phi_N(f) = 1 - f$. For this case the equilibrium concentrations are shown in the middle panel of Fig. 2. We use $U_\Delta(n_0) = -50\mathrm{MeV}$. There exists a region with multiple solutions for equilibrium concentrations at a given density, so the ρ^- condensation in our model appears by a $\mathrm{I^{st}}$ order phase transition (PT). Such an early and strong PT results in a substantial decrease of the NS mass down to $2.03\,M_\odot$ for $\Phi_{\rho,\min} = 0$, but the maximum NS mass constraint is still satisfied within the experimental error-bars.

If one assumes that $\Phi_\rho(f)$ has the same decrease rate as Φ_N at low scalar field values (corresponding to low densities), but saturates then at some value $\Phi_{\rho,\min}$ (see curves $\Phi_{\rho,\min} = 0.3, 0.5, 0.7$), the effect of the ρ^- condensation is diminished. The critical density for the ρ^-

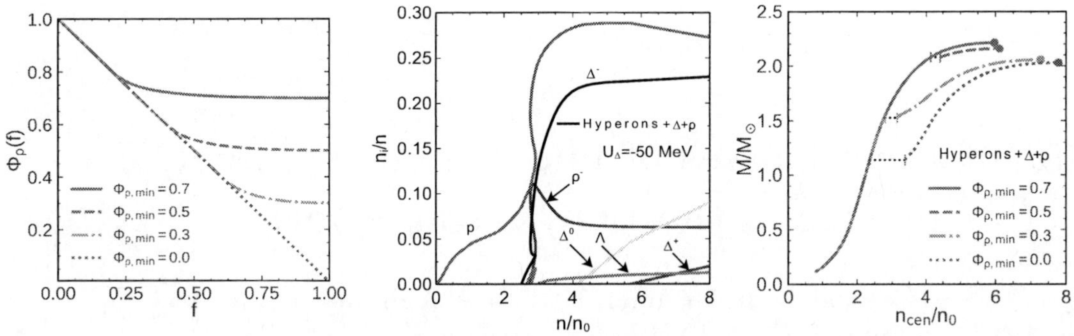

Figure 2. Left panel: Effective mass scaling function $\Phi_\rho(f)$ for $\Phi_{\rho,\min} = 0, 0.3, 0.5, 0.7$ in BEM. Middle panel: $p, \Delta, \Lambda, \rho^-$ fractions as functions of the density for the MKVOR* model with Δs and ρ^- included, $\Phi_{\rho,\min} = 0$. Right panel: NS mass as a function of central density for models with Δs and ρ^- included for various $\Phi_{\rho,\min}$. Horizontal dotted lines correspond to densities within Maxwell constructions. No stable stars with such central densities exist in these models.

condensation increases, and the density jump in the Maxwell construction decreases. This leads to an increase of the NS maximum mass up to 2.06, 2.16 and 2.21 $M_\odot$ (Fig. 2, right panel) for $\Phi_{\rho,\min} = 0.3, 0.5, 0.7$, respectively. For $\Phi_{\rho,\min} = 0.7$ there is already no ρ^- condensate in NSs.

5. Conclusion

We studied the possibility of appearance of Δ isobars and ρ^--meson condensate within a realistic RMF model with scaled hadron masses and coupling constants. We have shown that the Δs are energetically favorable to appear in NS matter, but do not lead to a substantial decrease of the maximum predicted NS mass, if one uses a realistic value for the Δ potential at saturation in ISM. Thus our model resolves the "Δ-resonance puzzle". The ρ^- condensation in our model appears by the Ist order PT and leads to a substantial decrease of the maximum NS mass, but it remains above the observational limit. The decrease can be lowered, if one assumes a lower rate of decrease of the ρ-meson effective mass than that of the nucleon at large densities.

Acknowledgments

The reported study was funded by the Russian Foundation for Basic Research (RFBR) according to the research project No 16-02-00023-A. The work was also supported by the Ministry of Education and Science of the Russian Federation within the state assignment, project No 3.6062.2017/BY, by the Slovak grant No. VEGA-1/0469/15 and partially by Russian Science Foundation, grant No. 17-12-01427. K.A.M. acknowledges the support by grant of the Foundation for the Advancement of Theoretical Physics "BASIS".

References

[1] Klähn T *et al.* 2006 *Phys. Rev.* C **74** 035802.
[2] Danielewicz P, Lacey R and Lynch W G 2002 *Science* **298** 1592.
[3] Antoniadis J *et al.* 2013 *Science* **340** 6131.
[4] Drago A *et al.* 2014 *Phys. Rev.* C **90** 065809.
[5] Voskresensky D N 1997 *Phys. Lett.* B **392** 262.
[6] Maslov K A, Kolomeitsev E E and Voskresensky D N 2015 *Phys. Lett.* B **748** 369.
[7] Maslov K A, Kolomeitsev E E and Voskresensky D N 2016 *Nucl. Phys.* A **950** 64.
[8] Kolomeitsev E E, Maslov K A and Voskresensky D N 2017 *Nucl. Phys.* A **961** 106.
[9] Kolomeitsev E E and Voskresensky D N 2005 *Nucl. Phys.* A **759** 373.
[10] Riek F, Lutz M F M and Korpa C L 2009 *Phys. Rev.* C **80** 024902.

IOP Conf. Series: Journal of Physics: Conf. Series **932** (2017) 012041 doi:10.1088/1742-6596/932/1/012041

The effective masses of nuceons in asymmetric nuclear matter in the LOCV framework

H Mazidabadi Farahani, H R Moshfegh and S Goudarzi

Department of Physics, University of Tehran, Tehran

E-mail: h.mazidabadi@ut.ac.ir, hr.moshfegh@yahoo.com

Abstract. The momentum dependence of single-particle potential (SPP) and effective masses of nuclei in asymmetric nuclear matter are studied in the framework of the lowest order constrained variational (LOCV) method at zero temperature. The Av18 interactions including two-body interactions and Urbana type three-body force (TBF) are considered as the input for the nucleon-nucleon potential. We investigate the TBF effect on the momentum-dependence of neutron and proton SPP. Also, we calculate the isospin splitting of the neutron and proton effective masses in neutron-rich nuclear matter.

1. Introduction

One of the major goals of nuclear physics is to obtain the equation of state (EOS) of nuclear matter based on the microscopic many-body approach. One can find the EOS by comparing the experimental data with the results of transport models [1]. The single-particle potential (SPP), which a nucleon feels in the nuclear matter, is one of the main components of transport models. Therefore, in order to obtain the EOS of the nuclear matter from experimental results on heavy ion collisions, we need to calculate the SPP. Note that not only the strength but also the variation of SPP around the Fermi energy is important [2]. Finally, we should solve a semi-classical equation instead of solving N-coupled equations by using the SPP [2].

The single-particle properties of nuclear matter have been studied by using various approaches, such as the *Brueckner–Hartree–Fock* (BHF) [3], *Dirac–Brueckner–Hartree–Fock* (DBHF) [4], and variational fermion hypernetted chain (FHNC) approaches [5]. In 1975, a group of scientists at the University of Manchester introduced a purely variational LOCV method for studying the nuclear matter [6]. Modarres used the LOCV method for investigating the asymmetric nuclear matter at finite temperature [7]. Also, Moshfegh *et al* extended this method for calculating the EOS of beta-stable matter at finite temperature [8]. The SPP and effective mass have been calculated for the symmetric nuclear matter [9]; the LOCV method has been generalized for three-body forces (TBFs) [10]. Goodarzi *et al* have shown that the input of TBFs plays an important role in the computational results [10]. Of course, it is well known that cold neutron stars contain asymmetric nuclear matter due to the presence of a large neutron fraction. Some important properties of neutron stars depend on the effective masses of nucleons [11]. Therefore, neutron stars can be considered as natural laboratories for studying the nuclear matter.

As far as we know, the TBFs have not been studied in the asymmetric nuclear matter in LOCV framework. Therefore, we intend to calculate the SPP and the effective mass of nucleons by analysing the TBFs for the asymmetric nuclear matter.

2. The single-particle potential in the LOCV framework

The nuclear matter Hamiltonian is usually written as

$$H = \sum_{i=1}^{A} \frac{p_i^2}{2m} + \sum_{i<j} V(ij), \tag{1}$$

where p_i is the momentum of i'th nucleon, m is the nucleon mass, $V(ij)$ is the bare two-body nucleon–nucleon potential operators, and A is the number of nucleons.

The SPP and the effective mass are defined as

$$e(k_i) = \frac{\hbar^2 k_i^2}{2m} + u(k_i), \qquad \frac{m^*}{m} = \frac{k\hbar^2}{m}\left[\frac{de(k)}{dk}\right]^{-1}, \tag{2}$$

where the single-particle potential $u(k_i)$ is determined by the equation [9]:

$$u(k_i) = \sum_j \langle ij|w(12)|ij - ji\rangle\,, \tag{3}$$

and the "effective interaction" operator is given by the following equation

$$w(12) = -\frac{\hbar^2}{2m}\left[F(12), \left[\nabla_{12}^2, F(12)\right]\right] + F(12)V(12)F(12). \tag{4}$$

Here, $F(12)$ is the two-body correlation operator [9] and $V(12)$ is the Argonne Av18 potential [12] supplemented by the UIX three-body force [10]. Algebraic calculations show that the single-particle potential at the lowest order level is

$$\begin{aligned} u(k_i) &= \frac{4\pi}{\Omega}\sum_{k_j}\sum_{M_T m_{Tj}}\sum_{LSJT}(2J+1)\left(\frac{1-(-1)^{L+S+T}}{2}\right) \\ &\times \left|\left\langle \frac{1}{2}m_{T1}\,\frac{1}{2}m_{T2}|TM_T\right\rangle\right|^2 \int r^2 dr W^{LSJT}(r)\left|j_L(kr)\right|^2, \end{aligned} \tag{5}$$

where Ω is the system volume, W^{LSJT} is the expectation value of the effective interaction in each $\{LSJT\}$ partial wave channel, $J_L(kr)$ is the standard spherical Bessel function of the order L and $k = 2^{-1}|\boldsymbol{k}_i - \boldsymbol{k}_j|$. Note that the correlation functions in the LOCV method have been obtained from the solution of the Euler-Lagrange equation [6].

3. Result and discussion

Figure 1 shows the nuclear matter binding energy per particle versus density for 3 values of the asymmetry parameter ($\beta = 0$ red, 0.2 blue, 0.8 black), where the asymmetry parameter is defined as

$$\beta = \frac{\rho_n - \rho_p}{\rho_n + \rho_p}. \tag{6}$$

Its limiting values (0 and 1) correspond to the symmetric nuclear matter and pure neutron matter, respectively. The solid and dashed lines in figure 1 refer to the Av18 interaction and the Av18+UIX interaction, respectively.

Figure 2 shows the SPP versus the momentum of neutrons and protons at saturation. According to figure 2a, the magnitude of the SPP increases with increasing the asymmetry parameter (adding neutrons to the system). Figure 2b shows that the SPP of neutrons decreases with increasing the asymmetry parameter (adding neutrons to the system). Of course, these results are expected because the asymmetry parameter shits to one (pure neutron matter) when the potential of the neutrons shits to the positive values (due to the instability). These results are in good agreement with the results of [13].

The effective mass ratio at the Fermi surface of the nuclear matter versus the asymmetry parameter (calculated with TBFs) are shown in figure 3. As we expect, the effective masses of protons and neutrons are equal for the symmetric nuclear matter. If the asymmetry parameter increases (neutrons are added to the system), the effective mass of the neutron increases and the effective mass of the proton decreases. As a result, in the framework of our computational method, in the asymmetric nuclear matter the effective mass of the neutron is greater than the effective mass of the proton. Our results are in line with the results obtained using the BHF+TBF (for K-mass) method [14]. The splitting of neutron and proton effective masses was a controversial issue, especially between relativistic and nonrelativistic treatments. The discrepancies were found to originate from the use of different definitions [15].

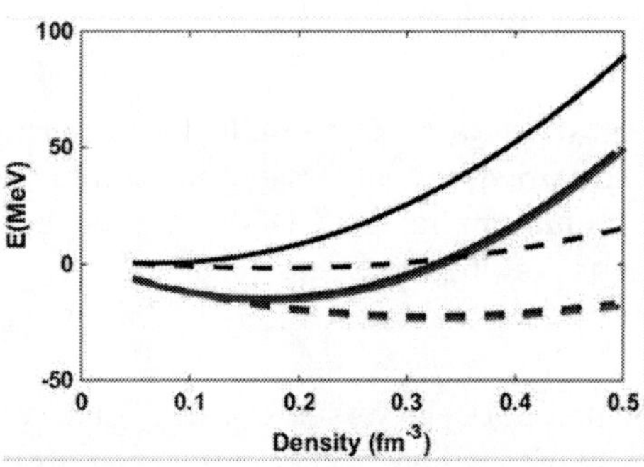

Figure 1. Density dependence of the energy per nucleon for 3 values of the asymmetry parameter ($\beta = 0$ red, 0.2 blue, 0.8 black). The dashed lines refer to the Av18 interaction and the solid lines refer to the Av18+UIX interaction.

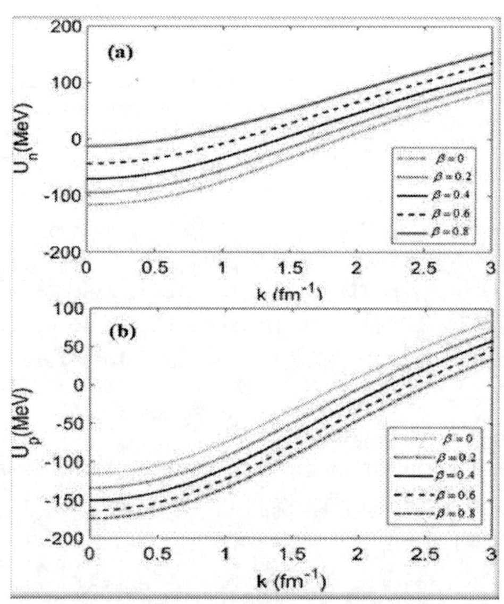

Figure 2. The SPP versus the momentum of nucleons at saturation density with TBFs for a) neutrons and b) protons.

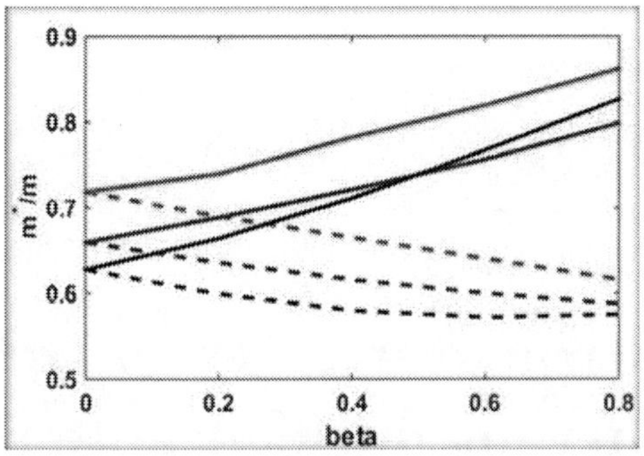

Figure 3. Isospin splitting of the effective mass at three densities. The red, blue and black lines refer to the densities of 0.08, 0.17 and 0.34, respectively. The solid lines show the neutron effective mass and the dashed line shows the proton effective mass.

4. Conclusion

The momentum and asymmetry parameter of SPP and effective mass of the asymmetric nuclear matter with and without including TBF have been studied in the LOCV framework. The results show that the SPP of proton (neutron) increases (decreases) with increasing the asymmetry parameter (by adding neutrons to the system). Also, as the asymmetry parameter increases, the effective mass of the neutron increases and the effective mass of the proton decreases.

References

[1] Baldo M 1999 *Nuclear methods and the nuclear equations of state* (World Scientific)
[2] Pines D and Nozieres P 1999 *Theory of quantum liquids* (Avalon Publishing)
[3] Baldo M, Burgio G F, Schulze H J and Taranto G 2014 *Phys. Rev.* C **89** 048801 (*Preprint* 1404.7031)
[4] Ma Z Y, Rong J, Chen B Q, Zhu Z Y and Song H Q 2004 *Physics Letters* B **604** 170–174 (*Preprint* nucl-th/0412030)
[5] Togashi H, Takano M, Sumiyoshi K and Nakazato K 2014 Nuclear equation of state with the variational method and its application to supernova simulations *International Journal of Modern Physics Conference Series* (*International Journal of Modern Physics Conference Series* vol 29) p 1460221
[6] Owen J C, Bishop R F and Irvine J M 1975 *Physics Letters* B **59** 1–3
[7] Modarres M 1993 *Journal of Physics G Nuclear Physics* **19** 1349–1358
[8] Modarres M and Moshfegh H R 2000 *Phys. Rev.* C **62** 044308 (*Preprint* nucl-th/0003028)
[9] Modarres M and Rajabi A 2011 *Nuclear Physics* A **867** 1–11
[10] Goudarzi S and Moshfegh H R 2015 *Phys. Rev.* C **91** 054320
[11] Yakovlev D G, Kaminker A D, Gnedin O Y and Haensel P 2001 *Physics Reports* **354** 1–155 (*Preprint* astro-ph/0012122)
[12] Wiringa R B, Stoks V G J and Schiavilla R 1995 *Phys. Rev.* C **51** 38–51 (*Preprint* nucl-th/9408016)
[13] Gan S X, Zuo W and Lombardo U 2012 *Chinese Physics Letters* **29** 042102
[14] Li A, Hu J N, Shang X L and Zuo W 2016 *Phys. Rev.* C **93** 015803
[15] van Dalen E N, Fuchs C and Faessler A 2005 *Physical Review Letters* **95** 022302 (*Preprint* nucl-th/0502064)

Transport coefficients in neutron star cores in BHF approach. Comparison of different nucleon potentials

P S Shternin[1], **M Baldo**[2] **and H-J Schulze**[2]

[1] Ioffe Institute, St. Petersburg, Russia
[2] INFN Sez. di Catania, Catania, Italy

E-mail: pshternin@gmail.com

Abstract. Thermal conductivity and shear viscosity of npeμ matter in non-superfluid neutron star cores are considered in the framework of Brueckner-Hartree-Fock many-body theory. We extend our previous work (Shternin *et al* 2013 *PRC* **88** 065803) by analysing different nucleon-nucleon potentials and different three-body forces. We find that the use of different potentials leads up to one order of magnitude variations in the values of the nucleon contribution to transport coefficients. The nucleon contribution dominates the thermal conductivity, but for all considered models the shear viscosity is dominated by leptons.

1. Introduction

Transport coefficients serve as an important part of the microphysical input required for modelling of various non-equilibrium processes in neutron stars (NSs). For instance, thermal conductivity is needed to study the NS cooling (especially at early stages) and thermal wave propagation from internal heating sources [1]. Viscosity coefficients regulate the damping and stability of the oscillation modes within the stars [2]. In these proceedings we consider the nucleon contribution to the thermal conductivity κ and the shear viscosity η in non-superfluid NS cores in equilibrium with respect to beta-processes with the simplest npeμ composition. Transport coefficients of a mixture are governed by the collisions between the species. In NS cores conditions, the lepton and nucleon sub-systems decouple and these sectors can be considered separately (e.g., [3]).

In [4], κ and η were calculated in the non-relativistic Brueckner-Hartree-Fock (BHF) approach for one particular nucleon-nucleon potential, Argonne v18 (Av18 in short) supplemented by the phenomenological Urbana IX (UIX in short) model for three-body forces. It is known, that purely two-body interactions in non-relativistic calculations can not correctly reproduce the empirical nuclear matter saturation point and the inclusion of three-body forces is necessary. Several models for the three-body forces are available on the market, and we investigate the model-dependence of the results obtained in [4]. In this respect, we employ the same microscopic potentials as in [5], where appropriate references and more detailed discussion can be found. Namely, in addition to the Av18+UIX model, we also use the CD-Bonn two-body potential with the corresponding adjustment of UIX three-body interaction. Both the Av18+UIX and CD-Bonn+UIX combinations reproduce the correct saturation point. In addition, we employ the three-body force obtained within the meson-nucleon theory, where the same meson-exchange parameters are used for two- and three- body interactions. As in [5], we use this model constructed for Argonne v18 potential and denote it as Av18+TBF.

2. Formalism

The contribution of each strongly degenerate species' in the mixture can be written as (we omit the species index, for brevity)

$$\kappa = C_\kappa \frac{\pi^2 k_{\rm B}^2 T n}{3m^*}\tau, \quad \eta = C_\eta \frac{p_{\rm F}^2 n}{5m^*}\tau, \tag{1}$$

where T is the temperature, $p_{\rm F}$ is the Fermi momentum, n is the number density, m^* is the effective mass on the Fermi surface, $k_{\rm B}$ is the Boltzmann constant, τ is the characteristic relaxation time and $C_{\kappa,\eta}$ are numerical constants. The latter values are found as a solution of the system of transport equations for the multicomponent Fermi-liquid. A method to obtain exact solution for these equations was developed in [6]. Below we adapt this formalism to the form convenient to perform the calculations for nuclear matter.

The necessary microscopic input is the values of m^* and the quasiparticle scattering amplitude $\hat{T}$. We obtain m^* and $\hat{T}$ within the BHF approximation, see [4] for details. Consider the scattering of two nucleon species c and i, $c + i \to c' + i'$. We define $\mathcal{Q}_{ci}(P, q) = 1/4 \sum_{\rm spins} |\langle ci|\hat{T}|c'i'\rangle|^2$, where all particles are placed on the Fermi surface and the momentum and energy conservation delta-functions are taken out. This quantity depends on the total nucleon pair momentum $\boldsymbol{P} = \boldsymbol{p}_c + \boldsymbol{p}_i$ and the transferred momentum $\boldsymbol{q} = \boldsymbol{p}_c' - \boldsymbol{p}_c$. In what follows we will need the angular averages $\mathcal{Q}_{ci}^{(kl)} = \langle \mathcal{Q}_{ci} P^k q^l \rangle$. The explicit definition for these averages which have the meaning of generalized transport cross-sections can be found in the appendix in [4]. When the nucleon interaction is expanded in partial waves, such representation is more convenient than the traditional use of the Abrikosov-Khalatnikov angles [4].

In transport theory, one expresses a non-equilibrium distribution function $F(\boldsymbol{p}, \epsilon)$ by introducing a linear (in external perturbation) correction to the local equilibrium distribution function as

$$F(\boldsymbol{p}, \epsilon) = f(\epsilon) + \tau \Psi(x) D(\boldsymbol{p}) \frac{\partial f}{\partial \epsilon}, \tag{2}$$

where $f(\epsilon)$ is the Fermi-Dirac distribution, $x = (\epsilon - \mu)/(k_{\rm B}T)$, μ is the chemical potential and $D(\boldsymbol{p})$ is the anisotropic part of the driving term. For example, $D(\boldsymbol{p}) = \boldsymbol{v}\nabla T$ for thermal conductivity. The transport coefficients are given then by (1) with

$$C_{\kappa/\eta} = \xi \int_{-\infty}^{+\infty} {\rm d}x\, \Xi(x) \Psi(x) f(x)(1 - f(x)). \tag{3}$$

For the shear viscosity, $\Xi(x) = 1$ and $\xi = 1$, while for thermal conductivity, $\Xi(x) = x$ and $\xi = 3/\pi^2$. The unknown functions $\Psi_c(x)$, where c now enumerates particle species, are determined from integral equations, derived by the linearization of the system of transport equations [6] using (2). Defining the characteristic relaxation time as

$$\tau_c = \frac{8\pi^2 \hbar^7 p_{{\rm F}c}}{m_c^*(k_{\rm B}T)^2} \mathcal{N}_c, \quad \mathcal{N}_c = \left(\sum_i m_i^{*2} \mathcal{Q}_{ci}^{(00)} \right)^{-1}, \tag{4}$$

one obtains the system of integral equations in a simple form

$$\Xi(x)(1 - f(x)) = \left(1 + \frac{x^2}{\pi^2}\right) \Psi_c(x) + \frac{2}{\pi^2} \int_{-\infty}^{+\infty} {\rm d}x'\, f(x') \frac{x + x'}{1 - {\rm e}^{x+x'}} \sum_i \lambda_{ci} \Psi_i(x'), \tag{5}$$

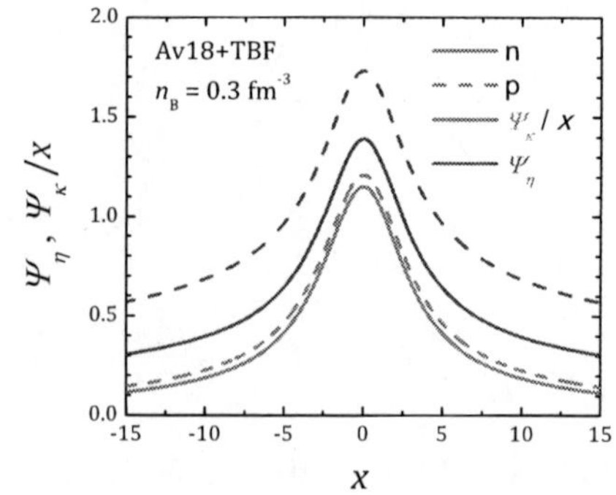

Figure 1. Functions $\Psi_\eta(x)$ and $\Psi_\kappa(x)/x$ for neutrons (——) and protons (- - - -) from the exact solutions of the system (5) for the Av18+TBF microscopic model and $n_{\rm B} = 0.3$ fm^{-3}. Thick red and thin blue lines correspond to thermal conductivity and shear viscosity problems, respectively.

where all information about the quasiparticle scattering is encapsulated in the matrix λ_{ci}. For κ and η problems, the matrices λ_{ci} are expressed through angular averages $\mathcal{Q}_{ci}^{(kl)}$ as

$$\lambda_{cc}^{\kappa} = \mathcal{N}_c \sum_i m_i^{*2} \left(\mathcal{Q}_{ci}^{(00)} - \frac{1}{2p_{\rm Fc}^2} \mathcal{Q}_{ci}^{(02)} \right), \tag{6}$$

$$\lambda_{ci}^{\kappa} = \frac{\tau_i}{p_{\rm Fc}^2 \tau_c} \mathcal{N}_c m_c^* m_i^* \left(\mathcal{Q}_{ci}^{(20)} - (p_{\rm Fc}^2 + p_{\rm Fi}^2)\, \mathcal{Q}_{ci}^{(00)} + \frac{1}{2} \mathcal{Q}_{ci}^{(02)} \right), \tag{7}$$

$$\lambda_{cc}^{\eta} = \mathcal{N}_c \sum_i m_i^{*2} \left(\mathcal{Q}_{ci}^{(00)} - \frac{1}{p_{\rm Fc}^2} \mathcal{Q}_{ci}^{(02)} + \frac{3}{8p_{\rm Fc}^4} \mathcal{Q}_{ci}^{(04)} \right), \tag{8}$$

$$\lambda_{ci}^{\eta} = \frac{3\tau_i}{4p_{\rm Fc}^4 \tau_c} \mathcal{N}_c m_c^* m_i^* \left(\mathcal{Q}_{ci}^{(22)} - (p_{\rm Fc}^2 + p_{\rm Fi}^2)\, \mathcal{Q}_{ci}^{(02)} + \frac{1}{2} \mathcal{Q}_{ci}^{(04)} \right). \tag{9}$$

Provided the matrix λ_{ci} is known, the solution of the system (5) can be found numerically by iteration method, as we do here. In [6] the general method is developed to express the solution in form of the rapidly converging series. The frequently used simplest variational solution corresponds to a first term in these series (namely, $\Psi(x) \propto \Xi(x)$). For carriers of one type, these variational solutions give $C_\kappa^{\rm Var} = 1.2(3 - \lambda_{\rm cc}^\kappa)^{-1}$ and $C_\eta^{\rm Var} = 0.75(1 - \lambda_{\rm cc}^\eta)^{-1}$ [6].

3. Results and conclusions

As an example, consider the Av18+TBF model at the total baryon density $n_{\rm B} = 0.3$ fm^{-3} and $T = 10^8$ K. The proton fraction of β-stable matter is $x_{\rm p} = 0.1$, the effective masses (in units of nucleon mass) are $m_{\rm n}^* = 0.81$ and $m_{\rm p}^* = 0.67$ [5]. The neutron and proton effective relaxation times (4) are similar, $\tau_{\rm n} = 1.2 \times 10^{-16}$ s and $\tau_{\rm p} = 0.79 \times 10^{-16}$ s, but since $n_{\rm p}$ is much smaller than $n_{\rm n}$, the proton contribution to transport coefficients is small [see (1)]. The exact solutions of the system (5) are shown in figure 1 giving $C_\kappa^{\rm n} = 0.67$ and $C_\eta^{\rm n} = 1.21$ for the neutrons. The total value of the thermal conductivity is $\kappa = \kappa_{\rm n} + \kappa_{\rm p} = 1.2 \times 10^{23}$ erg cm^{-1} s^{-1} K^{-1} and the total shear viscosity is $\eta = 2.8 \times 10^{18}$ g cm^{-1} s^{-1}. It is instructive to note that the influence of $\Psi_{\rm p}(x)$ on the equation for $\Psi_{\rm n}(x)$ via the non-diagonal terms in (5) is negligible, since $\lambda_{\rm nn}^\kappa = 0.07\lambda_{\rm np}^\kappa$. However, the inverse is not true, $\lambda_{\rm pn}^\kappa = 1.39\lambda_{\rm pp}^\kappa$ and the non-equilibrium neutron distribution strongly affects the proton one. Thus it seems a good starting approximation to consider only the neutron contribution to transport (but taking into account np collisions as well, see, e.g., [4]). In the present example, this gives $C_\kappa^{\rm n,\,Var} = 0.55$ and $C_\eta^{\rm n,\,Var} = 1.13$ where we additionally used the simplest variational solution. Comparing this approximation with the exact values, we get $\kappa/\kappa_{\rm n}^{\rm Var} = 1.36$ and $\eta/\eta_{\rm n}^{\rm Var} = 1.1$.

In figure 2 we present the final results of our calculations. We show the results for the Av18 potential on the two-body level and with addition of the UIX and TBF three-body forces. The

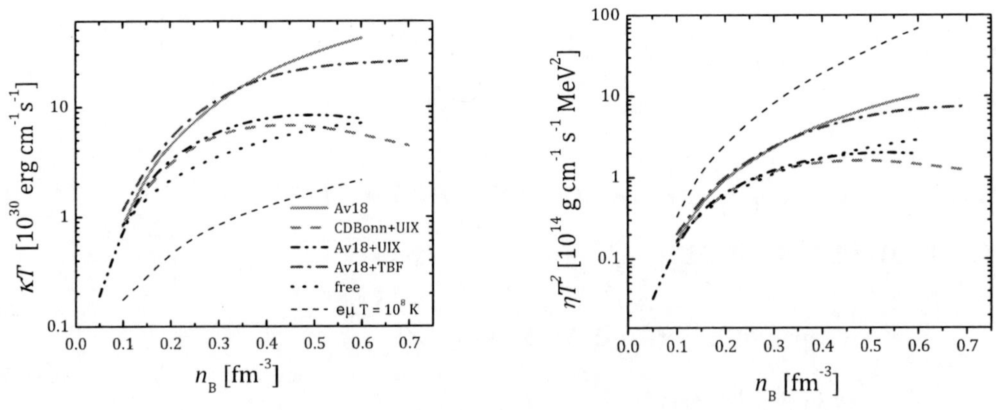

Figure 2. Temperature-independent combinations κT (*left*) and ηT^2 (*right*) as functions of baryon number density for the models considered in the text. Thin dashed lines show lepton contribution for $T = 10^8$ K and Av18+UIX EOS calculated according [7, 8].

CD-Bonn results on the two-body level give similar values as Av18 and are not shown, while the CD-Bonn+UIX results are shown with red dashed lines. For comparison, we show also the results obtained using free-space scattering probability and $m^* = 1$ (dotted lines) as well as the lepton contribution (thin dashed lines). Notice, that the lepton contribution has non-Fermi-liquid temperature dependence ($\eta_{e\mu} \propto T^{-5/3}$ and $\kappa_{e\mu} = \text{const}(T)$) [7, 8]. One clearly sees the importance of in-medium effects. At a two-body level, both η and κ in-medium values are much higher than the results based on the free-space interaction. The inclusion of UIX three-body force acts in the opposite direction, partially (but not fully) due to the increase in particles' effective masses [4, 5]. The effect of the meson-exchange three-body model TBF is much less pronounced, and the results stay approximately at the two-body level. In the future we plan to investigate the contribution of different partial waves to κ and η calculations in order to understand the difference in UIX and TBF effects.

The variations of κ and η found in our calculations are large. Fortunately, the shear viscosity is dominated by the lepton contribution, whose dependence on the equation of state is mainly through the particle fractions and effective masses and may be easily accounted for. For the thermal conductivity, the situation is reverse. However, the precise value of the thermal conductivity is usually not important since it is so large that the NS core is almost isothermal. In our study we did not consider the effects of the superfluidity that can be very important in NS cores. This is a good project for the future.

Acknowledgments

The work of PSS was supported by the Russian Foundation for Basic Research, grant 16-32-00507 mol_a and the "BASIS" Foundation.

References

[1] Yakovlev D G, Kaminker A D, Gnedin O Y and Haensel P 2001 *Phys. Rep.* **354** 1–155
[2] Haskell B 2015 *International Journal of Modern Physics* E **24** 1541007
[3] Flowers E and Itoh N 1979 *ApJ* **230** 847–58
[4] Shternin P S, Baldo M and Haensel P 2013 *Phys. Rev.* C **88** 065803
[5] Baldo M, Burgio G F, Schulze H J and Taranto G 2014 *Phys. Rev.* C **89** 048801
[6] Anderson R H, Pethick C J and Quader K F 1987 *Phys. Rev.* B **35** 1620–29
[7] Shternin P S and Yakovlev D G 2007 *Phys. Rev.* D **75** 103004
[8] Shternin P S and Yakovlev D G 2008 *Phys. Rev.* D **78** 063006

IOP Conf. Series: Journal of Physics: Conf. Series **932** (2017) 012043 doi:10.1088/1742-6596/932/1/012043

Evolution of the surface magnetic field of rotating proto-neutron stars

M Obergaulinger and M Á Aloy

Departamento de Astronomía y Astrofísica, Universidad de Valencia, C/ Dr. Moliner 50, 46100 Burjassot, Spain

E-mail: martin.obergaulinger@uv.es

Abstract. We study the evolution of the field on the surface of proto-neutron stars in the immediate aftermath of stellar core collapse by analyzing the results of self-consistent, axisymmetric simulations of the cores of rapidly rotating high-mass stars. To this end, we compare the field topology and the angular spectra of the poloidal and toroidal field components over a time of about one seconds for cores. Both components are characterized by a complex geometry with high power at intermediate angular scales. The structure is mostly the result of the accretion of magnetic flux embedded in the matter falling through the turbulent post-shock layer onto the PNS. Our results may help to guide further studies of the long-term magneto-thermal evolution of proto-neutron stars. We find that the accretion of stellar progenitor layers endowed with low or null magnetization bury the magnetic field on the PNS surface very effectively.

1. Introduction

The wide range of surface magnetic field strengths found across the population of neutron stars (e.g. [1]) is the result of a combination of processes operating at their formation during stellar core collapse and effects that modify their structure afterwards, such as cooling or accretion. Most observations pertain to relatively old neutron stars, and do not place tight constraints on the very early evolution of the field. Hence, it is difficult to draw conclusions on the fields of nascent proto-neutron stars (PNSs) from observations or, conversely, to connect theoretical results from models of supernova core collapse to evolved neutron stars. Nevertheless, the field is likely to bear at least a significant imprint of these early conditions.

A thorough theoretical study of the evolution of the magnetic field of PNSs should optimally be based on three-dimensional long-term simulations including magnetohydrodynamics and neutrino transport, which treat the global dynamics of core collapse and a possible supernova explosion and the generation of the PNS field in a self-consistent manner. The enormous computational costs of spectral neutrino transport mean that thus far only a small number of such simulations exist with the most sophisticated ones [2–4] run for only a fairly limited period.

Hence, less expensive axisymmetric models are still of considerable use for approaching this topic (e.g. [5–9]). Depending on the rotational energy and the seed field of the pre-collapse star, but also potentially on the input physics and the numerical method and grid resolution, their results range from minor modifications of non-magnetized versions of the same cores to magnetically driven explosions of a preferentially axial morphology. Depending on the global evolution, the PNS may accrete matter at all latitudes or only through narrow streams with a potential final collapse to a black hole (BH). The time evolution of the rate and geometry of mass accretion and the magnetization of the accreted layers have an important impact on the accumulation of magnetic flux at the PNS surface. The picture thus generated may show a shell of enhanced magnetic field strength, but it is also possible for the magnetic field to be buried underneath additional layers of weakly magnetized gas [10]. These processes are complemented by the possible amplification of magnetic fields in the interior of the PNS, in particular if rotation and convection constitute a dynamo.

Previously, we studied the evolution of magnetized core collapse for stars with an initial mass of $15\,M_{\odot}$ without rotation [11]. Varying the pre-collapse field strength between negligible values below 10^{10} G and dynamically relevant, yet from the point of view of stellar evolution rather overestimated, values of 10^{12} G, we found explosions driven by neutrino heating with potentially a strong contribution of magnetic forces. We were following the evolution of the PNSs for several hundreds of milliseconds and found the development of layers of strong field on their surfaces owing their origin to the accretion of magnetized gas. We now focus on cores of higher-mass stars with rapid rotation and study the geometry of the magnetic field during the first up to two seconds after PNS formation [12]. We note that our set of models includes both PNSs that collapse to a BH within this time and ones that will likely remain stable for much longer or potentially indefinitely.

We outline our simulations in section 2, discuss the result in section 3, and present our conclusions in section 4.

2. Simulations

Our numerical models of the stellar cores are based on special relativistic magnetohydrodynamics (MHD), an approximately general relativistic gravitational potential, and spectral neutrino transport. Neutrino physics is treated in the hyperbolic two-moment formulation, which provides a good approximation to a full solution of the Boltzmann equation of radiative transfer and is intrinsically multi-dimensional, and includes corrections due to the gas velocity (e.g., Doppler shifts) and gravity and all relevant interactions between neutrinos of all flavours and nucleons, nuclei, and electrons and positrons [13]. The transport solver, computationally the most expensive part of the simulations, is the main factor restricting us to axisymmetric rather than full three-dimensional simulations.

As reported in [12], we simulated the core collapse of several stars with a high initial mass of $35\,M_{\odot}$ and subsolar metallicity (pre-collapse models `35OB` and `35OC` of [14]). To test whether these properties make them viable progenitors of long GRBs, we combined the progenitors with different rotational profiles and magnetic fields. The results can be briefly summarized as follows:

- In spite of their high compactness, usually taken as an indication for a high resistance to shock revival [15], our cores develop successful supernova explosions predominantly driven either by neutrino heating or by magnetorotational stresses. All explosions have a bipolar geometry with the magnetically driven showing the highest degree of collimation of the ejecta and the highest explosion energies.
- While the ejecta expand asymmetrically, matter continues to fall onto the PNS, which therefore grows in mass. For most of our models, the PNS reaches the threshold for BH formation within less than 2 s after bounce. Some models within this class show conditions that make a later collapsar phase conceivable.
- Models without a BH collapse tend to contain a rapidly rotating and hence notably flattened PNS of relatively strong magnetization.

In the following, we will present a study of the magnetic fields on the surfaces of the PNSs of a selection of cores. We focus on both models that collapse to BHs and models where the

PNS remains stable throughout the simulation. Although the simulations end too early for observing the crust formation [16] and the late time fall back of magnetized gas [10], we intend to get insights on the field structure of the very young neutron stars. To this end, we focus on the time evolution of the strengths of the poloidal and toroidal components of the surface field and analyze their multipole expansion in spherical harmonics in order to distinguish large-scale structures from the field fluctuating on small scales. We refer to figure 1 for the pre-collapse state of the models, which, as the blue lines indicate, all rotate rapidly and differentially. We briefly describe their dynamics in the following:

Model `350C-RO` includes the original rotational profile and magnetic field distribution of the core as given by the stellar evolution models. The field has roughly similarly strong poloidal and toroidal components and is confined to the radiative zones of the core. About 200 ms after core bounce, the model develops an explosion driven by a combination of magnetic forces and neutrino heating. The ejecta expand in the form of bipolar outflows along the rotational axis while matter falls onto the PNS, whose mass and rotational energy continue to grow. The accretion, however, does not reach the threshold of collapse to a BH.

Model `350C-Rs` has a strong initial magnetic field based on the vector potential of [17] with the normalization set such that the profiles of toroidal and poloidal are equal and launches a magnetically powered, very energetic explosion almost immediately after bounce. The explosion quenches the accretion onto the PNS rather effectively, leading to a reduced PNS mass and possibly preventing BH formation indefinitely. Internal redistribution of angular momentum from the centre to the outer layers makes the PNS assume an extremely oblate shape.

Model `350B-RO` has the much weaker and strongly toroidally dominated magnetic field of the progenitor star `350B`, mostly eliminating the magnetic contribution to shock revival. The higher compactness of the pre-collapse core causes the PNS to exceed the instability threshold and collapse to a BH at $t \sim 1.5\,\mathrm{s}$.

The magnetic field distribution of the stellar evolution models `350B/C` is the result of a simple model for the amplification of the field by a dynamo operating only in radiative zones of the stars [18]. The cores do not include any magnetic field in convective zones. Consequently, there are large field-free regions in the cores, of which the ones outside the iron core of model `350B` are of particular importance, as we will show below.

3. Results

We illustrate the evolution of the surface field of the three models in figure 2. At three selected times between an early stage after bounce and one late in the evolution (close to BH collapse for model `350B-RO`), we identified the PNS surface with the angle-dependent radius of ν_e-sphere, $R_\nu(\theta)$ and focussed on a region centred at this radius, $r(\theta) \in [R_{\min}; R_{\max}] = [R_\nu(\theta) - \Delta R; R_\nu(\theta) + \Delta R]$ with $\Delta R = 4\,\mathrm{km}$. The top row of the figure shows the current density, $j = \left(\vec{\nabla} \times \vec{b}\right)^{\phi}$, and the toroidal field strength together with (poloidal) field lines in this region. Since the radii of the PNS surface decreases as time progresses, the three areas in each panel possess an ordering in time with the outermost and innermost being the first and last of the three times, respectively. In addition, we computed the angular spectra of the field at the same times (bottom row of panels) by expanding its volume-averaged poloidal and toroidal components in terms of spherical harmonics Y_{lm} ($m = 0$ because of axisymmetry):

$$b_l^{\mathrm{pol},\phi} = \int \mathrm{d}\theta\, Y_{lm}(\theta) \tilde{b}^{\mathrm{pol},\phi}(\theta), \tag{1}$$

$$\tilde{b}^{\mathrm{pol},\phi}(\theta) = \int \mathrm{d}V\, b^{\mathrm{pol},\phi}(r,\theta) \left(\int \mathrm{d}V\right)^{-1}, \tag{2}$$

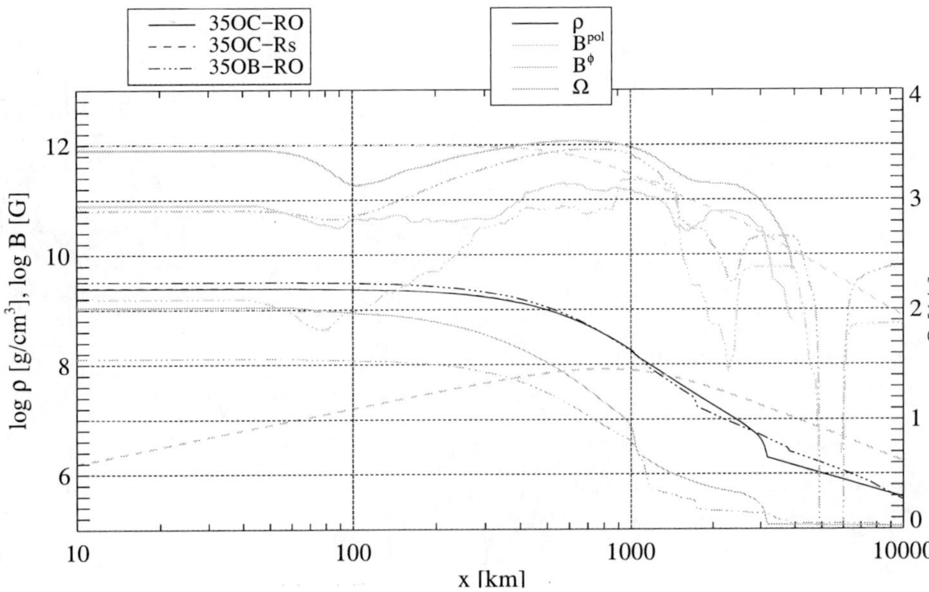

Figure 1. Profiles of density, angularly averaged poloidal and toroidal magnetic field, and angular velocity for all three models at the onset of collapse.

where the volume integration extends over the surface region defined above.

We first note that no crust has formed in any of the models. Hence, the dynamics of the surrounding region in which the PNS is immersed, characterized by radial as well as angular flows, leaves an imprint on the structure and evolution of the surface field. Consequently, we can find important variations of the strength and orientation of the field on intermediate length scales at all times for model 35OC-RO (left panels). At $t = 0.24$ s, regions where the poloidal field is mostly radial are situated close to ones where the θ-component is strongest, and the toroidal field shows reversals of polarity, though on average negative/positive in the northern/southern hemisphere. These patterns are reflected in rather strong current localized in domains of an angular extent around $\lesssim 20$ degrees. These general observations hold also at later times. In addition, we find an enhancement of the field strength at the north and south poles, where the PNS radius is lowest. There, the poloidal field is mostly radial. Variations on very small length scales lead to intense currents. We note that the gas ending up in the bipolar explosion is accelerated in the small volume immediately outside the polar PNS surface.

The angular spectra of the poloidal field are rather similar for all three times shown here, except for a growth of the mean field strength. Modes up to $l \lesssim 10$ possess a similar spectral power. Going to higher orders, the spectra decay according to a power law with an index slightly exceeding 1. The spectra of the toroidal field are somewhat steeper towards low order modes with a power-law component beginning already at $l < 10$. In agreement with the partition of the energy between poloidal and toroidal field, the toroidal spectrum is about an order of magnitude stronger than the poloidal one.

Model 35OC-Rs presents a similar case, albeit at a much higher axis ratio of the PNS and with a more equal distribution of energies of the poloidal and toroidal components. Compared to the above model, the enhancement of the radial field and the concentration of the currents at the polar caps are more pronounced. As before, the poloidal component shows a spectral plateau at $l \lesssim 10$, while the spectrum of the toroidal field has a power-law shape across the entire range of modes.

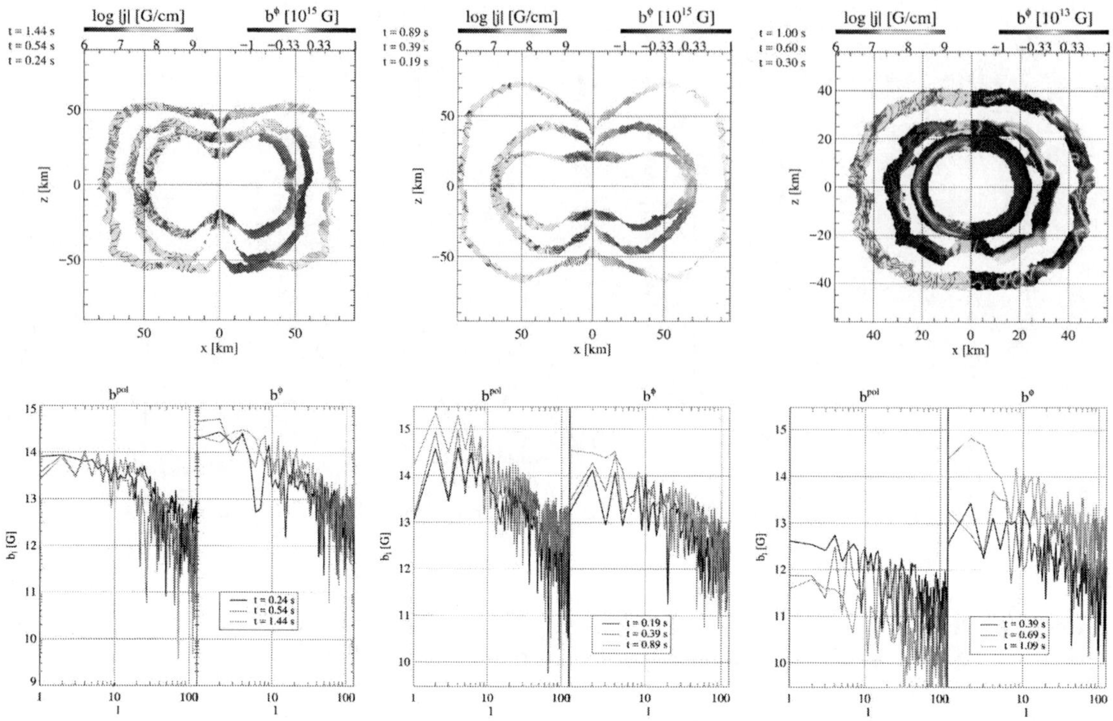

Figure 2. Top row: maps of the current density (colours in the left parts of the figures) and toroidal field component (right parts) overlayed by field lines of models (left to right) `350C-R0`, `350C-Rs`, and `350B-R0`. For each model, we only show the surface layer of the PNS at three different times as printed in the top left part of the figures. Due to the contraction of the PNS, the outermost, intermediate, and innermost PNS surfaces correspond to the sequence from the earliest to the latest times. Bottom row: angular spectra of the absolute value of the poloidal and the toroidal field components of the same models at the three times shown in the to row.

The evolution of model `350B-R0` differs from this pattern mostly due to the large unmagnetized shell that is accreted onto the PNS after the iron core. At an early stage of the model ($t = 0.39\,\mathrm{s}$), the qualitative results are similar to the other two models: variations on large and intermediate scales dominate the structure of both field components and the surface layer is filled by a highly variable pattern of currents. However, at later times, gas without a magnetic field is accreted, burying the field already present in the PNS. Therefore, the PNS surface at $t = 0.69\,\mathrm{s}$ is, except for the polar caps, almost devoid of magnetic field. At intermediate to low latitudes, poloidal field can be found only at its base. It does not reach the matter surrounding the PNS. The toroidal field is slightly less confined to the interior, but also rather weak outside the PNS. During the later phases of accretion and contraction of the PNS, this picture is modified slightly. At $t = 1.09\,\mathrm{s}$, the structure of the field is visible to a larger extent at the bottom of the surface layer. The poloidal field lines are now entirely closed within the PNS. Even the polar region where at $t = 0.69\,\mathrm{s}$ still some radial field lines were present is no longer threaded by field lines connected to the surrounding matter. Still, the toroidal field is slightly more extended, but also drops quickly with radius. Any speculation as to whether this mostly buried magnetic field might reemerge later is idle because of the quick collapse to a BH. For this model, the plateaus in the spectra of the poloidal component tend to be even more extended than for the two models discussed above. The toroidal component, which is by far stronger than, also shows notable power at high-order modes.

4. Conclusions

We investigated the structure and evolution of the magnetic field on the surface of rapidly rotating PNSs, which are the results of the axisymmetric simulations coupling magnetohydrodynamics with a spectral two-moment neutrino transport solver of [12]. We selected three representative models based on two different cores, one developing a magnetically supported explosion and leaving behind a PNS (`35OC-RO`), one exploding very energetically driven by the magnetic field and with an extremely oblate PNS at the centre (`35OC-Rs`), and a third one that explodes with little magnetic influence and produces a BH at the end of the simulation (`35OB-RO`).

For each of the models, we studied the morphology of the poloidal and toroidal magnetic field components on the PNS surface, which we approximately identified with the ν_e-sphere, at different times and computed the angular spectra of the two components by expanding their distribution in terms of spherical harmonics. We typically find a rather complex field geometry with field lines tangled on small to intermediate length scales. An enhancement of the field strength at the polar caps of the PNSs is particularly prominent for models `35OC-RO` and `35OC-Rs`. For both models, the surface fields are quite strong, reaching or exceeding 10^{14} G on average, which would correspond to magnetars when placed in the parameter space of old neutron stars. The angular spectra of the field can be described in terms of a combination of a plateau at low mode orders, l, and a power-law decay towards higher l. The former component is most clearly visible for the poloidal component, for which it extends to $l \gtrsim 10$, whereas the toroidal component shows only slight hints of its presence at lowest l. This behaviour means that the field cannot be described well in terms of a single low-order mode such as, e.g. a dipole. Model `35OB-RO` possesses a similar field early on, but later undergoes a transition to a very weakly magnetized surface as the magnetic field is buried underneath unmagnetized gas accreted at late times. The spectra of the field shows the same two components, but on a much weaker strength compared to the other two models strength.

We attribute this field structure to two main agents governing the field evolution: the internal dynamics of the PNS and the accretion of matter with a varying degree of magnetization. The PNSs develop convective activity quickly after their formation. While the surface usually is not part of the convective layer, it is nevertheless affected by overshoot from below. Convective eddies typically have sizes up to intermediate angles of few tens of degrees, which are reflected in the strength of the spectra at these angles. The field embedded in the gas falling onto the PNS surface is modulated by the non-radial flows in the post-shock layer on angular scales similar to those of the motions inside the PNS. At late times, after the onset of the explosions, these accretion flows take the shape of relatively narrow streams with an angular width that also corresponds to intermediate orders $l \sim 10$ impinging on the PNS at stochastically changing locations. Hence, the effect of both processes cannot be clearly disentangled. In the absence of a magnetic field in the accreted matter, the PNS surface rather quickly turns virtually non-magnetic, despite the ongoing activity of magneto-convection below. This fact suggests that accretion rather than the PNS convection is the process mainly responsible for generating and maintaining the surface field.

The main limitations of the current study, both caused by the restrictions of computational time, are the assumption of axisymmetry and the still fairly limited simulation time. The latter restriction prevents us from studying, e.g. the possible reemergence of buried magnetic fields in cores obtaining a similar structure to model `35OB-RO`. To remedy both drawbacks at the same time is currently not feasible. Shorter three-dimensional runs, on the other hand, can be done at reasonable costs, albeit not for a large number of models. Running the simulations until the crust formation would be desirable since the field evolution on the surface is slowed down considerably after that moment. So far, such a long simulation time of about 70 s has been achieved by [16] in spherical symmetry. To reach comparable times in axisymmetry, we would have to drastically reduce the computational costs by replacing the full neutrino transport by a much simpler treatment at a suitably chosen moment. Otherwise, computational costs range in several millions of core-hours per model. This problem is exacerbated by the fact that the main factor driving up the costs is the not the grid size, but the large number of time steps, against which parallelization does not offer a straightforward solution.

Acknowledgements
We acknowledge support from the European Research Council (grant CAMAP-259276) and from the Spanish Ministry of Economy and Finance and the Valencian Community grants under grants AYA2015-66899-C2-1-P and PROMETEOII/2014-069, resp. We thank Oliver Just and Thomas Janka for valuable help, in particular regarding the aspects of microphysics and neutrinos. The computations were performed under grants AECT-2016-1-0008, AECT-2016-2-0012, AECT-2016-3-0005, AECT-2017-1-0013, and AECT-2017-2-0006 of the Spanish Supercomputing Network on hosts *Pirineus* of the Consorci de Serveis Universitaris de Catalunya (CSUC), *Picasso* of the Universidad de Málaga, *FinisTerrae2* of the Centro de Supercomputación de Galicia, Santiago de Compostela, and *MareNostrum* of the Barcelona Supercomputing Centre, respectively, and on the clusters *Tirant* and *Lluisvives* of the Servei d'Informàtica of the University of Valencia.

References

[1] Kaspi V M 2010 *Proceedings of the National Academy of Science* **107** 7147–52
[2] Mösta P, Ott C D, Radice D, Roberts L F, Schnetter E and Haas R 2015 *Nature* **528** 376–9
[3] Mösta P, Richers S, Ott C D, Haas R, Piro A L, Boydstun K, Abdikamalov E, Reisswig C and Schnetter E 2104 *ApJL* **785** L29
[4] Winteler C, Käppeli R, Perego A, Arcones A, Vasset N, Nishimura N, Liebendörfer M and Thielemann F-K 2012 *ApJL* **750** L22
[5] Bisnovatyi-Kogan G S and Moiseenko S G 2008 *Progress of Theoretical Physics Supplement* **172** 145–55
[6] Burrows A, Dessart L, Livne E, Ott C D and Murphy J 2007 *ApJ* **664** 416–34
[7] Harikae S, Takiwaki T and Kotake K 2009 *ApJ* **704** 354–71
[8] Sawai H, Yamada S and Suzuki H 2013 *ApJL* **770** L19
[9] Takiwaki T, Kotake K and Sato K 2009 *ApJ* **691** 1360–79
[10] Torres-Forné A, Cerdá-Durán P, Pons J A and 2016 Font J A *MNRAS* **456** 3813–26
[11] Obergaulinger M, Janka H-T and Aloy M Á 2014 *MNRAS* **445** 3169–99
[12] Obergaulinger M and Aloy M Á 2017 *MNRAS* **469** L43–7
[13] Just O, Obergaulinger M and Janka H-T 2015 *MNRAS* **453** 3386–413
[14] Woosley S E and Heger A 2006 *ApJ* **637** 914–21
[15] O'Connor E and Ott C D 2011 *ApJ* **730** 70
[16] Suwa Y 2014 *Publications of the ASJ* **66** L1
[17] Suwa Y, Takiwaki T, Kotake K and Sato K 2007 *Publications of the ASJ* **59** 771–85
[18] Spruit H C 2002 *A&A* **381** 923–32

IOP Conf. Series: Journal of Physics: Conf. Series **932** (2017) 012044 doi:10.1088/1742-6596/932/1/012044

The apparent decay of pulsar magnetic fields

A Biryukov[1,2], A Astashenok[3], S Karpov[4,2] and G Beskin[4,2]

[1] Sternberg Astron. Inst. of MSU, 13 Universitetsky pr., Moscow 119234, Russia
[2] Kazan Federal University, 18 Kremlyovskaya str., Kazan 420008, Russia
[3] Phys., Math. & IT Dept., Baltic Fed. Univ., 14 A. Nevskogo str., Kaliningrad 236041, Russia
[4] Special Astrophysical Observatory RAS, Nizhnii Arkhyz, 369167, Russia

E-mail: ant.biryukov@gmail.com

Abstract.
Neutron stars are extremely strong cosmic magnets which fields are expected to decay with time. Here we report on the simple test of this process. Adopting a novel approach, we have estimated surface magnetic fields B for 76 radiopulsars (the most numerous subclass of the isolated neutron stars) which ages t are known independently. Focusing on the accurate evaluation of the precision of both quantities, we determined a significant power-law trend $B(t) \propto t^{-\beta}$ with index $\beta = 0.19^{+0.05}_{-0.06}$ at 95% C.L. The effects of the observational selection turn this value into the upper limit for the intrinsic field decay rate. If so, then neutron star crusts are close to the "impurity-free crystals", which results in a relatively slow magnetic fields decay.

1. Introduction

The large-scale magnetic field is the key feature of the neutron stars (NS). Being as strong as $\sim 10^{8-14}$ Gs at the star's surface, it to a large extent controls many physical properties and observational manifestations of these compact objects: their rotational evolution, cooling, x-ray flux etc.

Magnetic field strength B on the surface of an isolated NS unlikely remains constant, but is secularly decaying up to one order of magnitude throughout the entire lifetime of the star owing to the Ohmic decay of the electric currents located in the NS crust and/or core (e.g. [1, 2])

In spite of the number of indirect evidences for the NS magnetic fields decay (by analyzing and reproducing of pulsar rotational and/or cooling parameters, e.g. [3, 4, 5]) no direct observation of this process has been obtained so far. The main obstacle in this way is that ages and magnetic fields of individual *isolated* neutron stars are difficult to be estimated independently and with a confidently calculated precision at the same time.

In our research we focus on the well-justified estimation of radiopulsars magnetic fields, ages and especially the uncertainties of the both quantities. Analyzing the apparent pulsars $B(t)$ distribution directly, we, therefore, provide the reliable constraint on the evolution of isolated NSs magnetic fields.

2. Probing the magnetic field evolution

Up to date, observed pulsar periods P and their derivatives $\dot{P}$ remain the basic source for the massive and homogeneous magnetic fields estimations. The spin-down law $\dot{P}(B, P)$, which is assumed to be known from theory, provides a strict functional relationship between the timing

IOP Conf. Series: Journal of Physics: Conf. Series **932** (2017) 012044 doi:10.1088/1742-6596/932/1/012044

parameters and the strength of the field induction B on the star's surface. The general form of this law ($P\dot{P} \propto B^2$) is well fixed among theoreticians, while its details are dependent on the specific physical assumptions. However, recent achievements in numerical simulation of pulsar magnetospheres from "first principles" have shown that accurately calculated spin-down of a realistic isolated pulsar follows the simple equation

$$P\dot{P} = -\frac{4\pi^2 R^6(M)}{I(M)c^3} \cdot B^2 \cdot (k_0 + k_1 \sin^2 \alpha), \tag{1}$$

where α is the magnetic obliquity (the angle between the spin and the magnetic axes of the star), R, I and M are the star radius, moment of inertia and mass, respectively, c is the speed of light while $k_0 \approx 1$ and $k_1 \approx 1.4$ can be considered as universal constants for real NSs [6, 7]. Dependencies $R(M)$ and $I(M)$ here are determined by the neutron star equation of state (EOS or eos hereafter).

The value of the logarithm[1] of the magnetic field extracted from this equation thus depends on the actual values of a neutron star mass, magnetic obliquity, spin period and its derivative and, finally, the equation of state. Its estimation can be obtained within a probabilistic approach by substituting the distributions of masses and magnetic angles of isolated pulsars to $\log B(P, \dot{P}, M, \alpha \mid \text{EOS})$. As it has been shown in our previous research ([8], which we will hereafter refer to as BAB), the distribution of $\log B$ is nearly Gaussian for the representative subset of 22 EOSes with width $\sigma[\log B] \approx 0.06...0.09$ dex depending on the specific EOS, when state-of-the-art observational constraints to the pulsar masses and obliquities distributions are adopted (see BAB for details).

The value of $\sigma[\log B]$ being as small as ~ 0.1 dex represents the formal accuracy of the magnetic field estimate and, finally, makes possible the accurate probing of the isolated NS magnetic fields evolution.

Another component which is required for this procedure is the knowledge of confident pulsars ages. Notice that the most common (and simple to calculate) so-called characteristic age $\tau_{\text{ch}} = P/(2\dot{P})$ seems to be inapplicable here. Mostly because of its bias relative to the real ages of pulsars owing to nonzero initial spin period P_0.

Therefore, we have compiled a list of 76 normal rotation-powered pulsars, whose ages were estimated independently of the τ_{ch} and with more or less confidently known uncertainties. The list consists of: (*i*) 22 normal pulsars associated with young (up to 10^5 yr) supernova remnants (SNRs), ages of which have been discussed in the literature. This part is mostly the compilation of the data already prepared by other authors [9, 10] with some extensions. (*ii*) 36 pulsars with well-constrained kinematic ages t_{kin} mostly derived by Noutsos et al [11]. These data (along with t_{kin} uncertainties) have been extracted from the tracing of the galactic motion of particular objects back in time assuming that they had been born within the galactic disk. (*iii*) 18 pulsars, whose kinematic ages have been derived with relatively low precision (all obtained in [11]). These ones are quite old pulsars with typical $t_{\text{kin},0} \sim 3 \times 10^7$ years.

The empirical dependence of the calculated magnetic fields $\log B(P, \dot{P})$ on the ages $\log t$ of the described 76 pulsars is shown on the left plot of the figure 1. The values of the fields were calculated by substituting pulsar's P and $\dot{P}$ to eq (1)[2], while uncertainties of $\log B$ were modeled according to the distribution of $\Delta_{\text{B}}^{\text{BSk21}}$ obtained in BAB.

The clear linear gradient of $\log B$ with an increase of $\log \mathbf{t}$ can be seen there: the estimated magnetic fields of aged pulsars in average are one order of magnitude smaller than that of young neutron stars.

[1] We adopt decimal logarithms within the paper

[2] Both parameters for every pulsar were taken from the ATNF database [12] (version 1.54), `http://www.atnf.csiro.au/research/pulsar/psrcat/expert.html`

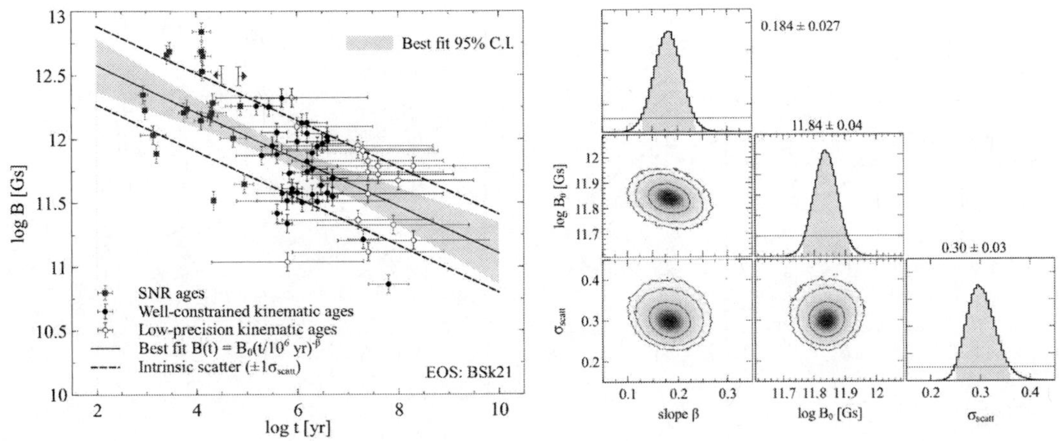

Figure 1. *Left plot:* The apparent dependence of the timing-based magnetic field values, calculated within the BSk21 equation of state, on the estimation of ages of 76 pulsars. The best fit by the power-law model is shown by the solid line with the corresponding 95 per cent C.L. filled by cyan. The dashed lines represent $\pm 1\sigma_{\text{scatt}}[\log B]$ of estimated intrinsic scatter in the observed $\log B$. The slope of the best fit $\beta \approx 1/5$ means the existence of a highly significant evolutionary trend in the magnetic fields of the considered pulsars. *Right plot:* The marginalized posterior distributions of the parameters of apparent pulsar magnetic fields evolution: the slope β, magnetic field B_0 at the age 10^6 years and the intrinsic scatter standard deviation $\sigma_{\text{scatt}}[\log B]$. The numbers given next to the plots are the averages and standard deviations for the corresponding distributions. Uniform priors adopted for these parameters are shown by red lines on the same plots. The solid-line contours on the 2-d plots represent 68, 95 and 99% credible areas respectively.

To check the significance of this gradient we undertook a full Bayesian fit of the data by a minimal power law $B(t) = B_0\,(t/10^6 \text{ years})^{-\beta}$, where B_0 is the average field strength at age 10^6 years, while β is the field decay index. As it can be also seen from the plot, the scatter of $\log B$ values at a given age is larger than the typical uncertainty of the values of $\log B$. It implies the existence of the intrinsic scatter in the physical parameters of the neutron stars. To quantify it the variance of the scatter $\sigma^2_{\text{scatt}}[\log B]$ has been also added in the Bayesian model. All the calculations have been performed by means the *Just Another Gibbs Sampler*[3] package (JAGS, [13]), which uses the Markov chain Monte Carlo simulations for producing the posterior distributions of model parameters. Within the calculations, the formal errors of $\log B$ and $\log t_{\text{SNR/kin}}$ have been taken into account.

The results of the calculations are shown on the right plot of the figure 1, where the posterior distributions of the β, $\log B_0$ and $\sigma_{\text{scatt}}[\log B]$ marginalized over individual pulsar parameters are plotted along with their correlations. The analysis of the apparent $\log B - \log t$ diagram shows that the magnetic fields of the considered pulsars follow a highly significant trend so that $B(t) \propto t^{-1/5}$.

3. Discussion and conclusions

The provided fit to the collected data is nothing more than a minimal phenomenological description of the apparent field evolution. The trend $B(t) \propto t^{-1/5}$ is likely to be significantly

[3] `http://mcmc-jags.sourceforge.net/`

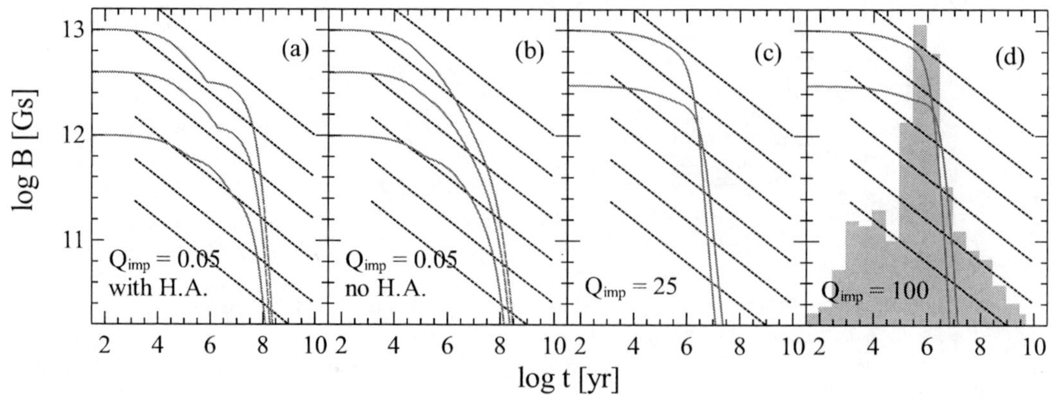

Figure 2. Theoretical curves of pulsar magnetic fields decay (red solid lines) for different initial $B_{\rm init}$ and amount of impurities within a NS crust ($Q_{\rm imp}$ parameter). The dashed lines are plotted with the slope $\beta = 0.2$ representing the most probable upper limit for the field decay rate obtained in the current work. (a) The phenomenological model proposed in [15]. (b) The same model as in the panel (a) but without the Hall attractor (H.A.); (c, d) The evolutionary tracks from [2] assuming larger amount of impurities. The empirical distribution of pulsar ages of the considered subset of 76 pulsars is also shown by the grey-filled histogram.

affected by the effect of the observational and analysis selection. Firstly, the systematic magnetic angle evolution mentioned above seems to be able to shift the estimated value of β. Indeed, average obliquity of older pulsars is expected to be less than that of younger ones. However, even within the most extremal case when *all* young pulsars have magnetic angles $\alpha \approx 90^\circ$, which further evolve to $\alpha \approx 0^\circ$ at the time interval of $10^3 \ldots 10^8$ years, one can get the maximal bias in determined β close to ≈ 0.04. This is comparable to the estimated accuracy of β estimation. Hence, the obliquity evolution unlikely can affect the apparent slope of $\log B - \log t$ distribution.

At the same time we undertook an accurate population synthesis of normal isolated radiopulsars to figure out the main components of the observational selection which affect the apparent slope in $\log B - \log t$ diagram. We mostly reproduced the model of pulsars evolution implemented in [14], but adopting the Spitkovsky's spin-down equation (1) along with the consistent model for the magnetic obliquity evolution [7].

As a result, it has been found, that even within the non-evolving magnetic fields, the distribution of "observed" pulsars (i.e. those of them which successfully passed the instrumental selection) on the $\log B - \log t$ plane is significantly inclined with the slope $\beta \approx 0.3$. (Here we assume that modeled pulsar ages t are unbaised relative to that estimated from PSR-SNR associations and NS motion within the Galaxy).

After a number of runs of the population synthesis (within different assumptions) we concluded that the obtained value $\beta \approx 0.2$ for the considered 76 pulsars represents *the upper limit* for the intrinsic rate of the magnetic field evolution. This means that our results likely reject possible rapid field decay, which, however, is predicted by a number of existing theoretical models.

NS fields dissipation history can be roughly divided into two epochs. The first one lasts up to $\sim 10^{4-6}$ years while the star's surface is hot enough and the electrons scattering from the phonons dominates. The magnetic field changes relatively weakly within this stage. In contrast, during the second epoch, the electrons scattering from the impurities of the NS crust crystalline lattice become important and can significantly increase the rate of the field decay. The amount of the impurities within the crust matter can be described by the parameter $Q_{\rm imp}$, which higher values lead to the faster magnetic field decay of middle-age and old pulsars [16]. Generally, there is no consensus about the value of $Q_{\rm imp}$ for the realistic neutron stars. Although the crusts of normal pulsars are likely to be close to "impurity-free crystals" ($Q_{\rm imp} \sim 0$) in contrast to that of the magnetars [2, 17].

In figure 2 four models of the pulsars magnetic field decay are overplotted on the lines representing the upper limit for the decay rate consistent with our results.[4]

It is clearly seen that the models proposed in [15], assuming the low $Q_{\rm imp}$, are much closer to the weak field decay scenario, which satisfy the empirical data definitely better. Additionally notice that rapidly decaying magnetic field makes it hard to explain the observation of active pulsars with ages greater than $\sim 10^6$ years which exist in the considered subset (see figure 2d).

Thus, the final results of our research can be summarized as follows: (*i*) Using the ages of 22 supernova remnants associated with young pulsars and 54 kinematic ages of older sources as well as the refined version of the timing-based estimator of pulsars magnetic fields, we have found significant apparent trend $B(t)$ in the magnetic fields of normal radiopulsars so that $B(t) \propto t^{-1/5}$ (*ii*) The ascertained trend $B(t)$ is significantly affected by the effects of the observational selections. The accurate population synthesis of the observed pulsars has shown that the slope β in the apparent $\log B - \log t$ distribution is likely to be the *upper limit* of the intrinsic one β_0. (*iii*) Thus, our result confirms that crusts of normal (rotation-powered) pulsars are likely to be close to "impurity-free crystals", which lead to a relatively slow surface magnetic field decay at the final stages of their active life. At the same time we reject the theoretical models of NS magnetic fields evolution that would assume high amount of nuclear impurities within their crusts.

Acknowledgments

We are grateful to Andrei Igoshev, Sergey Popov, Stefano Andreon, Yury Lyubarsky and Jérôme Pétri for their comments and fruitful discussions.The work is performed according to the Russian Government Program of Competitive Growth of Kazan Federal University. Data analysis and simulations used hardware and software supported by the Russian Science Foundation grant No. 14-50-00043. Artyom Astashenok thanks Russian Ministry of Education and Science for the support (project 2058/60).

References

[1] Aguilera D N, Pons J A and Miralles J A 2008 *Astron. and Astrophys.* **486** 255–71
[2] Viganò D, Rea N, Pons J A, Perna R, Aguilera D N and Miralles J A 2013 *MNRAS* **434** 123–41
[3] Pons J A, Link B, Miralles J A and Geppert U 2007 *Physical Review Letters* **98** 071101
[4] Chashkina A and Popov S B 2012 *New Astronomy* **17** 594–602
[5] Igoshev A P and Popov S B 2014 *MNRAS* **444** 1066–76
[6] Spitkovsky A 2006 *Astrophys. J. Lett.* **648** L51–4
[7] Philippov A A, Spitkovsky A and Cerutti B 2015 *Astrophys. J. Lett.* **801** L19
[8] Biryukov A, Astashenok A and Beskin G 2017 *MNRAS* **466** 4320–31
[9] Popov S B and Turolla R 2012 *Astrophys. and Space Sci.* **341** 457–64
[10] Gill R and Heyl J S 2013 *MNRAS* **435** 3243–50
[11] Noutsos A, Schnitzeler D H F M, Keane E F, Kramer M and Johnston S 2013 *MNRAS* **430** 2281–301
[12] Manchester R N, Hobbs G B, Teoh A and Hobbs M 2005 *AJ* **129** 1993–2006
[13] Plummer M 2012 JAGS: Just Another Gibbs Sampler Astrophysics Source Code Library *ascl*:1209.002
[14] Faucher-Giguère C A and Kaspi V M 2006 *Astrophys. J.* **643** 332–55
[15] Igoshev A P and Popov S B 2015 *Astronomische Nachrichten* **336** 831
[16] Shternin P S and Yakovlev D G 2006 *Phys. Rev. D* **74** 043004
[17] Gullón M, Miralles J A, Viganò D and Pons J A 2014 *MNRAS* **443** 1891–9

[4] We are grateful to Andrey Igoshev and Miguel Gullón for providing us with these theoretical curves of the magnetic field decay.

Magnetic field amplification by the r-mode instability

A I Chugunov[1], J L Friedman[2], L Lindblom[3] and L Rezzolla[4,5]
[1]Ioffe Institute, Politekhnicheskaya 26, 194021 St. Petersburg, Russia

E-mail: andr.astro@mail.ioffe.ru
[2]Leonard Parker Center for Gravitation, Cosmology and Astrophysics, University of Wisconsin-Milwaukee, P.O. Box 413, Milwaukee, Wisconsin 53201, USA

E-mail: friedman@uwm.edu
[3]Center for Astrophysics and Space Sciences, University of California at San Diego, La Jolla, CA 92093, USA

E-mail: llindblom@ucsd.edu
[4]Institute for Theoretical Physics, Max-von-Laue-Straße 1, 60438 Frankfurt, Germany

E-mail: rezzolla@itp.uni-frankfurt.de
[5]Frankfurt Institute for Advanced Studies, Ruth-Moufang-Str. 1, 60438 Frankfurt, Germany

Abstract. We discuss the magnetic field enhancement by unstable r-modes (driven by the gravitational radiation reaction force) in rotating stars. In the absence of a magnetic field, gravitational radiation exponentially increases the r-mode amplitude α, and accelerates differential rotation (secular motion of fluid elements). For a magnetized star, differential rotation enhances the magnetic field energy. Rezzolla et al (2000–2001) argued that if the magnetic energy grows faster than the gravitational radiation reaction force pumps energy into the r-modes, then the r-mode instability is suppressed. Chugunov (2015) demonstrated that without gravitational radiation, differential rotation can be treated as a degree of freedom decoupled from the r-modes and controlled by the back reaction of the magnetic field. In particular, the magnetic field windup does not damp r-modes. Here we discuss the effect of the back reaction of the magnetic field on differential rotation of unstable r-modes, and show that it limits the generated magnetic field and the magnetic energy growth rate preventing suppression of the r-mode instability by magnetic windup at low saturation amplitudes, $\alpha \ll 1$, predicted by current models.

1. Introduction

Neutron stars contain matter which is denser than that in atomic nuclei. Such a matter cannot be produced in terrestrial laboratories. Accordingly, neutron stars provide a unique opportunity to study the properties of super-dense matter by comparing observations of these stars with theoretical models. One such possibility is associated with the gravitational radiation driven instability of r-modes in rapidly rotating neutron stars [1–7]. In the absence of dissipation and magnetic fields, the r-modes are unstable in any (even slowly) rotating star [6, 7]. Dissipation (e.g., shear and bulk viscosities [8, 9], an Ekman layer [10, 11], mutual friction [12–16]) can suppress this instability to some extent. However if a neutron star spin frequency exceeds a (temperature dependent) critical value, it can be unstable (see e.g., [17] for recent review), and the instability will alter the star's evolution [5, 18–20]. Observations of rapidly rotating neutron stars (i.e., neutron stars in accreting low mass X-ray binaries and their descendants

IOP Conf. Series: Journal of Physics: Conf. Series **932** (2017) 012045 doi:10.1088/1742-6596/932/1/012045

– millisecond pulsars) put important constraints on the r-mode instability threshold and thus on properties of super-dense matter (c.f. [21–26]). However, these constraints assume that the internal fluid dissipation is the main damping mechanism for the r-modes; such constraints can be irrelevant if the r-mode instability is, in fact, suppressed by an alternate mechanism like a magnetic field windup [27–29]. These publications argue that the growth of unstable r-modes should be accompanied by differential rotation and associated bending of the magnetic field lines powered by the r-mode energy. If the r-mode energy loss exceeds the energy supplied by the gravitational radiation instability, the r-modes should be damped out. Here we summarize our analysis of the back reaction of bent magnetic lines on the differential rotation, and obtain an upper limit on the the magnetic field enhancement by the r-mode instability. We show that magnetic windup cannot suppress the r-mode instability if the nonlinear saturation amplitude is low, as predicted by current theoretical models, e.g. [30–32]. More details can be found in [33].

2. R-modes and differential rotation

Differential rotation which suppresses the gravitational radiation driven instability in the magnetic windup scenario [27–29], is a second-order effect in the r-mode amplitude. Before considering the full problem (including the effects of gravitational radiation and magnetic fields), we outline previously published results on two simplified problems: (1) the r-modes in non-magnetized stars and (2) stable r-modes (i.e. neglecting gravitational radiation) in magnetized stars.

The solution for stable r-modes in non-magnetized stars to second-order in amplitude α was obtained in [34]. The axisymmetric part of this solution describes differential rotation and contains a gauge dependent part; the general solution includes contribution, corresponding to the second order differential rotation allowed in non-oscilating star (i.e., an arbitrary time independent azimuthal flow, stratified on cylinders). A particular solution with vanishing drift velocity (no secular motion of fluid elements) exists. However, the latter solution does not mean that the axisymmetric oscillation averaged second order velocity perturbations are absent. On the contrary, according to [34] these terms are not zero and are not cylindrically stratified for any choice of initial conditions. The apparent contradiction between the absence of secular motion of the fluid elements and non-vanishing velocity perturbations is superficial. It is associated with Stokes drift [36, 37] – second order secular motion of fluid elements associated with the oscillating velocity profile in first order of perturbation theory. The Stokes drift contributes to the secular motion along with the oscillation averaged velocity perturbations and resolves the contradiction (see [35] for details).

Recently, an analytical second order solution for the unstable r-modes in non-magnetized rotating stars was obtained [38], neglecting internal fluid dissipation. This solution demonstrates an exponential growth of the r-mode amplitude, $\alpha(t) = \alpha(0)\exp(t/\tau_{\rm GR})$ (with the time scale $\tau_{\rm GR} > 0$ determined previously in [8]). The growth is accompanied by the exponential growth of differential rotation (with a particular spatial profile determined by the equation of state). Like a stable r-mode, this solution has a gauge freedom – the general solution includes an arbitrary second order time independent differential rotation profile. The secular motion of fluid elements is still cylindrically stratified, but the solution with vanishing drift is absent; the exponential growth of velocity perturbations leads to an exponential growth of fluid element displacements.

For a magnetized star with perfect conductivity, the secular motion of fluid elements leads to bending of magnetic field lines. A solution with time independent non-vanishing drift velocity becomes impossible. This was confirmed in [35] where the general second order solution for the stable r-modes in a magnetized neutron star was obtained. As with non-magnetized stars, the general solution has gauge freedom; it is associated with perturbations of a non-oscillating star and can be described as an ensemble of Alfvén modes. A solution with the vanishing drift of fluid elements exists for specific initial conditions. The general second order solution can be presented as a superposition of two solutions: (a) a solution which describes the evolution of differential rotation in a non-oscillating magnetized star (i.e. an ensemble of Alfvén modes, determined by initial conditions) and (b) the r-mode solution with vanishing drift. This solution demonstrates that stable r-modes are not damped by magnetic windup.

The criteria for the gravitation radiation driven instability was generalized to magnetized stars in [39]. However this result cannot be directly applied to check the magnetic windup scenario. This scenario assumes the r-mode instability is not suppressed initially, but it can be suppressed by the growth of the unstable r-mode leading to windup of the magnetic field and subsequent suppression of the instability.

The magnetic field evolution in the presence of an unstable r-mode is analyzed in our recent paper [33] using second order perturbation theory for an infinitely conducting rotating star including gravitational radiation reaction forces. The results are illustrated there by a simple toy model, and confirmed by a detailed analysis of the full problem using the symplectic product formalism [2, 40]. They can be summarized as follows. In second order perturbation theory, the gravitational radiation reaction and magnetic forces have axisymmetric components with non-vanishing oscillation averages that directly affect the secular motion of fluid elements. These components are (a) a gravitational radiation reaction force, with magnitude (per unit mass) $f_{\rm GR} \sim \alpha^2(t)\,\tau_{\rm GR}^{-1}\,\Omega R$, where $\Omega = 2\pi\nu$ is the spin frequency of the star, and (b) a force associated with the second order magnetic terms in the perturbation equations, with magnitude (per unit mass) $f_{\rm m} \sim \alpha^2(t)\,\omega_{\rm A}\Omega$ [33], assuming a non-superconducting stellar interior. Here, $\omega_{\rm A} = \sqrt{\pi B^2/\rho R^2}$ is a typical frequency of Alfvén modes, B is an initial magnetic field, ρ is a typical density, and R is the stellar radius. Before saturation, when the mode amplitude $\alpha(t)$ is exponentially growing, $\alpha(t) = \alpha(0)\exp(t/\tau_{\rm GR})$, the displacements of fluid elements in the azimuthal direction can be estimated in the same way as the displacement of an "Alfvén" harmonic oscillator with frequency $\omega_{\rm A}$ evolving under the action of an exponentially growing external force $f_{\rm GR} + f_{\rm m}$ (with timescale $\tau_{\rm GR}/2$). The result is [33]

$$\xi^{\hat\phi} \sim \frac{f_{\rm GR} + f_{\rm m}}{\left(2\tau_{\rm GR}^{-1}\right)^2 + \omega_{\rm A}^2} \sim \alpha^2(t)\,\Omega R\,\frac{\max(\omega_{\rm A},\,\tau_{\rm GR}^{-1})}{4\tau_{\rm GR}^{-2} + \omega_{\rm A}^2} \le \alpha^2(t)\,R\,\frac{\Omega}{\omega_{\rm A}}, \qquad (1)$$

where the last inequality follows from a detailed analysis of the terms. This estimate also allows us to estimate the magnetic field energy density,

$$\epsilon_{\rm m} \sim \rho\,\omega_{\rm A}^2\left(\xi^{\hat\phi}\right)^2 \le \alpha^4(t)\,\rho\,\Omega^2 R^2 \qquad (2)$$

and its growth rate $\dot\epsilon_{\rm m} \sim 4\alpha^4(t)\,\tau_{\rm GR}^{-1}\,\rho\,\Omega^2 R^2$, again using the harmonic oscillator analogy, or a more detailed mathematical analysis of the full problem [33]. The energy density of r-modes can be estimated as $\epsilon_{\rm r} \sim \alpha^2(t)\,\rho\Omega^2 R^2$ (c.f. [8]), and its rate as $\dot\epsilon_{\rm r} \sim 2\alpha^2(t)\,\tau_{\rm GR}^{-1}\,\rho\,\Omega^2 R^2 \sim \alpha^{-2}(t)\dot\epsilon_{\rm m} \gg \dot\epsilon_{\rm m}$. Therefore, the magnetic wind up cannot suppress the r-mode instability before saturation (unless the saturation amplitude $\alpha(t) \sim 1$, as assumed in early papers [27–29]).

In the simplest model of nonlinear r-mode saturation, the amplitude growth simply stops and remains frozen once $\alpha(t) = \alpha_{\rm sat}$. In this case, $\xi^{\hat\phi}$ can be estimated as the displacement of a harmonic Alfvén oscillator under the influence of the constant force $f_{\rm GR} + f_{\rm m}$ with mode amplitude $\alpha = \alpha_{\rm sat}$,

$$\xi^{\hat\phi}_{\rm sat} \sim \frac{f_{\rm GR} + f_{\rm m}}{\omega_{\rm A}^2} \sim \alpha_{\rm sat}^2\,\Omega R\,\frac{\max(\omega_{\rm A},\,\tau_{\rm GR}^{-1})}{\omega_{\rm A}^2}. \qquad (3)$$

It can significantly exceed the displacement in the exponential growth phase, $\xi^{\hat\phi}$, if $\omega_{\rm A} \ll \tau_{\rm GR}^{-1}$. The growth rate of the magnetic energy can be estimated as $\dot\epsilon_{\rm m} \sim \omega_{\rm A}\epsilon_{\rm m}$, with $\epsilon_{\rm m}$ given by

IOP Conf. Series: Journal of Physics: Conf. Series **932** (2017) 012045 doi:10.1088/1742-6596/932/1/012045

(2) with $\xi^{\hat{\phi}} = \xi^{\hat{\phi}}_{\rm sat}$. Formally, $\dot{\epsilon}_{\rm m}$ is comparable to $\dot{\epsilon}_{\rm r}$ if $\omega_{\rm A} \sim \alpha^2_{\rm sat}\tau^{-1}_{\rm GR} \ll \tau^{-1}_{\rm GR}$. However, even for an unexpectedly large $\alpha_{\rm sat} \sim 10^{-3}$ in a non-superconducting neutron star, this will affect the r-mode instability only if the initial magnetic field is $\lesssim 100\,{\rm G}$. Alternatively, the magnetic windup cannot suppress the r-mode instability. Equation (3) can also be used to estimate the maximum azimuthal magnetic field, generated by a magnetic windup in a non-superconducting star,

$$\delta B^{\hat{\phi}} \sim B\frac{\xi^{\hat{\phi}}}{R} \lesssim 4\times 10^8 \left(\frac{\alpha}{10^{-4}}\right)^2 \frac{\max(\omega_{\rm A},\,\tau^{-1}_{\rm GR})}{\omega_{\rm A}} \frac{\nu}{500\,{\rm Hz}} \frac{R}{10^6\,{\rm cm}} \frac{\rho}{4\times 10^{14}\,{\rm g\,cm^{-4}}}\,{\rm G}. \quad (4)$$

3. Summary and conclusions

We have outlined nonlinear (second order in perturbation theory) unstable r-modes in magnetized stars composed of a perfect fluid with infinite conductivity. Previously, the second order r-modes were studied for non-magnetized and magnetized Newtonian stars ([34] and [35], respectively), and for non-magnetized stars with gravitational radiation reaction [38]. Our approach here includes both gravitational radiation reaction and magnetic forces and, therefore, provides a self-consistent description of the r-mode instability in magnetized stars. This problem is important because of the magnetic windup scenario [27–29] in which the r-mode instability leads to magnetic field enhancement that suppresses the r-mode instability. Our results confirm the suggestion [27–29], that the growing r-modes induce the differential drift of fluid elements and generation of magnetic fields. However, before nonlinear saturation is achieved [i.e., at the stage of exponentially growing $\alpha(t)$], the magnetic energy growth rate is a factor of $\alpha^2(t)$ smaller than the r-mode energy growth rate by radiation reaction. Thus magnetic windup cannot affect the r-mode instability if $\alpha(t) \ll 1$. After the saturation, the enhancement of the magnetic energy is also strongly restricted. Current models of nonlinear saturation [30–32] suggest that the r-mode amplitude is limited by $\alpha \sim \alpha_{\rm sat} \lesssim 10^{-4}$. Therefore, the magnetic windup cannot suppress the instability in this case. We have also estimated the maximal magnetic field that could be generated by the r-mode instability. For an unexpectedly large saturation amplitude $\alpha_{\rm sat} = 10^{-3}$, an initial magnetic field $10^8\,{\rm G}$ could be amplified up to about 10^{11} G, which can be important for the magnetic field generation in neutron stars. However if the initial magnetic field is $\sim 10^{10}$ G, it would not be significantly affected by the r-mode magnetic windup. The assumption that the magneto-rotational instability does not operate during magnetic windup should be carefully checked in the future.

Acknowledgments

We thank anonymous referee for useful suggestions, which improve the paper and Ruxandra Bondarescu, Mikhail E. Gusakov, Branson Stephens and Ira Wasserman for helpful conversations. The work by AIC consisted in supporting consideration of B-field amplification (supported by the Russian Science Foundation, grant No. 14-12-00316). LL was supported in part by NSF grants PHY 1604244 and DMS 1620366 to the University of California at San Diego. LR was supported in part by "NewCompStar", COST Action MP1304, from the LOEWE-Program in HIC for FAIR, the European Union's Horizon 2020 Research and Innovation Programme under grant agreement No. 671698 (call FETHPC-1-2014, project ExaHyPE), from the ERC Synergy Grant "BlackHoleCam - Imaging the Event Horizon of Black Holes" (Grant 610058), and from JSPS Grant-in-Aid for Scientific Research(C) No. 26400274.

References

[1] Chandrasekhar S 1970 Phys. Rev. Lett. **24** 611–5
[2] Friedman J L and Schutz B F 1978 ApJ **221** 937–57
[3] Friedman J L and Schutz B F 1978 ApJ **222** 281–96

[4] Friedman J L 1978 *Communications in Mathematical Physics* **62** 247–78
[5] Chugunov A I 2017 PASA **34** e046
[6] Andersson N 1998 ApJ **502** 708–13
[7] Friedman J L and Morsink S M 1998 ApJ **502** 714–20
[8] Lindblom L, Owen B J and Morsink S M 1998 Phys. Rev. Lett. **80** 4843–6
[9] Lindblom L, Mendell G and Owen B J 1999 Phys. Rev. D **60** 064006
[10] Levin Y and Ushomirsky G 2001 MNRAS **324** 917–22
[11] Glampedakis K and Andersson N 2006 Phys. Rev. D **74** 044040
[12] Lindblom L and Mendell G 2000 Phys. Rev. D **61** 104003
[13] Andersson N, Glampedakis K and Haskell B 2009 Phys. Rev. D **79** 103009
[14] Haskell B, Andersson N and Passamonti A 2009 MNRAS **397** 1464–85
[15] Gusakov M E, Chugunov A I and Kantor E M 2014 Phys. Rev. Lett. **112** 151101
[16] Gusakov M E, Chugunov A I and Kantor E M 2014 Phys. Rev. D **90** 063001
[17] Haskell B 2015 *International Journal of Modern Physics E* **24** 1541007
[18] Owen B J, Lindblom L, Cutler C, Schutz B F, Vecchio A and Andersson N 1998 Phys. Rev. D **58** 084020
[19] Levin Y 1999 ApJ **517** 328–33
[20] Chugunov A I, Gusakov M E and Kantor E M 2014 MNRAS **445** 385–91
[21] Ho W C G, Andersson N and Haskell B 2011 Phys. Rev. Lett. **107** 101101
[22] Haskell B, Degenaar N and Ho W C G 2012 MNRAS **424** 93–103
[23] Alford M G and Schwenzer K 2014 *Physical Review Letters* **113** 251102
[24] Schwenzer K, Boztepe T, Güver T and Vurgun E 2017 MNRAS **466** 2560–9
[25] Chugunov A I, Gusakov M E and Kantor E M 2017 MNRAS **468** 291–304
[26] Bhattacharya S, Heinke C O, Chugunov A I, Freire P C C, Ridolfi A and Bogdanov S 2017 MNRAS **472** 3706–21
[27] Rezzolla L, Lamb F K and Shapiro S L 2000 ApJ **531** L139–42
[28] Rezzolla L, Lamb F K, Marković D and Shapiro S L 2001 Phys. Rev. D **64** 104013
[29] Rezzolla L, Lamb F K, Marković D and Shapiro S L 2001 Phys. Rev. D **64** 104014
[30] Bondarescu R, Teukolsky S A and Wasserman I 2007 Phys. Rev. D **76** 064019
[31] Bondarescu R, Teukolsky S A and Wasserman I 2009 Phys. Rev. D **79** 104003
[32] Haskell B, Glampedakis K and Andersson N 2014 MNRAS **441** 1662–8
[33] Friedman J L, Lindblom L, Rezzolla L and Chugunov A I 2017 *Preprint* arXiv:1707.09419
[34] Sá P M 2004 Phys. Rev. D **69** 084001
[35] Chugunov A I 2015 MNRAS **451** 2772–9
[36] Stokes G G 1847 *Transactions of the Cambridge Philosophical Society* **8** 441–55
[37] Longuet-Higgins M S 1953 *Phil. Trans. R. Soc. Long.* A **245** 535–81
[38] Friedman J L, Lindblom L and Lockitch K H 2016 *Phys. Rev.* D **93**(2) 024023
[39] Glampedakis K and Andersson N 2007 MNRAS **377** 630–44
[40] Dyson J and Schutz B F 1979 *Proceedings of the Royal Society of London Series* A **368** 389–410

IOP Conf. Series: Journal of Physics: Conf. Series **932** (2017) 012046 doi:10.1088/1742-6596/932/1/012046

Heat blanketing envelopes vs cooling of neutron stars

M V Beznogov[1], A Y Potekhin[2,3], M Fortin[4], P Haensel[4], J L Zdunik[4] and D G Yakovlev[2]

[1] Institute of Astronomy, National Autonomous University of Mexico, Mexico D.F. 04510, Mexico

[2] Ioffe Institute, Politekhnicheskaya 26, St Petersburg 194021, Russia

[3] Central Astronomical Observatory at Pulkovo, Pulkovskoe Shosse 65, St Petersburg 196140, Russia

[4] Nicolaus Copernicus Astronomical Center of the Polish Academy of Sciences, Bartycka 18, Warsaw 00-716, Poland

E-mail: mikavb89@gmail.com

Abstract. We discuss the effect of the chemical composition of heat blanketing envelopes of neutron stars (NSs) on the interpretation of the observations of these stars. First we analyze the diffusive fluxes of ions in non-isothermal and non-ideal Coulomb plasmas. Then we outline models of diffusively-equilibrated heat blanketing envelopes composed of binary ionic mixtures and finally we study their effect on the cooling of isolated NSs.

1. Introduction

The interpretation of the observations of isolated NSs is a difficult task for many reasons. One of them is our poor knowledge of the chemical composition of the outer heat blanketing NS envelopes. The uncertainties in their chemical composition create an obstacle in the theoretical inference of the internal structure of NSs from the observations of their thermal radiation.

Let us study different chemical compositions of heat blanketing envelopes. Such studies have been mostly conducted relying on the simplified "onion"-like (PCY97, [1]) model of envelopes (several shells of pure chemical H, He, C, and Fe with abrupt boundaries between the shells). To extend it, we constructed new models with a proper treatment of ion diffusion [2]. In this latter work (which was based on our previous work on diffusion in isothermal dense stellar plasmas [3]) we have considered non-isothermal systems and diffusively-equilibrated or non-equilibrated heat blanketing envelopes composed of binary ionic mixtures (BIMs), H–He, He–C or C–Fe, with different amounts of lighter ion species. Then we have applied the new models to investigate their effect on the cooling of isolated NSs [4]. Here we summarize these results.

2. Diffusion in non-isothermal dense stellar plasmas

Consider a multicomponent Coulomb plasma of several fully ionized ion species ($\alpha = j$, $j = 1, 2, 3 \ldots$) immersed in a neutralizing electron background ($\alpha = e$) (neglecting magnetic fields). The detailed derivation of the general expressions for the diffusive fluxes of ions is given in [2]; here we present a brief sketch.

The generalized thermodynamic forces acting on given particles α can be written as

$$\widetilde{\boldsymbol{f}}_\alpha = \boldsymbol{f}_\alpha - \left(\boldsymbol{\nabla}\mu_\alpha - \left.\frac{\partial \mu_\alpha}{\partial T}\right|_P \boldsymbol{\nabla} T \right), \tag{1}$$

where $\boldsymbol{f}_\alpha$ is the total force acting on the particles α, μ_α is their chemical potential and T is the temperature. In the outer envelope,

$$\boldsymbol{f}_\alpha = Z_\alpha e \boldsymbol{E} + m_\alpha \boldsymbol{g}, \tag{2}$$

m_α and $Z_\alpha e$ being the mass and charge of the particles α; $Z_e = -1, e > 0$; $\boldsymbol{g}$ is the surface gravity. Possible deviations from the diffusion equilibrium are characterized by the quantities

$$\boldsymbol{d}_\alpha = \frac{\rho_\alpha}{\rho} \sum_\beta n_\beta \widetilde{\boldsymbol{f}}_\beta - n_\alpha \widetilde{\boldsymbol{f}}_\alpha, \tag{3}$$

where ρ_α is the mass density of the particles α and ρ is the total mass density; then $\sum_\alpha \boldsymbol{d}_\alpha = 0$. Using (1) and (2), the Gibbs-Duhem relation $\sum_\alpha n_\alpha \boldsymbol{\nabla}\mu_\alpha = \boldsymbol{\nabla}P - S\boldsymbol{\nabla}T$ (S being the entropy density) and the electric neutrality condition, we obtain

$$\sum_\alpha n_\alpha \widetilde{\boldsymbol{f}}_\alpha = \rho \boldsymbol{g} - \boldsymbol{\nabla} P. \tag{4}$$

If we assume that the envelope as a whole is in hydrostatic equilibrium (i.e. $\rho\boldsymbol{g} = \boldsymbol{\nabla}P$), then $\boldsymbol{d}_\alpha = -n_\alpha \widetilde{\boldsymbol{f}}_\alpha$ and

$$\boldsymbol{d}_\alpha = -\frac{\rho_\alpha}{\rho} \boldsymbol{\nabla} P - Z_\alpha n_\alpha e \boldsymbol{E} + n_\alpha \left(\boldsymbol{\nabla}\mu_\alpha - \left.\frac{\partial \mu_\alpha}{\partial T}\right|_P \boldsymbol{\nabla} T \right). \tag{5}$$

Using the adiabatic (Born-Oppenheimer) approximation we can factor the electrons out of the problem of ion transport, $\boldsymbol{d}_e = 0$, implying that the electrons are in mechanical quasi-equilibrium in response to the motion of ions. Assuming also $m_e \to 0$, we obtain $\widetilde{\boldsymbol{f}}_{\rm e} = 0$ and

$$e\boldsymbol{E} = -\left(\boldsymbol{\nabla}\mu_{\rm e} - \left.\frac{\partial \mu_{\rm e}}{\partial T}\right|_P \boldsymbol{\nabla} T \right). \tag{6}$$

The final phenomenological expression for the diffusive fluxes becomes

$$\boldsymbol{J}_\alpha = \frac{n m_\alpha}{\rho k_{\rm B} T} \sum_{\beta \neq \alpha} m_\beta D_{\alpha\beta} \boldsymbol{d}_\beta - D_\alpha^T \frac{\boldsymbol{\nabla} T}{T}, \tag{7}$$

where $D_{\alpha\beta}$ is the generalized diffusion coefficient for the particles α with respect to the particles β, D_α^T is the thermal diffusion coefficient of the particles α, and the coefficient before the sum is chosen so as to match the conventional definition of $D_{\alpha\beta}$, e.g., [2, 3, 5].

3. Diffusively-equilibrated heat blanketing envelopes of neutron stars

Now we apply (7) to model heat blanketing envelopes. Consider envelopes composed of the BIMs ^{1}H – ^{4}He, ^{4}He – ^{12}C or ^{12}C – ^{56}Fe. In all examples we employ a model of 'canonical' NS (with a mass $M = 1.4 M_\odot$ and radius $R = 10$ km). We neglect the thermal diffusion (it is usually slower compared to ordinary diffusion [2]) and assume that the electrons have a little impact on the transport of ions (as ions are much heavier [6]), so $\boldsymbol{J}_e \to 0$ and $\boldsymbol{J}_2 = -\boldsymbol{J}_1$. From (7) we obtain

$$\boldsymbol{J}_2 = -\boldsymbol{J}_1 = \frac{n m_1 m_2}{\rho k_{\rm B} T} D_{12} \boldsymbol{d}_1. \tag{8}$$

The diffusion equilibrium implies the absence of diffusive fluxes, $\boldsymbol{J}_1 = -\boldsymbol{J}_2 = 0$, which leads to $\boldsymbol{d}_1 = \boldsymbol{d}_2 = 0$. Then from (5) we have a system of three linear first-order differential equations,

$$\widetilde{\boldsymbol{\nabla}}\mu_e = -e\boldsymbol{E}, \quad \widetilde{\boldsymbol{\nabla}}\mu_j = m_j\boldsymbol{g} + Z_je\boldsymbol{E}, \tag{9}$$

where

$$\widetilde{\boldsymbol{\nabla}}\mu_\alpha \equiv \sum_j \frac{\partial\mu_\alpha}{\partial n_j}\boldsymbol{\nabla}n_j + \frac{\partial P}{\partial T}\sum_j n_j\frac{\partial\mu_\alpha}{\partial n_j}\left(\sum_k n_k\frac{\partial P}{\partial n_k}\right)^{-1}\boldsymbol{\nabla}T; \tag{10}$$

$j, k = 1, 2$ enumerate the ion species. The unknowns are $\boldsymbol{\nabla}n_1$, $\boldsymbol{\nabla}n_2$ and $\boldsymbol{E}$. Note that here we do not need an explicit expression for D_{12}, owing to the assumptions made in the beginning of this section. To solve (9), one needs to add the heat transport equation, e.g. [1, 2].

Fig. 1 shows a schematic cross-section of the heat blanketing envelope in the plane-parallel approximation as well as the expected distribution of ion species. The effective density ρ^* of the transition between the light and heavy ions is determined by the total mass ΔM of light ions in the envelope [2].

Fig. 2 presents the internal temperature (at the bottom of the envelope) T_b as a function of the effective transition density ρ^* for the He – C mixture and a fixed surface temperature $T_s = 1.47$ MK. The different curves correspond to different possible densities ρ_b at the bottom of the envelope; after Fig. 4 in [2].

4. Cooling of neutron stars with diffusively-equilibrated envelopes

Now, following [4], we outline how the chemical composition of the outer envelopes affects the cooling of isolated NSs. To this aim, we employ the envelope models from the previous section.

Consider the famous Vela pulsar as an example. Its characteristic age is $t \approx 11$ kyr and the redshifted effective surface temperature (inferred using the 'canonical' NS model) $T_s^\infty = 0.68 \pm 0.03$ MK [7]; its non-redshifted temperature is $T_s = T_s^\infty/\sqrt{1-(2GM)/(Rc^2)} \approx 0.89$ MK. Using

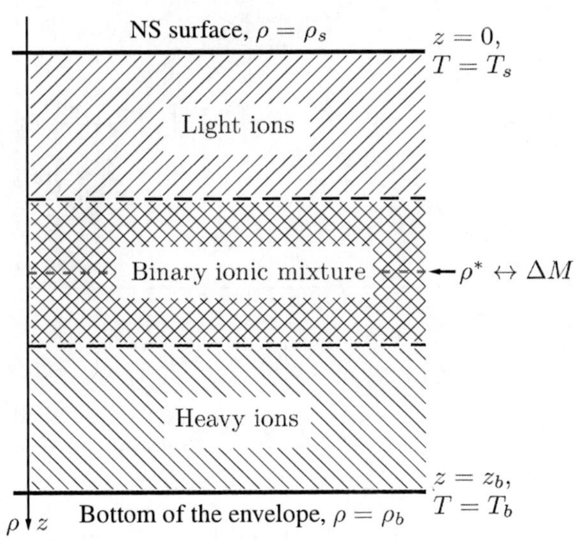

Figure 1: Schematic cross-section of a heat blanketing envelope composed of a BIM in the plane-parallel approximation vs depth z from the surface.

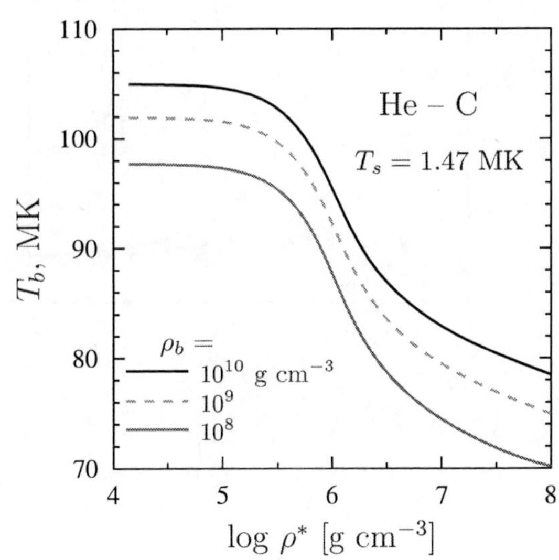

Figure 2: Internal temperature T_b vs effective transition (He $\leftrightharpoons$ C) density ρ^* in a He – C mixture with a fixed surface temperature T_s at three values of ρ_b.

$T_s - T_b$ relations from the previous section, we can calculate T_b and $\widetilde{T}_b = T_b\sqrt{1-(2GM)/(Rc^2)}$ (the redshifted internal temperature that should be constant over the Vela's interior). The pulsar has a magnetic field $B \sim 3 \times 10^{12}$ G which has no strong influence on the $T_s - T_b$ relations [8].

Fig. 3 demonstrates the dependence of the redshifted internal Vela's temperature $\widetilde{T}_b$ on the chemical composition of the blanketing envelope. $\widetilde{T}_b$ is presented as a function of the accumulated mass ΔM of lighter ion species in He – C and C – Fe envelopes. The solid black curve is for the PCY97 model, where ΔM is the total accumulated mass of all light elements (H+He+C). One sees that the variations in the chemical composition cause up to ~ 2.5 times changes in the internal temperature; the maximum variation occurs in the PCY97 model which takes into account a wider range of elements than the binary envelope models. The different maximum values of ΔM for the various mixtures are explained in [2].

It is widely accepted that the Vela pulsar is cooling mainly via neutrino emission from its core (the so-called 'neutrino cooling stage', e.g., [9, 10]). The thermal evolution of such NSs is mostly regulated by the neutrino cooling function $\ell(\widetilde{T}_b) = L_\nu^\infty(\widetilde{T}_b)/C(\widetilde{T}_b)$, where L_ν^∞ is the total redshifted neutrino luminosity of the star and C is its total heat capacity (e.g., [11]). The function $\ell(\widetilde{T}_b)$ is determined by the internal temperature; $\ell(\widetilde{T}_b)$ is a fundamental quantity of superdense matter in NS cores. It is convenient to introduce the so-called 'standard candle' $\ell_{\rm SC}$, which corresponds to a non-superfluid star cooling via the modified Urca process. Then one can express ℓ in units of $\ell_{\rm SC}$, $f_\ell = \ell/\ell_{\rm SC}$. Using the fits to f_ℓ from [12] which are almost independent of the equation of state (EOS) for nucleon NS cores, one can calculate $f_\ell(\widetilde{T}_b)$ for the Vela pulsar. As shown in [4], a factor of ~ 2 uncertainty in $\widetilde{T}_b$ translates into a factor of $\sim 10^2$ uncertainty in f_ℓ, because $f_\ell \propto \widetilde{T}_b^7$ (mainly due to the strong temperature dependence of the neutrino emissivity; see [13]).

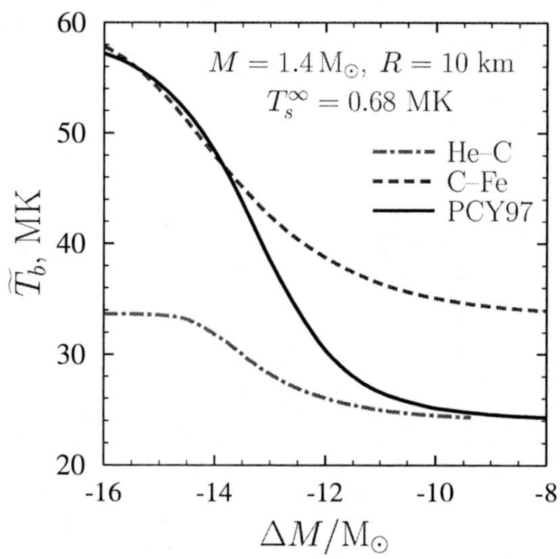

Figure 3: Vela's redshifted internal temperature as a function of the accumulated mass of light elements in different blanketing envelopes.

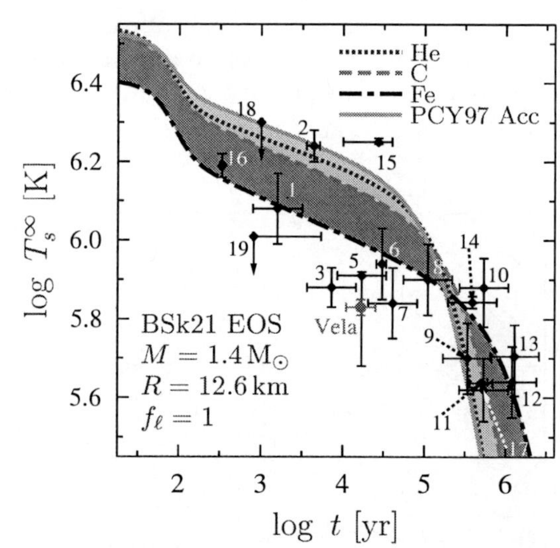

Figure 4: Cooling of a standard NS candle with different blanketing envelopes as compared with observations; after Fig. 4 in [4].

Fig. 4 presents cooling curves $T_s^\infty(t)$ for a standard candle ($f_\ell = 1$) with the BSk21 EOS [14], $M = 1.4\mathrm{M}_\odot$ and different chemical compositions of the heat blankets (for the BSk21 EOS, the radius $R = 12.6$ km is larger than the canonical one, which corresponds to a lower inferred non-redshifted surface temperature $T_s = 0.83$ MK for the Vela pulsar). Such cooling curves are almost universal, independent of the EOS and M. The theoretical curves are compared with the observational data on known isolated NSs. The thick curves correspond to the envelopes composed of pure He, C, Fe and for the fully accreted PCY97 envelope. The filled areas are covered with cooling curves obtained for envelopes with different ΔM: the C – Fe envelopes between the Fe and C curves, the He – C envelopes between the C and He curves, and the PCY97 envelopes between the Fe and PCY97 ones. One can also see that despite the 'broadening' of the curve for the iron envelope owing to the presence of light elements, it is impossible to explain all the data. In order to explain the sources that are warmer or cooler than the presented cooling curves one needs to consider deviations from standard neutrino candles [4] (see also [9, 10] and references therein).

5. Conclusions

We have derived general expressions for the diffusive fluxes in non-isothermal Coulomb plasmas with an arbitrary Coulomb coupling of ions (in gaseous and liquid states). We have employed these expressions to construct models of diffusively-equilibrated heat blanketing envelopes of NSs and studied how the chemical composition of these envelopes affects the thermal evolution of isolated NSs. We have shown that for a fixed surface temperature (known from observations) the internal temperature can vary up to ~ 2.5 times depending on the chemical composition of the heat blanket. This uncertainty translates into a factor of ~ 200 uncertainty in the neutrino cooling function, creating a strong obstacle to study the internal structure of NSs. More work should be done to overcome this obstacle [4].

Acknowledgments

The work of MB was partially supported by a Conacyt grant (CB-2014-1, #240512). The work of AP was supported by Program P13 of Presidium of Russian Academy of Sciences and by RFBR grant 16-29-13009-ofi-m and the work of MF, PH, and LZ was supported by the Polish NCN research grant no. 2013/11/B/ST9/04528.

References

[1] Potekhin A Y, Chabrier G and Yakovlev D G 1997 *Astron. Astrophys.* **323** 415–28
[2] Beznogov M V, Potekhin A Y and Yakovlev D G 2016 *MNRAS* **459** 1569–79
[3] Beznogov M V and Yakovlev D G 2013 *Phys. Rev. Lett.* **111** 161101
[4] Beznogov M V, Fortin M, Haensel P, Yakovlev D G and Zdunik J L 2016 *MNRAS* **463** 1307–13
[5] Hirschfelder J O, Curtiss C F and Bird R B 1954 *Molecular Theory of Gases and Liquids* (Wiley, New York)
[6] Paquette C, Pelletier C, Fontaine G and Michaud G 1986 *Astrophys. J. Suppl.* **61** 177–95
[7] Pavlov G G, Zavlin V E, Sanwal D, Burwitz V and Garmire G P 2001 *Astrophys. J. Lett.* **552** L129–33
[8] Potekhin A Y, Yakovlev D G, Chabrier G and Gnedin O Y 2003 *Astrophys. J.* **594** 404–18
[9] Yakovlev D G and Pethick C J 2004 *Annu. Rev. Astron. Astrophys.* **42** 169–210
[10] Potekhin A Y, Pons J A and Page D 2015 *Space Sci. Rev.* **191** 239–91
[11] Yakovlev D G, Ho W C G, Shternin P S, Heinke C O and Potekhin A Y 2011 *MNRAS* **411** 1977–88
[12] Ofengeim D D, Kaminker A D, Klochkov D, Suleimanov V and Yakovlev D G 2015 *MNRAS* **454** 2668–76
[13] Yakovlev D G, Kaminker A D, Gnedin O Y and Haensel P 2001 *Phys. Rep.* **354** 1–155
[14] Potekhin A Y, Fantina A F, Chamel N, Pearson J M and Goriely S 2013 *Astron. Astrophys.* **560** A48

IOP Conf. Series: Journal of Physics: Conf. Series **932** (2017) 012047 doi:10.1088/1742-6596/932/1/012047

Neutron stars with outbursts from superfluid crust

A D Kaminker[1,2], E A Chaikin[2], A A Kaurov[3] and D G Yakovlev[1]

[1] Ioffe Institute, Politechnicheskaya 26, St. Petersburg, 194021, Russia

[2] Peter the Great St. Petersburg Polytechnic University, Politechnicheskaya 29, St. Petersburg, 195251, Russia

[3] Institute for Advanced Study, 1 Einstein Drive Princeton, New Jersey 08540, USA

E-mail: kam.astro@mail.ioffe.ru

Abstract. We model heat propagation and the thermal surface luminosity $L_s^\infty(t)$ of a neutron star after an internal outburst in its crust. Simulations take into account superfluidity of free neutrons and the thickness of the outbursting layer (heater) in the crust. Crustal superfluidity can shorten and intensify variations of $L_s^\infty(t)$.

1. Introduction and simulations

Here we continue to study the effects of crustal heaters on the variable surface thermal emission [1, 2] of neutron stars (NSs). Our results can be useful for the interpretation of observations of old transiently accreting NSs in low-mass X-ray binaries (LMXBs) and of warm bursting magnetars. It is currently thought [3] that NSs in LMXBs are reheated by nuclear transformations in the accreted crust [4, 5]. Magnetars, which are mainly hot, middle-aged and active NSs are most likely reheated by the processes associated with strong magnetic fields, e.g., [6].

To simulate NS thermal evolution, we use a one-dimensional generally relativistic code [2] for a spherically symmetric NS. We solve the thermal evolution equations at $\rho \geq \rho_b \sim 10^9 - 10^{10}$ g cm^{-3} (to the bottom of the heat blanketing envelope – a thin surface layer with a strong thermal insulation [7]). For the densities $\rho < \rho_b$, we use separately calculated $T_b - T_s$ relations, where T_b is the temperature at $\rho = \rho_b$ and T_s is the local effective temperature of the stellar surface. We consider two NS models with the BSk21 equation of state [8]. The masses of these stars are $M = 1.4$ and $1.85\ M_\odot$; their radii are $R = 12.6$ and 12.25 km, respectively. For simplicity, we neglect the effects of superfluidity (SF) in NS cores (but consider SF of free neutrons in the crust). The $1.4M_\odot$ NS demonstrates standard neutrino cooling via modified direct Urca process; the $1.85M_\odot$ star undergoes much faster neutrino cooling via the direct Urca process in its central kernel.

The heater will be treated as a spherical layer where a given heat power $Q(\rho, t)$ [erg s^{-1} cm^{-3}] is released. Integrating $Q(\rho, t)$ over the heater, we calculate the total redshifted heat power $L_h^\infty(t)$ [erg s^{-1}]. Using the cooling code, we find the redshifted effective surface temperature $T_s^\infty(t)$ and thermal luminosity $L_s^\infty(t)$. The problem is if the $L_s^\infty(t)$ [or $T_s^\infty(t)$] variations are observable.

2. Thin heating shell and crustal superfluidity

In this section, following [2], we treat the heater as a thin spherical shell located at $\rho_1 < \rho < \rho_2$; ρ_1 and ρ_2 being the lowest and highest densities in the shell, respectively. We start with the case of non-superfluid NS. Initially, we let a young and hot NS to cool freely. Then we turn on

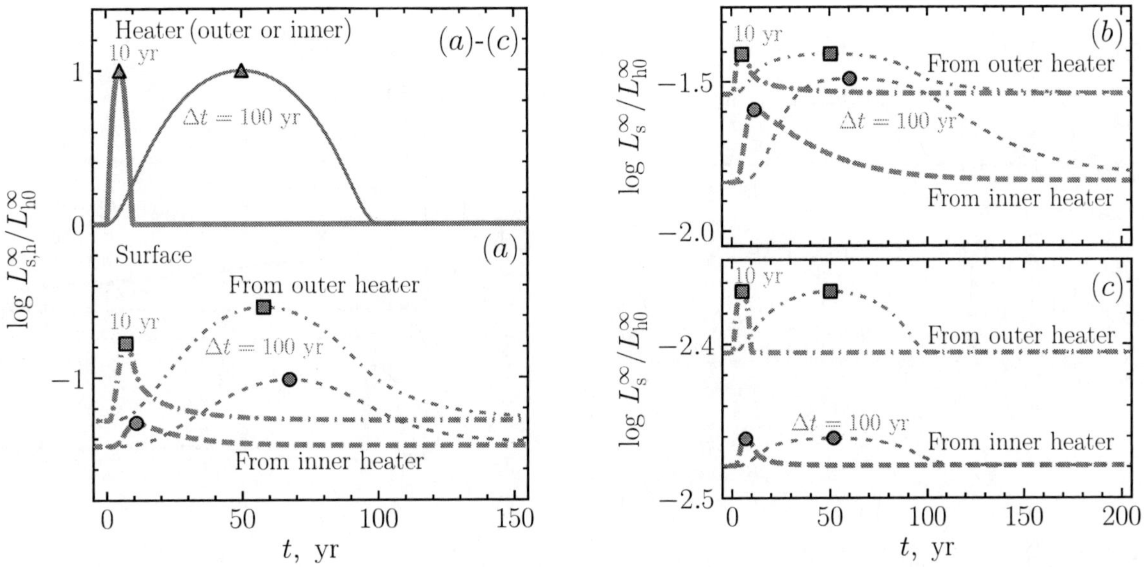

Figure 1. The heater's power $L_\mathrm{h}^\infty(t)$ and the surface luminosities $L_\mathrm{s}^\infty(t)$ of a 1.4 $M_\odot$ non-superfluid NS for three steady heat powers L_h0^∞ (a)–(c), outer and inner heaters, and two heat durations (Δt=10 and 100 yr) at $H_\mathrm{max} = 10\,H_\mathrm{h0}$. Symbols show maxima of $L_\mathrm{h}^\infty(t)$ and $L_\mathrm{s}^\infty(t)$.

a time-independent heat power $Q(\rho,t) \equiv Q_0(\rho)$, $L_\mathrm{h}^\infty \equiv L_\mathrm{h0}^\infty$, and wait until a stationary state within the star is established. After that we vary $Q(\rho,t)$ (imitating an internal outburst) and follow the outburst dynamics of $L_\mathrm{h}^\infty(t)$ and $L_\mathrm{s}^\infty(t)$.

We set $Q(\rho,t)$ to be uniform inside the heater, with the time dependence [2] $Q(t) = H_0 + (H_\mathrm{max} - H_0)\sin^2(\pi t/\Delta t)$, $0 \le t \le \Delta t$, where t is the time from the beginning of the outburst, Δt is the internal outburst duration and H_max is the maximum of $Q(t)$ reached at $t = \Delta t/2$ (assuming $H_\mathrm{max} > H_0$). The associated total heat power varies as $L_\mathrm{h}^\infty(t) = L_\mathrm{h0}^\infty + (L_\mathrm{h,max}^\infty - L_\mathrm{h0}^\infty)\sin^2(\pi t/\Delta t)$, where $L_\mathrm{h,max}^\infty$ is the maximum of $L_\mathrm{h}^\infty(t)$. When t exceeds Δt, we return the heater to the stationary state, with $Q(\rho,T) = H_0$ and $L_\mathrm{h}^\infty = L_\mathrm{h0}^\infty$. The NS also begins returning to its state before the outburst.

We consider two heaters. The *outer* heater is located between $\rho_1 = 10^{11}$ and $\rho_2 = 10^{12}$ g cm^{-3}. For the *inner* heater, we take $\rho_1 = 10^{12}$ g cm^{-3}, with $\rho_2 = 1.27 \times 10^{13}$ and 1.23×10^{13} g cm^{-3} for the 1.4 and 1.85 $M_\odot$ stars, respectively (to have equal $L_\mathrm{h}^\infty(t)$ at the same H_0, H_max and Δt for both heaters).

Figure 1 compares the generated heat power $L_\mathrm{h}^\infty(t)$ with the NS surface luminosity $L_\mathrm{s}^\infty(t)$ at $L_\mathrm{h,max}^\infty = 10\,L_\mathrm{h0}^\infty$. The cases (a), (b) and (c) refer to the three values of $H_0 = 5 \times 10^{17}$, 5×10^{18} and 5×10^{19} erg cm^{-3} s^{-1}, respectively. The corresponding steady-state heat powers are $L_\mathrm{h0}^\infty = 1.7 \times 10^{35}$ (a), 1.7×10^{36} (b) and 1.7×10^{37} erg s^{-1} (c). The figure refers to two outburst durations, Δt=10 and 100 yr, produced by the outer and inner heaters. The two solid curves on the upper-left part (a)–(c) show $L_\mathrm{h}^\infty/L_\mathrm{h0}^\infty$ which are the same for the outer and inner heaters. Any two pairs of curves on panels (a), (b) and (c) show $L_\mathrm{s}^\infty(t)/L_\mathrm{h0}^\infty$ produced either by the outer (dash-dot lines) or the inner (dashed lines) heater. At the same L_h^∞ and Δt, the surface luminosities $L_\mathrm{s}^\infty(t)$ depend on the heater's position; the deeper the heater, the lower $L_\mathrm{s}^\infty(t)$. In all the cases, only a small fraction of the generated heat is emitted through the NS surface ($L_\mathrm{s}^\infty \ll L_\mathrm{h}^\infty$), while the rest is emitted by neutrinos [2].

Now we notice that NSs are likely superfluid inside [9]. We consider only singlet-state SF of free neutrons in the inner crust (at $\rho \gtrsim 4 \times 10^{11}$ g cm^{-3}). Let $T_\mathrm{c}(\rho)$ be the critical temperature

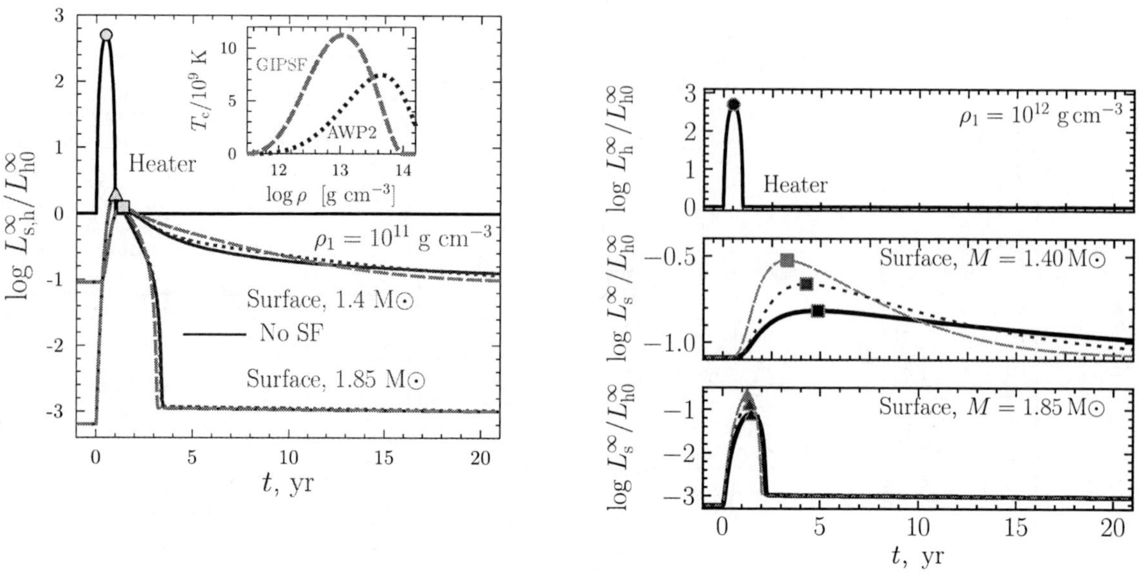

Figure 2. The heater's power $L_{\rm h}^\infty(t)$ ('Heater', the same for the $1.4M_\odot$ and $1.85M_\odot$ NSs) and the surface luminosity $L_{\rm s}^\infty(t)$ ('Surface') for two models [GIPSF – long dashes, AWP2 – dots] of crustal SF as well as for non-superfluid crust (No SF) at $L^\infty_{\rm h,max}/L^\infty_{\rm h0} = 500$ and $\Delta t = 1$ yr. The left and right panels correspond to the outer and inner heaters, respectively. The circles show maxima of $L_{\rm h}^\infty(t)$; the squares and triangles position maxima of $L_{\rm s}^\infty(t)$ for the 1.4 and $1.85M_\odot$ NSs, respectively. *Inset:* Critical temperature $T_{\rm c}(\rho)$ for the two models of crustal SF.

that depends on ρ. Calculations of $T_{\rm c}(\rho)$ give a large scatter of results depending on microphysics of the matter. For example, the inset in figure 2 shows $T_{\rm c}(\rho)$ for two SF models. The 'GIPSF' model [10] is close to basic the BCS model and gives a high and broad peak of $T_{\rm c}(\rho)$. The 'AWP2' model [11] is stronger affected by the in-medium effects; the $T_{\rm c}(\rho)$ peak is lower and shifted to higher ρ.

SF strongly reduces the heat capacity in the inner crust, $C(\rho, T)$ [12]. In the absence of SF ($T > T_{\rm c}$), the main contribution to the heat capacity of the inner crust is provided by free neutrons, but at $T \ll T_{\rm c}$ the neutron heat capacity is suppressed, and the heat capacity is determined by much smaller contribution of ions and electrons.

Figure 2 is similar to figure 1 and compares the variations of $L_{\rm h}^\infty(t)$ and $L_{\rm s}^\infty(t)$ (for the same 1.4 and $1.85M_\odot$ stars) but for the two models of crustal SF (see the inset) in a powerful outburst, $L^\infty_{\rm h,max}/L^\infty_{\rm h0} = 500$ and $\Delta t = 1$ yr. With the suppressed crustal heat capacity, one needs less energy to warm up the heater and its vicinity. Then more energy is directed to heat the surface. Therefore, for the heaters with $\rho_1 > \rho_{\rm drip}$ (the neutron drip density), the $L_{\rm s}^\infty(t)$ peaks are more pronounced in the presence of superfluid neutrons. The initial luminosity $L^\infty_{\rm s0}$ for the $1.85M_\odot$ star with fast neutrino cooling is about two orders of magnitude lower than for the $1.4M_\odot$ star.

3. Deep crustal heating and afterburst relaxation

Here we outline the crust-core relaxation in NSs which enter quasi-persistent LMXBs (six sources known at the moment [13]). Such an NS transiently accretes matter from its low-mass companion during long (months–years) and intense (about the Eddington level) accretion episodes (outbursts). During an outburst, the crustal heater is so strong that it makes the crust hotter than the core, destroying the isothermality of NS interiors.

After the outburst, the surface accretion energy release stops and the crust undergoes gradual (a few years) relaxation with the core. An afterburst (quiescent) stage is characterized by much weaker, but observable surface emission which is likely supported by deep crustal heating. It

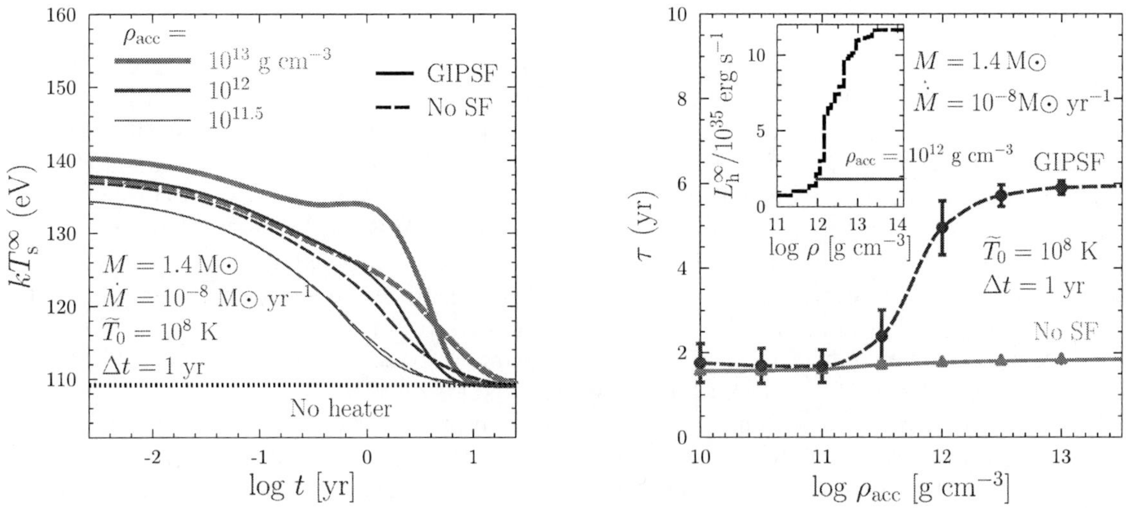

Figure 3. *Left:* Temporal relaxation $T_s^\infty(t)$ after an outburst in the $1.4M_\odot$ NS with the pre-outburst redshifted internal temperature $\widetilde{T}_0 = 10^8$ K; t is time after the *return to quiescence.* Solid curves correspond to $\rho_{acc} = 10^{11.5}$, 10^{12} and 10^{13} g cm^{-3} and to the GIPSF superfluidity. The dashed curves refer to the same ρ_{acc} but without SF. Dotted line – no heater. *Right:* Long-term relaxation time τ versus ρ_{acc} extracted from the theoretical cooling curves. Six filled circles and triangles with their errorbars correspond to a non-superfluid and superfluid crust, respectively. *Inset:* Heat power $L_h^\infty(\rho)$, integrated over a spherical layer from the surface up to a current density ρ, in the crust at $\dot{M} = 10^{-8}$ $M_\odot$ yr^{-1}. The grey horizontal line shows the modification of this curve for the partly accreted crust with $\rho_{acc} = 10^{12}$ g cm^{-3}.

operates in the NS crust during outburts, when the freshly accreted matter compresses the underlying crust. Since the accreted matter is different from the fully equilibrium cold catalyzed matter, the compression triggers nuclear transformations in the accreted matter (e.g., [4, 5, 14]). As a result, the energy is generated deeply in the crust with the rate $Q(\rho, t)$ distributed from the outer NS layer to the maximum density ρ_{acc} occupied by the accreted matter in the crust.

One can consider the deep crustal heating with fully accreted and partly accreted crusts (e.g., [15]). For example, we simulate the long-term cooling of the $1.4M_\odot$ NS in a quasi-persistent LMXB. We start with a hot and passively cooling NS. In about one century, it becomes thermally relaxed inside, with the same redshifted internal temperature $\widetilde{T}(t)$ over the internal region. When $\widetilde{T}(t)$ reaches the value $\widetilde{T}_0 = 10^8$ K, we switch on the distributed heat sources within the accreted crust (the inset in figure 3) assuming a mass accretion rate $\dot{M} = 10^{-8}$ $M_\odot$ yr^{-1}. In $\Delta t = 1$ yr after the outburst onset, we abruptly switch off the deep crustal heating $[L_h^\infty(t) = 0]$ to simulate the long-term cooling in quiescence for the cases of normal (No SF) and highly superfluid (GIPSF) crust.

The left panel of figure 3 shows the cooling curves $T_s^\infty(t)$ (in energy units) for six models of long-term relaxation after the return to quiescence. The three solid lines are calculated for the GIPSF superfluidity assuming the partially accreted crust with $\rho_{acc} = 10^{11.5}, 10^{12}$ and 10^{13} g cm^{-3}. The other three dashed curves are the same but neglecting neutron SF. A thinner accreted crust reduces T_s^∞. SF considerably shortens the relaxation by reducing the heat capacity in the inner crust; this is seen as a faster decay of $T_s^\infty(t)$ at large t.

In first few months of quiescence the relaxation can be affected by poorly known shallow heating (e.g. [13, 16]), which we neglect. In late long-term relaxation this effect is unimportant so that the observable and calculated cooling curves $T_s^\infty(t)$ are well described (e.g., [17, 18]) by a function $T_s^\infty(t) = A\exp(-t/\tau) + B$, where τ, A and B are the fit parameters, with τ being a relaxation time. Taking the sets of our theoretical cooling curves for normal and SF crusts, we infer two branches of $\tau(\rho_{\rm acc})$ plotted on the right panel of figure 3. If $\rho_{\rm acc} \gtrsim 10^{11}$ g cm^{-3}, τ is greatly reduced by SF and independent of $\rho_{\rm acc}$. Our preliminary analysis demonstrates that the reduction agrees with the observations of quasi-persistent LMXBs and gives evidence for crustal SF. The same conclusion has been made earlier from more complicated simulations based on similar microphysics and SF effects; see, e.g., [16, 19–21] and references therein. If $\rho_{\rm acc} \lesssim 10^{11}$ g cm^{-3}, the superfluid and normal cooling curves merge into one curve.

To summarize, we confirm the results of [2] that only a small amount of heat is emitted by photons through the surface. The surface outburst $L_s^\infty(t)$ is delayed, broadened and asymmetric with respect to the internal outburst $L_h^\infty(t)$. The long-term afterburst relaxation in NSs entering quasi-persistent LMXBs depends on crustal SF and on the maximum density $\rho_{\rm acc}$ occupied by the accreted matter. At $\rho_{\rm acc} \gtrsim 10^{11}$ g cm^{-3} superfluidity can greatly shorten the relaxation and amplify $T_s^\infty(t)$.

These results are important for theoretical interpretation of observed thermal evolution of NSs in quasi-persistent LMXBs. They can also be useful for modeling of afterburst relaxation in magnetars where the relaxation can have much in common with that in LMXBs but microphysics input is greatly complicated by strong magnetic fields (e.g. [22]).

Acknowledgments

The work of AK and DY was supported by the Russian Science Foundation (grant 14-12-00316). The authors are indebted to P Shternin for strong criticism.

References

[1] Kaminker A D, Kaurov A A, Potekhin A Y and Yakovlev D G 2014 *MNRAS* **442** 3484–94
[2] Chaikin E A, Kaminker A D, Kaurov A A and Yakovlev D G 2017 *Europhysics Letters* **108** 31002
[3] Brown E F, Bildsten L and Rutledge R E 1998 *Astrophys. J.* **504** L95–8
[4] Haensel P and Zdunik J L 1990 *Astron. Astrophys.* **227** 431–6
[5] Haensel P and Zdunik J L 2008 *Astron. Astrophys.* **480** 459–64
[6] Kaspi V M and Beloborodov A M 2017 *Ann. Rev. Astron. Astrophys.* **55** 261–301
[7] Potekhin A Y, Pons J A and Page D 2015 *Space Sci. Rev.* **191** 239–91
[8] Potekhin A Y, Fantina A F, Chamel N, Pearson J M and Goriely S 2013 *Astron. Astrophys.* **560** A48
[9] Page D, Lattimer J M, Prakash M and Steiner A W 2014 *Novel Superfluids* ed Bennemann K H and Ketterson J B (Oxford Univ. Press, Oxford) p 550
[10] Gandolfi S, Illarionov A Y, Pederiva F, Schmidt K E and Fantoni S 2009 *Phys. Rev. C* **80** 045802
[11] Ainsworth T L, Wambach J and Pines D 1989 *Physics Letters B* **222** 173–8
[12] Yakovlev D G, Kaminker A D, Gnedin O Y and Haensel P 2001 *Phys. Rep.* **354** 1–155
[13] Homan J, Fridriksson J K, Wijnands R, Cackett E M, Degenaar N, Linares M, Lin D and Remillard R A 2014 *Astrophys. J.* **795** 131
[14] Haensel P, Potekhin A Y and Yakovlev D G 2007 *Neutron Stars. 1. Equation of State and Structure* (Springer, New York)
[15] Wijnands R, Degenaar N and Page D 2013 *MNRAS* **432** 2366–77
[16] Brown E F and Cumming A 2009 *Astrophys. J.* **698** 1020–32
[17] Cackett E M, Wijnands R, Linares M, Miller J M, Homan J and Lewin W H G 2006 *MNRAS* **372** 479–88
[18] Degenaar N, Wijnands R, Brown E F, Altamirano D, Cackett E M, Fridriksson J, Homan J, Heinke C O, Miller J M, Pooley D and Sivakoff G R 2013 *Astrophys. J.* **775** 48
[19] Shternin P S, Yakovlev D G, Haensel P and Potekhin A Y 2007 *MNRAS* **382** L43–7
[20] Turlione A, Aguilera D N and Pons J A 2015 *Astron. Astrophys.* **577** A5
[21] Merritt R L et al 2016 *Astrophys. J.* **833** 186
[22] An H, Kaspi V M, Tomsick J A, Cumming A, Bodaghee A, Gotthelf E V and Rahoui F 2012 *Astrophys. J.* **757** 68

IOP Conf. Series: Journal of Physics: Conf. Series **932** (2017) 012048 doi:10.1088/1742-6596/932/1/012048

Looking for Hall attractor in astrophysical sources

S B Popov[1], A P Igoshev[2], R Taverna[3] and R Turolla[3]

[1] Sternberg Astronomical Institute, Lomonosov Moscow State University, Russia
[2] Department of Astronomy/IMAPP Radboud University Nijmigen, Netherlands
[3] Department of Physics and Astronomy, University of Padua, Italy

E-mail: polar@sai.msu.ru

Abstract. Recently, numerical calculations of the magnetic field evolution in neutron stars demonstrated the possible existence of a Hall attractor, a stage at which the evolution of the field driven by the Hall cascade ends. The existence of such a stage in neutron star magnetic evolution is very important, and can be potentially probed by observations. Here we discuss three types of objects which could have reached this stage. First, we briefly describe the evolution of normal radio pulsars with ages about a few hundred thousand years. Then we analyse in more detail observations of RX J1856.5-3754, one of the Magnificent Seven, focusing on the surface temperature distribution and comparing model predictions with the temperature map inferred from X-ray observations. Finally, we discuss the necessity of the Hall attractor stage to explain the hypothetical existence of accreting magnetars. We conclude that at the moment there is no direct confirmation of the Hall attractor stage in known sources. However, more detailed observations in the near future can demonstrate existence (or absence) of this stage of the crustal magnetic field evolution.

1. Introduction

Magnetic field evolution is one of the most important features in the biography of a neutron star (NS). The most natural behavior of the field is to decay, due to the progerssive dissipation of the supporting electric currents (see for a review, see e.g. [1]). However, in some instances the external field can increase due to the re-emergence of the surface field buried by an accretion episode (see recent calculations in [2]).

The evolution of the crustal component of the NS magnetic field occurs on two distinct time-scales: $\tau_{\rm Ohm}$, related to Ohmic resistivity (i.e. to a finite conductivity of the crust) and $\tau_{\rm Hall}$, related to the Hall cascade. The Hall evolution is in principle non-dissipative. Nevertheless, it redistributes the magnetic energy from the larger spatial scales (e.g that of the dipole field) to the smaller ones (those of higher order multipoles) over which Ohmic decay acts faster. Ultimately, this causes the decay of the dipole component and an enhanced release of magnetic energy.

The timescale of the Hall evolution is

$$\tau_{\rm Hall} = \frac{4\pi e n_{\rm e} L^2}{cB(t)}, \qquad (1)$$

with $n_{\rm e}$ the electron density, e the elementary charge, $B(t)$ the magnetic field, L the typical spatial scale of electric currents and c the speed of light [3].

The Hall cascade stops when (and if) the so-called *Hall attractor* is reached. The existence of this stage was proposed by [4, 5] and then independently confirmed by [6] (see also [7], where the authors studied stability of Hall equilibria in NS crust). In [4] it was shown that this stage is reached after a few initial Hall time-scales. Relaxation of the Hall process after early very active phase was also noted before by [8]. Recently, the attractor stage was confirmed by [9] in simulations which include the core magnetic field evolution.

Up to now there are no direct observational evidences in favour of the Hall attractor stage. In this note we discuss three types of sources, and outline three different strategies to identify the presence of the Hall attractor.

2. Hall attractor and normal radio pulsars

The large number of known normal radio pulsars gives an opportunity to probe the magneto-rotational evolution of NSs at least on time scales $\lesssim 10$ Myr (for older sources selection effects become very important). However, till present, population synthesis or/and studies of individual objects of this type did not provide a clear evidence for magnetic field decay for NSs with initial field $\sim 10^{12} - 10^{13}$ G during the radio pulsar phase.

In [10] we applied a modified pulsar current method to study the magnetic field evolution of radio pulsars. It was shown that, over several hundred thousand years, the fields of radio pulsars decay by a factor two with a time-scale $\sim 4 \times 10^5$ yrs. However, for ages $\gtrsim 10^6$ yrs no significant decay is seen. To identify the cause responsible for this decay we plot in figure 1 the electron density $n_{\rm e}$ against the characteristic spatial scale of currents L. The Hall time-scale with the value 4×10^5 yrs corresponds to the black dashed line, see eq (1). In the crust of a real NS each electron density corresponds to a unique spatial scale due to the pressure gradient [3]. This relation is shown with red and blue lines depending on the exact relation between the pressure height gradient H and L (dashed lines corresponds to $L = 3H$), and the electron fraction $Y_{\rm e}$.

Our region of interest is at the intersection of the black dashed line and one of the colored lines. From figure 1 one can see that the Hall timescale $\sim 4 \times 10^5$ yrs corresponds to the case in which electric currents are mostly situated deep in the crust at electron densities about $(0.5 - 2.0) \times 10^{36}$ cm^{-3}, that corresponds to a mass density about $(2.4 - 9.5) \times 10^{13}$ g cm^{-3}. This is a realistic range of densities. However, the onset of the Hall attractor in this case occurs at an age ~ 2 Myrs, which is too large to fit the data. Thus, a fine-tuning of the parameters is required in order to match the predictions of the Hall cascade and attractor model with the findings from [10] (see [11]).

A different possibility to explain time-scales $\sim 4 \times 10^5$ yrs is related to the Ohmic decay due to phonons. The desired value is obtained for quite realistic conditions. Also, this type of decay stops naturally after ~ 1 Myr due to cooling when the NS temperature drops below the critical value, $T_{\rm U}$.

Therefore, we think that the second possibility is more probable. It may be that the Hall effect also contributes to the field decay in normal radio pulsars, but the main contribution, according to our analysis [11], is due to Ohmic decay on phonons. Thus, there are no clear signs of the Hall attractor in the case of normal radio pulsars.

3. Hall attractor and the Magnificent Seven

Modern scenarios of magnetic field evolution with the Hall attractor can make important, and potentially testable, predictions about NS properties. Calculations made under the assumption of axial symmetry by [5, 4] have shown that Hall evolution saturates after a few $\tau_{\rm Hall}$. The field then reaches some stable configuration, and the successive evolution is driven mainly by the relatively slow Ohmic dissipation. The stage when the Hall cascade stops – the Hall attractor – is reached in $\lesssim 1$ Myr for magnetar-like initial fields. This picture was confirmed through 3D numerical simulations by [6], and recently generalized by [9] (however, different time-scales were obtained in this investigation). The Hall cascade moves the crustal field towards the crust-core boundary, where it dissipates [5], so that the field looks more core-centered. According

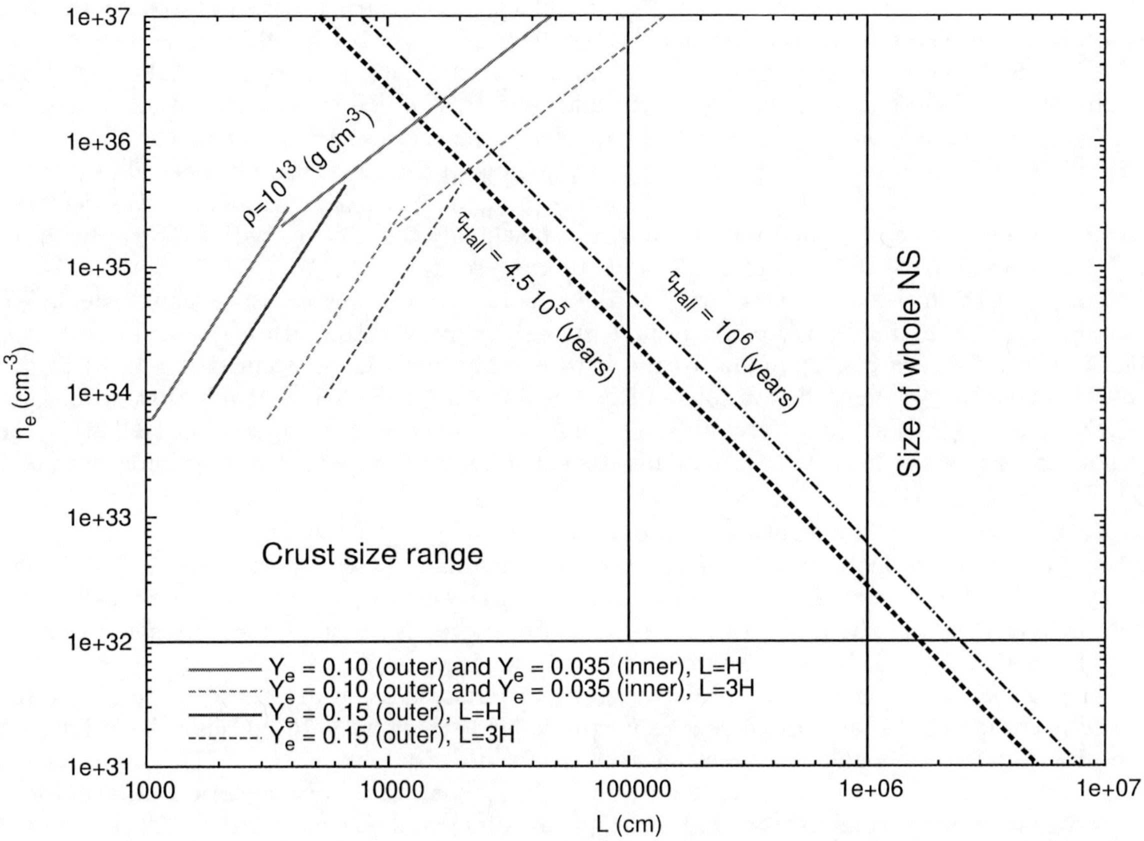

Figure 1. Spatial scale and electron density. Black thick lines show the locus where the Hall timescale is $\tau_{\rm Hall} = 4.5 \times 10^5$ yr (dotted line) and $\tau_{\rm Hall} = 10^6$ years (dot-and-dashed line). The red thick lines correspond to taking the spatial scale equal to the local pressure scale height in the crust. The break corresponds to the density $\rho = 10^{13}$ g cm^{-3}. The thinner dotted red line corresponds to the same relation but when the spatial scale is three times larger than the local pressure scale height.

to [5] the Hall attractor has a well-defined property: when the field structure is stabilized, its poloidal part mainly consists of dipole and octupole components (with a small $l = 5$ multipole in addition). Since the surface thermal distribution in cooling, magnetized INSs is determined by the structure of their crustal field, the analysis of their spectral properties can provide a direct test for the Hall attractor scenario. We used this approach in [12].

In order to compare the predictions on the field evolution [5, 4] with observations, we have to select a sample of INSs which are presumably close enough to the Hall attractor stage. In particular, these sources need to be of the right age ($\lesssim 1$ Myr) and possess initially high fields; moreover, the emission from their cooling surface should be not polluted by strong contributions of non-thermal magnetospheric emission. The most promising candidates belong to a small group of near-by cooling isolated NSs dubbed the Magnificent Seven (M7). They are characterized by stable, purely thermal emission with temperatures $\sim$ 50-100 eV. Typically, their spectra can be fitted by one or two blackbody components with the addition of a broad absorption feature at few hundred eVs. The pulsed fraction is usually low ($\lesssim 20\%$) and their luminosities are in the range $L \sim 10^{31}$–10^{32} erg s^{-1} (see a review in [13]).

Among the Magnificent Seven RX J1856.5-3754 looks particularly promising, since, in the first place, it can be closer to the Hall attractor stage, having a lower blackbody temperature ($\sim$ 50–60 eV), and hence luminosity, than other sources. Furthermore, its dipolar B-field, $\sim 10^{13}$ G, is one of the weakest among the Seven. In addition, this source is very well studied, e.g. in comparison with RX J0420.0-5022, which also seems to be close to the attractor stage. RX J1856.5-3754 is the prototype and the brightest member of the class, and its spectral properties are very well-characterized: the pulsed fraction is the lowest, $\sim 1\%$, no variability was detected so far and the X-ray spectrum is well fitted by two blackbody components with $kT_1^\infty \sim$ 61–62 eV, $R_1^\infty \sim$ 4.5–5 km and $kT_2^\infty \sim$ 39 eV, $R_2^\infty \sim$ 11–16 km (see [14]).

Details of the methods we used are described in [12]. Here we present the main results. We found that, at the Hall attractor stage and fitting the multi-temperature spectrum with two blackbodies, the area corresponding to the hard component is larger than that related to the soft one, contrary to what observations of RX J1856.5-3754 show. We conclude that in the case of RX J1856.5-3754 (and, most probably, also in most or all other M7 sources) the Hall attractor is not reached, or the field structure at this stage is different from what current models predict.

4. Hall attractor and accreting magnetars

Many NSs have been observed in X-ray binaries. In many cases, the magnetic field of the compact object can be measured via cyclotron line observations, or estimated by modeling the spin behavior or/and the properties of the accretion flow. All reliable measurements provide values $\sim 10^{12} - 10^{13}$ G. Indirect estimates result in a wider range.

Among NS X-ray binaries of special interest for our discussion are those containing a so-called accreting magnetar. For these NSs magnetic field strenghts $\sim 10^{14}$ G have been inferred [15, 16, 17, 18], although no direct measurements are available so far. The existence of such systems can potentially contradict the standard scenario for NS magnetic field evolution, according to which magnetar-scale fields are typically dissipated on time-scales $\lesssim 10^6$ yrs, shorter than the inferred age of the neutron star. In [19] we explored different possibilities to form middle-aged magnetars varying the parameters which control the magnetic field evolution.

We constructed a model of the magnetic field evolution which includes the Hall cascade, Ohmic dissipation due to phonons and impurities, and the Hall attractor. Varying the model parameters, we tried to fit the properties of two accreting magnetar candidates: ULX M82 X-2 and 4U 0114+65. For the latter one we also presented a new estimate of its age based on backward integration of its trajectory in the Galactic potential within the Bayesian statistics.

The bottom line of this study is the following: to form an accreting magnetar with an age $\gtrsim$ few Myrs it is necessary to include three main ingredients: (i) the field should have reached the Hall attractor stage; (ii) scatterings on phonons should become negligible after $\sim$few hundred thousand years (i.e. a relatively rapid cooling to the critical temperature when phonons are not important anymore should occur); and (iii) impurities have to be negligible (i.e. the parameter Q which characterizes the role of impurities should assume low values, $\lesssim$ a few). Notice that the latter option (low Q) is considered to be untypical for magnetars, see [20]. In addition, new approaches to accretion physics allow to explain parameters of many proposed magnetar candidates with standard fields $\lesssim$ few 10^{13} G (see [21] and references therein). Multiple contradicting magnetic field estimates prevent us from concluding that the Hall attractor definitely exists in the accreting magnetar candidates.

5. Discussion and conclusions

Typically, when we are dealing with high mass X-ray binaries, we do not know the precise age of the system. This complicates the comparison between model predictions and observations. However, there are several important exceptions, such as the source SXP 1062 observed in a supernova remnant (see recent data and references in [22]). This object has been already proposed as an accreting magnetar candidate [17]. On the other hand, its properties can be explained in the framework of a decaying field model [23], with the present day value of B below the magnetar scale. The existence of different accretion models does not allow at the moment to make a definite estimate of the magnetic field in SXP1062. More direct methods – especially cyclotron line measurements – might be very important in this case.

Up to a very recent time the most popular models of magnetic field evolution in the case of magnetars were based on the properties of crustal fields. However, the magnetic field in the core can play an important role. Recently this question has been studied in [9]. Accounting for the field behavior in the core might modify the standard set of assumptions about field evolution, especially in the case of magnetars. In particular, the time scales of field variations can become larger for some processes. If this is indeed the case, then other samples of sources might be used to test the hypothesis of the Hall attractor, especially for standard field values. Recently, a new approach to describe field evolution in the core has been proposed [24]. Enhanced activity of theoreticians in analysing the behavior of magnetic fields in NS interiors can produce fruitful results, hopefully testable with observations.

Taking altogether, we conclude that different types of astrophysical sources can provide an opportunity to test predictions of magnetic field evolution scenarios and so to probe the existence of the Hall attractor. However, up to now there are no strong arguments in favor of this hypothesis.

Acknowledgements
SBP was supported by the grant from the Russian Science Foundation 14-12-00146.

References

[1] Geppert U 2009 *Astrophysics and Space Science Library* (*Astrophysics and Space Science Library* vol 357) ed Becker W p 319
[2] Igoshev A P, Elfritz J G and Popov S B 2016 *MNRAS* **462** 3689–702
[3] Cumming A, Arras P and Zweibel E 2004 *ApJ* **609** 999–1017
[4] Gourgouliatos K N and Cumming A 2014 *Physical Review Letters* **112** 171101
[5] Gourgouliatos K N and Cumming A 2014 *MNRAS* **438** 1618–29
[6] Wood T S and Hollerbach R 2015 *Physical Review Letters* **114** 191101
[7] Marchant P, Reisenegger A, Alejandro Valdivia J and Hoyos J H 2014 *ApJ* **796** 94
[8] Pons J A, Miralles J A and Geppert U 2009 *A&A* **496** 207–16
[9] Bransgrove A, Levin Y and Beloborodov A 2017 *Preprint* 1709.09167
[10] Igoshev A P and Popov S B 2014 *MNRAS* **444** 1066–76
[11] Igoshev A P and Popov S B 2015 *Astronomische Nachrichten* **336** 831
[12] Popov S B, Taverna R and Turolla R 2017 *MNRAS* **464** 4390–8
[13] Turolla R 2009 *Astrophysics and Space Science Library* (*Astrophysics and Space Science Library* vol 357) ed Becker W p 141
[14] Sartore N, Tiengo A, Mereghetti S, De Luca A, Turolla R and Haberl F 2012 *A&A* **541** A66
[15] Doroshenko V, Santangelo A, Suleimanov V, Kreykenbohm I, Staubert R, Ferrigno C and Klochkov D 2010 *A&A* **515** A10
[16] Reig P, Torrejón J M and Blay P 2012 *MNRAS* **425** 595–604
[17] Fu L and Li X D 2012 *ApJ* **757** 171
[18] Ho W C G, Klus H, Coe M J and Andersson N 2014 *MNRAS* **437** 3664–9
[19] Igoshev A P and Popov S B 2017 *Preprint* 1709.10385
[20] Pons J A, Viganò D and Rea N 2013 *Nature Physics* **9** 431–4
[21] Postnov K A, Shakura N I, Kochetkova A Y and Hjalmarsdotter L 2014 *European Physical Journal Web of Conferences* (*European Physical Journal Web of Conferences* vol 64) p 02002
[22] Gonzalez-Galan A, Oskinova L M, Popov S B, Haberl F, Kuehnel M, Gallagher III J S, Schurch M P E and Guerrero M A 2017 *Preprint* 1706.05002
[23] Popov S B and Turolla R 2012 *MNRAS* **421** L127–31
[24] Gusakov M E, Kantor E M and Ofengeim D D 2017 *Preprint* 1705.00508

IOP Conf. Series: Journal of Physics: Conf. Series **932** (2017) 012049 doi:10.1088/1742-6596/932/1/012049

Cooling status of three neutron stars

D D Ofengeim and D G Yakovlev

Ioffe Institute, 26 Politekhnicheskaya Street, St. Petersburg 194021, Russia

E-mail: ddofengeim@gmail.com

Abstract. We briefly review a simple method to analyse observations of cooling neutron stars (developed in our previous works) and discuss its applications to XMMU J173203.3–244518, RX J1856.5–3754 and the Vela pulsar.

1. Introduction

The cooling theory of neutron stars (NSs), e.g. [1], is important for exploring superdense NS interiors. An NS cooling rate is regulated by the properties of the superdense matter, such as equation of state and baryon superfluidity (SF), as well as by chemical composition of an outer heat blanketing envelope (HBE). Here, we combine the cooling theory of NSs (including, particularly, the results of [2–5]) with the X-ray spectral analysis. Spectra of several NSs have been studied for a wide range of M and R that allows one to constrain M and R. We consider three NSs, XMMU J173203.3–344518 (hereafter XMMU J1732), RX J1856.5–3754 (RX J1856), and PSR B0833–45 (Vela).

2. Analytic description of neutron star cooling

In $\lesssim 100$ yr after an NS birth its interior becomes isothermal except for a thin outer HBE [1]. Then the main cooling equation is $C\mathrm{d}\widetilde{T}/\mathrm{d}t = -L_\nu^\infty - L_\gamma^\infty$, where $\widetilde{T} = T\sqrt{g_{00}}$ is the redshifted internal temperature that is constant over the thermally relaxed stellar interior at any moment t; T is a local temperature, g_{00} is the metric tensor component; C, L_ν^∞, and L_γ^∞ are the total NS heat capacity, neutrino and photon luminosities, respectively. The superscript '∞' denotes quantities in the restframe of a distant observer.

In a thermally relaxed NS, the main contribution into C and L_ν^∞ comes from the NS core, with $C(\widetilde{T}) \propto \widetilde{T}$ and $L_\nu^\infty(\widetilde{T}) \propto \widetilde{T}^n$; typically, n is either 6 or 8. We have approximated [4] these quantities as $L_\nu^\infty = \Lambda(M,R)\widetilde{T}^n$ and $C = \Sigma(M,R)\widetilde{T}$, where Λ and Σ are analytic fit functions of M and R. They appear to be the same for a wide class of nucleonic equations of state in NS cores [4]. L_ν^∞ is the sum of contributions from several neutrino mechanisms in the core, and C is the sum of heat capacities of different particle species (n, p, e or μ). All these partial contributions have been calculated (neglecting SF) and fitted by the expressions which have similar structure [4]. Below we will mainly consider the case in which the main contribution to L_ν^∞ comes either from the modified Urca process ($\Lambda = \Lambda_{\mathrm{MU}}$) or from neutron-neutron neutrino bremsstrahlung ($\Lambda = \Lambda_{\mathrm{nn}}$). The former case is for non-superfluid NSs (the standard cooling), and the latter is for strong proton superfluidity but normal neutrons (the slowest neutrino cooling [6]). In both cases $n = 8$. In what follows, we need C for the same two cases. The heat capacity

IOP Conf. Series: Journal of Physics: Conf. Series **932** (2017) 012049 doi:10.1088/1742-6596/932/1/012049

of a non-superfluid NS contains the contributions of all particles in the core ($\Sigma = \Sigma_{\mathrm{npe}\mu}$); strong proton SF fully suppresses the contribution of protons ($\Sigma = \Sigma_{\mathrm{ne}\mu}$).

The photon luminosity $L_\gamma^\infty = 4\pi\sigma R_\infty^2 (T_\mathrm{s}^\infty)^4$ (σ being the Stefan-Boltzmann constant and $R_\infty = R/\sqrt{1-2GM/(Rc^2)}$) can be determined using a relation between the surface temperature T_s^∞ and $\widetilde{T}$ from the theory of HBEs [7, 8]. If HBE consists of one ion species and the effects of magnetic fields are neglected, $(T_\mathrm{s}^\infty)^4 \propto \widetilde{T}^\alpha$, where $\alpha \sim 2$ slightly depends on the HBE composition. For onion-like accreted HBEs and/or magnetized HBEs, this scaling is modified [7].

Assuming $L_\gamma^\infty \propto \widetilde{T}^\alpha$, the cooling equation is approximately solved analytically [5],

$$\widetilde{T}(t) \approx (s/q)^{1/(n-\alpha)} \left[\left((\alpha-2) s^k q^{-\gamma} t + 1 \right)^{k/\gamma} - 1 \right]^{-1/(n-2)}, \qquad (1)$$

where $k = (n-2)/(n-\alpha)$, $\gamma = (\alpha-2)/(n-\alpha)$; $q = \Lambda/\Sigma$ and $s = L_\gamma^\infty/(\Sigma\widetilde{T}^\alpha)$ are the functions of M and R. For the iron HBE, $\alpha \approx 2.2$, and the maximum relative fit error of (1) is $\delta_\mathrm{max} \lesssim 7\%$. For the carbon carbon HBE, $\alpha \approx 2.3$ and $\delta_\mathrm{max} \lesssim 11\%$. At the neutrino cooling stage (1) reduces to a well-known formula $\widetilde{T} \approx [(n-2)qt]^{-1/(n-2)}$.

The effects of proton or neutron SFs in the core can be described by phenomenological cooling factors (e.g., [5]) $f_\ell = q/q_\mathrm{SC}$ and $f_C = C/C_{\mathrm{npe}\mu} = s_\mathrm{SC}/s$. Here, q_SC and s_SC refer to the case of standard cooling. Strictly speaking, the cooling factors are time-dependent. The factor f_C is mainly affected by triplet-state SF of neutrons and can vary from $\lesssim 2$ (if $\widetilde{T}$ is just below the maximum critical temperature for neutron SF) to $\Sigma_{\mathrm{e}\mu}/\Sigma_{\mathrm{npe}\mu} \sim 0.1$ (if neutron and proton SFs are strong). In the absence of neutron SF, proton SF affects f_C much weaker, by a factor from ~ 0.75 to ~ 1.25. The factor f_ℓ can vary from $\Lambda_\mathrm{nn}/\Lambda_\mathrm{MU} \sim 0.01$ for strong proton SF and normal neutrons to ~ 100 if triplet-state SF of neutrons strongly enhances L_ν^∞. In what follows we will treat f_ℓ and f_C as constants.

If strong proton SF greatly suppresses the neutrino cooling, it is convenient to introduce another cooling factor $f_{\ell p}$ [3] as $q = f_{\ell p} q_\mathrm{SC} + \Lambda_\mathrm{nn}/\Sigma_{\mathrm{ne}\mu}$. It is better to use $f_{\ell p}$ instead of f_ℓ at $f_{\ell p} \ll 1$ but it is simple co calculate f_ℓ if $f_{\ell p}$ is known.

Then we can take $\widetilde{T}$ from (1) or its simplification at the neutrino cooling stage, use the appropriate q, s and $T_\mathrm{s} - \widetilde{T}$ relation to obtain a star's cooling curve $T_\mathrm{s}^\infty(t)$.

3. Cooling status of RX J1856, XMMU J1732, and Vela

Detailed analysis of XMMU J1732 is given in [3]; the spectral and cooling analyses of RX J1856 are described in [9] and [5], respectively. The previous spectral analysis of Vela was performed [10, 11] at $M = 1.4\,M_\odot$ and $R = 10$ km; we have reanalyzed (Zyuzin & Ofengeim, in prep.) these results taking the Vela's X-ray spectrum from archival *Chandra* observations (ObsID 127, 131 and 1852). We have employed a magnetic hydrogen (`nsmax`) atmosphere model from the `Sherpa` package for a wide ranges of M and R at the distance $D = 290$ kpc [12]. However there are indications [11] that the Vela' surface has a small hot region. It cannot be properly described by the atmosphere model we use and the problem deserves further studies.

The derived confidence levels on the $M - R$ plane for Vela (figure 1a) show that the preferable values of Vela's M and R lie far from the values assumed in [10]. By comparing these confidence contours with the constraints from [13], we conclude that Vela is likely to be a massive NS.

Figure 1b shows $T_\mathrm{s}^\infty(t)$ for all three NSs compared to the standard cooling scenario (black line). XMMU J1732 is much hotter than for the standard cooling, while Vela and RX J1856 are much colder. Other colored curves give examples of possible explanations of the data. Most suitable atmosphere models allow XMMU J1732 to have a carbon HBE and allow RX J1856 to have an iron HBE (but hydrogen atmosphere). Thus the red cooling curve cannot explain

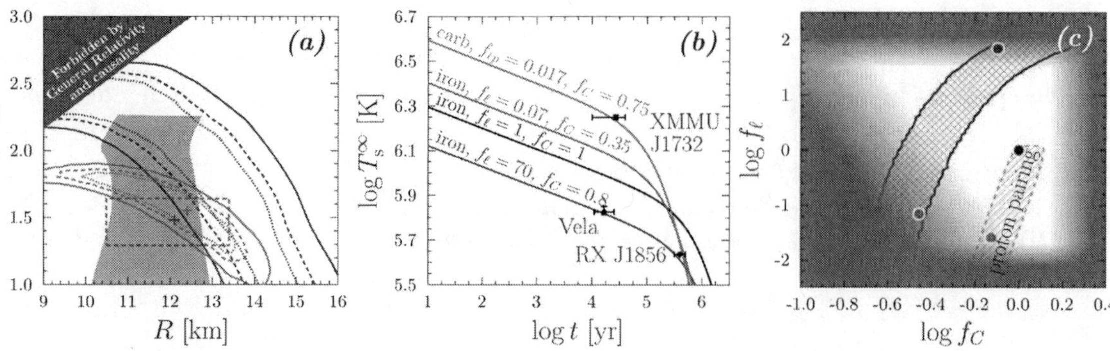

Figure 1. *(a)* $M - R$ confidence levels for the Vela (blue), XMMU J1732 (red), and RX J1856 (1σ, green). Solid, dashed and dotted lines are for the 90%, 68%, and 50% confidence levels, respectively. Crosses mark the best-fits. The greyed shaded area is the area of M and R that was claimed to be the 90% confidence area in [13]. *(b)* Cooling curves $T_s^\infty(t)$; the black line shows the standard candle cooling with the iron HBE; the red curve shows how to hit the XMMU J1732 error box with fully carbon HBE and strong proton SF in the core; the blue and green curves illustrate different possibilities to hit the error box for RX J1856 assuming the iron HBE (the green curve is for the slow neutrino emission and the blue one is for the enhanced one). The blue curve also hits the Vela's error box. Corresponding values of f_ℓ or $f_{\ell p}$ and f_C are presented next to the curves. *(c)* $f_\ell - f_C$ diagram. The colored points correspond to the cooling curves plotted in the middle panel. The singly hatched area is a schematic $(f_\ell - f_C)$ domain for NSs with normal neutrons and partially SF protons. The double-hatched strip is the restriction on RX J1856 cooling factors from the observational ranges of T_s^∞ and t.

RX J1856 cooling though it hits its error box. Vela is too cold for its age for having a large fraction of accreted matter in its HBE; in addition, it should have enhanced neutrino emission.

Each cooling scenario with constant cooling factors is equivalent to a point on the $f_\ell - f_C$ diagram (figure 1c). Particularly, each curve on the middle panel is represented by a point of corresponding color. Note that the magenta point that refers to XMMU J1732 lies in the (f_ℓ, f_C)-domain which is determined by proton SF without neutron SF. Grey-shaded regions with smoothed boundaries are forbidden on some theoretical grounds [5].

We have no access to the results of the RX J1856 spectral analysis. Thus we have analysed its cooling in a simple way. We have determined all those (f_ℓ, f_C) values which allow us to hit the error box on the $T_s^\infty - t$ plane, adopting the best-fit M and R (figure 1c, a double-hatched strip) [5]. It indicates that RX J1856 has neutron SF in the core.

The two other NSs are at the neutrino cooling stage ($L_\nu^\infty \gg L_\gamma^\infty$). Their cooling is independent of f_C that simplifies our study. XMMU J1732 is a slowly cooling NS with strong proton SF and normal neutrons. It is conveniently described by the factor $f_{\ell p}$. For each pair of M and R, we derive T_s^∞ from spectral analysis. Then by employing a carbon HBE model [8] with fixed amount of accreted carbon, one can derive $\widetilde{T}$ and $f_{\ell p}$. Such a plot is given in figure 2. The mass ΔM of accreted carbon can be parameterized by a maximum density ρ_C near the bottom of the carbon layer. For instance, figure 2 employs $\rho_C = 3 \times 10^9$ g cm^{-3}. Since proton SF vanishes at high densities, we take $f_{\ell p} \geqslant 1/60$ [3]. Thus, the outer region of the curve $f_{\ell p} = 1/60$ on figure 2 becomes forbidden. In addition, one has $\rho_C \lesssim 3 \times 10^9$ g cm^{-3} (e.g. [3]). Therefore, the inner area of the $f_{\ell p} = 1/60$ curve on figure 2 is the widest and possible $M - R$ values are located at the intersection of this domain and the confidence levels. Thus the cooling theory sets additional constraints on M and R of XMMU J1732.

Similar analysis has been done for Vela to constrain f_ℓ using the 'onion' model of the HBE [7] with the accreted mass ΔM. Figure 3 shows our results for $\Delta M = 10^{-14}\ M_\odot$. Since Cooper pairing neutrino emission can provide $f_\ell < 100$ only, the inner region of the thick black curve on

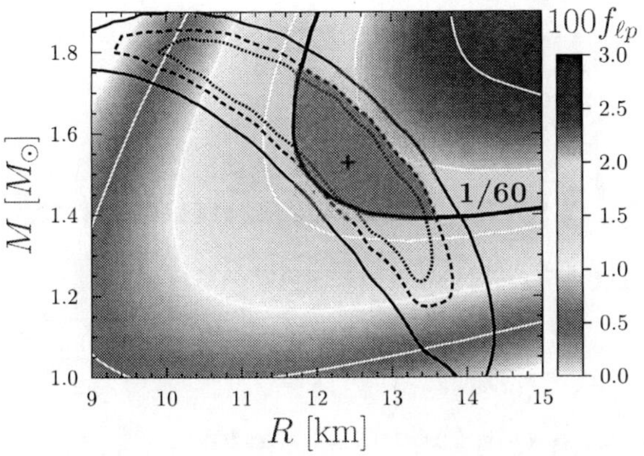

Figure 2. The map of $f_{\ell p}$ over the $M-R$ plane for XMMU J1732 with $\rho_C = 3\times10^9$ g cm^{-3}. Thin black lines are the same confidence contours as the red ones on figure 1a. Thick black line shows $f_{\ell p} = 1/60$ that we treat as the minimum value. The intersection of the 68% confidence region and the $f_{\ell p} > 1/60$ domain is grey-shaded. White lines show levels of constant $f_{\ell p} = 0.005, 0.010, \ldots, 0.025$.

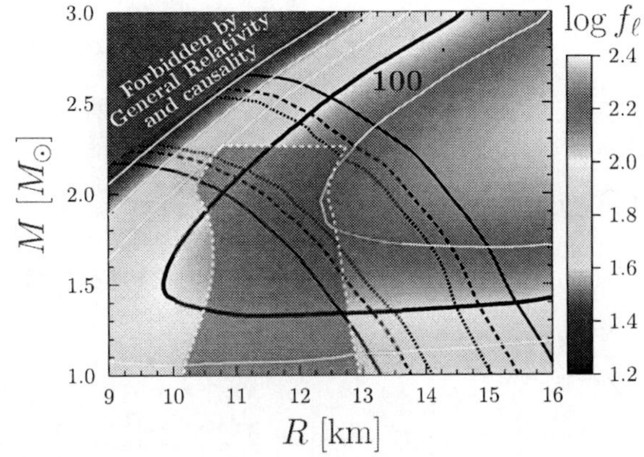

Figure 3. The map f_ℓ over the $M-R$ plane for the Vela pulsar with $\Delta M = 10^{-14}\ M_\odot$. Thin black lines are the same confidence contours as the blue ones on figure 1a. Thick black lines show $f_\ell = 100$ which we treat as the maximum value. The grey-shaded area in the white-dashed contour is the $M-R$ area which was considered as the 90% confidence level in [13]. White lines show levels of constant $\log f_\ell = 1.4, 1.6, \ldots, 2.2$.

figure 3 is prohibited. If ΔM increases, the forbidden area increases too; $\Delta M = 3\times10^{-13}\ M_\odot$ can prohibit all intersections of the Vela confidence contours and the constraints from [13]. Thus, ΔM in Vela's HBE does not exceed this value.

4. Conclusions

We have outlined a method to analyse cooling NSs. The method is based on simple analytic approximations valid for a wide class of nucleonic equations of state. Introducing the phenomenological cooling factors, we have made this analysis free of any specific SF model. The method has been applied to the three cooling NSs, XMMU J1732, RX J1856, and the Vela pulsar. The Vela's spectrum has been reanalysed.

Acknowledgments

Authors are indebted to Dmitry Zyuzin who performed the Vela's spectral analysis.We thank the anonymous referee for useful comments on the Vela's spectrum interpretation. The work was supported by the Russian Science Foundation (grant 14-12-00316).

References

[1] Potekhin A Y, Pons J A and Page D 2015 *Sp. Sci. Rev.* **191** 239–91
[2] Yakovlev D G, Ho W C G, Shternin P S, Heinke C O and Potekhin A Y 2011 *MNRAS* **411** 1977–88
[3] Ofengeim D D, Kaminker A D, Klochkov D, Suleimanov V and Yakovlev D G 2015 *MNRAS* **454** 2668–76
[4] Ofengeim D D, Fortin M, Haensel P, Yakovlev D G and Zdunik J L 2017 *Phys. Rev. D* **96** 043002
[5] Ofengeim D D and Yakovlev D G 2017 *MNRAS* **467** 3598–603
[6] Klochkov D, Suleimanov V, Pühlhofer G, Yakovlev D G, Santangelo A and Werner K 2015 *A&A* **573** A53
[7] Potekhin A Y, Yakovlev D G, Chabrier G and Gnedin O Y 2003 *ApJ* **594** 404–18
[8] Beznogov M V, Potekhin A Y and Yakovlev D G 2016 *MNRAS* **459** 1569–79
[9] Ho W C G, Kaplan D L, Chang P, van Adelsberg M and Potekhin A Y 2007 *MNRAS* **375** 821–30
[10] Pavlov G G, Zavlin V E, Sanwal D, Burwitz V and Garmire G P 2001 *ApJL* **552** L129–33
[11] Manzali A, De Luca A and Caraveo P A 2007 *ApJ* **669** 570
[12] Dodson R, Legge D, Reynolds J E and McCulloch P M 2003 *ApJ* **596** 1137–41
[13] Nättilä J, Steiner A W, Kajava J J E, Suleimanov V F and Poutanen J 2016 *A&A* **591** A25

IOP Conf. Series: Journal of Physics: Conf. Series **932** (2017) 012050 doi:10.1088/1742-6596/932/1/012050

Toward understanding the physical underpinnings of spatial and spectral morphologies of pulsar wind nebulae

O Kargaltsev[1], N Klingler[1], S Chastain[1] and G G Pavlov[2]

[1] Department of Physics, The George Washington University, Washington, DC 20052, USA

[2] Department of Astronomy & Astrophysics, Pennsylvania State University, University Park, PA 16802, USA

E-mail: kargaltsev@gwu.edu

Abstract. We discuss the observational properties of pulsar wind nebulae (PWNe) linking them to the injected (at the termination shock) electron spectral energy distribution and parameters of pulsar magnetospheres. In particular, we (1) present spatially-resolved *Chandra* ACIS spectral maps of twelve PWNe and measure the slopes of the uncooled PWN spectra just downstream of the termination shock obtained from these maps, (2) consider the connections between PWN morphologies and predictions of the magnetospheric emission models and (3) discuss the limits on the maximum energies of particles in PWNe from X-ray and γ-ray observations.

1. Introduction

Pulsars are among nature's most powerful particle accelerators, capable of producing and accelerating particles up to PeV energies. For rotation-powered pulsars (RPPs), most of the neutron star's (NS's) rotational energy is imparted into a magnetized relativistic pulsar wind (PW) which emits nonthermal radiation from radio to TeV γ-rays, forming a pulsar wind nebula (PWN). The particle wind is ultra-relativistic immediately beyond the pulsar magnetosphere (light cylinder), but the interaction with the surrounding medium causes the bulk flow to slow down abruptly at a termination shock (TS). The shocked magnetized PW is confined between the termination shock (TS) and contact discontinuity (CD). It produces synchrotron radiation observed from radio to GeV γ-rays and inverse Compton radiation observed from GeV to TeV energies (see, e.g., reviews [1, 2, 3, 4]).

The morphologies and spectra of PWNe depend on the anisotropy of the wind, its magnetization, particle acceleration efficiency, magnetic field strength and pulsar velocity. While most pulsars have reliably measured spin-down energy loss rates ($\dot{E}$), it is much more difficult to determine the other parameters responsible for the diverse morphologies, radiative efficiencies and spectra of PWNe. It is natural to assume that the angle ζ between the pulsar's spin axis and the observer's line of sight and the inclination angle α between the spin and magnetic dipole axes (if the field is indeed largely dipolar) would leave an imprint on the PWN properties. However, these angles are notoriously difficult to measure. Constraints on these angles can be obtained by comparing theoretical models of pulsar magnetospheric emission with the *Fermi* LAT GeV light curves and radio light curves (e.g., [5]). However, different magnetosphere models can lead to differing predictions. The PWN spectra are also affected by cooling, which is expected to become progressively more important with increasing distance from the pulsar. Therefore, to determine the particle spectral energy distribution (SED) injected at the TS, one must obtain spatially-resolved PWN spectra.

For a stationary or slow-moving pulsar, the CD (and therefore the PWN) will expand as more energy and particles are injected by the pulsar, until the radiative and adiabatic expansion losses balance the rate of energy input. For an isotropic PW, one would expect the CD/PWN to have a spherical shape; however, no spherical PWNe are observed which shows that pulsar winds are intrinsically anisotropic. Many PWNe exhibit prominent equatorial and polar components (e.g., tori and jets) which reflect symmetry with respect to the pulsar spin axis (see, e.g., the Crab and Vela PWNe in figure 1). In such cases it becomes possible to measure the angle between the line of sight and the pulsar's spin axis and compare it with the predictions of magnetosphere emission models. Some PWNe display nearly-axial symmetry, but identifying the equatorial and polar outflows can be challenging (e.g., G21.5–0.9 and MSH 11–62 in figure 1). The anisotropic winds can lead to anisotropic PWN spectra, flow speeds, magnetic fields, and cooling trends, requiring high-resolution imaging, spatially-resolved spectroscopy, and spatially-dependent models accounting for the wind anisotropy.

It is even more challenging to comprehend the effects of the magnetic inclination α on PWN parameters. One can hypothesize that lower values of α would result in a highly magnetized wind with a stronger polar outflow (compared to the equatorial torus) and also perhaps a reduced particle acceleration efficiency and lower PWN luminosities (see modeling by [6]).

The "natal kicks" that pulsars receive in supernova explosions result in large pulsar velocities with the average 3D velocity $v_{\rm psr} \sim 400$ km s^{-1} [7]. This implies that most pulsars remain inside their host SNRs only for a few tens of kyrs; after that they travel through the interstellar medium (ISM) where the speed of sound $c_{\rm ISM} \ll v_{\rm psr}$ ($c_{\rm ISM} \sim 3-30$ km s^{-1}, depending on the ISM phase). In this environment, the pulsar motion becomes supersonic (Mach number $\mathcal{M} \equiv v_{\rm psr}/c_{\rm ISM} > 1$), and the ambient medium ram pressure strongly modifies the PWN appearance leading to formation of extended pulsar tails (see [3, 8] for reviews on supersonic PWNe). Although in supersonic PWNe (SPWNe) the structures formed by the anisotropic post-TS wind (e.g., jets and tori) are deformed by the ram pressure, in some cases the torus-jet structure can still be identified in high-resolution images (see, e.g., J1509–5850, Geminga, B1706–44, and B0355+54 in figure 2 of [8]).

Great progress in measuring the physical properties of PWNe has been made with the Advanced CCD Imaging Spectrometer (ACIS) onboard the *Chandra X-ray Observatory* (*CXO*), whose unprecedented angular resolution combined with the low ACIS background makes it possible to resolve (spatially and spectrally) the complex anisotropic structures of PWNe. On smaller scales (∼0.01–0.1 pc), structures such as bow shocks, wisps, knots, rings, jets, arcs and combinations thereof have been observed (see figures 1–3). On larger scales (∼0.1–10 pc), structures such as tori and long diffuse tails directed opposite the pulsar's motion are often seen, and in a few cases (associated with particularly fast-moving pulsars), puzzling linear structures strongly misaligned with the pulsar's direction of motion have been discovered (see, e.g., figures 2 and 9 in [8]). Many PWNe include various combinations of the above structures but sometimes can have amorphous morphologies.

In this paper we focus on investigating (1) the slopes of the particle SEDs injected at the TS using adaptively binned spectral maps, (2) the correspondences between the angles α and ζ inferred from PWN morphologies with those predicted by magnetospheric emission models and the possible dependencies of PWN parameters on these angles and (3) the constraints on maximum PW particle energies.

IOP Conf. Series: Journal of Physics: Conf. Series **932** (2017) 012050 doi:10.1088/1742-6596/932/1/012050

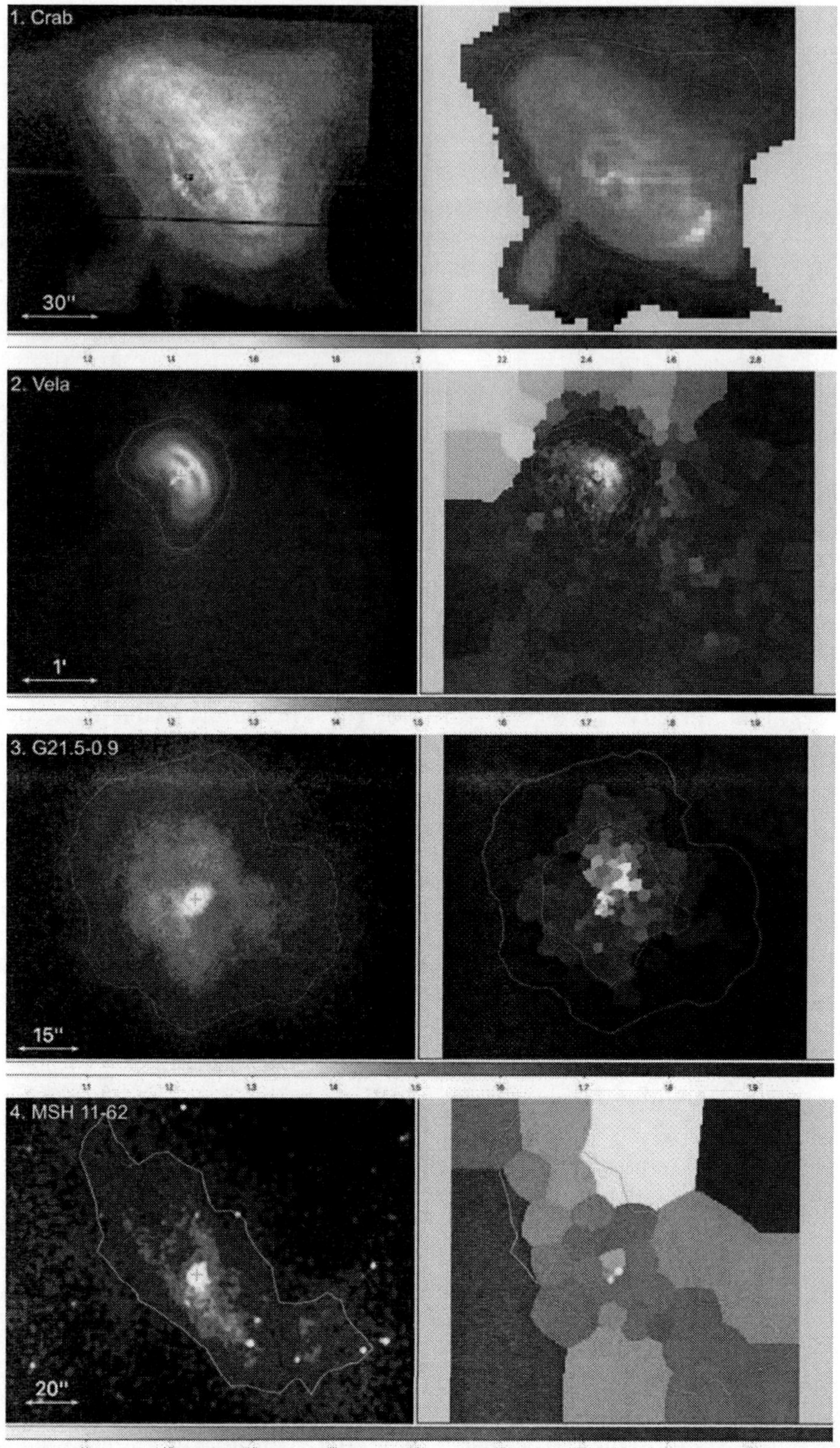

Figure 1. *Left panels:* *CXO* ACIS images (in 0.5–8 keV) of bright, well-resolved PWNe from table 1. *Right panels:* Adaptively-binned spatially-resolved spectral maps for the PWNe shown on the left. The color bars represent the photon index Γ measured in the 0.5–8 keV band. The green contours are shown for illustrative purposes, and the green crosses mark the pulsar positions. The gray-colored areas in panels 1 and 2 were excluded from mapping. The Crab spectral map is adopted from [9].

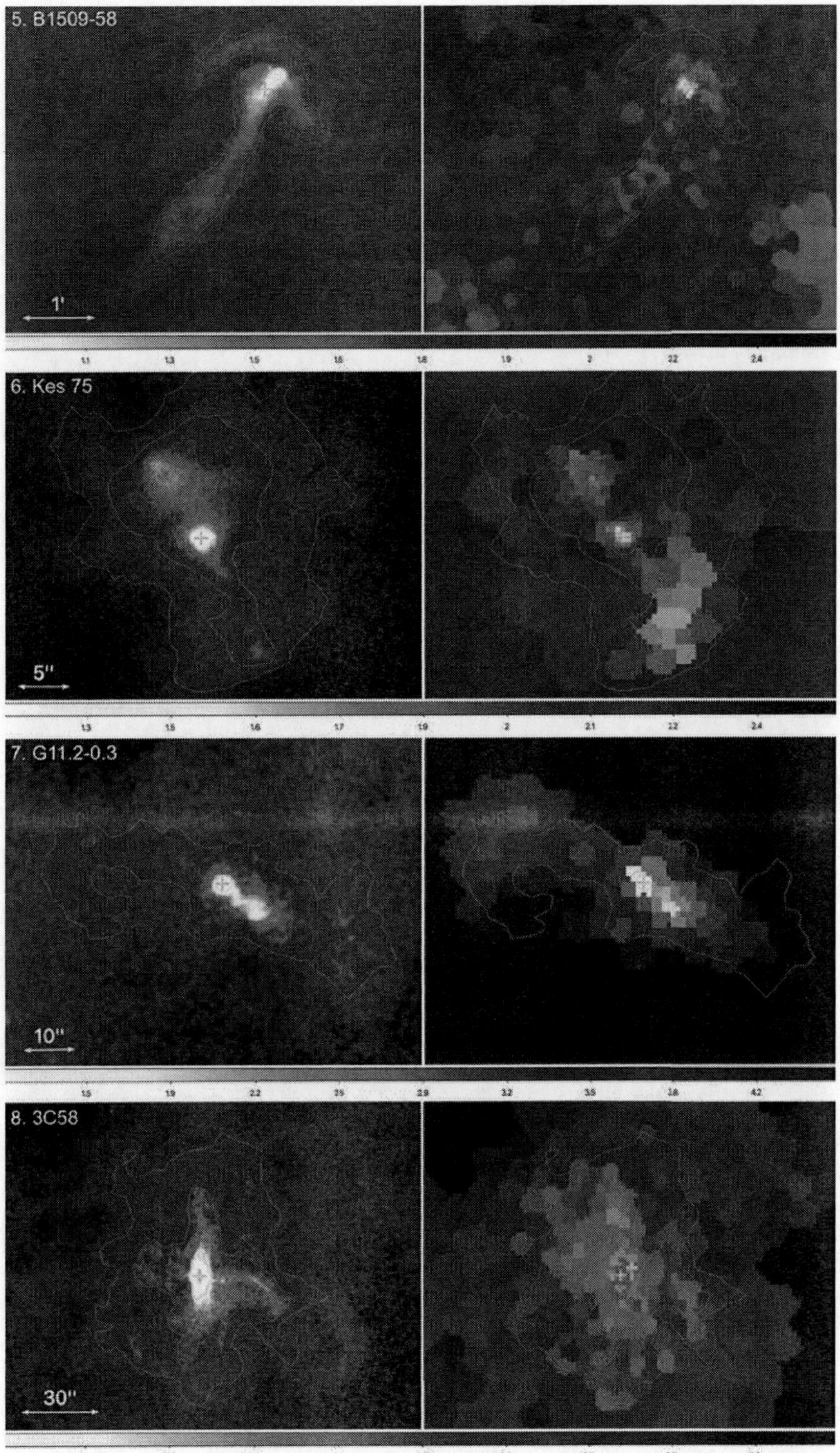

Figure 2. ACIS images and spectral maps of bright, well-resolved PWNe (continued from figure 1).

IOP Conf. Series: Journal of Physics: Conf. Series **932** (2017) 012050 doi:10.1088/1742-6596/932/1/012050

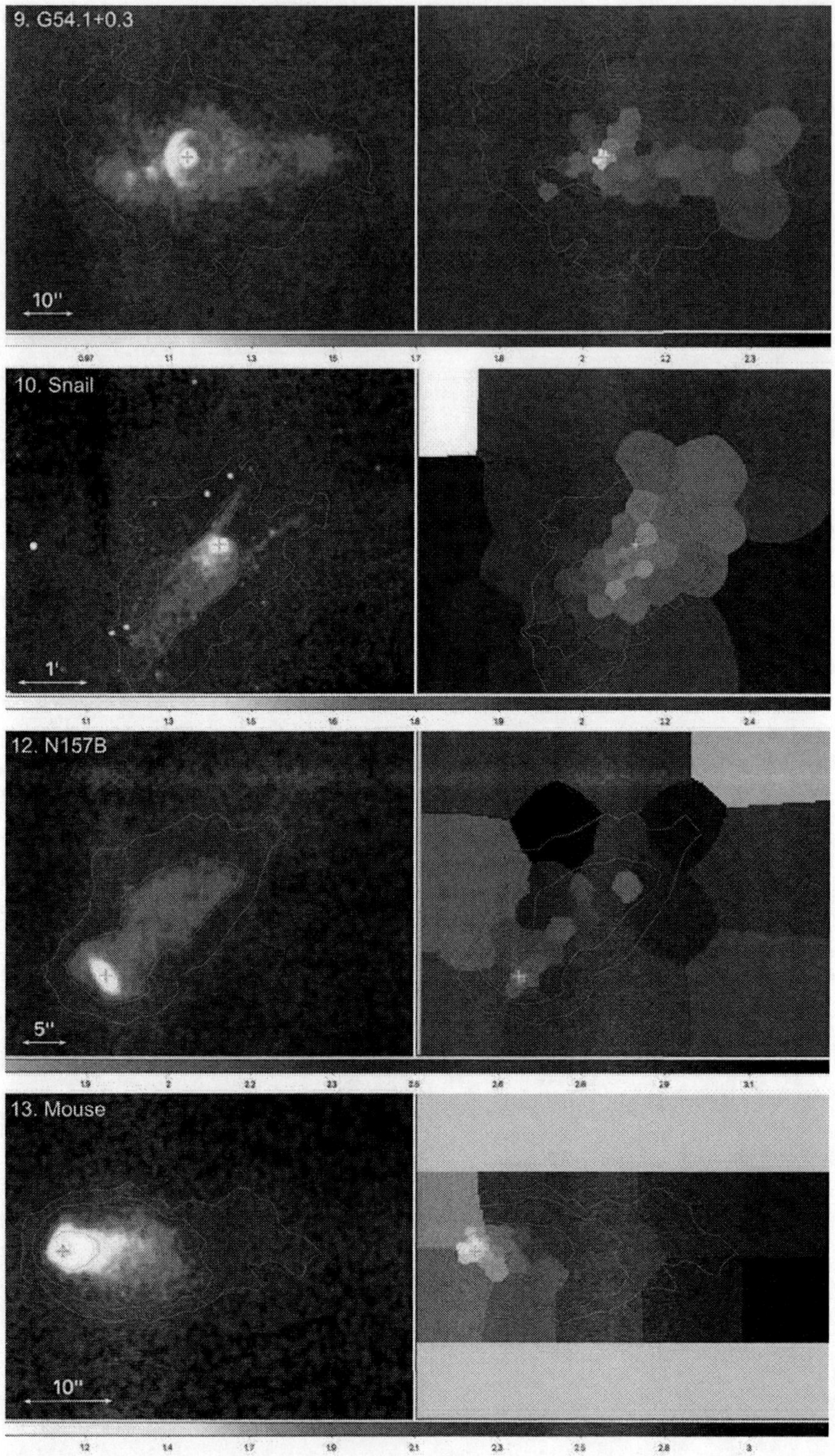

Figure 3. ACIS images and spectral maps of bright, well-resolved PWNe (continued from figures 1 and 2).

IOP Conf. Series: Journal of Physics: Conf. Series **932** (2017) 012050 doi:10.1088/1742-6596/932/1/012050

Table 1. *Chandra* ACIS observations of bright X-ray PWNe suitable for spatially resolved spectroscopy. Column 3 lists the on-axis exposure times of imaging ACIS observations (the number of separate observations is in parentheses). Column 4 gives the total observed (absorbed) X-ray flux (0.5–8 keV). Column 5 lists prominent features found in the PWN. The first group of PWNe (above the horizontal line) is located inside SNRs, the second group is supersonic PWNe. † – the scientific exposure for the Crab PWN is much smaller (107 ks).

Number	Name	Exposure (ks)	F_X (10^{-11} cgs)	Features
1	Crab	565† (57)	2000	ring, torus, jets
2	Vela	630 (22)	6.1	arcs, jets
3	G21.5–0.9 (J1833-1034)	738 (82)	3.8	tori, jets
4	MSH 11–62	472 (9)	3	torus?, 2 arcs(?)
5	G320.4–1.2 (B1509–58)	331 (10)	5.3	arcs, jet, tail(?)
6	Kes 75 (J1846-0258)	383 (8)	1.5	jets, torus
7	G11.2–0.3 (J1811-1925)	480 (12)	0.6	jets, torus(?)
8	3C58 (J0205+6449)	392 (4)	0.7	jets, loops, torus
9	G54.1+0.3 (PSR J1930+1852)	326 (5)	0.5	torus, jets
10	Snail/G327.1–1.1	386 (4)	0.3	tail, prongs
11	MSH 15–56 (G326.3–1.8)	56 (1)	0.83	amorphous morphology
12	N157B (J0537–6910)	48 (1)	0.6	torus, tail/trail
13	Mouse (J1747–2958)	154 (5)	0.7	equatorial outflow, tail
14	Lighthouse (J1101–6101)	302 (7)	0.2	misaligned outflow
15	Geminga	680 (14)	0.075	equatorial & lateral outflows
16	CTB 80 (B1951+32)	85 (1)	0.6	torus, tail(?)
17	J1509–5850	413 (5)	0.05	bow shock, jets, tail
18	Mushroom (B0355+54)	461 (9)	0.03	equatorial outflow, jets, tail
19	J1741–2054	331 (7)	0.03	bow shock, tail
20	J0357+3205	134 (4)	0.04	(disconnected) tail
21	Guitar (B2224+65)	195 (6)	0.006	bow shock, misaligned outflow

2. Spatially-resolved X-ray Spectra of PWNe

Over the last 18 years of its operation, the *CXO* has accumulated a substantial amount of ACIS data on bright, well-resolved PWNe with different morphologies. At present, there are over 20 sufficiently bright PWNe with deep ACIS exposures (see table 1), suitable for detailed spatially-resolved spectroscopy. For the brightest of these PWNe (shown in figures 1–3), this can be done by producing adaptively-binned spectral maps with spatial resolutions ranging from a few arc seconds to about an arc minute.

A number of previous studies have investigated the spectral properties of PWNe observed by the *CXO* ([10] and references therein). However, even for relatively bright, extended PWNe, most authors report spatially-averaged spectral indices from differently defined regions (e.g., based on morphology, brightness contours, number of counts, etc). This often does not allow one to cleanly separate the effects of varying injection spectra (measurable just downstream the TS) from cooling trends or to look for azimuthal spectral dependences associated with the polar and equatorial outflows. The lack of uniformity also complicates the comparison between different PWNe. An example of complications arising from the measurements of spatially-averaged spectra can be seen in the analysis of deep observations of the Mouse PWN (Klingler et al., in prep.). The high-S/N spectrum of the fairly compact (50″-long) X-ray tail can be fit well by a power-law (PL) model with photon index $\Gamma = 2.09 \pm 0.03$. Yet, our spatially-resolved analysis shows that the spectral slope varies dramatically with distance from the pulsar along the same region, with Γ increasing from $1.65{\pm}0.06$ to $3.0{\pm}0.1$. Thus, to study the dynamics of pulsar winds and the particle acceleration mechanisms, one needs to create accurate, high-resolution spectral maps following a uniform approach.

An adaptive binning methods using weighted Voronoi tessellations (WVTs) were developed in [11, 12] to analyze galaxy cluster emission. We adopted the technique to PWNe and incorporated a method to utilize an arbitrary number of ACIS observations by applying observation-specific calibration corrections, calculating region- and observation-specific detector responses, and performing joint fitting of spectra extracted from the individual observations for each spatial bin of the spectral map.

These spectral maps can be used to (1) measure the slope of the injected electron SED, (2) characterize and visualize synchrotron cooling trends[1] which vary among PWNe and, possibly, among different features of the same PWN (e.g., jets vs. tori), (3) constrain the flow velocities, magnetic fields and particle diffusion, (4) search for in-situ acceleration sites located downstream of the termination shock (e.g., due to magnetic turbulence and reconnection), (5) understand the origin of the PWN features (e.g., puzzling asymmetric structures seen in several PWNe associated with fast-moving pulsars). Recently, Porth et al. [13] presented calculations of spatially-resolved, radial (i.e., azimuthally-averaged) spectra which take into account particle cooling, advection and diffusion using 3D MHD. The results suggest that the competition between the advection and diffusion can result in different spectral properties of PWNe with different magnetic fields and ages. The latitudinal dependence of the diffusion transport can cause anisotropy in the particle SED within the same PWN (e.g., toroidal vs. polar directions) which has not yet been investigated. Once synthetic spatially-resolved spectra are calculated for 3D models, they can be compared to the spectral maps to obtain further constraints on the diffusion coefficients, flow speeds and magnetic field strengths of PWNe.

The spectral maps shown in figures 1–3 (right panels) reveal varying injection spectra (see table 2), drastically different cooling trends, and other interesting features. For example, the Vela PWN map offers two surprises. On a large scale, there appears to be an enveloping region of spectral hardening outside the inner nebula (in all directions but the south; see figure 1). On small scales (within the inner compact nebula) the southeast inner jet appears to be brighter than the northwest jet suggesting that the former jet is approaching while the latter one is receding [14, 15]. Interestingly, the spectra of the jets appear to differ, with the fainter northwest inner jet appearing harder ($\Gamma = 1.25 \pm 0.02$) than the brighter southeast inner jet ($\Gamma = 1.36 \pm 0.04$). Although Doppler boosting can change the apparent brightnesses of jets, it should not affect the jets' spectra. Moreover, in the G11.2–0.3 and G54.1+0.3 maps, the apparently approaching brighter jets appear to be harder. The map of 3C58 shows that the edge-on torus and western jet are the hardest features of the PWN embedded into a softer emission. Therefore, any averaged radial profiles of photon index will not reflect the actual anisotropic dependences. Overall, in the cases when both polar (jets) and equatorial (torus) components can be identified, their spectra appear to be similar. In the context of the Komissarov & Lyubarsky model [16] where the jets form from the equatorial outflow diverted by magnetic field hoop stress, it means that the particle SED does not change in this process. In G21.5–0.9, the PWN brightness distribution appears roughly symmetric along a line running northeast to southwest, but the spectral map reveals a bulbous lobe with a harder spectrum extending northwest from the pulsar, which reveals the otherwise invisible additional anisotropy of the PW (figure 2). The distribution of the photon indices Γ_i measured from the spectral maps for the innermost regions of 17 PWNe and the SED slopes inferred from them ($p_{\rm i} = 2\Gamma_i - 1$, assuming negligible cooling[2]) is shown in figure 4. The p_i values span the range from 1.4 (for Vela and G320.4–1.2) to 3.3 (in N157B). However, the latter PWN is very remote (it is in the LMC), and the outflow can be affected by cooling even for the smallest region resolvable by *CXO*. Therefore, a more plausible range of $p_i = 1.8$–2.5 with a maximum of the distribution between 2.0 and 2.4 (see figure 4). We have also looked for correlations between Γ_i and luminosity (L_X) or radiative efficiency ($\eta_X = L_X/\dot{E}$) of the innermost regions of the brightest PWNe (taken from table 2 of [17]). Despite the significant scatter, there appears to be a hint of a positive correlation between Γ_i and L_X, suggesting that PWNe with harder injected SEDs are less luminous (see figure 5, left panel). In the Γ_i–η_X

[1] The characteristic cooling length scales depend on the bulk flow velocity, the magnetic field strength, the presence of in-situ acceleration, the geometry of the outflow and ISM entrainment.

[2] This may not hold for PWNe at large distances (e.g., in the LMC) where small angular scales correspond to large distances.

IOP Conf. Series: Journal of Physics: Conf. Series **932** (2017) 012050 doi:10.1088/1742-6596/932/1/012050

Table 2. Injection spectra of selected bright PWNe, obtained from spatially-resolved spectral mapping (objects above the horizontal line; shown in figures 1–3) and spatially-resolved spectroscopy (below the horizontal line). Note that four PWNe (MSH 11–62, the Lighthouse nebula, J0357+3205, and B2224+65) listed in table 1 were excluded, as the injection spectrum can not be reliably determined in those cases. Here, we define $p_i \equiv 2\Gamma_i - 1$, assuming an uncooled electron SED described by a PL with a slope p_i.

Number	Name	Γ_i	p_i
1	Crab	1.80 ± 0.05	2.60 ± 0.10
2	Vela	1.33 ± 0.06	1.67 ± 0.12
3	G21.5–0.9	1.43 ± 0.09	1.86 ± 0.18
4	MSH 11–62	1.10 ± 0.05	1.20 ± 0.10
5	B1509–58 (G320.4–1.2)	1.19 ± 0.13	1.38 ± 0.27
6	Kes 75 (J1846–0258)	1.85 ± 0.10	2.70 ± 0.20
7	G11.2–0.3	1.75 ± 0.1	2.79 ± 0.46
8	3C58	1.97 ± 0.07	2.94 ± 0.15
9	G54.1+0.3	1.7 ± 0.1	2.30 ± 0.21
10	Snail (G327.1–1.1)	1.66 ± 0.06	2.32 ± 0.11
12	N157B (J0537–6910)	2.17 ± 0.11	3.34 ± 0.23
13	Mouse (J1747–2958)	1.57 ± 0.06	2.14 ± 0.12
15	Geminga	1.4 ± 0.1	1.8 ± 0.2
16	CTB 80	1.7 ± 0.1	2.4 ± 0.2
17	J1509–5850	1.43 ± 0.18	1.86 ± 0.36
18	Mushroom (B0355+54)	1.54 ± 0.05	2.08 ± 0.10
19	J1741–2054	1.5 ± 0.15	2.0 ± 0.3

Table 3. Viewing angles ζ and magnetic inclination angles α for γ-ray pulsars estimated in different ways: $\zeta_{\rm PWN}$ – obtained from fitting of the PWN geometry in [22, 23, 24, 25] (unless marked with †, in which case it was estimated by us based on the appearance of the tori/jets); $\alpha_{\rm pred}$ – the predicted value of α based on $\zeta_{\rm PWN}$ from figure 2 of [5] for the two pole caustic (TPC) and outer gap (OG) magnetosphere emission models; $\zeta_{\rm mod}$ and $\alpha_{\rm mod}$ – allowed values of ζ and α obtained from figure 2 of [5] for the pulsar γ-ray and radio light curve properties taken from the 2nd Fermi Pulsar Catalog [26]. The object numbers correspond to those in table 1, and the objects first presented in this table (PSR B1706–44 and the Dragonfly PWN with PSR J2021+3651) are numbered sequentially (starting with 22). ^{a} – The OG model simulations can not produce two γ-ray peaks with a phase separation > 0.5 and the lack of a radio pulse. ^{b} – The model can not produce the observed light curves for the inferred $\zeta_{\rm PWN}$, so we list the α values for the ζ value closest to the observed one (which we note in parentheses in the subscript of ζ). ‡ – Indicates pulsars not seen in radio.

#	Name	$\zeta_{\rm PWN}$	$\alpha_{\rm pred,TPC}$	$\alpha_{\rm pred,OG}$	$\zeta_{\rm mod,TPC}$	$\zeta_{\rm mod,OG}$	$\alpha_{\rm mod,TPC}$	$\alpha_{\rm mod,OG}$
1	Crab	61.3 ± 1.2	63-86	78-90 $(\zeta_{68})^b$	45-90	68-90	40-90	55-90
2	Vela	63.6 ± 0.7	61-86	78-90 $(\zeta_{68})^b$	45-90	68-90	40-90	55-90
3	G21.5-0.9	$85.4^{+0.2}_{-0.3}$	60-90	60-90	45-90	68-90	40-90	55-90
5	B1509-58	≈ 50	27-54	51-70	22-60	38-69	15-60	42-90
6	Kes 75‡	62 ± 5	10-25	18-25 $(\zeta_{70})^b$	10-90	13-31	0-90	55-90
8	3C58	91.6 ± 2.7	72 $(\zeta_{75})^b$	$-^a$	47-75	$-^a$	42-48; 72-83	$-^a$
13	Mouse	$\sim 70^\dagger$	41-57	50-67	40-71	58-78	39-70	47-80
15	Geminga‡	$\sim 90^\dagger$	36-46	13-55	13-26; 42-62; 85-90	76-90	35-45; 80-90	6-55
17	J1509–5850	$\sim 90^\dagger$	41 $(\zeta_{62})^b$	43-53 $(\zeta_{70})^b$	36-62	50-71	40-65	41-61
22	B1706–44	53.3 ± 4.5	47-54	54-60	36-62	50-71	40-65	41-61
23	Dragonfly	79 ± 3	53-90	60-90	45-90	68-90	40-90	55-90

plot (figure 5, right panel) the correlation is not seen, but there seems to be a deficit of objects with high efficiencies and hard spectra. Some of the PWNe (e.g., the Mouse, Klingler et al., in prep.; the Lighthouse [18]; N157B [19]) show strong cooling trends (the spectra is softening rapidly with increasing distance from the pulsar) while other PWNe (e.g., those associated with PSRs J1509–5850 and B0355+54) do not show any substantial cooling along their long tails (see [20, 21] for spatially-resolved spectral measurements). Interestingly, the misaligned outflows (see [8] and also below) do not show spectral softening with distance from the pulsar (see figure 7 of [18] for the adaptively binned spectral map of the Lighthouse PWN).

3. Connection Between PWN Morphologies and Pulsar Light Curves

It is natural to expect that the magnetic inclination angle α (between the spin and magnetic dipole axes) and the viewing angle ζ (between the line of sight and the spin axis) will leave an imprint on both the pulsar light curves and the compact PWN morphologies. The viewing angle can be inferred from the PWN morphology if the PWN has an identifiable torus or ring(s) associated with the TS or, to a lesser extent, if there are two jets of unequal brightness. In the former case, the ellipticity of the torus or ring(s) provides a direct measurement of ζ while in the latter case, the ratio of the jet brightnesses can be related to the ζ-dependent Doppler factor as $f_b = [(1+\beta\cos\zeta)/(1-\beta\cos\zeta)]^{\Gamma+2}$ where $\beta = v_{\rm flow}/c$ [22]. However, the Doppler factor depends on the flow speed, $v_{\rm flow}$, which has been directly[3] measured only in a few cases (Vela PWN outer jet: $0.3c$–$0.6c$, [27]; Crab PWN torus: $0.25c$–$0.55c$, [28, 29]; J1509–5850 PWN jets: possibly $0.2c$–$0.3c$, [21]). Therefore, unlike the torus ellipticity fitting, any estimates based purely on the Doppler brightening are highly uncertain. Inferring α from PWN morphologies is even harder, but some constraints can still be obtained if the prediction that PWNe with pronounced toroidal components and weak polar outflows (jets) are likely to be associated with pulsars having large α [6] finds observational support.

Models of pulsar magnetospheric emission (e.g., [30, 31, 5, 32, 33, 34, 35]) provide solutions for α and ζ; however, the predicted values of these angles are often not well constrained, and widely separated solutions are possible. Moreover, competing models provide different predictions (below we only compare the Two Pole Caustic (TPC) and Outer Gap (OG) magnetosphere emission models – see [35] for a review). Table 3 lists the values suggested by the PWN morphologies ($\zeta_{\rm PWN}$), and the values ($\alpha_{\rm mod}$ and $\zeta_{\rm mod}$) predicted by the pulsar light curve modeling (from [5, 32]). To determine plausible ranges of $\zeta_{\rm PWN}$ from PWN morphologies, we used the values from [22, 23] obtained by careful fitting of tori geometry (whenever available), and estimated plausible ranges of $\zeta_{\rm PWN}$ from the appearance of the tori/jets in the other cases. Table 3 shows that even when ζ is reasonably well constrained from PWN morphology, the $\alpha_{\rm pred}$ value(s) allowed by the magnetospheric emission models for $\zeta = \zeta_{\rm PWN}$ can still span a wide range, which, however, shrinks substantially compared to what could be predicted from the magnetospheric emission models alone. The values of $\alpha_{\rm pred}$ show that PWNe with weaker toroidal components indeed tend to have smaller values of α (see, e.g., the B1509–58, Kes 75, and G11.2–0.3 PWNe in figure 2). The TPC model seems to be performing somewhat better compared to the OG model, in the sense that it has less trouble finding an allowed $\alpha_{\rm pred}$ for a given $\zeta_{\rm PWN}$. The next step in this direction will be to increase the sample of PWNe with interpretable morphologies powered by γ-ray and radio pulsars. It would also be interesting to look at the properties of PWNe whose pulsars lack γ-ray emission. With a sufficiently large sample, one could also start looking into correlations between α and other PWN parameters such as $\Gamma_{\rm i}$ and luminosity.

4. Maximum Particle Energy in Pulsar Wind

The maximum energy of a synchrotron photon can be estimated by equating the synchrotron losses in a magnetic field B to the energy gain in the electric field of strength $E \simeq B$ (for typical, highly-conducting astrophysical plasmas[4] $E < B$), which leads to the maximum synchrotron photon energy $\epsilon_{\gamma,{\rm max}} = \hbar mc^3/e^2 \sim 100$ MeV. For electrons radiating close to this energy the acceleration must occur on a gyro-radius length scale, which means that any stochastic acceleration (e.g., Fermi type; see [38] and references therein) would result in a substantially lower energy [39]. An estimate of the corresponding maximum electron energy requires knowing the magnetic field within the particle acceleration region, which is challenging because the acceleration region location and the acceleration process itself are still a matter of debate (see [40, 38] and references therein).

[3] If the torus inclination is known and it is Doppler brightened on one side, the contrast with the other side can be used to estimate the flow speed in the equatorial region.

[4] Flares in the Crab PWN suggest that this condition may be violated in certain regions where reconnection occurs (see e.g., [37]).

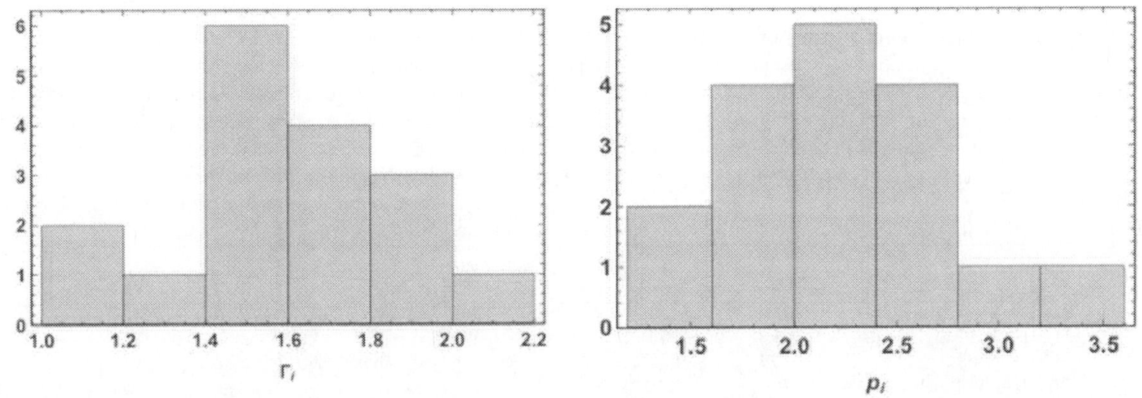

Figure 4. Histograms of $\Gamma_{\rm i}$ and $p_{\rm i} = 2\Gamma_{\rm i} - 1$ for the 17 PWNe listed in table 2 (note that pulsars 11 and 14 are omitted from table 2; see the caption). The bin width corresponds to the average measurement uncertainty.

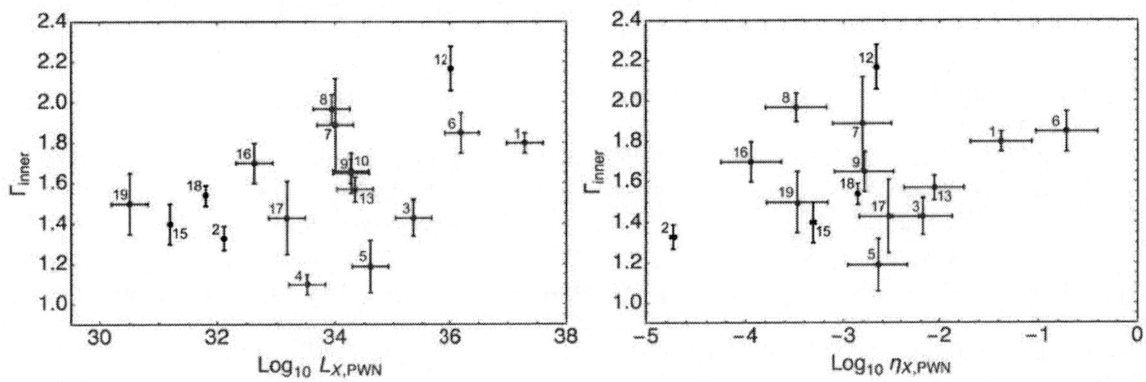

Figure 5. *Left:* Photon index of the injection spectrum vs. X-ray luminosity of the central regions of PWNe (from table 2 of [17]; the values for B0355, J1509, J1747, and J1741 are from [20, 21, 36], respectively). *Right:* Photon index of the injection spectrum vs. PWN radiative efficiency $\eta = L_X/\dot{E}$. Pulsars/PWNe whose distances are known accurately (e.g., via parallax, and for N157B in the LMC) are shown in black. For the rest of the pulsars (shown in blue) 30% uncertainties in the distances are assumed.

On the other hand, if the energy for particle acceleration in PW ultimately comes from the potential drop available in a pulsar magnetosphere, $\Phi \sim (\dot{E}/c)^{1/2}$, then the maximum $\gamma_{\rm max} = \Phi e/m_e c^2 \sim 1 \times 10^{10}(\dot{E}/10^{37}\ {\rm erg/s})^{1/2}$. For the Crab, Vela, Geminga and the Guitar (B2224+65) pulsars, the corresponding values of $\gamma_{\rm max}$ are, respectively, $\sim 7\times10^{10}$, 9×10^{9}, 6×10^{8} and 1×10^{8}. The dependence on $\dot{E}$ suggests that by observing older (lower $\dot{E}$) pulsars in X-rays, one can probe the fraction of the available magnetospheric potential drop used to accelerate PW particles. The largest uncertainty in such estimates comes from the PWN magnetic field (or PW magnetization). Since it is difficult to measure the field directly, it must be inferred from other PWN properties (e.g., from synchrotron brightness; see e.g., [27]). This difficulty can possibly be circumvented for PWNe of supersonic pulsars which exhibit *misaligned outflows* (see below).

The recently launched *NuSTAR* observatory was able to detect and resolve the Crab (up to 78 keV; [41]) and the B1509–58 (up to 40 keV; [42]) PWNe. Such energies correspond to electron Lorentz factors of $\gamma_{\rm Crab} \sim 10^8$ and $\gamma_{\rm B1509} \sim 4\times10^8$ for plausible magnetic fields within these PWNe ($B_{\rm Crab} \sim 300\ \mu$G [29] and $B_{\rm B1509} \sim 10\mu$G [42]). The lack of sensitive instruments with

Table 4. Comparison of various properties for four supersonic pulsars with misaligned outflows. For the apex stand-off distance r_s we used the observed values for B0355+54 [20] and B2224+65 [45], and estimated r_s from the pressure balance (for ISM number density $n = 1$ cm^{-3}) for J1509–5850 and J1101–6101. The lower limits on the Lorentz factors, $\gamma_{\rm esc}$, of electrons escaping into the ISM from the PWN apex are estimated for $B_{\rm ISM} = 5$ μG and synchrotron photon energy of 8 keV. The maximum Lorentz factors, $\gamma_{\rm max}$, are calculated from the full potential drop available in the pulsar magnetosphere (see text). $B_{\rm apex}$ is the PWN magnetic field near the PWN apex. The gyro-radii, $r_{\rm g,max}$, of electrons with Lorentz factor $\gamma_{\rm max}$ are calculated using the upper limit on $B_{\rm apex}$ for the PWN magnetic field. See [46, 18, 21, 20] for the estimates of pulsar velocities, $v_{\rm PSR}$.

Name	$\dot{E}$ (10^{35} erg s^{-1})	P (ms)	$v_{\rm PSR}$ (km s^{-1})	$r_{\rm s}$ (10^{15} cm)	$\gamma_{\rm esc}$ (10^8)	$\gamma_{\rm max}$ (10^8)	$B_{\rm apex}$ (μG)	$r_{\rm g,max}$ (10^{15} cm)
B2224+65 (Guitar)	0.012	682	~1600	2.4	6	1	<125	1.4
J1101–6101 (Lighthouse)	5.1	88.9	~2000	6	6	39	<57	117
J1509–5850	14	62.8	~800	10	6	24	<34	120
B0355+54 (Mushroom)†	0.45	156	61^{+12}_{-9}	54	6	7.1	<6	202

good angular resolution in the MeV range makes it challenging to determine, from synchrotron emission, if electrons with even higher γ are present. Perhaps, the Crab PWN is the only PWN with reliable MeV spectrum measurements showing that the (quiescent) synchrotron spectrum breaks sharply above ~20 MeV [43], which corresponds to $\gamma_{\rm Crab} \sim 4 \times 10^9$ for the assumed $B_{\rm Crab} \sim 300$ μG. This falls substantially short of the $\gamma_{\rm max,Crab} \approx 7 \times 10^{10}$ value based on the full magnetospheric potential drop, but it is close to the limit that is expected to be imposed by the synchrotron losses (see above).

The inverse Compton (IC) emission from background photons up-scattered by PW electrons has been seen by HAWC (e.g., [44]) up to energies ($\epsilon_{\rm IC}$) of a few tens of TeV. If the up-scattered photons come primarily from Cosmic Microwave Background (CMB), then $\gamma \sim 3 \times 10^8 (\epsilon_{\rm IC}/60\ {\rm TeV})^{1/2} (\epsilon_{\rm CMB}/6 \times 10^{-4}\ {\rm eV})^{-1/2}$. This is close to the maximum γ values inferred from synchrotron emission (see above) and may exceed $\gamma_{\rm max}$ for older low-$\dot{E}$ pulsars which are believed to power relic PWNe (e.g., $\gamma_{\rm max} \sim 6 \times 10^8$ for Geminga, whose relic PWN is likely associated with the HAWC source seen above 20 TeV [44]). We note that although the TeV emitting particles can be produced earlier in pulsar's life (when $\gamma_{\rm max}$ was higher), those particles that radiate at 60 TeV have cooling times of around 10 kyr (e.g., Eqn. 4 in [10]), which is significantly smaller than the 340 kyr spin-down age of Geminga. Therefore, these particles experienced an accelerating potential corresponding to a nearly modern-day $\dot{E}$.

An alternative approach to the estimation of $\gamma_{\rm max}$ has become possible with the discovery of misaligned outflows produced by a few supersonically moving pulsars (B2224+65, J1101–6101, J1509–5850, and possibly B0355+54; see [8] and references therein). These puzzling structures can be explained by the kinetic leakage of the most energetic particles near the apex of the bow shock, which becomes possible when the bow shock stand-off distance, $r_{\rm s} \simeq (\dot{E}/4\pi c m_H n v_p^2)^{1/2}$ (n is the ISM baryon number density; see e.g., [8]), is comparable to the gyro-radius, $r_g = \gamma m_e c^2/eB_{\rm apex}$, in the PWN magnetic field $B_{\rm apex}$ near the bow-shaped PWN apex [47]. This condition can be used to put a lower limit on the energies of particles escaping into the ISM, if the ISM magnetic field is known (or assumed). Since supersonically moving pulsars with ages $\gtrsim 10$ kyr are expected to leave their host SNRs and move in a relatively unperturbed ISM with typical $B_{\rm ISM} \simeq 5$ μG [48], the Lorentz factor of the electrons that escaped into the ISM can be estimated as $\gamma_{\rm esc} \gtrsim 6 \times 10^8 (\epsilon/8\,{\rm keV})^{1/2} (B_{\rm ISM}/5\,\mu{\rm G})^{-1/2}$, where ϵ is the observed energy of synchrotron photon (8 keV is the upper boundary of the *CXO* band). From the escape condition, $r_g \gtrsim r_{\rm s}$, one can also set an upper limit on the PWN field near the apex, $B_{\rm apex} \lesssim 34 (\epsilon/1\,{\rm keV})^{1/2} (B_{\rm ISM}/5\,\mu{\rm G})^{-1/2} (r_{\rm s}/10^{16}\ {\rm cm})^{-1}$ μG (here we adopted 1 keV as the lower boundary of the *CXO* energy band because the typical PWN spectra are substantially absorbed below this energy). As one can see from table 4, for B0355+54 the upper limit on $B_{\rm apex}$ approaches the ISM field and therefore provides a good constraint suggesting that $B_{\rm apex}$

is a few μG. Table 4 also shows a comparison of $\gamma_{\rm max}$ and $\gamma_{\rm esc}$ for four supersonic pulsars with misaligned outflows. In the case of the Guitar nebula (PSR B2224+65), the energies of X-ray radiating electrons in the misaligned outflow appear to exceed the energy available in the magnetospheric potential drop (due to the space limitations, we refer the reader to [38, 49] for discussion and possible explanation). If a substantial number of particles are able to reach $\gamma_{\rm max}$ (a plausible situation for the old PSR B0355+54 given that it apparently happens in the somewhat older PSR B2224+65), they should be able to leak into the ambient ISM very easily since the gyro-radii of electrons with Lorentz factors $\gamma_{\rm max}$ would greatly exceed the stand-off distance r_s.

5. Summary

We discussed the observational properties of PWNe, linking them to the injected (post-TS) electron SED and parameters of pulsar magnetospheres. The main findings can be summarize as follows.

- The photon indices of PWNe measured in X-rays for the innermost resolved regions of bright, extended PWNe vary in the range of $\Gamma_i = 1.1 - 2.2$ with the majority (80%) being in the $\Gamma_i = 1.3 - 2.0$ range. The corresponding ranges of the uncooled electron SED slopes, $p_i = 2\Gamma_i - 1$, are 1.2–2.9 and 1.8–2.5, respectively. Effects of cooling are obvious in many high-resolution spectral maps of PWNe, hence inferring p_i from the spatially-averaged Γ measured from a large PWN region does not provide accurate diagnostics of particle SED injected at the TS. The spectra of the inner regions of the tori are similar to those of the inner jets, in the same PWN.
- Measuring the viewing angle ζ from approximately axisymmetric morphologies of PWNe with tori and/or jets allows one to test pulsar magnetosphere models which are now capable of predicting ζ and the magnetic inclination α based on γ-ray and radio light curves. A correlation between α and PWN properties can be expected from general principles and is supported by recent numerical modeling [6]. We used the α values predicted by magnetospheric models to look for a correlation between the PWN spatial morphologies, luminosities, or spectra. We find some observational support for a hypothesis that pulsars with smaller α power PWNe with relatively weaker toroidal components (compared to their jet components).
- The capabilities of modern X-ray and TeV γ-ray observatories allow one to infer (under some assumptions) PW particle energies. These energies approach, and in one case (B2224+65) exceed, the full potential drop available in pulsar magnetosphere.

Acknowledgements: We thank Andery Bykov and Demos Kazanas for stimulating discussions. We acknowledge partial support from *CXO* awards GO3-14082, GO3-14084, GO3-14057 and AR5-6006.

References

[1] Gaensler B M and Slane P O 2006 *ARAA* **44** 17–47
[2] Kargaltsev O, Klingler N, Rangelov B and Pavlov G 2015 X-ray view of pulsar wind nebulae: What we can learn with Athena? *Exploring the Hot and Energetic Universe: The first scientific conference dedicated to the Athena X-ray observatory* ed Ehle M p 60
[3] Reynolds S P, Pavlov G G, Kargaltsev O, Klingler N, Renaud M and Mereghetti S 2017 *SSR* **207** 175–234
[4] Slane P 2017 *Preprint* 1703.09311
[5] Watters K P, Romani R W, Weltevrede P and Johnston S 2009 *ApJ* **695** 1289–301
[6] Bühler R and Giomi M 2016 *MNRAS* **462** 2762–76
[7] Hobbs G, Lorimer D R, Lyne A G and Kramer M 2005 *MNRAS* **360** 974–92
[8] Kargaltsev O, Pavlov G G, Klingler N and Rangelov B 2017 *Preprint* 1708.00456
[9] Mori K, Burrows D N, Hester J J, Pavlov G G, Shibata S and Tsunemi H 2004 *ApJ* **609** 186–93

IOP Conf. Series: Journal of Physics: Conf. Series **932** (2017) 012050 doi:10.1088/1742-6596/932/1/012050

[10] Kargaltsev O, Rangelov B and Pavlov G G 2013 *ArXiv e-prints*
[11] Cappellari M and Copin Y 2003 *MNRAS* **342** 345–54
[12] Diehl S and Statler T S 2006 *MNRAS* **368** 497–510
[13] Porth O, Vorster M J, Lyutikov M and Engelbrecht N E 2016 *MNRAS* **460** 4135–49
[14] Helfand D J, Gotthelf E V and Halpern J P 2001 *ApJ* **556** 380–91
[15] Durant M, Kargaltsev O, Pavlov G G, Kropotina J and Levenfish K 2013 *ApJ* **763** 72
[16] Komissarov S S and Lyubarsky Y E 2004 *MNRAS* **349** 779–92
[17] Kargaltsev O and Pavlov G G 2008 Pulsar Wind Nebulae in the Chandra Era *40 Years of Pulsars: Millisecond Pulsars, Magnetars and More* (*American Institute of Physics Conference Series* vol 983) ed Bassa C, Wang Z *et al* pp 171–85
[18] Pavan L *et al* 2016 *A&A* **591** A91
[19] Chen Y, Wang Q D, Gotthelf E V, Jiang B, Chu Y H and Gruendl R 2006 *ApJ* **651** 237–49
[20] Klingler N *et al* 2016 *ApJ* **833** 253
[21] Klingler N, Kargaltsev O, Rangelov B, Pavlov G G, Posselt B and Ng C Y 2016 *ApJ* **828** 70
[22] Ng C Y and Romani R W 2004 *ApJ* **601** 479–84
[23] Ng C Y and Romani R W 2008 *ApJ* **673** 411-7
[24] Ng C Y, Slane P O, Gaensler B M and Hughes J P 2008 *ApJ* **686** 508-19
[25] Yatsu Y, Kawai N, Shibata S and Brinkmann W 2009 *PASJ* **61** 129–35
[26] Abdo A A *et al* 2013 *ApJS* **208** 17
[27] Pavlov G G, Teter M A, Kargaltsev O and Sanwal D 2003 *ApJ* **591** 1157–71
[28] Bietenholz M F, Frail D A and Hester J J 2001 *ApJ* **560** 254–60
[29] Hester J J 2008 *ARAA* **46** 127–55
[30] Romani R W 1996 *ApJ* **470** 469
[31] Muslimov A G and Harding A K 2003 *ApJ* **588** 430–40
[32] Romani R W and Watters K P 2010 *ApJ* **714** 810–24
[33] Kalapotharakos C, Harding A K and Kazanas D 2014 *ApJ* **793** 97
[34] Pierbattista M, Harding A K, Grenier I A, Johnson T J, Caraveo P A, Kerr M and Gonthier P L 2015 *A&A* **575** A3
[35] Harding A K 2016 *Journal of Plasma Physics* **82** 635820306
[36] Auchettl K *et al* 2015 *ApJ* **802** 68
[37] Kroon J J, Becker P A, Finke J D and Dermer C D 2016 *ApJ* **833** 157
[38] Bykov A M, Amato E, Petrov A E, Krassilchtchikov A M and Levenfish K P 2017 *SSR* **207** 235–90
[39] Lyutikov M 2010 *MNRAS* **405** 1809–15
[40] Lyutikov M, Sironi L, Komissarov S and Porth O 2016 *Preprint* 1603.05731
[41] Madsen K K *et al* 2015 *ApJ* **801** 66
[42] An H *et al* 2014 *ApJ* **793** 90
[43] Kuiper L, Hermsen W, Cusumano G, Diehl R, Schönfelder V, Strong A, Bennett K and McConnell M L 2001 *A&A* **378** 918–35
[44] Abeysekara A U and HAWC Collaboration 2017 *ApJ* **843** 40
[45] van Kerkwijk M H and Ingle A 2008 *ApJL* **683** L159
[46] Chatterjee S, Cordes J M, Vlemmings W H T, Arzoumanian Z, Goss W M and Lazio T J W 2004 *ApJ* **604** 339–45
[47] Bandiera R 2008 *A&A* **490** L3–6
[48] Opher M, Bibi F A, Toth G, Richardson J D, Izmodenov V V and Gombosi T I 2009 *Nature* **462** 1036–8
[49] Pétri J 2016 *Journal of Plasma Physics* **82** 635820502

Polarized synchrotron X-ray emission from supernova shells: XIPE/IXPE perspective

A Bykov[1,2] **and Y Uvarov**[1]

[1] Ioffe Institute, St. Petersburg, Russia
[2] SPbPU, St. Petersburg, Russia

E-mail: byk@astro.ioffe.ru, uv@astro.ioffe.ru

Abstract. Young supernova remnants are the sources of a broadband nonthermal synchrotron continuum emission from radio to X-rays produced by accelerated electrons. Diffusive shock acceleration (DSA) of relativistic particles is accompanied by strong amplification of the turbulent magnetic field. This is an intrinsic feature of an efficient DSA. Synchrotron X-rays are produced by the multi-TeV electrons which are concentrated in a narrow region near the shock due to the strong energy losses. The confinement of the X-ray emitting electrons in this region filled with highly fluctuating magnetic field results in a complex time dependent clumpy structure of the observed X-ray images of supernova shells. The lack of Faraday depolarization in X-rays makes X-ray polarimetry a powerful tool to study synchrotron structures in the young galactic shell type SNRs Cas A, Tycho, SN 1006, RX J1713.7−3946. We present here simulated synchrotron images of SNR shells for different models of isotropic and anisotropic magnetic turbulence in a shell and argue that polarized X-ray imaging can be constructed for Tycho and Cas A SNRs with the expected sensitivity and the angular resolution of the upcoming missions XIPE (ESA) and IXPE (NASA).

1. Introduction

A young supernova remnant that supersonically expands in the interstellar medium generates a strong collisonless shock wave. This shock is an effective accelerator of charged particles due to the DSA mechanism. This mechanism produces a power law particle spectrum in a broad energy range. In the vicinity of the shock a strong super-adiabatic amplification of the fluctuating magnetic field likely takes place due to the cosmic-ray driven plasma instabilities. This turbulent magnetic field is an important part of the DSA itself and is responsible for particle diffusion [1, 2, 3, 4]. In this process electrons can be accelerated up to an energy of tens (or even hundreds) TeVs. These energetic particles are immersed in a magnetic field and efficiently emit nonthermal synchrotron radiation up to keV energies [5]. Modern observations confirm this model with detection of nonthermal radiation in a broad spectral range from radio to X-rays.

Recent high resolution synchrotron X-ray images of some SNRs (Cas A, Tycho, RX J1713.72−3946 etc) obtained with Chandra observatory revealed a complicated small-scale structure that includes thin filaments, clumps and dots. In the case of SNR RX J1713.72−3946 it was shown that small X-ray structures can be time-variable on a timescale of few years [6]. The nature of filaments can be explained as a combination of a geometry projection effect together with the limited width of the emission region. This limitation can be a result of a strong electron synchrotron looses [7, 8] or a rapid magnetic field decay [9] in the downstream region.

IOP Conf. Series: Journal of Physics: Conf. Series **932** (2017) 012051 doi:10.1088/1742-6596/932/1/012051

Time variability and clumpy structure can be naturally explained by the turbulent nature of magnetic field in the vicinity of a shock front. The influence of the stochastic turbulent magnetic field on spectra and images was studied in [10, 11] where some of the observed features of clumpy structures were indeed reproduced by simulated images. It occurs that intensity and polarization maps are sensitive to the turbulent magnetic field spectrum. The isotropic turbulent spectrum was assumed in that model. In this work we also consider an anisotropic turbulent field that can be formed naturally in the downstream medium after shock compression of the incoming isotropic magnetic turbulence. Polarization can be significantly higher in the case of the anisotropic turbulence and it can be observed with the future generation of X-ray polarimeters such as XIPE [12] or IXPE. The preferred direction of polarization in this case is radial.

2. The model

2.1. Stochastic magnetic field

The stochastic properties of a turbulent magnetic field with an axis of symmetry is $\langle B_\perp^2\rangle = \xi\langle B_\parallel^2\rangle = \langle B^2\rangle\xi/\,(\xi+1)$, where $\xi = 2$ in the isotropic case and $\xi > 2$ in the case of the compressed anisotropic turbulence. We introduce here the Cartesian coordinate system with axis $\|$ directed along the symmetry axis. We simulate isotropic and anisotropic turbulence according to our previous works [10], [11] that use method based on [13]. We consider a summation of a big number of harmonics with random wave vectors and phases:

$$\mathbf{B}(\mathbf{r},t) = \sum_{n=1}^{N_m}\sum_{\alpha=1}^{2} \mathbf{A}^{(\alpha)}(k_n)\cos(\mathbf{k}_n\cdot\mathbf{r} - \omega_n(\mathbf{k}_n)\cdot t + \phi_n^{(\alpha)})$$

Polarizations $\mathbf{A}^{(\alpha)}(k_n)$ $(\alpha = 1,2)$ of magnetic field are orthogonal to each other and to the wave vector $\mathbf{A}^{(\alpha)}(k_n)\perp\mathbf{k}_n$ in order to satisfy $\nabla\cdot\mathbf{B} = 0$ condition. For the 1st polarization we chose direction in a plane formed by the symmetry axis and the wave vector. The direction of the 2nd polarization is orthogonal both to the symmetry axis and to the wave vector. The 2nd polarization contributes only to the value of $B_\perp$ while the 1st polarization contributes both to $B_\perp$ and $B_\parallel$. If $A_1(k) = A_2(k)$ this method simulates an isotropic turbulence [13]. If $A_1^2(k) = (1-a)\,A^2(k)$, $A_2^2(k) = (1+a)\,A^2(k)$, where $1 > a > 0$, this method simulates anisotropic turbulence with stochastic properties $\langle B_\perp^2\rangle = \xi\langle B_\parallel^2\rangle$, where $\xi = (2+a)\,/\,(1-a)$. The amplitude is connected with turbulence spectrum energy density by the relation $W(k)\sim k^{-\delta}\sim k^2A^2(k)$, where δ is a spectral index.

2.2. Particle distribution function and SNR geometry

We use an electron distribution function obtained in the one-dimensional DSA model of particle acceleration including synchrotron looses [14],[15]. High energy electrons are concentrated in a thin layer near the shock (see Figure 1) due to the synchrotron looses.

We simulated SNR with radius ~ 3 pc at a distance of 2.5 kpc, so the SNR angular radius was 4 arcmin. These parameters are similar to those of the Tycho SNR. For the field simulations we used a rectangular region $(16\times 8\times 4)\cdot 10^{18}$ cm consisting of $(4\times 2\times 1)$ cubic boxes with size $D = 4\cdot 10^{18}$ cm (Figure 1). The minimal and maximal characteristic scales of turbulence that we used were $L_{min}\sim 2.5\cdot 10^{15}$ cm and $L_{max}\sim 1.2\cdot 10^{18}$ cm (~ 0.07 and ~ 30 arcsec in angular units). We calculated synchrotron emission for isotropic and anisotropic turbulence with spectral index $\delta = 1; 5/3$.

2.3. Synchrotron radiation properties

Emission and polarization properties of radiation can be fully described by the Stokes parameters. We consider magnetic field fluctuations of scales larger than the synchrotron

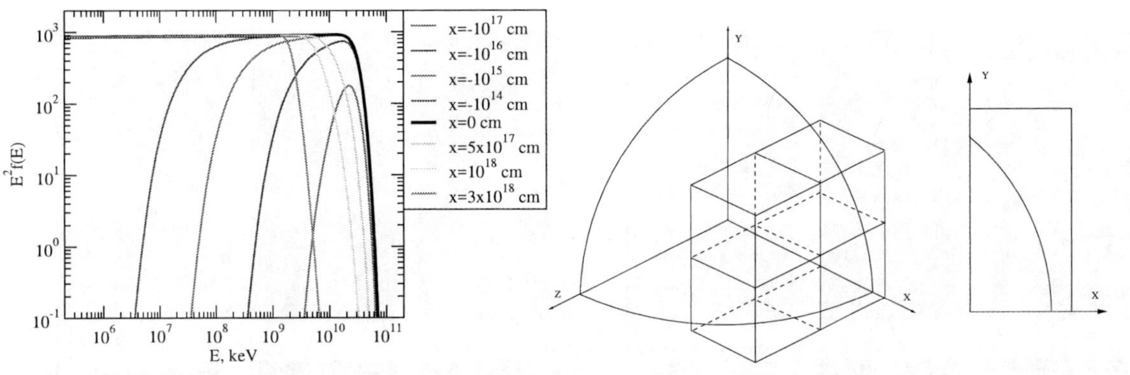

Figure 1. In the left panel the electron distribution function at different distances from the shock front is shown. Negative values correspond to the upstream region while positive to the downstream flow. In the middle panel the geometry of the model is shown: 4 cubic boxes with simulated magnetic field (one half of 8 boxes we used) and the spherical SNR shock. In the right panel the schematic 2D image is shown that is obtained by projection along the line of sight (z-axis).

radiation formation length $l_f \approx mc^2/eB \approx 1.7 \times 10^9 B_{\mu G}^{-1}$ cm, so we can use standard relations for the synchrotron radiation in a homogeneous magnetic field [16].

$$\widetilde{I}(\mathbf{r},t,\nu) = \frac{\sqrt{3}e^3}{mc^2r^2}\frac{\nu}{\nu_c}B_\perp(\mathbf{r})\int\limits_{\nu/\nu_c}^{\infty}K_{5/3}(\eta)d\eta,\ \widetilde{Q}(\mathbf{r},t,\nu) = \frac{\sqrt{3}e^3}{mc^2r^2}\frac{\nu}{\nu_c}B_\perp(\mathbf{r})K_{2/3}\left(\frac{\nu}{\nu_c}\right)cos\,(2\chi),$$

$\widetilde{U}(\mathbf{r},t,\nu) = \frac{\sqrt{3}e^3}{mc^2r^2}\frac{\nu}{\nu_c}B_\perp(\mathbf{r})K_{2/3}\left(\frac{\nu}{\nu_c}\right)sin\,(2\chi)$, $\mathbf{B}_\perp$ is a magnetic field projection to a plane transverse to the line of sight, χ is an angle between the selected direction in this plane and the main axis of polarization ellipse. The projected maps can be obtained by integrating over the line of sight taking into account the isotropic distribution of emitting particles $f_e(E,\mathbf{r})$:

$$I(\nu) = \frac{\sqrt{3}e^3}{mc^2}\int dl\,dE\frac{\nu}{\nu_c}B_\perp(\mathbf{r})f_e(E,\mathbf{r})\int\limits_{\nu/\nu_c}^{\infty}K_{5/3}(\eta)d\eta,$$

$$Q(\nu) = \frac{\sqrt{3}e^3}{mc^2}\int dl\,dE\frac{\nu}{\nu_c}cos\,(2\chi)B_\perp(\mathbf{r})f_e(E,\mathbf{r})K_{2/3}\left(\frac{\nu}{\nu_c}\right),$$

$$U(\nu) = \frac{\sqrt{3}e^3}{mc^2}\int dl\,dE\frac{\nu}{\nu_c}sin\,(2\chi)B_\perp(\mathbf{r})f_e(E,\mathbf{r})K_{2/3}\left(\frac{\nu}{\nu_c}\right),\ \nu_c = \frac{3eB_\perp}{4\pi mc}\gamma^2,\ \int f_e(E,\mathbf{r})\,dE = \frac{n(\mathbf{r})}{4\pi}.$$

Here I, Q, U are normalized so that the radiation flux near Earth can be obtained as $dF(\nu) = d\Omega I(\nu) = \frac{dS}{r^2}I(\nu)$ and so on, parameters ν_c,χ depend on location. The degree of polarization can be obtained as: $\Pi = \sqrt{U^2+Q^2}/I$. The main axis of the polarization ellipse of the synchrotron radiation is perpendicular to the direction of the local magnetic field projection $B_\perp(\mathbf{r})$. If the rotation angle of the polarization ellipse χ is counted clockwise from the negative direction of the x-axis then $cos\,(\chi) = B_{\perp y}/B_\perp$ and $sin\,(\chi) = B_{\perp x}/B_\perp$, so

$$\widetilde{Q}(\mathbf{r},t,\nu) = \frac{\sqrt{3}e^3}{mc^2r^2}\frac{\nu}{\nu_c}B_\perp(\mathbf{r})K_{2/3}\left(\frac{\nu}{\nu_c}\right)\frac{B_{\perp y}^2-B_{\perp x}^2}{B_\perp^2},\ \widetilde{U}(\mathbf{r},t,\nu) = \frac{\sqrt{3}e^3}{mc^2r}\frac{\nu}{\nu_c}B_\perp(\mathbf{r})K_{2/3}\left(\frac{\nu}{\nu_c}\right)\frac{2B_{\perp y}\cdot B_{\perp x}}{B_\perp^2}.$$

It is evident that in the case of the isotropic turbulence $\left\langle\widetilde{Q}(\mathbf{r},t,\nu)\right\rangle = 0$, $\left\langle\widetilde{U}(\mathbf{r},t,\nu)\right\rangle = 0$. In reality the stochastic ensemble is not full so the average is not zero but it is substantially suppressed. If in the case of the anisotropic turbulence the axis of symmetry coincides with axis Ox (or Oy) then $\left\langle B_{\perp y}^2\right\rangle \neq \langle B_{\perp x}^2\rangle$ and $\left\langle\widetilde{Q}(\mathbf{r},t,\nu)\right\rangle \neq 0$ while $\left\langle\widetilde{U}(\mathbf{r},t,\nu)\right\rangle = 0$. If $B_\perp^2/B_\parallel^2$ is high then $\left\langle\widetilde{Q}(\mathbf{r},t,\nu)\right\rangle$ can be close to the value for the case of the homogeneous magnetic field and polarization can be close to the maximum theoretical limit.

IOP Conf. Series: Journal of Physics: Conf. Series **932** (2017) 012051 doi:10.1088/1742-6596/932/1/012051

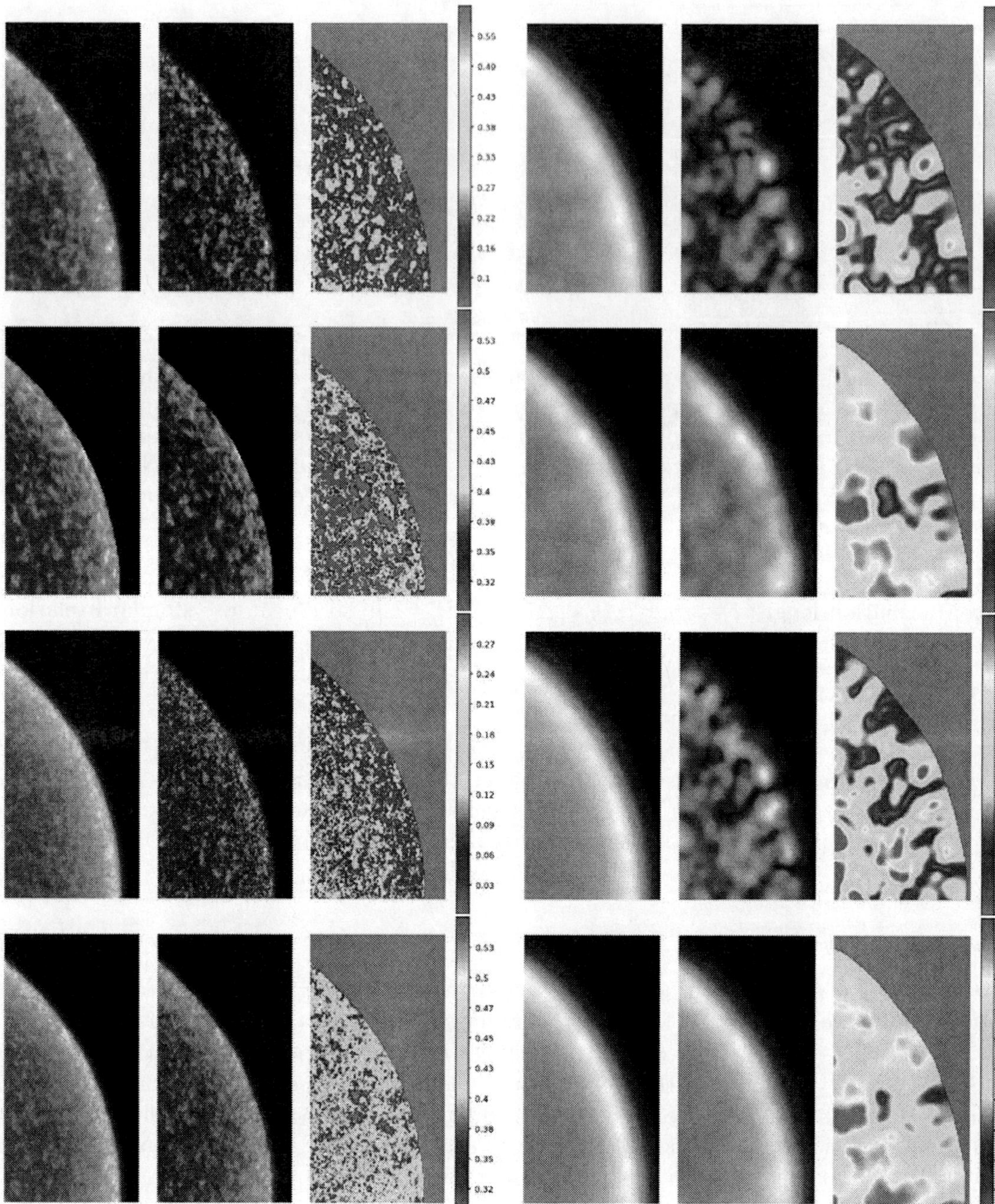

Figure 2. The model synchrotron images at 5keV. The total emission is in the left panel, the polarized emission is in the middle panel while the polarization degree is in the right panel. In the left column the images have the angular resolution of 1.2 arcsec while in the right column the images are shown after the convolution with XIPE PSF. The first and second rows show the maps for an isotropic turbulence case and anisotropic turbulence case with $\xi = 5$ and value of the spectral index of magnetic field fluctuations $\delta = 5/3$. The third and forth rows show maps for an isotropic turbulence case and anisotropic turbulence case with $\xi = 5$ and $\delta = 1$.

3. Discussion

Based on the model described above we constructed X-ray intensity and polarization maps with angular resolution of 1.2 arcsec. We also constructed maps with the lower angular resolution expected for the next generation X-ray polarimeters such as XIPE or IXPE. To do so we used the PSF of polarimeter XIPE [17]. Simulations were made for both the isotropic and the anisotropic magnetic turbulence with $\xi = 5$ and the spectral indexes $\delta = 1$ (strong CR-driven turbulence) and $\delta = 5/3$ (Kolmogorov turbulence). The simulated synchrotron X-ray images at the photon energy 5 keV are shown in Figure 2 for different values of δ. The spectral index determines the relative sizes of the magnetic structures at the high resolution intensity and polarization images. The influence of anisotropy on the degree of polarization is also clearly seen in Figure 2.

The XIPE PSF width exceeds the smallest sizes of the image structure, so after convolution the difference in the small scale structure becomes much less prominent, and the intensity images obtained for turbulent fields with different power spectrum look almost the same. On the other hand the difference in the polarization degree is still present after the convolution, so future X-ray polarimeters could constrain the level of anisotropy of the magnetic turbulence in SNRs.

We show that in the case of the high anisotropy of the magnetic field turbulence the polarization degree can be close to the maximum possible theoretical limit attainable in the case of the homogeneous magnetic field. The difference between the case of the highly anisotropic turbulence and the case of the homogeneous magnetic field is that in the latter case the total emission from SNR is highly polarized. While in the case of the turbulent field only some polarized patches of various scales can be seen in the spatially resolved X-ray images.

Acknowledgments

The authors acknowledge support from RSF grant 16-12-10225. Most calculations were done on computers of the RAS JSCC, St. Petersburg department of the RAS JSCC and on the Tornado cluster of SPbGTU.

References

[1] Blandford R and Eichler D 1987 Phys. Rep. **154** 1–75
[2] Malkov M A and Drury L O 2001 *Reports on Progress in Physics* **64** 429–81
[3] Bell A R 2004 MNRAS **353** 550–8
[4] Marcowith A *et al* 2016 *Reports on Progress in Physics* **79** 046901
[5] Reynolds S P and Chevalier R A 1981 ApJ **245** 912
[6] Uchiyama Y, Aharonian F A, Tanaka T, Takahashi T and Maeda Y 2007 Nature **449** 576–8
[7] Vink J and Laming J M 2003 ApJ **584** 758–69
[8] Bamba A, Yamazaki R, Yoshida T, Terasawa T and Koyama K 2005 ApJ **621** 793–802
[9] Pohl M, Yan H and Lazarian A 2005 ApJ **626** L101–4
[10] Bykov A M, Uvarov Y A and Ellison D C 2008 ApJ **689** L133–6
[11] Bykov A M, Uvarov Y A, Bloemen J B G M, den Herder J W and Kaastra J S 2009 MNRAS **399** 1119–25
[12] Soffitta P *et al* 2016 *Space Telescopes and Instrumentation 2016: Ultraviolet to Gamma Ray* (Proc. SPIE vol 9905) p 990515
[13] Giacalone J and Jokipii J R 1999 ApJ **520** 204–14
[14] Bykov A M and Uvarov Y A 1999 *Soviet Journal of Experimental and Theoretical Physics* **88** 465–75
[15] Bykov A M, Chevalier R A, Ellison D C and Uvarov Y A 2000 ApJ **538** 203–16
[16] Ginzburg V L and Syrovatskii S I 1965 *Annual Review of Astronomy and Astrophysics* **3** 297
[17] Fabiani S *et al* 2014 ApJS **212** 25

IOP Conf. Series: Journal of Physics: Conf. Series **932** (2017) 012052 doi:10.1088/1742-6596/932/1/012052

Confinement of pulsar wind nebulae by their supernova remnants and magnetic dissipation

S J Tanaka

Department of Physics, Faculty of Science and Engineering, Konan University, 8-9-1 Okamoto, Kobe, Hyogo 658-8501, Japan

E-mail: sjtanaka@center.konan-u.ac.jp

Abstract. The standard model of the pulsar wind nebula states that the pulsar wind should be low-magnetization just upstream the termination shock, called the 'sigma-problem'. Low-magnetization is required in order to confine the pulsar wind nebula inside the slowly expanding supernova remnant whose expansion velocity is much smaller than the speed of light. Although the standard model is based on the ideal magnetohydrodynamic approximation, the recent studies indicate that the non-ideal magnetohydrodynamic effects would be important to resolve the sigma-problem. In this study, we extend the standard model including the turbulent magnetic field and the magnetic dissipation. The conversion of the toroidal magnetic field to the turbulent magnetic field and the magnetic dissipation terms are treated phenomenologically. We find that the conversion of the toroidal magnetic field to the turbulent magnetic field decelerates the nebular flow to fit the observed value of the expansion velocity. On the other hand, the magnetic dissipation hardly contributes the flow deceleration, although the magnetic dissipation is important to reproduce the observed emission properties of the PWNe.

1. Introduction

Pulsar wind nebulae (PWNe) are composed of relativistic magnetized plasma supplied by the central pulsar. The relativistic magnetized outflow is called pulsar wind and PWNe are the shocked pulsar wind. The pulsar wind is originally the rotational energy of the central pulsar and spins down the pulsar. Because the extraction of the rotational energy is mediated by the strong magnetic field of the pulsar itself [1], the magnetization (the ratio of the magnetic to particle energy fluxes) of the pulsar wind should be much larger than unity [2, 3].

The observed luminosity of the pulsed emission from the pulsar has only a fraction of spin-down power, while the PWN around the pulsar has the luminosity comparable with the spin-down power, i.e., the PWN is more luminous and larger astrophysical object than the pulsar itself. Important observational properties of the Crab Nebula, which is a prototypical example of PWNe, are the small expansion velocity $v_{\rm PWN} \ll c$ and the large ratio of the synchrotron to the inverse Compton fluxes. These observations indicate that the pulsar wind of the Crab Nebula has the magnetization of much less than unity at the immediately upstream the pulsar wind shock and also at the post shock region [4–7]. This mismatch of the magnetization at the inner magnetosphere and that at around the shock is called the 'sigma-problem'.

A lot of efforts have been made to resolve the sigma-problem. Most works consider the magnetic dissipation at the pre-shock pulsar wind [8–10] or that at the shock [11, 12]. The recent 3D relativistic MHD simulation of a PWN states that the magnetic dissipation and the randomization of the magnetic field at the post shock (PWN) region are important [13]. Recently, the magnetic dissipation at the PWN region is also studied analytically [14, 15].

Here, we extend the model of PWN by [5] including the magnetic dissipation and the randomization of the magnetic field. It will be found that only the randomization of the magnetic filed is enough to reproduce the observed property $v_{\rm PWN} \ll c$. In section 2, we describe the our model of PWN outflow. In section 3, we show the flow profile based on our model and demonstrate that the outflow is decelerated by the randomization of the magnetic field. In section 4, we discuss the obtained results and summarise this paper. We are preparing a paper by adding greater details of our model and more results than the contents of this proceeding. The readers who are interested in this study will be able to find our paper in near future.

2. Model

Here, we extend the model of [5] (KC model) by including the turbulent component of the magnetic field and the magnetic dissipation. The upstream properties of the nebular flow is set to exactly the same as the KC model. We assume that the pulsar spin-down power is the same as the power of the pulsar wind,

$$L_{\rm spin} = L_{\rm wind} = \kappa \dot{N}_{\rm GJ} \gamma_{\rm w} m_{\rm e} c^2 (1 + \sigma_{\rm w}), \tag{1}$$

where κ is the pair multiplicity (the particle number flux normalized by the Goldreich-Julian number flux $\dot{N}_{\rm GJ} \equiv \sqrt{6cL_{\rm spin}}/e$), $\gamma_{\rm w}$ is the wind bulk Lorentz factor and $\sigma_{\rm w}$ is the wind magnetization parameters (the ratio of the Poynting to the kinetic energy fluxes). At the termination shock radius $r_{\rm TS}$, we impose the Rankin-Hugoniot relations for the strong perpendicular shock derived by [5]. The termination shock is set to the standing shock in the observer (pulsar) frame.

The nebular flow of the KC model is based on the following assumptions; (i) steady state, (ii) spherical symmetry, (iii) ideal MHD, (iv) pure radial flow velocity, and (v) pure toroidal magnetic field. Adopting the same assumptions, we add the following two effects inside the nebula; (vi) the conversion of the toroidal magnetic field to the turbulent one and (vii) the decay of the turbulent magnetic field. We adopt phenomenological descriptions of these effects referring to the studies by [16, 17].

The nebular flow has six variables, the radial four-velocity u, the proper density n, the proper enthalpy density w, the pressure p, the proper toroidal magnetic field $\bar{b}$ and the proper turbulent magnetic field δb. The turbulent magnetic field is set to isotropic in the proper frame and then the magnetic field three-vector in the proper frame would be written as $\boldsymbol{b} = \bar{b}\boldsymbol{e}_\phi + \delta\boldsymbol{b}$ satisfying $<\boldsymbol{b}>= \bar{b}\boldsymbol{e}_\phi$ and $<\boldsymbol{b}^2>= \bar{b}^2 + \delta b^2$, where $<>$ represents ensemble average. The corresponding six equations are two algebraic equations,

$$w = nm_{\rm e}c^2 + \frac{\hat{\Gamma}}{\hat{\Gamma}-1}p, \tag{2}$$

$$4\pi nucr^2 = \kappa \dot{N}_{\rm GJ} = {\rm const.}, \tag{3}$$

and four differential equations,

$$\frac{d}{dr}\left[r^2\gamma u\left(w + \bar{b}^2 + \frac{2}{3}\delta b^2\right)\right] = -r^2 n\gamma\frac{p_{\rm syn}}{c}, \tag{4}$$

$$\frac{d}{dr}(w-p) + w\frac{d}{dr}\ln ur^2 = \frac{\delta b^2/2}{uc\tau_{\rm diss}} - \frac{n}{u}\frac{p_{\rm syn}}{c}, \tag{5}$$

$$\frac{d}{dr}\frac{\bar{b}^2}{2} + \frac{\bar{b}^2}{2}\frac{d}{dr}\ln u^2r^2 = -\frac{\bar{b}^2/2}{uc\tau_{\rm conv}}, \tag{6}$$

Table 1. Summary of the parameters.

$L_{\rm spin}$[erg s^{-1}]	κ	$r_{\rm TS}$[pc]	$r_{\rm PWN}$[pc]	$\sigma_{\rm w}$	$\tau_{\rm diss}$[yr]	$\tau_{\rm conv}$[yr]	$\eta_{\rm syn}$
4.6 $\times 10^{38}$	10^4	0.1	2.0	10^3	10	0.228	0.026
				10		0.308	0.026
				0.1		6.51	0.037
				0.003	∞	∞	0.359
				10^3		0.223	$< 10^{-3}$
				10		0.301	$< 10^{-3}$
				0.1		7.56	0.167

$$\frac{d}{dr}\frac{\delta b^2}{2} + \frac{2}{3}\delta b^2 \frac{d}{dr}\ln ur^2 = -\frac{\delta b^2/2}{uc\tau_{\rm diss}} + \frac{\bar{b}^2/2}{uc\tau_{\rm conv}}. \tag{7}$$

Equations (2) and (3) correspond to the equation of state and the conservation of the particle number, respectively. If we ignore the right-hand sides of Equations (4) – (7), these equations represent the conservation of the total energy, the internal energy of the fluid, the toroidal magnetic energy and the turbulent magnetic energy, respectively, and exactly correspond to the equations of the KC model for $\delta b = 0$. We use the adiabatic index of $\hat{\Gamma} = 4/3$ throughout this paper.

In the right-hand sides of Equations (4) – (7), the terms proportional to the synchrotron power per particle $p_{\rm syn}$, to the inverse of the conversion time-scale $\tau_{\rm conv}^{-1}$ and to the inverse of the dissipation time-scale $\tau_{\rm diss}^{-1}$ represent the synchrotron cooling effect, the conversion of the toroidal magnetic field to the turbulent magnetic field, and the dissipation of the turbulent magnetic field, respectively. Introducing the thermal Lorentz factor $\gamma_{\rm th} \equiv (w-p)/(nm_{\rm e}c^2)$, for simplicity, we assume

$$p_{\rm syn} = \frac{4}{3}\sigma_{\rm T} c \gamma_{\rm th}^2 \frac{\bar{b}^2 + \delta b^2}{2}, \tag{8}$$

where $\sigma_{\rm T}$ is the Thomson cross section. $\tau_{\rm conv}$ and $\tau_{\rm diss}$ are the parameters of the basic equations and are set to constants in the proper frame.

Evolution of the four-velocity is obtained from Equations (2) – (7),

$$\epsilon(\beta^2 - \beta_{\rm c}^2)\frac{du}{dr} = \frac{2u}{r}\left(\hat{\Gamma}p + \frac{2}{9}\delta b^2\right) + (\hat{\Gamma}-1)n\frac{p_{\rm syn}}{c} + \frac{\bar{b}^2}{3c\tau_{\rm conv}} + \left(\frac{4}{3} - \hat{\Gamma}\right)\frac{\delta b^2}{2c\tau_{\rm diss}}, \tag{9}$$

where $\beta_{\rm c} = (\hat{\Gamma}p + \bar{b}^2 + 2/9\delta b^2)/(w + \bar{b}^2 + 2/3\delta b^2)$ is the characteristic velocity corresponding to a phase velocity of a fast magnetosonic wave for the case of the pure toroidal magnetic field. The right-hand side of Equation (9) is positive so that the nebular flow always decelerates with r for the post shock flow ($\beta < \beta_{\rm c}$).

3. Results

We apply the model described in section 2 to the Crab Nebula. We set $L_{\rm spin} = 4.6 \times 10^{38}$ erg s^{-1}, $\kappa = 10^4$, $r_{\rm pc} = 0.1$ pc and $r_{\rm PWN} = 2.0$ pc. We consider three different values of the wind magnetization $\sigma_{\rm w} = (0.1, 10, 10^3)$ and two different dissipation time-scale $\tau_{\rm diss} = (10\ {\rm yr}, \infty)$, and look for the values of $\tau_{\rm conv}$ which satisfies $v_{\rm PWN} = 1500$ km s^{-1}. The parameters are also summarised in Table 1.

Figure shows the resultant flow profiles. The four-velocity hardly depends on $\tau_{\rm diss}$, i.e., the thick and thin lines are overlapped in the top panel and the obtained $\tau_{\rm conv}$ are almost the same

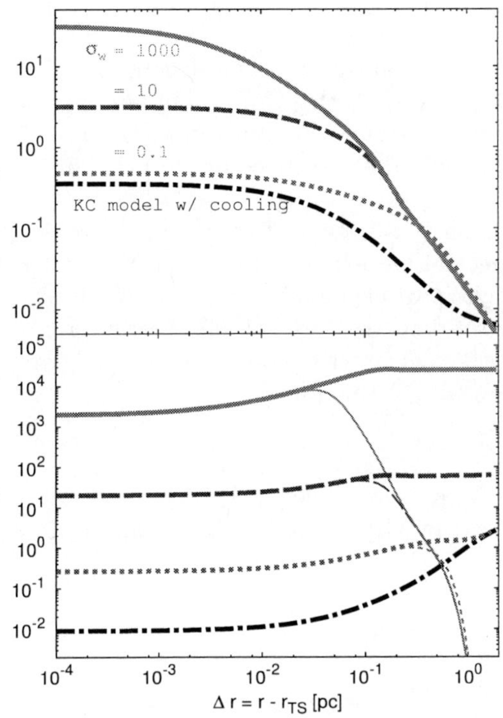

Figure 1. The nebular flow profiles (Equations (2) – (7)) which satisfy $v_{\rm PWN} = 1500$ km s^{-1} are plotted. The profiles of the four velocity u (top) and the magnetization σ (bottom) are plotted. The horizontal axis is the distance from the termination shock and not from the pulsar. Adopted parameters are summarised in Table 1 and combinations of $(\sigma_{\rm w}, \tau_{\rm diss})$ are different for each line, where $\tau_{\rm diss} = 10$ yr, (thin lines) ∞ (thick lines) and $\sigma_{\rm w} = 10^3$ (solid red lines), 10 (dashed blue lines), and 0.1 (dotted green lines). For reference, the KC model with the cooling effect are plotted (dot-dashed black), where we set $(\sigma_{\rm w}, \kappa) = (0.003, 10^4)$ and $\tau_{\rm conv} = \tau_{\rm diss} = \infty$.

for the cases of $\tau_{\rm diss} = 10$ yr and ∞. On the other hand, σ in the bottom panels are different between thick and thin lines because the magnetic energy is converted to the internal energy of the plasma by $\tau_{\rm diss}$. The definition of the magnetization is

$$\sigma(r) \equiv \frac{\bar{b}^2 + 2/3\ \delta b^2}{w}, \tag{10}$$

and this is different from the definition of $\sigma_{\rm w}$, while those are almost the same for the case of $\gamma_{\rm w} \gg 1$.

The magnetic dissipation heats up the plasma and changes the synchrotron cooling efficiency $\eta_{\rm syn}$ (the fraction of the energy flux radiated away from the system by the synchrotron cooling $p_{\rm syn}$). However, the synchrotron cooling hardly changes the flow profile even for the KC model ($\eta_{\rm syn} \sim 0.36$). The thick and thin lines of the top panel (four-velocity) is not exactly the same but are overlapped.

4. Discussion & Conclusions

We extend the KC model including the turbulent magnetic field and the magnetic dissipation. The conversion of the toroidal magnetic field to the turbulent magnetic field and the magnetic dissipation terms are treated phenomenologically. The time-scales of the conversion and dissipation are the parameters of the system and we take them as constants in the proper frame for simplicity. We find that the conversion of the toroidal magnetic field to the turbulent magnetic field decelerates the nebular flow to fit the observed value of the expansion velocity. On the other hand, the magnetic dissipation hardly contributes the flow deceleration.

Here, we did not consider the emission spectrum of the PWN. Equation (8) is too simple description of the radiation effect. The low-magnetization of the Crab Nebula is required in both

IOP Conf. Series: Journal of Physics: Conf. Series **932** (2017) 012052 doi:10.1088/1742-6596/932/1/012052

dynamical and spectral point of views, i.e., we should reproduce not only the observed expansion velocity of the Crab Nebula but also the observed synchrotron to the inverse Compton flux ratio. In order to reproduce the flux ratio, $\tau_{\rm diss}$ plays an important role. In addition, the values of $\eta_{\rm syn}$ in this study are much smaller than the observations of the Crab Nebula. $\tau_{\rm diss}$ is also used to control the synchrotron cooling efficiency and it should be studied with a broadband emission model. The studies of the radial profiles of the radiation spectrum are also important to constrain the parameters [18].

In addition to $\tau_{\rm diss}$, κ changes the synchrotron cooling efficiency. In this paper, we adopted $\kappa = 10^4$, although the studies of the broadband spectral modeling of the Crab Nebula suggest $\kappa > 10^6$ [19–21], however, see also [22]. The larger particle number κ leads to the smaller temperature $\gamma_{\rm th}$ of the post shock flow because the total energy flux is limited by Equation (1). We need a larger $\gamma_{\rm th}$ (smaller κ) in order to make the synchrotron cooling more effective than this study. Adopting $\kappa > 10^4$ does not change the results of this paper.

Acknowledgments

The author would like to thank K. Toma and N. Tominaga. They will be co-authors of the paper which is in preparation. This work is supported by Grants-in-Aid for Scientific Research Nos 15H05440 and 17K18270.

References

[1] Goldreich P and Julian W H 1969 *Astrophys. J.* **157** 869
[2] Daugherty J K and Harding A K 1982 *Astrophys. J.* **252** 337–37
[3] Hibschman J A and Arons J 2001 *Astrophys. J.* **560** 871–84
[4] Rees M J and Gunn J E 1974 *Mon. Not. R. Astron. Soc.* **167** 1–12
[5] Kennel C F and Coroniti F V 1984 *Astrophys. J.* **283** 694–709
[6] Kennel C F and Coroniti F V 1984 *Astrophys. J.* **283** 710–30
[7] Atoyan A M and Aharonian F A 1996 *Mon. Not. R. Astron. Soc.* **278** 525–41
[8] Coroniti F V 1990 *Astrophys. J.* **349** 538–45
[9] Lyubarsky Y and Kirk J G 2001 *Astrophys. J.* **547** 437–48
[10] Kirk J G and Skjæraasen O 2003 *Astrophys. J.* **591** 366–79
[11] Komissarov S S 2013 *Mon. Not. R. Astron. Soc.* **428** 2459–66
[12] Amano T and Kirk J G 2013 *Astrophys. J.* **770** 18
[13] Porth O, Komissarov S S and Keppens R 2014 *Mon. Not. R. Astron. Soc.* **438** 278–306
[14] Lyutikov M 2010 *Mon. Not. R. Astron. Soc.* **405** 1809–15
[15] Zrake J and Arons J 2016 *Preprint* 1612.02430
[16] Drenkhahn G 2002 *Astron. Astrophys.* **387** 714–24
[17] Drenkhahn G and Spruit H C 2002 *Astron. Astrophys.* **391** 1141–53
[18] Ishizaki W, Tanaka S J, Asano K and Terasawa T 2017 *Astrophys. J.* **838** 142
[19] Tanaka S J and Takahara F 2010 *Astrophys. J.* **715** 1248–57
[20] Bucciantini N, Arons J and Amato E 2011 *Mon. Not. R. Astron. Soc.* **410** 381–98
[21] Torres D F, Cillis A, Martín J and de Oña Wilhelmi E 2014 *J. High Energy Astrophys.* **1** 31–62
[22] Tanaka S J and Asano K 2017 *Astrophys. J.* **841** 78

IOP Conf. Series: Journal of Physics: Conf. Series **932** (2017) 012053 doi:10.1088/1742-6596/932/1/012053

Search for the time evolution of the synchrotron X-ray spectrum of the youngest Galactic supernova remnant G1.9+0.3 using *Suzaku*

A Sezer[1], R Yamazaki[2], Y Ohira[2], S Tanaka[3] and S Kisaka[2]

[1] Department of Electrical-Electronics Engineering, Avrasya University, 61250, Trabzon, Turkey

[2] Department of Physics and Mathematics, Aoyama Gakuin University, 5-10-1 Fuchinobe, Sagamihara 252-5258, Japan

[3] Department of Physics, Faculty of Science and Engineering, Konan University, 8-9-1 Okamoto, Kobe, Hyogo 658-8501, Japan

E-mail: aytap.sezer@avrasya.edu.tr

Abstract. G1.9+0.3 is the youngest known Galactic supernova remnant (SNR) and dominated by X-ray synchrotron emission. Synchrotron X-rays can be a useful tool to study the electron acceleration in young SNRs. The X-ray spectra of young SNRs give us information about the particle acceleration at the early stages of evolution of SNRs. In this work, we investigate the time evolution of roll-off frequency of the synchrotron spectrum from SNR G1.9+0.3 using *Suzaku*. For this analysis, we use ∼101 ks (2011) and ∼92 ks (2015) observations with the X-ray Imaging Spectrometer. We find that there is no significant differences in the spectral parameters and interpret our results.

1. Introduction

The X-ray spectra of several young SNRs are dominated by non-thermal emission: e.g., SN 1006 [1], G347.3−0.5 [2, 3] and G266.2−1.2 [4]. These discoveries in shell-like SNRs suggest that relativistic electrons are accelerated by SNR shocks. Their spectra have a power-law form (the synchrotron flux density at frequency ν is $F_\nu \propto \nu^{-\alpha} \propto \nu^{1-\Gamma_{\rm x}}$, where α is the spectral index and $\Gamma_{\rm x}$ is the photon index). *Chandra* X-ray spectra showed that G1.9+0.3 is a young and X-ray synchrotron-dominated shell type SNR [5-7]. The age of this youngest known Galactic supernova remnant is estimated to be ∼110 yr [6]. Using *Chandra* and *NuStar* data, Zoglauer et al. [8], performed a detailed X-ray study of the SNR and found a spectral index of $\alpha \sim 0.63$, and a roll-off frequency of $\nu_{\rm rolloff} \sim 3.1 \times 10^{17}$ Hz. From *Suzaku* 2011 observation, Gök and Ergin [9], obtained a spectral index of ∼0.61 and a roll-off frequency of $\sim 3.1 \times 10^{17}$ Hz.

The νF_ν spectrum of synchrotron X-ray emission has a peak around the roll-off frequency, $\nu_{\rm roll}$, which is related to the maximum electron energy of accelerated electrons ($E_{\rm max,e}$) and the magnetic field (B), $\nu_{\rm roll} \propto E_{\rm max,e}^2\ B$ [10, 11]. The value of $\nu_{\rm roll}$ decreases as SNRs evolve (e.g., [12]). The SNR G1.9+0.3 is one of the best studied SNRs to search for the time variability. In this work, we investigate the time evolution of roll-off frequency of the synchrotron spectrum from G1.9+0.3. For this study, we use 2011 and 2015 observations with the X-ray Imaging Spectrometer (XIS; [13]) on board *Suzaku* [14], which has a high spectral resolution. In Section 2,

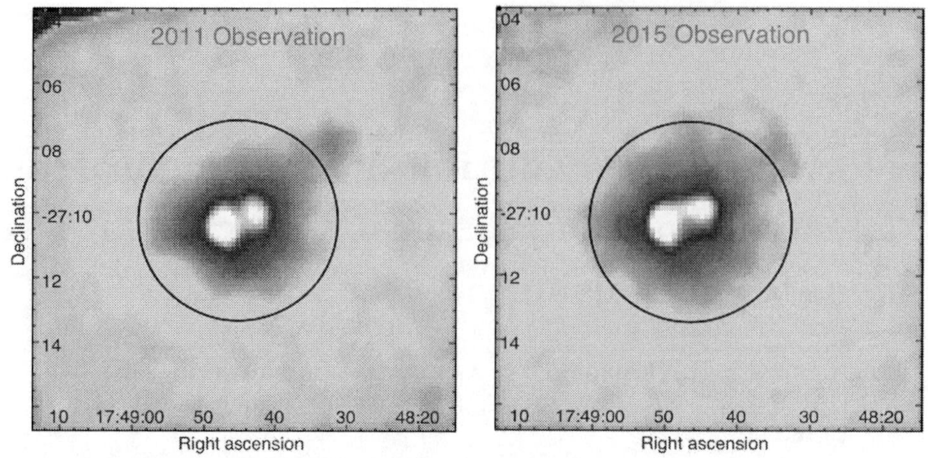

Figure 1. XIS images of G1.9+0.3 in the 0.3−10.0 keV energy band. The circles indicate the extraction regions used for the spectral analysis.

we describe the *Suzaku* XIS observations and data reduction. The spectral analysis is presented in Section 3. Finally, in Section 4, we discuss our results.

2. Observations and Data Reduction

SNR G1.9+0.3 observed with XIS on 2011 March (Obs ID: 505053010) and 2015 March (Obs ID:509003010) for $\sim$ 101 ks and $\sim$ 92 ks, respectively. A detailed spectral analysis of *Suzaku* observation in 2011 was previously presented in [9]. The XIS consists of four X-ray charge-coupled devices (CCDs). Three of them (XIS0, 2 and 3) are front-illuminated CCDs, and the other (XIS1) is a back-illuminated CCD. After 2006 November 9, the XIS2 was out of operation. Therefore, we use only XIS0, 1 and 3 data.

We retrieved the data from the public *Suzaku* science data archive through the DARTS

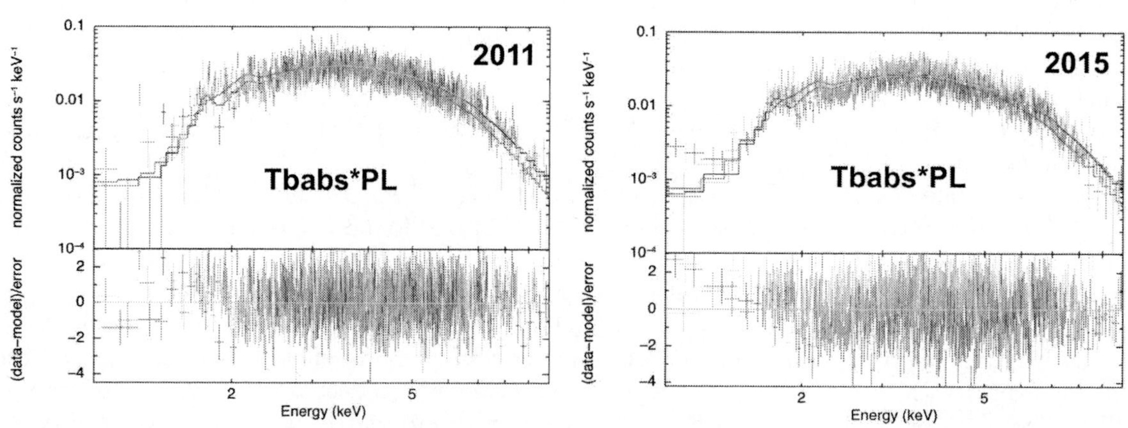

Figure 2. Background-subtracted XIS0 (black), XIS1 (red) and XIS3 (green) spectra of G1.9+0.3.

Table 1. Results of Spectral Fitting

Parameter	Observation 2011	Observation 2015
TBABS*PL		
N_H ($\times 10^{22}$ cm^{-2})	$9.8^{+0.4}_{-0.4}$	$9.1^{+0.3}_{-0.3}$
Photon Index (Γ)	$2.52^{+0.07}_{-0.07}$	$2.49^{+0.06}_{-0.06}$
Flux[a] ($\times 10^{-11}$ ergs cm^{-2} s^{-1})	$2.33^{+0.11}_{-0.07}$	$2.01^{+0.09}_{-0.05}$
χ^2/dof	1848.3/1801	1596.5/1373
TBABS*SRCUT		
N_H ($\times 10^{22}$ cm^{-2})	$9.2^{+0.3}_{-0.4}$	$8.6^{+0.3}_{-0.2}$
α	$0.608^{+0.013}_{-0.008}$	$0.616^{+0.011}_{-0.007}$
$\nu_{\rm rolloff}$ ($\times 10^{17}$ Hz)	$3.12^{+1.14}_{-0.51}$	$3.33^{+1.04}_{-0.51}$
norm (Jy at 1 GHz)	1.24 (fixed)	1.24 (fixed)
χ^2/dof	1842.5/1801	1592.9/1373
TBABS*SRCUT (α=0.62)		
N_H ($\times 10^{22}$ cm^{-2})	$8.9^{+0.2}_{-0.2}$	$8.6^{+0.3}_{-0.1}$
α	0.62 (fixed)	0.62 (fixed)
$\nu_{\rm rolloff}$ ($\times 10^{17}$ Hz)	$3.71^{+0.08}_{-0.09}$	$3.59^{+0.06}_{-0.09}$
norm (Jy at 1 GHz)	1.24 (fixed)	1.24 (fixed)
χ^2/dof	1844.6/1802	1603.4/1374

[a] Unabsorbed fluxes in the 0.5−10.0 keV energy range.

interface[1]. Data reduction and analysis were made using HEASOFT package[2] version 6.20, XSPEC version 12.9.1 and AtomDB version 3.0.8. The redistribution matrix file and ancillary response file were produced by XISRMFGEN and XISSIMARFGEN [15], respectively.

3. Spectral Analysis

Figure 1 shows the *Suzaku* XIS images of 2011 and 2015 observations in the 0.3−10.0 keV. In order to investigate the time variability of X-ray emission from G1.9+0.3, we extracted XIS spectra from a circular region at the source with a radius of 3.1 arcmin centered on the remnant for both data sets. To estimate the background, we extracted data from source-free region on the same field of view. We subtracted the non-X-ray background (NXB) from both observations. The NXB spectra were made by using XISNXBGEN [16].

First, we fitted both spectra with a power-law (PL) model modified by an absorption model (TBABS; [17]) with free parameters of the column density (N_H), the photon index (Γ) and the normalization. Then, we tried an absorbed SRCUT model instead of the PL model. SRCUT model describes synchrotron radiation from a PL distribution of electrons with an exponential cut-off [11]. The flux (at 1 GHz) for SRCUT is fixed at the value 1.24 Jy similar to previous studies [8, 9]. PL and SRCUT models are well fitted to the data. The parameters of both models for 2011 and 2015 observations are given in table 1, where the uncertainties quoted are

[1] https://darts.isas.jaxa.jp/astro/suzaku/
[2] https://heasarc.nasa.gov/lheasoft/

IOP Conf. Series: Journal of Physics: Conf. Series **932** (2017) 012053 doi:10.1088/1742-6596/932/1/012053

the 90% confidence limits. As a next step, the SRCUT model was also applied with the spectral index fixed to 0.62 at 1 GHz as reported by [18]. This model also well reproduces the spectra with best-fit values shown in table 1. XIS spectra for 2011 and 2015 observations are shown in figure 2.

4. Results and Discussion

In this work, we focus on the time evolution of roll-off frequency of the synchrotron spectrum from SNR G1.9+0.3 using archival *Suzaku* data. We found that the range of the roll-off frequencies of both observations is $\sim(3.1 - 3.7)\times10^{17}$ Hz, which are consistent with the roll-off frequencies derived by [8, 9]. The roll-off frequency has been measured in several young SNRs (e.g., SN 1006 [19]). They found that the non-thermal component of X-ray spectra has a roll-off frequency of $\sim 5.7 \times 10^{16}$ Hz, which is lower than the roll-off frequency found in our analysis.

From our spectral analysis, as seen table 1, we find no significant differences in the spectral parameters between 2011 and 2015 observations. The roll-off frequency can be written as

$$\nu_{\rm roll} \sim 0.97 \times 10^{17}\ {\rm Hz}\ \xi^{-1} (V_{\rm s}/10^8\ {\rm cm\ s^{-1}})^2 \qquad (1)$$

where ξ is the gyro-factor and $V_{\rm s}$ is the shock velocity (e.g., [20, 21]). The shock velocity of SNRs decreases with time after the free expansion phase [22]. Our results imply that the shock velocity is almost unchanged between 2011 and 2015. This may indicate that G1.9+0.3 is still in the free expansion phase.

Acknowledgments

AS is supported by the Scientific and Technological Research Council of Turkey (TÜBİTAK) through the BİDEB-2219 fellowship program. This work is supported in part by grant-in-aid from the Ministry of Education, Culture, Sports, Science, and Technology (MEXT) of Japan, No.15K05088(RY), No.16K17702(YO), No.17K18270 (ST) and 16J06773(SK).

References

[1] Koyama K, Petre R, Gotthelf E V, Hwang U, Matsura M, Ozaki M and Holt S S 1995 *Nature* **378** 255
[2] Koyama K, Kinugasa K, Matsuzaki K, Nishiuchi M, Sugizaki M, Torii K, Yamauchi S and Aschenbach B 1997 *Pub. Astron. Soc. Japan* **49** L7
[3] Slane P, Gaensler B M, Dame T M, Hughes J P, Plucinsky P P and Green A 1999 *ApJ* **525** 357
[4] Slane P, Hughes J P, Edgar R J, Plucinsky P P, Miyata E, Tsunemi H and Aschenbach B 2001 *ApJ* **548**, 814
[5] Reynolds S P, Borkowski K J, Green D A, Hwang U, Harrus I, Petre R 2008 *ApJ* **680** L41
[6] Carlton A K, Borkowski K J, Reynolds S P, Hwang U, Petre R, Green D A, Krishnamurthy K and Willett R 2011 *ApJ* **737** L22
[7] Borkowski K J, Reynolds S P, Hwang U, Green D A, Petre R, Krishnamurthy K and Willett R 2013 *ApJ* **771** L9
[8] Zoglauer A *et al* 2015 *ApJ* **798** 98
[9] Gök F and Ergin T 2015 *Adv Space Res* **56** 1793
[10] Reynolds S P 1998 *ApJ* **493** 375
[11] Reynolds S P and Keohane J W 1999 *ApJ* **525** 368
[12] Bamba A, Yamazaki R, Yoshida T, Terasawa T and Koyama K 2005 *ApJ* **621** 793
[13] Koyama K *et al* 2007 *Pub. Astron. Soc. Japan* **59** 23
[14] Mitsuda K *et al* 2007 *Pub. Astron. Soc. Japan* **59** 1
[15] Ishisaki Y *et al* 2007 *Pub. Astron. Soc. Japan* **59** 113
[16] Tawa N *et al* 2008 *Pub. Astron. Soc. Japan* **60** 11
[17] Wilms J, Allen A and McCray R 2000 *ApJ* **542** 914
[18] Green D A, Reynolds S P, Borkowski K J, Hwang U, Harrus I and Petre R 2008 *MNRAS* **387** L54
[19] Bamba A *et al* 2008 *Pub. Astron. Soc. Japan* **60** 153
[20] Aharonian F A and Atoyan A M 1999 *Astron. Astrophys.* **351** 330
[21] Yamazaki R, Kohri K, Bamba A, Yoshida T, Tsuribe T and Takahara F 2006 *MNRAS* **371** 1975
[22] Yamazaki R, Ohira Y, Sawada M and Bamba A 2014 *Research in Astronomy and Astrophys.* **14** 165

IOP Conf. Series: Journal of Physics: Conf. Series **932** (2017) 012054 doi:10.1088/1742-6596/932/1/012054

Confirming the nature of the knot near the pulsar B1951+32

D A Zyuzin[1], Yu A Shibanov [1,2], G G Pavlov[3] and A A Danilenko[1]
[1] Ioffe Institute, Politekhnicheskaya 26, 194021, Saint Petersburg, Russia
[2] Peter the Great Saint Petersburg Polytechnic University, Politekhnicheskaya 29, 195251, Saint Petersburg, Russia
[3] Pennsylvania State University, 525 Davey Lab., University Park, PA 16802, USA

E-mail: da.zyuzin@gmail.com

Abstract.
The energetic and fast-moving radio and γ-ray pulsar B1951+32 is associated with the supernova remnant CTB 80. It powers a complex pulsar wind nebula detected in the radio, Hα and X-rays (Moon *et al* 2004 *ApJ* **610** L33).A puzzling optical knot was detected about $0''.5$ from the pulsar in the optical and near-IR (Moon *et al* 2004 *ApJ* **610** L33; Hester 2000 *Bulletin of the AAS* **32** 1542).It is reminiscent of the unique "inner optical knot" located $0''.6$ from the Crab pulsar. Until now there has been no evidence that B1951+32 knot is indeed associated with the pulsar.

We observed the pulsar field with the Gemini-North telescope in 2016 to check the association. We performed first near-IR high spatial resolution imaging in the K_s band using the NIRI+Altair instrument and deep optical imaging in the *gr* bands using the GMOS instrument. Our observations showed that the current knot position is shifted by $\approx 0''.6$ from the position measured with the HST in 1997. This is consistent with the known pulsar proper motion and is direct evidence of the pulsar–knot connection. We compared the spectral energy distribution of the knot emission with that of the Crab knot. Possible implications of the results are discussed.

1. Introduction

Pulsar wind nebulae (PWNe) are formed as a result of interaction of a relativistic pulsar wind with ambient matter. The wind consists of particles accelerated in magnetospheres of rotation-powered pulsars. It propagates freely until it experiences a collisionless termination shock (TS). The shocked wind streams out at $\sim 0.3 - 0.5c$, producing synchrotron emission observed from the radio to X-rays [1]. The PWNe generally show two morphological types. Around young pulsars they usually appear as X-ray tori with polar jets. The classical examples are the Crab [2] and the Vela [3]. Subarcsecond optical imaging of the Crab [4] revealed puzzling knots and wisps. The so-called inner knot located at only $0''.6$ from the Crab pulsar is projected onto the free wind zone upstream of the TS, where no emission from the unshocked wind is expected [5]. More distant wisps are interpreted as Doppler-boosted emission from the equatorial zones of the shocked wind [6]. Extensive studies of the knot and wisps in the Crab revealed their temporal and spectral variability providing insight into the complex dynamics of non-stationary relativistic flows. Another type is bullet-shaped PWNe with long tails seen around pulsars moving through the ambient medium with a supersonic velocity. In this case, the overall PWN geometry is governed by the ram pressure of the oncoming medium flow, and both the TS and

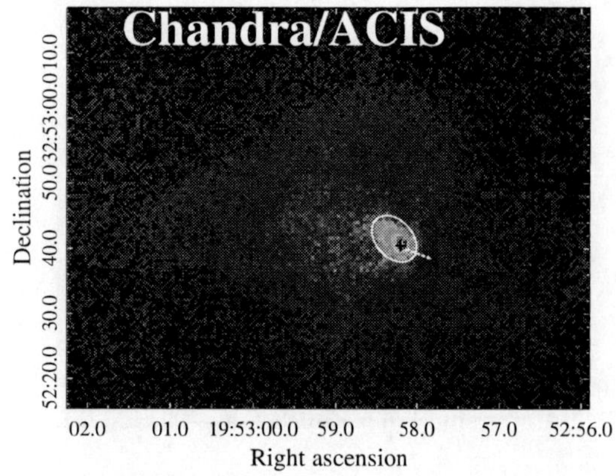

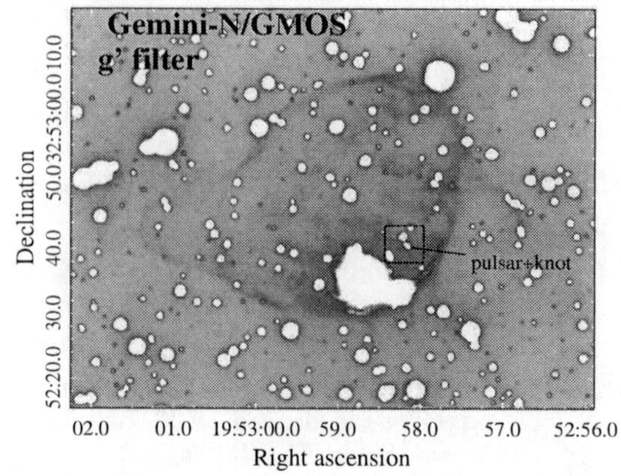

Figure 1. *Left*: X-ray image of the PWN associated with SNR CTB 80. The pulsar position is marked by the cross. The arrow shows the pulsar proper motion direction. The X-ray spectrum of the "knot+pulsar" (figure 3) was extracted from the white ellipse. *Right:* g$'$ band image of the PWN obtained with Gemini-N. The PWN emission is dominated by the Hβ Balmer line, excited by the interaction of the pulsar wind with ambient matter. The pulsar and the knot are within the black dashed rectangle region, which is enlarged in figure 2.

the forward shock (FS) in the ambient medium acquire a bow-shock shape ahead of the moving pulsar. Intermediate types are possible for transonic motion.

An example of the latter type is the PWN around the 51 kyr old, fast-moving ($\mu = 32.3 \pm 0.9$ mas yr^{-1} [7]; $v_\perp \approx 225$ km s^{-1} for $d = 1.5$ kpc), energetic ($\dot{E} \approx 3.7 \times 10^{36}$ erg s^{-1}) PSR B1951+32, associated with the supernova remnant (SNR) CTB 80. The age is consistent with the pulsar proper motion (p.m.) shift from the SNR centre [7], and we accept the distance of 1.5 kpc recently estimated from HI studies [8]. The PWN demonstrates a bow-shock morphology in the radio, X-rays, and Hα, allowing one to determine the TS and FS positions [9]. The nebula is superimposed on the SNR limb, and the ambient medium may thus be over-pressured with respect to the general ISM. The unique feature is a compact optical knot at $0\farcs5$ from the pulsar detected with the HST in the F547M band [10]. It has dimensions of $0\farcs8 \times 1\farcs3$ and, as the Crab inner knot, it is projected onto the pulsar proper-motion path. It was confirmed by ground-based observations in the r and K$_s$ bands in 2001 and 2003 [9]. The data suggest a flat energy spectrum with $F_\nu = 40 \pm 10$ μJy, implying non-thermal nature of the knot emission. However, the fluxes in different bands were obtained at different epochs. Based on the Crab knot studies, the knot temporal variability is expected, and the spectral flatness needs a confirmation by simultaneous observations in different bands. There still is no direct evidence that the knot is not a background object. The evidence can be obtained by measuring its p.m. and comparing it with the known p.m. of the pulsar. Here we briefly report results of the new near-infrared and optical observations of the PSR B1951+32 field.

2. Observations with Gemini-North

The pulsar B1951+32 field was observed in June-July 2016 with the ALTtitude conjugate Adaptive optics for the InfraRed (ALTAIR) connected with the Near InfraRed Imager and Spectrometer (NIRI), and with the Gemini Multi-Object Spectrograph (GMOS) mounted on the Gemini-North telescope. The observations were carried out in the K$_s$, g$'$ and r$'$ bands.

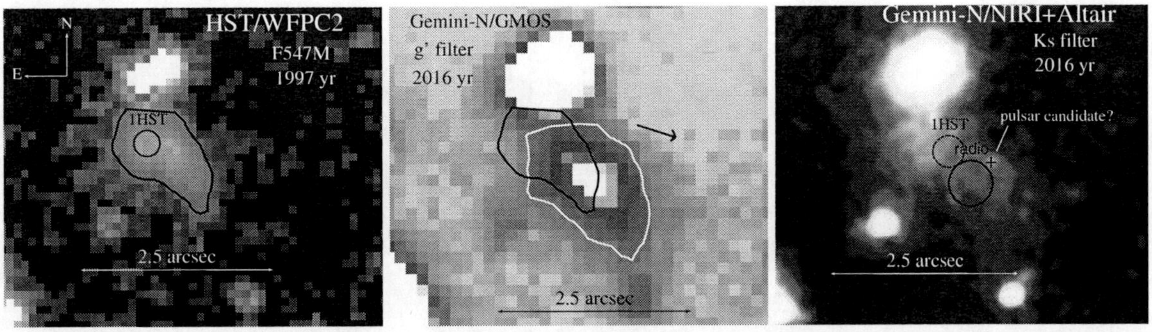

Figure 2. *Left:* HST image of the knot region obtained in 1997 in the F547M band. The "1HST" circle within the knot extended emission marks the position of the possible pulsar optical counterpart reported in [10]. *Middle:* Gemini-N image of the same field obtained in 2016 in the g′ band. The black and white contours in this and the left panel show the knot boundaries defined at 3σ level in 1997 and 2016, respectively. The black arrow demonstrates the pulsar p.m. shift between 1997 and 2016. As seen, the knot contour shift is consistent with that of the pulsar. *Right:* Gemini-N high spatial resolution image of the field obtained in 2016 in the K_s band. The small circle demonstrates the expected position of the "1HST" source in 2016 if it would be the real pulsar counterpart. However, no point source is detected there. The large black circle labelled as "radio" is the pulsar radio position error circle in 2016. A new pulsar counterpart candidate, marked by the cross, is detected near the edge of the radio circle.

3. Results

The main results are presented in figures 1–3. The PWN shock structure in the optical is quite complex (figure 1), likely due to the interaction with the SNR limb. Comparing the images obtained in the g' and HST/F547M bands (figure 2), whose band-passes are close to each other, we found that the knot was shifted by $\approx 0\farcs6$ between the epochs of the observations with the HST (1997) and Gemini-N (2016). The shift value and direction approximately coincide with the pulsar p.m. shift of $0\farcs61 \pm 0\farcs02$ with the position angle of $249^\circ \pm 2^\circ$. This is direct evidence of the pulsar-knot association. In K_s the knot is more extended and brighter in the north–east direction than in g′. In general, it is much less compact then the Crab inner knot. A new near-IR pulsar counterpart candidate is detected in the K_s band near the edge of the radio pulsar position error circle, while the previous candidate proposed in [10] is not detected (figure 2).

The knot optical-IR spectral energy distribution (SED) is almost flat, with the spectral index $\alpha_\nu \approx 0.14$ (figure 3), which is different from the steep SED of the Crab inner knot [11]. This can be related to different compactness and physical nature of the knots. To compare the optical and X-ray data, we extracted the X-ray spectrum of the knot+pulsar from the ellipse in figure 1, using *Chandra* archive, fitted it with the thermal + non-thermal model, and adopted the non-thermal component of the pulsar X-ray spectrum from [12]. It is hard to distinguish the non-thermal pulsar and knot contributions in X-rays (figure 3). Nevertheless, it is obvious that the spectrum of the knot should have a break between the optical and X-rays with $\Delta\alpha_\nu \geq 0.5$. The flux of the new pulsar counterpart candidate lies significantly below the long-wavelength extrapolation of the presumed pulsar nonthermal X-ray spectrum, as is typically observed for pulsars detected in both ranges. The thermal component is negligible in the optical.

4. Conclusions

We resolved a complex structure of the PSR B1951+32 bow-shock PWN in the optical and found strong evidence that its puzzling knot is indeed associated with the pulsar. In the near-IR

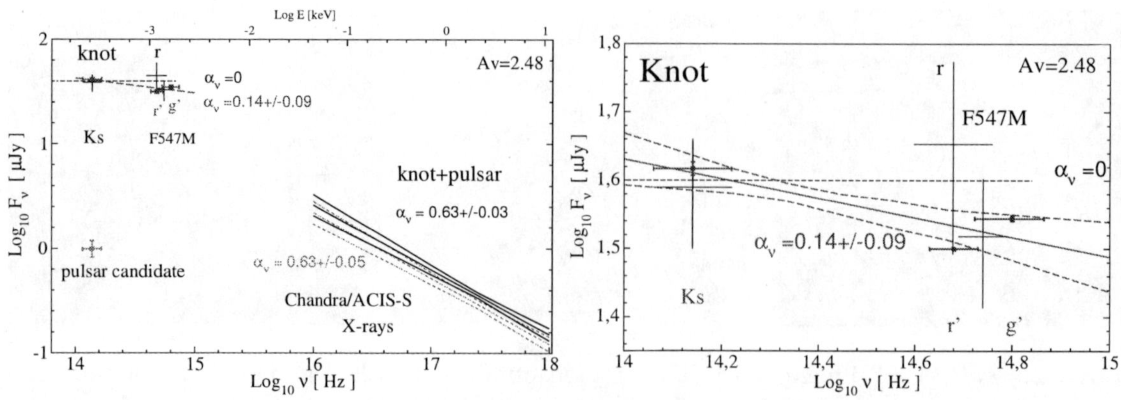

Figure 3. *Left:* Near-IR–X-ray SEDs of the knot and the pulsar corrected for interstellar absorption $A_V = 2.48$. Black optical-near-IR data points are taken from [9], while blue ones are obtained in this work. The knot+pulsar and the presumed pulsar X-ray spectra with their uncertainties are shown by black and red lines. The near-IR pulsar counterpart candidate flux is shown by red. *Right:* The closeup view of the optical–near-IR knot SED. Blue lines show the power-law fit to the fluxes obtained in this work and its uncertainties.

the knot is more extended than in the optical. Its SED is almost flat in the optical–near-IR, which together with the X-ray data implies a spectral break between these ranges. We did not detect the possible pulsar optical counterpart suggested in [10]; however, we found a new near-IR pulsar counterpart candidate near the edge of the pulsar radio position error circle.

Acknowledgments

DAZ work was partially supported by RF Presidential Programme MK-2566.2017.2. This work is based on observations obtained at the Gemini Observatory (Program ID: GN-2016A-Q-46), which is operated by the Association of Universities for Research in Astronomy, Inc., under a cooperative agreement with the NSF on behalf of the Gemini partnership.

References

[1] Kargaltsev O and Pavlov G G 2008 *40 Years of Pulsars: Millisecond Pulsars, Magnetars and More* (*American Institute of Physics Conference Series* vol 983) ed Bassa C, Wang Z, Cumming A and Kaspi V M pp 171–85 (*Preprint* `0801.2602`)
[2] Weisskopf M C *et al* 2000 *ApJ* **536** L81–4 (*Preprint* `astro-ph/0003216`)
[3] Pavlov G G, Kargaltsev O Y, Sanwal D and Garmire G P 2001 *ApJ* **554** L189–92 (*Preprint* `astro-ph/0104264`)
[4] Hester J J *et al* 1995 *ApJ* **448** 240
[5] Kennel C F and Coroniti F V 1984 *ApJ* **283** 694–709
[6] Komissarov S S and Lyubarsky Y E 2003 *MNRAS* **344** L93–6 (*Preprint* `astro-ph/0306162`)
[7] Zeiger B R, Brisken W F, Chatterjee S and Goss W M 2008 *ApJ* **674** 271–7 (*Preprint* `0801.0244`)
[8] Leahy D A and Ranasinghe S 2012 *MNRAS* **423** 718–24
[9] Moon D S, Lee J J, Eikenberry S S, Koo B C, Chatterjee S, Kaplan D L, Hester J J, Cordes J M, Gallant Y A and Koch-Miramond L 2004 *ApJ* **610** L33–6 (*Preprint* `astro-ph/0406240`)
[10] Hester J 2000 *American Astronomical Society Meeting Abstracts* (*Bulletin of the American Astronomical Society* vol 32) p 1542
[11] Sandberg A and Sollerman J 2009 *A&A* **504** 525–30 (*Preprint* `0906.2065`)
[12] Li X H, Lu F J and Li T P 2005 *ApJ* **628** 931–7 (*Preprint* `astro-ph/0504293`)